# Lecture Notes in Computer Science

# Lecture Notes in Computer Science

Edited by G. Goos and J. Hartmanis

## 344

J. van Leeuwen (Ed.)

# Graph-Theoretic Concepts in Computer Science

International Workshop WG '88
Amsterdam, The Netherlands, June 15–17, 1988
Proceedings

Springer-Verlag

Berlin Heidelberg New York London Paris Tokyo

**Editor**

J. van Leeuwen
Department of Computer Science, University of Utrecht
P.O. Box 80.089, NL-3508 TB Utrecht, The Netherlands

CR Subject Classification (1987): G.2.2, F.2.2, F.1.2, E.1, I.3.5, D.4.1, H.2.1, B.7.2

ISBN 3-540-50728-0 Springer-Verlag Berlin Heidelberg New York
ISBN 0-387-50728-0 Springer-Verlag New York Berlin Heidelberg

Printing and binding: Druckhaus Beltz, Hemsbach/Bergstr.
2145/3140-543210

# PREFACE

The 14th International Workshop on Graph-Theoretic Concepts in Computer Science (WG'88) was held at the Centre for Mathematics and Computer Science (CWI) in Amsterdam, June 15-17, 1988. The workshop was intended to provide a forum for researchers and other parties interested in the study and application of graph-theoretic concepts in Computer Science. The Workshop featured sessions on structural graph theory, parallel graph algorithms, graph-based modeling (e.g. in database theory and VLSI), computational geometry and applied graph theory.

In the call for papers contributions were solicited describing original results in the study and application of graph-theoretic concepts in the tradition of the Workshop, including e.g. the design and analysis of graph algorithms, applied graph theory in computer science, operations research algorithms on graphs and networks, parallel and distributed algorithms on graphs, graph-theoretic modeling (e.g. in hardware and software design), graph grammars and graph replacement systems, graph-theoretic aspects of computer graphics (e.g. in solid modeling and CAD), computational geometry, efficient data structures and complexity-theoretic aspects of programming and communication structures, and related fields.

The organizational committee for the Workshop consisted of

J. van Leeuwen    (University of Utrecht),
D.W. Matula       (Southern Methodist University, Dallas),
M. Nagl           (RWTH Aachen),
H. Noltemeier     (University of Würzburg),
H.J. Schneider    (University of Erlangen),
E. Welzl          (Free University of Berlin),
S. Zaks           (Technion, Haifa).

Out of the submissions the organizational committee selected thirty-one papers for presentation in the Workshop. The selection reflects several current directions of research that are representative for the area of graph-based algorithms, although certainly not all aspects could be covered in the three-day Workshop.

The present volume contains the revised version of all papers presented in the Workshop. The revised versions are based on the comments and suggestions received by the authors during and after the Workshop. Several papers are in the form of preliminary reports on continuing research, and it is expected that more elaborate versions of these papers will eventually appear in standard scientific journals. We hope that the papers in this volume give a good impression of the current work in graph-based algorithms and will stimulate further research.

The organizational committee is grateful to the Centre for Mathematics and Computer Science in Amsterdam for allowing the use of its excellent facilities for the Workshop, to the Department of Computer Science at the University of Utrecht for supporting the organization of the Workshop, and to Geraldine Leebeek (Department of Computer Science, University of Utrecht) for her invaluable assistance in all matters related to the Workshop.

Utrecht, November 1988.                                                    J. van Leeuwen

# CONTENTS

**Session 6 (Chairman: Jan van Leeuwen)**

# NC-ALGORITHMS
# FOR GRAPHS WITH SMALL TREEWIDTH

Hans L. Bodlaender
Department of Computer Science, University of Utrecht
P.O.Box 80.012, 3508 TA Utrecht, the Netherlands

### Abstract

In this paper we give a parallel algorithm for recognizing graphs with treewidth $\leq k$, for constant $k$, and building the corresponding tree-decomposition, that uses $O(\log n)$ time and $O(n^{3k+4})$ processors on a CRCW PRAM. Also, we give a parallel algorithm that transforms a given tree-decomposition of a graph $G$ with treewidth $k$ to another tree-decomposition of $G$ with treewidth $\leq 3k + 2$, such that the tree in this tree-decomposition is binary and has logarithmic depth. The algorithm uses a linear number of processors and $O(\log n)$ time. Many NP-complete graph problems are known to be solvable in polynomial time, when restricted to graphs with treewidth $\leq k$, $k$ constant. From the results in this paper, it follows that most of these problems are also in NC, when restricted to graphs with treewidth bounded by a constant.

## 1 Introduction

The class of graphs with treewidth $\leq k$ has the property that many graph problems, which are NP-complete for arbitrary graphs, become solvable in polynomial time, when restricted to this class [4,3,6,9,18,17]. Arnborg, Corneil and Proskurowski gave an $\mathcal{O}(n^{k+2})$ algorithm to recognize graphs with treewidth $\leq k$, and find the corresponding tree-decompositions [2]. Deep results from Robertson and Seymour on graph minors show that there exist $\mathcal{O}(n^2)$ algorithms to recognize graphs with treewidth $\leq k$ [16]. Recently, we were able to use this result to show the existence of $\mathcal{O}(n^2)$ algorithms that find the corresponding tree-decompositions. The non-constructive elements in the result of Robertson and Seymour can often, and also in this case, be avoided with a technique of Fellows and Langston [11].

In this paper we consider the parallel complexity of the problems. To be precise: we show that the problems are in NC, i.e. they can be solved on a CRCW PRAM, using a polynomial number of processors and polylogarithmic time. Chandrasekharan and Sitharama Iyengar [8] considered the related problem of recognizing $k$-trees, and showed that this can be done in $\mathcal{O}(\log n)$ time on a CRCW PRAM with $\mathcal{O}(n^4)$ processors. A related result on graphs with bounded treewidth and bounded degree was obtained by Engelfriet, Leih and Welzhl [10]. A special case of these problems is considered in [13].

This paper is organized as follows. In section 2 a number of fundamental definitions are given and some basic graph theoretic results are derived. In section 3 we show that recognizing graphs with treewidth $\leq k$ and finding the corresponding tree-decompositions is in NC, for constant $k$. In section 4 we give a parallel algorithm that transforms a given tree-decomposition of a graph $G$ with treewidth $k$ to another tree-decomposition of $G$ with treewidth $\leq 3k + 2$, such that the tree in this tree-decomposition is binary and has logarithmic depth. From this result, it follows that many sequential polynomial time algorithms for graphs with bounded treewidth can be transformed to NC-algorithms. All problems considered in [3] and [6] can be dealt with in this way.

## 2 Definitions and graph-theoretic results

First we give the definition of the treewidth of a graph, introduced by Robertson and Seymour [15]. Some alternative definitions of the same class of graphs can be found in [1].

**Definition.**
Let $G = (V, E)$ be a graph. A tree-decomposition of G is a pair $(\{X_i \mid i \in I\}, T = (I, F))$, with $\{X_i \mid i \in I\}$ a family of subsets of $V$ and $T$ a tree, with the following properties:

- $\bigcup_{i \in I} X_i = V$

- For every edge $e = (v, w) \in E$, there is a subset $X_i$, $i \in I$ with $v \in X_i$ and $w \in X_i$

- For all $i, j, k \in I$, if $j$ lies on the path in $T$ from $i$ to $k$, then $X_i \cap X_k \subseteq X_j$.

The treewidth of a tree-decomposition $(\{X_i \mid i \in I\}, T)$ is $\max_{i \in I} |X_i| - 1$. The treewidth of $G$, denoted treewidth$(G)$ is the minimum treewidth of a tree-decomposition of $G$, taken over all possible tree-decompositions of $G$.

For a set $S$, clique$(S)$ denotes the graph $(S, \{(v, w) \mid v, w \in S, v \neq w\})$. For graphs $G = (V, E)$, $H = (W, F)$, $G \cup H$ denotes the (possibly non-disjoint) union $(V \cup W, E \cup F)$. For $W \subseteq V$, $G[W]$ denotes the subgraph of $G = (V, E)$ induced by $W$: $G[W] = (W, \{(v, w) \mid v, w \in W \text{ and } (v, w) \in E\})$.

Next we give some graph-theoretic results, which will be used in later sections.

**Lemma 2.1**
Let $(\{X_i \mid i \in I\}, T = (I, F))$ be a tree-decomposition of $G = (V, E)$. Suppose $W \subseteq V$ forms a clique in $G$. Then $\exists i \in I : W \subseteq X_i$.

**Proof.**
Use induction to the clique size $|W|$. For $|W| \leq 2$, the result follows directly from the definition of tree-decomposition. Suppose the lemma holds up to clique size $l - 1$, $l \geq 3$. Consider a clique $W \subseteq V$, with $|W| = l$, and suppose the lemma does not hold for $W$. Choose a vertex $w \in W$, and let $W' = W - \{w\}$. Let $I' \subseteq I$ be the set $\{i \in I \mid W' \subseteq X_i\}$. By induction $I' \neq \emptyset$. Note that $w \in X_i \Rightarrow i \notin I'$. Now choose a node $i' \in I'$, and a node $i \in I$ with $w \in X_i$. Consider the path in $T$ from $i$ to $i'$. Let $i''$ be the last node on this path with $i'' \in I'$, and let $i'''$ be the next node on this path. Now, for every $w' \in W'$, there must be a node $j_{w'}$, with $\{w, w'\} \subseteq X_{j_{w'}}$. Consider the path from $i''$ to $j_{w'}$. There are two cases. *Case 1*: This path does not use $i'''$. In this case, the path in $T$ from $i$ to $j_{w'}$ uses $i''$. Now $w \in X_i$, $w \in X_{j_{w'}}$, hence $w \in X_{i''}$, contradiction. *Case 2*: This path uses $i'''$. Now $w' \in X_{i''}$ and $w' \in X_{j_{w'}}$, hence $w' \in X_{i'''}$. It follows that for all $w' \in W' : w' \in X_{i'''}$, hence $i''' \in I'$, which contradicts the assumption that $i''$ was the last node on the path from $i$ to $i'$, that was in $I'$. $\qquad\square$

**Definition.**
A tree-decomposition $(\{X_i \mid i \in I\}, T = (I, F))$ of a graph $G = (V, E)$ is called *full*, iff

(i) $\forall i, j \in I : |X_i| = |X_j|$, and

(ii) $\forall (i, j) \in F : X_i \not\subseteq X_j$ and $X_j \not\subseteq X_i$.

**Lemma 2.2**
Let $G = (V, E)$ be a graph with treewidth$(G) \leq k$ and $|V| \geq k + 1$. Then $G$ has a full tree-decomposition with treewidth $k$.

**Proof.**
Start with any tree-decomposition of $G$ with treewidth $\leq k$, and repeat the following operations, until a full tree-decomposition is obtained:

1. If there are $(i_0, i_1) \in F$ with $X_{i_0} \subseteq X_{i_1}$ or $X_{i_1} \subseteq X_{i_0}$, then we make a new tree-decomposition by merging $i_0$ and $i_1$. Take $(\{X_i \mid i \in I - \{i_1\}\}, T' = (I - \{i_1\}, \{(i, j) \mid (i, j) \in I - \{i_1\} \text{ and } (i, j) \in F\})$ or $(i = i_0 \text{ and } (i_1, j) \in F)$ or $(j = i_1 \text{ and } (i_1, i) \in F)\})$. This is again a tree-decomposition of $G$ with treewidth $\leq k$, but with a smaller index set $I$.

2. If there is an $i_0 \in I$ with $|X_{i_0}| \leq k$, then either operation 1 can be applied, or there is an adjacent node $i_1 \in I$ with $\exists v \in X_{i_1} : v \notin X_{i_0}$. Make a new tree-decomposition by adding $v$ to $X_{i_0}$: $(\{X_i' \mid i \in I\}, T = (I, F))$, with $X_{i_0}' = X_{i_0} \cup \{v\}$, and $X_i' = X_i$ for $i \neq i_0$. This is again a tree-decomposition of $G$ with treewidth $\leq k$. In this case the size of the index set $I$ does not change, but $\sum_{i \in I} |X_i|$ is increased by one.

Operation 1 can be applied less than $|I|$ times, operation 2 can be applied less than $(k+1)\cdot|I|$ times. So after applying operation 1 and 2 a finite number of times, we obtain a tree-decomposition of $G$ with treewidth $\leq k$, such that neither operation 1 or operation 2 can be applied. This is a a full tree-decomposition of $G$ with treewidth $k$. $\qquad\square$

# 3 An NC-algorithm for recognizing graphs with small treewidth

In this section we show that recognizing graphs with treewidth $\leq k$, and finding the corresponding tree-decomposition, are in NC, for constant $k$. The algorithm is quite inefficient in the use of processors, as it uses $\mathcal{O}(n^{3k+4})$ processors. The algorithm uses $\mathcal{O}(\log n)$ time on a CRCW PRAM.

Suppose $G = (V, E)$ is the input-graph.

First all $(k + 1)$-element vertex sets, which are a separator of $G$, are computed, and numbered $S_1, S_2, \ldots, S_i, \ldots$ For each such $S_i$, the connected components of $G[V - S_i]$ are numbered $S_i^1, S_i^2, \ldots, S_i^j, \ldots$ (Note the difference with the algorithm of Arnborg, Corneil and Proskurowski [2], where $k$-element vertex sets were considered, instead of $(k + 1)$-element sets.)

For each pair of $(k + 1)$-element separators $S_i, S_j, i \neq j$, let $R_{i,j}$ denote the set of vertices $v$, such that $v$ has a path to a vertex in $S_i - S_j$, which avoids $S_j$, and a path to a vertex in $S_j - S_i$, which avoids $S_i$.

## Definition

(i) A pair $(S_i, S_i^j)$ is called *good*, iff $G[S_i \cup S_i^j]\cup$ clique$(S_i)$ has treewidth $\leq k$.

(ii) A triple $(S_i, S_j, R_{i,j})$ is called *good*, iff $G[S_i \cup S_j \cup R_{i,j}]\cup$ clique$(S_i)\cup$ clique$(S_j)$ has treewidth $\leq k$.

The next three lemma's give the essential steps of the algorithm.

## Lemma 3.1
Let $|V| \geq k + 3$. Then: treewidth$(G) \leq k$, if and only if there exists a $(k + 1)$-vertex separator $S_i$, with all $(S_i, S_i^j)$ are good.

## Proof.
($\Rightarrow$) Consider a full tree-decomposition $(\{X_i \mid i \in I\}, T = (I, F))$ of $G$. Take $S_i = X_r$ for an arbitrary internal node $r \in I$.
($\Leftarrow$) For each $G[S_i \cup S_i^j]\cup$ clique$(S_i)$ there exists a tree-decomposition with treewidth $\leq k$. By lemma 2.1, each of these tree-decompositions contains an $X_{i_0}$, with $S_i \subseteq X_{i_0}$, so $S_i = X_{i_0}$. We now can compose a tree-decomposition of $G$ of the tree-decompositions of the $G[S_i \cup S_i^j]\cup$ clique$(S_i)$ graphs, by identifying the nodes $i_0$ with $X_{i_0} = S_i$. $\qquad\square$

## Lemma 3.2
Consider $S_i, S_k, R_{i,k}$ with $R_{i,k} \neq \emptyset$.
$(S_i, S_k, R_{i,k})$ is good, if and only if there exists a $(k + 1)$-vertex cutset $S_j \subseteq S_i \cup S_k \cup R_{i,k}$, such that

(i) $S_j \supseteq (S_i \cup R_{i,j}) \cap (S_k \cup R_{j,k})$

(ii) $(S_i, S_j, R_{i,j})$ and $(S_j, S_k, R_{j,k})$ are good

(iii) $|R_{i,j}| \leq \frac{1}{2}|R_{i,k}|; |R_{j,k}| \leq \frac{1}{2}|R_{i,k}|$

(iv) For all $m$ with $S_j^m \cap (S_i \cup S_k \cup R_{i,j} \cup R_{j,k}) = \emptyset$ and $S_j^m \subseteq R_{i,k} : (S_j, S_j^m)$ is good.

## Proof.
($\Rightarrow$) Consider a full tree-decomposition $(\{X_i \mid i \in I\}, T = (I, F))$ of $G' = G[S_j \cup S_k \cup R_{i,k}]\cup$ clique$(S_i)\cup$ clique$(S_j)$, with treewidth $k$. Note that there must be $i_0 \in I$ with $X_{i_0} = S_i$ and $i_1 \in I$ with $X_{i_1} = S_k$, by lemma 2.1. One may suppose that $i_0$ and $i_1$ are leaves in the tree $T$, else one can remove some nodes from $T$ and still have a full tree-decomposition of $G'$ with treewidth $k$. For each node $i' \in I$ on the path from $i_0$ to $i_1, i' \neq i_0, i' \neq i_1$, we have that $X_{i'}$ is a $k + 1$-vertex cutset of $G'$ (and hence of $G$). As $R_{i,k} \neq \emptyset, |I| \geq 3$, and so there is at least one such node $i'$. For each such $i' \in I$, let $X_{i'} = S_{\alpha(i')}$, and $s(i') = \max(|R_{i\alpha(i')}|, |R_{k\alpha(i')}|)$. Suppose $i_2$ has minimal $s(i_2)$ over all $i'$ on the path from $i_0$ to

$i_1, i' \notin \{i_0, i_1\}$. We claim that $s(i_2) \leq \frac{1}{2}|R_{i,k}|$. Note that $X_{i'} \cap R_{i\alpha(i_2)} \neq \emptyset \Leftrightarrow i'$ belongs to the tree in $T - \{i_2\}$ which also contains $S_i$, and similarly, $X_{i'} \cap R_{k(i_2)} \neq \emptyset \Leftrightarrow i'$ belongs to the tree in $T - \{i_2\}$ which also contains $S_k$. W.l.o.g. suppose $s(i_2) = |R_{k\alpha(i_2)}|$. Let $i_3$ be the next node on the path from $i_2$ to $i_1$. Now $R_{k\alpha(i_3)} \subset R_{k\alpha(i_2)}$ and $R_{k\alpha(i_3)} \neq R_{k\alpha(i_2)}$; and $v \in R_{k\alpha(i_2)} \Rightarrow v \notin R_{i\alpha(i_3)}$. This follows from the definition of tree-decomposition. It follows that $|R_{i\alpha(i_3)}| \geq s(i_2)$ and $|R_{i\alpha(i_3)}| \leq |R_{ik}| - |R_{k\alpha(i_2)}| = |R_{ik}| - s(i_2)$, hence $s(i_2) \leq \frac{1}{2}|R_{ik}|$.

Now let $S_j = X_{i_2}$. Property (i) follows from the observations, made before and the definition of tree-decomposition. Properties (ii) and (iv) follow because the corresponding graphs are subgraphs of $G^1 \cup$ clique$(S_j)$, which clearly has treewidth $\leq k$.

($\Leftarrow$) Note that if $S_j^m \cap (S_i \cup S_k \cup R_{i,j} \cup R_{j,k}) \neq \emptyset$, then $S_j^m \cap R_{i,k} \subseteq R_{i,j} \cup R_{j,k}$. Further, note that if a vertex belongs to two or more of the sets $S_i \cup R_{i,j} \cup S_j, S_j \cup R_{j,k} \cup S_k, S_j \cup S_j^m$, for any $m$ with $S_j^m \cap (S_i \cup S_k \cup R_{i,j} \cup R_{j,k}) = \emptyset$ and $S_j^m \subseteq R_{i,k}$, then it belongs to $S_j$.

Now make tree-decompositions with treewidth $\leq k$ of $G[S_i \cup S_j \cup R_{i,j}] \cup$ clique$(S_i) \cup$ clique$(S_j)$, $G[S_j \cup S_k \cup R_{j,k}] \cup$ clique$(S_j) \cup$ clique$(S_k)$, and $G[S_j \cup S_j^m] \cup$ clique$(S_j)$, for all $m$ as before. Each of these tree-decompositions contains an $i'$ with $X_{i'} = S_j$. By identifying all these $i'$ and so "glueing" the tree-decompositions together we obtain a new tree-decomposition with treewidth $\leq k$. By the two observations made above, it follows that this is indeed a correct tree-decomposition of $G[S_i \cup S_k \cup R_{i,k}] \cup$ clique$(S_i) \cup$ clique$(S_k)$. □

## Lemma 3.3

Consider $(S_i, S_i^j)$ with $|S_i^j| \geq k+1$. $(S_i, S_i^j)$ is good, if and only if there exists a $(k+1)$-vertex cutset $S_j$, such that

(i) $(S_i, S_j, R_{i,j})$ is good

(ii) for all $m$, with $S_j^m \cap (R_{i,j} \cup S_i) = \emptyset$ and $S_j^m \subseteq S_i^j$ : $(S_j, S_j^m)$ is good and $|S_j^m| \leq \frac{1}{2}|S_i^j|$.

(iii) $|R_{i,j}| \leq \frac{1}{2}|S_i^j|$.

## Proof.

($\Rightarrow$) Consider a full tree-decomposition $(\{X_i \mid i \in I\}, T = \{I, F\})$ of $G' = G[S_i \cup S_i^j] \cup$ clique$(S_i)$. For each $i' \in I$ and each component $T' = (I', F')$ of $T - \{i'\}$, let $s(T', i') = |\{v \in X_{i''} \cap S_i^j | i'' \in I'$ and $v \notin X_{i'}\}|$. For all $i' \in I$, define $s(i')$ to be the maximum of $s(T', i')$ over all connected components $T'$ of $T - \{i'\}$. Let $i_0 \in I$ be the node, such that $s(i_0)$ is minimal over all $i' \in I$, and $|X_{i_0} \cap S_i^j|$ is minimal over all $i'$ with minimal $s(i')$. From $|S_i^j| \geq k+1$, it easily follows that $i_0$ is an internal node of $G'$, and hence $X_{i_0}$ is a $(k+1)$-vertex cutset of $G'$, and hence also of $G$. Take $S_j = X_{i_0}$.

We claim that $s(i_0) \leq \frac{1}{2}|S_i^j|$. Let $i_1$ be the node that is adjacent to $i_0$ and in the component $T'$ of $T - \{i_0\}$ with $s(T', i_0) = s(i_0)$. Consider the component $T''$ of $T - \{i_1\}$, that contains $i_0$. From the definition of tree-decomposition it follows that for all $v \in S_i^j : v \in X_{i''}$ for some $i''$ in $T'$ and $v \notin X_{i_0} \Rightarrow v \notin X_{i''}$ for all $i''$ in $T''$ or $v \in X_{i_1}$. So $s(T', i_0) + s(T'', i_1) \leq |S_i^j|$. If there is an $v \in S_i^j$, with $v \in X_{i_0}$ and $v \notin X_{i_1}$, then the result follows. All components $T''''$ of $T - \{i_1\}$, except $T''$, are contained in $T'$, and for each of these components we have $s(T''', i_1) < s(T', i_0)$. So $s(T'', i_1) \geq s(T', i_0)$, and hence $s(i_0) = s(T', i_0) \geq \frac{1}{2}|S_i^j|$.

Also, if $s(i_1) > s(i_0)$, one easily derives that $s(i_0) \leq \frac{1}{2}|S_i^j|$. So suppose $s(i_1) = s(i_0)$ and $v \in S_i^j \cap X_{i_0} \Rightarrow v \in X_{i_1}$, i.e. $|S_i^j \cap X_{i_0}| \geq |S_i^j \cap X_{i_1}|$. By definition of $i_0 : |S_i^j \cap X_{i_0}| = |S_i^j \cap X_{i_1}|$. It follows that $S_i^j \cap X_{i_0} = S_i^j \cap X_{i_1}$. One can derive that $S_i \cap X_{i_0} = S_i \cap X_{i_1}$, and hence $X_{i_0} = X_{i_1}$. So the tree-decomposition was not full. Contradiction. So the claim $s(i_0) \leq \frac{1}{2}|S_i^j|$ follows.

One can now check without difficulty that conditions (i) - (iii) are fulfilled, when taking $S_j = X_{i_0}$.
($\Leftarrow$) Similar as in lemma 3.2. □

With help of these 3 lemma's, an NC-algorithm, using $\mathcal{O}(n^{3k+4})$ processors and $\mathcal{O}(\log n)$ time on a CRCW PRAM can be derived.

First, determine the set of $(k+1)$-vertex cutsets $S_i$.

Secondly, determine which $(S_i, S_i^j)$ are good for $|S_i^j| \leq k$, and which $(S_i, S_j, R_{i,j})$ are good, for $R_{i,j} = \emptyset$. This can be done with $\mathcal{O}(n^{2k})$ processors in $\mathcal{O}(1)$ time.

Then, in $\log n$ phases, one can determine for all $(S_i, S_i^j)$ and $(S_i, S_k, R_{i,k})$ whether they are good, with lemma 3.2 and 3.3; in phase $l$ one considers $S_i^j$ and $R_{i,k}$ with $|S_i^j|, |R_{i,k}| \in \{2^{l-1} + 1, \ldots, 2^l\}$.

Finally, verifying whether treewidth$(G) \leq k$ can be done in $\mathcal{O}(1)$ time, with lemma 3.1, with $\mathcal{O}(n^{k+2})$ processors.

We note that one can also find the corresponding tree-decomposition in $\mathcal{O}(\log n)$ time, with $\mathcal{O}(n^{3k+4})$ processors, using the construction method for the tree-decompositions, indicated by the proofs of lemma 3.1 - 3.3.

**Theorem 3.4**
For each constant $k$, there exists an algorithm that uses $\mathcal{O}(\log n)$ time and $\mathcal{O}(n^{3k+4})$ processors on a CRCW PRAM that determines whether a given input graph has treewidth $\leq k$, and if so, finds a corresponding tree-decomposition.

The algorithm is quite inefficient in the use of processors. By parallizing the algorithm of Arnborg, Corneil and Proskurowski [2] one obtains without much difficulty the following result.

**Theorem 3.5**
For each constant $k$, there exists an algorithm that uses $\mathcal{O}(n)$ time and $\mathcal{O}(n^{k+1})$ processors on a CRCW PRAM, that determines whether a given input graph has treewidth $\leq k$, and if so, finds a corresponding tree-decomposition.

# 4 NC-algorithms for NP-complete problems, restricted to graphs with small treewidth

In this section we give a method to design NC-algorithms, for a large number of graph problems, that are NP-complete for arbitrary graphs, when restricted to the class of graphs with treewidth $\leq k$. Our main result is the following: given a tree-decomposition of $G$ with constant treewidth, one can find (using an NC-algorithm) another tree-decomposition of $G$ with larger, but still constant treewidth, such that the tree $T$, appearing in this tree-decomposition is a binary tree with logarithmic depth.

For example, consider the sequential algorithms, proposed in [6]. The sequential algorithms are of the following form: we suppose a tree-decomposition of $G$ is given. For each node $i \in I$ we compute a table TABLE(i). For computing TABLE(i) one needs TABLE(j) for all sons $j$ of $i$ in $T$. The time to compute such a table is in several cases $\mathcal{O}(\#$ sons of $j$ $)$, in other cases it is polynomial in $n$. A close observation of the algorithms in [6] learns, that if $i$ has $\mathcal{O}(1)$ sons, then either TABLE(i) can be computed in $\mathcal{O}(1)$ time, or TABLE(i) can be computed with an NC-algorithm. We will not give the details here, but refer the reader to [6]. A large number of problems can be dealt with in this manner.

Each of these problems can be solved in NC, when restricted to graph with treewidth $\leq k$. The following two theorems give the main idea.

**Theorem 4.1**
Every binary tree $T = (V, E)$ has a tree-decomposition $\{\{X_i \mid i \in I\}; T' = (I, F)\}$ with treewidth $\leq 3$, and the depth of $T'$ is at most $2\lceil \log_{\frac{4}{3}}(|V|)\rceil$, and $T'$ is a binary tree.

**Proof.**
Our result is based upon the method of parallel tree-contraction of Miller and Reif [14]. We will obtain a series of (rooted) trees $T = T_0 = (V_0, E_0), T_1 = (V_1, E_1), T_2 = (V_2, E_2), \ldots, T_r = (V_r, E_r)$, with $|V_r| = 1$. To each $v \in V_i$ we assign a set $\varphi(v, i) \subseteq V$ representing the set of "vertices that are contracted to $v$". Define $\varphi(v, 0) = \{v\}$.

Each $T_{i+1}$ is obtained from $T_i$ by applying the following two operations in parallel:

1. RAKE. For each node $v \in V_i$, with at least 1 child of $v$ is a leaf in $T_i$ : remove the children from $v$ that are a leaf, and take $\varphi(v, i+1) = \bigcup\{\varphi(w, i) | w = v$ or $w$ is a child of $v$, and $w$ is a leaf $\}$.

2. COMPRESS. A sequence of nodes $v_1, \ldots, v_k$ is a chain if $v_{j+1}$ is the only child of $v_j$, and $v_k$ has exactly one child and that child is not a leaf. Now, in each maximal chain, identify $v_j$ and $v_{j+1}$ for $j$ odd and $1 \leq j < k$. Let $w_i$ be the new node. We take $\varphi(w, i+1) = \varphi(v_j, i) \cup \varphi(v_{j+1}, i)$.

Miller and Reif [14] showed that after $\lceil \log_{\frac{5}{4}} n \rceil$ simultaneous applications of RAKE and COMPRESS, $T$ is reduced to a single vertex. So it follows that $r \leq \lceil \log_{\frac{5}{4}} n \rceil$.

Each $\varphi(v, i)$ represents the set of vertices that are contracted to $v$ in $i$ contractions. Note that each $\varphi(v, i)$ induces a connected subtree of $T$ and that for each $i$, all $\varphi(v, i)$ are disjunct and partition $V$. Furthermore, if $(v, w) \in E$, then either $\exists x \in V_i$ with $v, w \in \varphi(x, i)$ or $\exists x, y \in V_i$ with $(x, y) \in E_i$ and $v \in \varphi(x, i)$ and $w \in \varphi(y, i)$.

Now define $\beta(v, i) = \{w \in \varphi(v, i) \mid \exists w' \in V$ with $(w, w') \in E$ and $w' \notin \varphi(v, i)\}$. $\beta(v, i)$ represents the vertices that are at the "border" of $\varphi(v, i)$. The following properties hold:

1. If $v \in V_i$, and the degree of $v$ in $T_i$ is 3, then $|\varphi(v, i)| = |\beta(v, i)| = 1$.

2. If $v \in V_i$, then $|\beta(v, i)|$ is at most the degree of $v$ in $T_i$.

To make $V_0, \ldots, V_r$ disjoint we label all $v \in V_i$ with $i$. We now give a "first version" of a tree-decomposition $\{\{X_i \mid i \in I\}, T' = (I, F)\}$ of $T$, which "almost" satisfies the constraints. We take $I = \bigcup_{i=0}^{r} V_i$. If vertices $v^i, w^i$, or $v^i, w^i, x^i \in V_i$ are contracted to $y^{i+1}$, then $y^{i+1}$ is the father of $v^i, w^i$ (and $x^i$) in $T'$. If $v$ is unchanged by going from $T_i$ to $T_{i+1}$, the $v^{i+1}$ is the father of $v^i$ in $T'$. Further we take $X_w = \bigcup \{R_x \mid x \text{ is a son of } W \text{ in } T'\}$.

We claim that this is a correct tree-decomposition of $T$ with treewidth $\leq 4$, and no node in $T'$ has more than 3 children.

First, for each edge $(v, w) \in E$, there must be an $i$, such that $v \in \varphi(x^i, i), w \in \varphi(y^i, i)$ and $v, w \in \varphi(z^{i+1}, i+1)$, for some $x^i, y^i, z^{i+1}$. Now $v \in \beta(x^i, i), w \in \beta(y^i, i)$, so $v, w \in X_{z^{i+1}}$.

Secondly, for each $v \in V$ : on each level $i$ of the tree $T'$, there is exactly one $w^i \in V_i$ with $v \in \varphi(w^i, i)$, and hence at most one $w^i \in V_i$ with $v \in \beta(w^i, i)$, so also at most one $w^i \in V_i$ with $v \in X_{w^i}$. Furthermore, if $v \in X_{w^i}$, and $x^{i+1}$ is the father of $w^i$ in $T'$, then either $v \in X_{x^{i+1}}$, or $v$ does not belong to any $X_y$ for $y$ on a level, higher than $i+1$. It follows that we have a correct tree-decomposition.

For all $w \in I, |X_w| \leq 4$ : either two vertices with degree $\leq 2$ are contracted, or a vertex with degree 3 is contracted with one or two leaves. Hence, the treewidth of the tree-decomposition is at most 4. As there are never more than 3 vertices contracted to a single node during a single step, it directly follows that no node in $T'$ has more than 3 children.

We show now that this tree-decomposition can be slightly modified, such that $T'$ is binary, and the treewidth $\leq 3$.

For a node with 3 children, use the transformation in figure 4.1. If $|X_{w^{i+1}}| = 4$, then $w^{i+1}$ is obtained

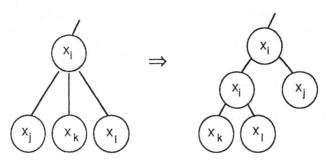

Figure 4.1

by contracting two nodes with degree 2, say $x^i$ and $y^i$. Suppose $\beta(x^i, i) = \{v_0, v_1\}; \beta(y^i, i) = \{v_2, v_3\}$ and $(v_1, v_2) \in E$. Then transform as in figure 4.2. Note that $\beta(w^{i+1}, i+1) = \{v_0, v_3\}$. A new correct tree-decomposition with treewidth $\leq 3$ results with the depth of $T'$ increased by at most a factor 2, and $T'$ is a binary tree. $\quad\square$

**Theorem 4.2**

Let $G = (V, E)$, with $|V| = n$, and treewidth$(G) \leq k$. Then $G$ has a tree-decomposition $(\{X_i \mid i \in I\}, T = (I, F))$ with $T$ a binary tree with depth $\leq 2\lceil \log_{\frac{5}{4}}(2n) \rceil$, and with treewidth of this decomposition $\leq 3k + 2$.

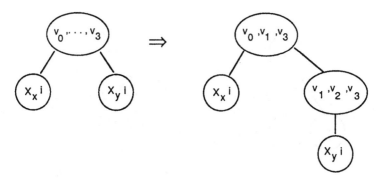

Figure 4.2

**Proof.**
Let $(\{X_i \mid i \in I_1\}, T_1 = (I_1, F_1))$ be a tree-decomposition of $G$ with treewidth $\leq k$, and $|I_1| \leq n$. By transforming nodes in $T_1$ as in figure 4.3, one obtains a new tree-decomposition $(\{X_i \mid i \in I_2\}, T_2 = (I_2, F_2))$ of $G$ with treewidth $\leq k$ and $|I_2| \leq 2n$, and $T$ is a binary tree. Let $(\{Y_i \mid i \in I_3\}, T_3 = (I_3, E_3))$ be a tree-decomposition of $T_2$, with $T_3$ a binary tree with depth $\leq 2\lceil \log_{\frac{5}{4}} |I_2| \rceil \leq 2\lceil \log_{\frac{5}{4}}(2n) \rceil$, and treewidth of this tree-decomposition $\leq 3$, (cf. theorem 4.1.). Then $(\{Z_i \mid i \in I_3\}, T_3 = (I_3, E_3))$ with $Z_i = \bigcup\{X_j \mid j \in Y_i\}$ is a tree-decomposition of $G$, with the required properties. $\qquad\square$

More-over, the tree-decompositions, indicated in theorem 4.1 and 4.2 can be found in logarithmic parallel time. Using the same technique as Miller and Reif [14], one can carry out the construction indicated in the proof of theorem 4.1 in $\mathcal{O}(\log n)$ time on a CRCW PRAM using a linear number of processors. Using similar techniques as in [14], one can also obtain a probabilistic algorithm, that uses $\mathcal{O}(\log n)$ time and $\mathcal{O}(n/\log n)$ processors. One easily sees that the construction, indicated in the proof of theorem 4.2 can be carried out within the same time. Thus, we have the following result.

**Theorem 4.3**
Given a graphs $G = (V, E)$ with treewidth$(G) \leq k = \mathcal{O}(1)$, we can find a tree-decomposition $(\{X_i \mid i \in I\}, T = (I, F)$ of $G$, with treewidth $\mathcal{O}(1)$, and $T$ a binary tree with depth $\mathcal{O}(\log n)$, with an NC$_1$-algorithm.

Hence, the sequential algorithms, proposed in [6] can be transformed to NC-algorithms in the following way. The TABLE's can be computed level by level: first compute the tables for all $i \in I$ with maximum distance to the root of $T$, then for all $i \in I$ with distance one smaller, etc. Each step either takes $\mathcal{O}(1)$ time, or can be carried out in NC. After $\mathcal{O}(\log n)$ such steps, we have found the table for the root of $T$. Finding the answer to the query then costs $\mathcal{O}(1)$ time, or is easily seen to be in NC.

In a similar way, the (sequential) polynomial and linear time algorithms of Arnborg, Lagergren and Seese [3] can be transformed to NC-algorithms, using theorem 4.3 as a first step. We summarize the results in the following theorem. We use the terminology of Garey and Johnson [12]. When vertices and/or edges have weights, these are assumed to be given in unary notation.

**Theorem 4.4**
Each of the following problems is in NC, when restricted to graphs with treewidth $\leq K$, for constant $K$: vertex cover [GT1], dominating set [GT2], domatic number [GT3], chromatic number [GT4], monochromatic triangle [GT5], feedback vertex set [GT7], feedback arc set [GT8], partial feedback edge set [GT9], minimum maximal matching [GT10], partition into triangles [GT11], partition into isomorphic subgraphs

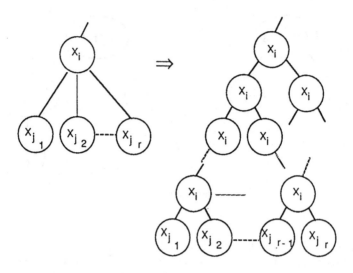

Figure 4.3.

for fixed $H$ [GT12], partition into Hamiltonian subgraphs [GT13], partition into forests [GT14], partition into cliques [GT15], partition into perfect matchings [GT16], clique [GT19], independent set [GT20], induced subgraph with property P (for monadic second order properties P) [GT21], induced connected subgraph with property P (for monadic second order properties P) [GT22], induced path [GT23], balanced complete bipartite subgraph [GT24], bipartite subgraph [GT25], degree bounded connected subgraph for fixed $d$ [GT26], planar subgraph [GT27], transitive subgraph [GT29], uniconnected subgraph [GT30], minimum $k$-connected subgraph for fixed $k$ [GT31], cubic subgraph [GT32], minimum equivalent digraph [GT33], Hamiltonian completion [GT34], Hamiltonian circuit [GT37], directed Hamiltonian circuit [GT38], Hamiltonian path (and directed Hamiltonian path) [GT39], subgraph isomorphism for fixed $H$, subgraph isomorphism for connected $H$ with bounded valence [GT 48], graph contractability for fixed $H$ [GT51], graph homomorphism for fixed $H$ [GT52], path with forbidden pairs for fixed $n$ [GT54], multiple choice matching for fixed $J$ [GT55], graph grundy numbering for graphs with bounded valence [GT56], kernel [GT57], $k$-closure [GT58], path distinguishers [GT60], degree constrained spanning tree [ND1], maximum leaf spanning tree [ND2], bounded diameter spanning tree [ND3], $k$'th best spanning tree for fixed $k$ [ND9], bounded component spanning forest for fixed $k$ [ND10], multiple choice branching for fixed $m$ [ND11], Steiner tree in graphs [ND12], max cut [ND16], minimum cut into bounded sets [ND17], rural postman [ND27], longest circuit [ND28], longest path [ND29], shortest weight-constrained path [ND30], $k$'th shortest path for fixed $k$ [ND31], disjoint connecting paths for fixed $k$ [ND40], maximum length-bounded disjoint paths for fixed $J$ [ND41], maximum fixed-length disjoint paths for fixed $J$ [ND42], chordal graph completion for fixed $k$, chromatix index, spanning tree parity problem, distance $d$ chromatic number for fixed $d$ and $k$, thickness $\leq k$ for fixed $k$, membership for each class $C$ of graphs, which is closed unded minor taking.

## 5   Remarks and open problems

One practical disadvantage of the sequential algorithms on graphs with small treewidth [3,4,7,6,18] is that of the large constants involved in the algorithms. For instance, Arnborg and Proskurowski gave a linear algorithm for Hamiltonian circuit (among others) [4], but consider their algorithm unfeasible for $k \geq 8$. The NC-algorithms, given in this paper will only add to this problem of the large constant factor. However, parallelism will help in a very straightforward way to decrease the running time, as the large constant is in a large extent due to a large number of actions which can be carried out in parallel. So, in many cases, using a large, but constant number of processors may decrease the running time by a large, but again constant factor. Thus, the results in this paper seem to be of mainly theoretical interest.

However, the theoretical importance of the results in this paper are stressed by the fact that a large number of classes of graphs have associated a constant $c$ with them, such that each graph in the class has treewidth $\leq c$. Examples of such classes of graphs are: graphs with bandwidth $\leq k$, graphs with cutwidth $\leq k$, the series-parallel graphs, the outerplanar graphs, the $k$-outerplanar graphs, Halin graphs, chordal graphs with maximum clique size $k$, graphs with genus $\leq d$ and disk dimension $\leq k(d, k$ constants). For an overview of several results of this type, see [5]. The class of partial $k$-trees equals the class of graphs with treewidth $\leq k$.

An interesting open problem is whether there exists a parallel variant of Robertson and Seymours algorithm [16], that finds a branch-decomposition (and hence also a tree-decomposition) of a graph with constant branchwidth or treewidth, that has again constant, but not necessarily optimal branchwidth or treewidth. Their algorithm uses $\mathcal{O}(n^2)$ time. A parallel variant of this algorithm could be used to determine whether a graph has treewidth $\leq k$, and as first step for the algorithms in section 4, using perhaps a smaller number of processors as the algorithm of section 3.

# References

[1] S. Arnborg. Efficient algorithms for combinatorial problems on graphs with bounded decomposability – A survey. *BIT*, 25:2–23, 1985.

[2] S. Arnborg, D. G. Corneil, and A. Proskurowski. Complexity of finding embeddings in a $k$-tree. *SIAM J. Alg. Disc. Meth.*, 8:277–284, 1987.

[3] S. Arnborg, J. Lagergren, and D. Seese. Problems easy for tree-decomposable graphs (extended abstract). In *Proc. 15 th ICALP*, pages 38–51, Springer Verlag, Lect. Notes in Comp. Sc. 317, 1988.

[4] S. Arnborg and A. Proskurowski. *Linear time algorithms for NP-hard problems on graphs embedded in k-trees*. TRITA-NA-8404, Dept. of Num. Anal. and Comp. Sci., Royal Institute of Technology, Stockholm, Sweden, 1984.

[5] H. L. Bodlaender. *Classes of Graphs with Bounded Treewidth*. Technical Report RUU-CS-86-22, Dept. Of Comp. Science, University of Utrecht, Utrecht, 1986.

[6] H. L. Bodlaender. *Dynamic programming algorithms on graphs with bounded tree-width*. Tech. Rep., Lab. for Comp. Science, M.I.T., 1987. Extended abstract in proceedings ICALP 88.

[7] H. L. Bodlaender. Polynomial algorithms for Graph Isomorphism and Chromatic Index on partial $k$-trees. In *Proc. 1st Scandinavian Workshop on Algorithm Theory*, pages 223–232, Springer Verlag LNCS 318, 1988.

[8] N. Chandrasekharan and S. S. Iyengar. *NC Algorithms for Recognizing Chordal Graphs and k-Trees*. Tech. Rep. 86-020, Dept. of Comp. Science, Louisiana State University, 1986.

[9] B. Courcelle. *Recognizability and Second-Order Definability for Sets of Finite Graphs*. Preprint, Universite de Bordeaux, 1987.

[10] J. Engelfriet, G. Leih, and E. Welzl. Characterization and complexity of boundary graph languages. 1987. Manuscript.

[11] M. R. Fellows and M. A. Langston. On seach, decision and the efficiency of polynomial-time algorithms. 1988. Extended abstract.

[12] M. R. Garey and D. S. Johnson. *Computers and Intractability, A Guide to the Theory of NP-Completeness*. W.H. Freeman and Company, New York, 1979.

[13] A. M. Gibbons, A. Israeli, and W. Rytter. Parallel $o(\log n)$ time edge-coloring of trees and halin graphs. *Inform. Proc. Letters*, 27:43–52, 1988.

[14] G. Miller and J. Reif. Parallel tree contraction and its application. In *Proc. of the 26th Annual IEEE Symp. on the Foundations of Comp. Science*, pages 478–489, 1985.

[15] N. Robertson and P. Seymour. Graph minors. II. Algorithmic aspects of tree-width. *J. of Algorithms*, 7:309–322, 1986.

[16] N. Robertson and P. Seymour. Graph minors. XIII. The disjoint paths problem. 1986. Manuscript.

[17] P. Scheffler. *Linear-time algorithms for NP-complete problems restricted to partial k-trees*. Report R-MATH-03/87, Karl-Weierstrass-Institut Für Mathematik, Berlin, GDR, 1987.

[18] P. Scheffler and D. Seese. A combinatorial and logical approach to linear-time computability. 1986. Extended abstract.

# GRAPH–THEORETIC PROPERTIES

# COMPATIBLE WITH GRAPH DERIVATIONS

**Annegret Habel**

Fachbereich Mathematik und Informatik

Universität Bremen

D–2800 Bremen 33

## Abstract

A graph–theoretic property is compatible with the rewriting process of hyperedge–replacement graph grammars if for each graph and each derivation of it the property holds just in case the property (or a related property) holds for some specific subgraphs determined by the fibres of the derivation. On the one hand, this leads to proper tests of compatible properties. On the other hand, compatible properties turn out to be decidable for the corresponding graph languages, i.e., the questions

    (1) Is there a graph in the generated language having the property?

    (2) Do all graphs in the generated language have the property?

are decidable for all hyperedge–replacement graph grammars as inputs. In this paper, we introduce the concept of compatible properties, show the compatibility of connectedness, existence of Eulerian paths and cycles, and edge–colorability, and apply the decidability result to these distinguished properties.

## 1. Introduction

Graph grammars and graph languages are motivated from a wide spectrum of applications so that, for the last 20 years, many researchers have felt encouraged to introduce and to investigate an enormous variety of graph–rewriting mechanisms (see the proceedings of the three graph grammar workshops [CER 79], [ENR 83], and [ENRR 87] as a survey on theory and applications). As a class of grammars defines and generates a class of languages, but does not tell much about the languages in general, it is hard to find a class of graph grammars with nice decidability properties and desirable structural results.

At least two approaches are more promising: Boundary node–label–controlled graph grammars [RW 86 a+b] and edge–replacement grammars [Co 87, HK 83+85]. In both cases, the generative mechanisms and powers are of context–free nature, and several — especially graph-theoretic — properties are decidable for the generated graph languages.

In [HKV 87] the investigation of decision problems concerning graph languages (generated by edge– resp. hyperedge–replacement grammars) is continued and a quite general criterion for decidability is given. Roughly speaking: whenever a graph–theoretic property is "compatible" with the derivation process of hyperedge–replacement grammars in a certain way, then, for each given hyperedge–replacement grammar, the questions

    (1) Is there a graph in the generated language having the property?

    (2) Do all graphs in the generated language have the property?

are decidable. By this criterion, compatibility implies decidability. It remains the task to prove the compatibility.

This paper is mainly concerned with the compatibility of graph–theoretic properties with the derivation process. A graph–theoretic property is termed to be compatible with the derivation process of hyperedge–replacement grammars if it can be tested for each graph $G$ and each derivation of the form $A^\bullet \Longrightarrow R \stackrel{*}{\Longrightarrow} G$ by testing the property (or related properties) for the graphs derived from the hyperedges of $R$ and composing the results of the tests to a result for $G$. In this way, compatible properties play an important role for their own: they allow proper tests for deciding whether a given graph in a generated language has the considered property.

The paper is organized as follows: In Section 2 we recall the basic notions of edge– and hyperedge–replacement grammars. In Section 3, graph–theoretic properties like total disconnectedness, connectedness, existence of Eulerian paths and cycles, and $k$–edge–colorability are discussed and it is shown that these properties can be easily tested for a graph provided that a derivation of the graph is known. In Section 4, a formal definition of compatibility is given and the graph properties investigated in Section 3 are shown to be compatible. Moreover, the set of compatible graph properties turns out to be closed under Boolean operations. Finally, a metatheorem is stated saying that the compatibility of a graph property yields the decidability of the questions (1) and (2) mentioned above. In an appendix, the basic notions of (hyper)graphs and hyperedge replacement are summarized as far as they are needed in this paper.

# 2. Hyperedge–Replacement Grammars

In this section we give a summary of the basic notions on hyperedge–replacement grammars generalizing edge–replacement grammars as investigated e.g. in [HK 83+85] and context–free string grammars. They can be seen as hypergraph–manipulating and hypergraph–language–generating devices and turn out to be closely related to other concepts in computer science. Details and examples can be found in [HK 87a+b].

Hyperedge replacement is a simple mechanism for replacing hyperedges by multi–pointed hypergraphs. Based on this concept, one can derive multi–pointed hypergraphs from multi–pointed hypergraphs by applying productions of a simple form. [The notions and notations of hypergraphs and hyperedge replacements are explained in the appendix.]

## 2.1 Definition (productions and derivations)

1. Let $N \subseteq C$. A *production* over $N$ is an ordered pair $p = (A, R)$ with $A \in N$ and $R \in \mathcal{H}_C$. $A$ is called *left–hand side* of $p$ and is denoted by $lhs(p)$, $R$ is called *right–hand side* and is denoted by $rhs(p)$. The *type* of $p$ is given by the type of $R$.

2. Let $H \in \mathcal{H}_C$, $B \subseteq E_H$, and $P$ be a set of productions over $N$. A mapping $prod : B \to P$ is called a *production base* in $H$ if $l_H(b) = lhs(prod(b))$ and $type(b) = type(rhs(prod(b)))$ for all $b \in B$.

3. Let $H, H' \in \mathcal{H}_C$ and $prod : B \to P$ be a production base in $H$. Then $H$ *directly derives* $H'$ through $prod$ if $H'$ is isomorphic to $REPLACE(H, repl)$ where $repl : B \to \mathcal{H}_C$ is given by $repl(b) = rhs(prod(b))$ for all $b \in B$. We write $H \underset{prod}{\Longrightarrow} H'$, $H \underset{P}{\Longrightarrow} H'$, or $H \Longrightarrow H'$ in this case.

4. A sequence of direct derivations $H_0 \underset{prod_1}{\Longrightarrow} H_1 \underset{prod_2}{\Longrightarrow} \ldots \underset{prod_k}{\Longrightarrow} H_k$ is called a *derivation* from $H_0$ to $H_k$ (of length $k$). Additionally, the case $H_0 \cong H_0'$ is called a *derivation* from $H_0$ to $H_0'$ of length 0. A derivation from $H$ to $H'$ is shortly denoted by $H \underset{P}{\overset{*}{\Longrightarrow}} H'$ or $H \overset{*}{\Longrightarrow} H'$. If the length of the derivation shall be stressed, we write $H \underset{P}{\overset{k}{\Longrightarrow}} H'$ or $H \overset{k}{\Longrightarrow} H'$. The set of all derivations (over a given set of productions) is denoted by $DER$.

5. A direct derivation through $prod : \emptyset \to P$ is called a *dummy*. [1] A derivation is said to be *valid* if at least one of its steps is not a dummy.

**Remark:** The application of a production $p = (A, R)$ of type $(m, n)$ to a multi–pointed hypergraph $H$ requires the following two steps only:

(1) Choose a hyperedge $e$ of type $(m, n)$ with label $A$.

(2) Replace the hyperedge $e$ in $H$ by $R$.

---

[1] A production base $prod : B \to P$ may be *empty*, i.e., $B = \emptyset$. In this case, $H \Longrightarrow H'$ through $prod$ implies $H \cong H'$, and there is always a trivial direct derivation $H \Longrightarrow H$ through $prod$.

Using the introduced concepts of productions and derivations hyperedge–replacement grammars and languages can be introduced in a straightforward way.

**2.2 Definition** (hyperedge–replacement grammars and languages)

1. A *hyperedge–replacement grammar* is a system $HRG = (N, T, P, Z)$ where $N \subseteq C$ is a set of *nonterminals*, $T \subseteq C$ is a set of *terminals*, $P$ is a finite set of *productions* over $N$, and $Z \in \mathcal{H}_C$ is the *axiom*.

2. $HRG$ is said to be *typed* if there is a mapping $ltype : N \cup T \rightarrow I\!N \times I\!N$ such that, for each production $(A, R) \in P$, $ltype(A) = type(R)$ and $ltype(l_R(e)) = type(e)$ for all $e \in E_R$ and $ltype(l_Z(e)) = type(e)$ for all $e \in E_Z$. It is said to be *well–formed* if the right–hand sides of the productions are well–formed and all hyperedges in $Z$ are well–formed.

3. The *hypergraph language $L(HRG)$ generated by $HRG$* consists of all hypergraphs which can be derived from $Z$ applying productions of $P$ and which are terminally labeled:

$$L(HRG) = \{H \in \mathcal{H}_T \,|\, Z \overset{*}{\underset{P}{\Longrightarrow}} H\}.$$

**Remarks:** 1. Even if one wants to generate graph languages rather than hypergraph languages, one may use nonterminal hyperedges because the generative power of hyperedge–replacement grammars increases with the maximum number of tentacles of a hyperedge involved in the replacement (see [HK 87b]).

2. Without effecting the generative power, we assume in the following that $N$ and $T$ are finite, $N \cap T = \emptyset$, and $Z$ is a singleton the label of which is in $N$. Furthermore, we assume that the hyperedge–replacement grammars considered in this paper are typed and well–formed.

The results presented in the following sections are mainly based on some fundamental aspects of hyperedge–replacement derivations. Roughly speaking, hyperedge–replacement derivations cannot interfere with each other as long as they handle different hyperedges: On the one hand, given a hypergraph $R$, a collection of derivations of the form $e^{\bullet} \overset{*}{\Longrightarrow} H(e)$ with $e \in E_R$ can be simultaneously embedded into $R$ leading to a single derivation $R \overset{*}{\Longrightarrow} H$.[2] On the other hand, restricting a derivation $R \overset{*}{\Longrightarrow} H$ to the handle $e^{\bullet}$ induced by the hyperedge $e \in E_R$, one obtains a so–called "restricted" derivation $e^{\bullet} \overset{*}{\Longrightarrow} H(e)$ where $H(e) \subseteq H$. Finally, restricting a derivation to the handles induced by the hyperedges, and subsequently embedding them again returns the original derivation. In other words, hyperedge–replacement derivations can be distributed to (the handles of) the hyperedges without loosing information.

**2.3 Theorem**

1. *EMBED:* Let $H \in \mathcal{H}_C$, $B \subseteq E_H$, and *fibre* $: B \rightarrow DER$ be a mapping where $fibre(e)$ for each $e \in B$ has the form $H(e)_0 \Longrightarrow H(e)_1 \Longrightarrow \ldots \Longrightarrow H(e)_k$ $(k \geq 0)$ and $type(H(e)_0) = type(e)$. Then there is a derivation $H_0 \Longrightarrow H_1 \Longrightarrow \ldots \Longrightarrow H_k$, denoted by $EMBED(H, fibre)$, where, for $i = 0, \ldots, k$, $H_i = REPLACE(H, repl_i)$ and $repl_i : B \rightarrow \mathcal{H}_C$ is defined by $repl_i(e) = H(e)_i$ for all $e \in B$.

2. *RESTRICT:* Let $der : H_0 \Longrightarrow H_1 \Longrightarrow \ldots \Longrightarrow H_k$ be a derivation and $H'_0 \subseteq H_0$. Then there is a derivation $H'_0 \Longrightarrow H'_1 \Longrightarrow \ldots \Longrightarrow H'_k$, denoted by $RESTRICT(der, H'_0)$, where, for $i = 0, \ldots, k$, $H'_i \subseteq H_i$.

3. The operations *EMBED* and *RESTRICT* are related in the following sense:
(a) Let $der : H \overset{*}{\Longrightarrow} H'$ be a derivation. Then there is a set $B \subseteq E_H$ such that, for $fibre : B \rightarrow DER$ given by $fibre(e) = RESTRICT(der, e^{\bullet})$ for $e \in B$, $EMBED(H, fibre) = der$ (up to isomorphisms), i.e., the old derivation can be reconstructed from its fibres.

---

[2] Given $R \in \mathcal{H}_C$, each hyperedge $e \in E_R$ induces a handle, denoted by $e^{\bullet}$, given by $(V, \{e\}, s, t, l, s_R(e), t_R(e))$ where $V$ is the set of nodes occurring in $s_R(e)$ or $t_R(e)$, $s$, $t$ are the restrictions of the mappings $s_R$, $t_R$ to the sets $\{e\}$ and $V^{\bullet}$, and $l$ is the restriction of $l_R$ to $\{e\}$. Obviously, $e^{\bullet} \subseteq R$ for all $e \in E_R$.

(b) Let $H \in \mathcal{H}_C$, $B \subseteq E_H$, and $fibre : B \to DER$ a mapping so that $EMBED(H, fibre)$ is defined. Then (up to isomorphisms) $RESTRICT(EMBED(H, fibre), e^{\bullet}) = fibre(e)$ for all $e \in B$, i.e., the embedded fibres can be extracted from the embedding derivation.

**Remarks:** 1. The common–length condition of the derivations composed by $EMBED$ is not a serious restriction because each derivation can be lengthened by dummy steps.
2. The explicit construction of the restricted derivation works as follows: Let $H_0 \Longrightarrow H_1$ through $prod_1 : B_0 \to P$ be the first derivation step. Consider $B_0' = B_0 \cap E_{H_0'}$ and the restriction $prod_1' : B_0' \to P$ of $prod_1$ to $B_0'$. Then $prod_1'$ is a production base in $H_0'$ and $H_0' \Longrightarrow H_1'$ can be constructed such that $H_1' \subseteq H_1$. Hence this restriction step can be iterated.
3. The result stated in 3(a) is called Context–Freeness Lemma in [HK 85]. It says that derivations in hyperedge–replacement grammars can be decomposed into "thin fibres" (where each starts from the handle induced by a hyperedge) without loosing information.

Theorem 2.3 is the basis of our considerations. We use it in the following recursive version concerning terminal hypergraphs which are derivable from handles.

## 2.4 Corollary

Let $HRG = (N, T, P, Z)$ be a typed and well–formed hyperedge–replacement grammar, $A \in N \cup T$, and $H \in \mathcal{H}_T$.

Then there is a derivation $A^{\bullet} \Longrightarrow R \overset{k}{\Longrightarrow} H$ for some $k \geq 0$ [3] if and only if

$A^{\bullet} \Longrightarrow R$ and for each $e \in E_R$ there is a derivation $l_R(e)^{\bullet} \overset{k}{\Longrightarrow} H(e)$ with $H(e) \subseteq H$ such that $H \cong REPLACE(R, repl)$ with $repl(e) = H(e)$ for $e \in E_R$.

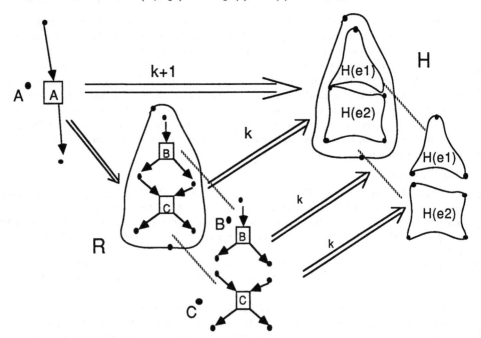

Note that the derivation $l_R(e)^{\bullet} \overset{k}{\Longrightarrow} H(e)$ may be valid or not. In the first case, it has the same form as the original derivation, but it is shorter as the original one. In the latter case, $H(e)$ is isomorphic to $e^{\bullet}$ (resp. $l_R(e)^{\bullet}$) and hence a terminal handle.

---

[3] For a symbol $A \in N \cup T$ with $ltype(A) = (m, n)$, $A^{\bullet}$ denotes an $(m, n)$–handle induced by $A$. [Note that $(m, n)$–handles induced by a symbol $A$ are isomorphic].

# 3.Compatibility of Some Graph–Theoretic Properties

A hyperedge–replacement grammar as a generating device specifies a (hyper)graph language. Unfortunately, the generating process produces never more than a finite section of the language explicitly (and even this may consume much time). Hence one may wonder what the hyperedge–replacement grammar can tell us about the generated language.

As a matter of fact, by Corollary 2.4, we have the following nice situation. Given a hyperedge–replacement grammar and an arbitrary terminal (hyper)graph $H$ with derivation $A^{\bullet} \Longrightarrow R \overset{*}{\Longrightarrow} H$, we get a decomposition of $H$ into "smaller" components which are derivable from the handles of the hyperedges in $R$. If one is interested in graph–theoretic properties of derived (hyper)graphs, one may ask how a certain property of a derived (hyper)graph depends on properties of the components and of the involved (hyper)graph $R$. If a property is "compatible" with the derivation process of hyperedge–replacement grammars, i.e., if it can be tested for each derived (hyper)graph $H$ by testing the property (or related properties) for the components and composing the results to a result for $H$, then a hyperedge–replacement grammar can tell us whether (1) there is a (hyper)graph in the generated language having the property and whether (2) all (hyper)graphs in the generated language have the property.

In this section, we pick up several graph–theoretic properties and show that they are "compatible" with the replacement process of hyperedges. A formal definition of compatibility as well as a metatheorem saying that compatibility implies decidability of the questions (1) and (2) is given in the next section. We are going to have a close look at connectedness, existence of Eulerian paths and cycles, bounded degreeness, and edge–colorability. To accustom to our kind of investigations, we first consider the question whether or not a given hypergraph $H$ is totally disconnected, i.e., whether or not $E_H = \emptyset$.

## 3.1 Total Disconnectedness

Given a typed and well–formed hyperedge–replacement grammar $HRG = (N, T, P, Z)$, a hypergraph $H \in \mathcal{H}_T$, and a derivation of the form $A^{\bullet} \Longrightarrow R \overset{*}{\Longrightarrow} H$ in $HRG$, there is a simple method for testing whether $H$ is totally disconnected or not. The proposed method makes use of the fact that $H$ can be obtained from $R$ by replacing each hyperedge $e \in E_R$ by the result $H(e)$ of its fibre.

### Theorem

Let $HRG = (N, T, P, Z)$ be a typed and well–formed hyperedge–replacement grammar, $A^{\bullet} \Longrightarrow R \overset{*}{\Longrightarrow} H$ a derivation in $HRG$ with $A \in N \cup T$ and $H \in \mathcal{H}_T$, and, for $e \in E_R$, $l_R(e)^{\bullet} \overset{*}{\Longrightarrow} H(e)$ the fibre of $R \overset{*}{\Longrightarrow} H$ induced by $e$. Then $H$ is totally disconnected if and only if $H(e)$ is totally disconnected for all $e \in E_R$.

**Proof:** Without loss of generality, we can assume that $H = REPLACE(R, repl)$ with $repl(e) = H(e) \subseteq H$ for each $e \in E_R$. If $H$ is totally disconnected, i.e., $E_H = \emptyset$, then $E_{H(e)} = \emptyset$ for all $e \in E_R$. Conversely, if $H(e)$ is totally disconnected for all $e \in E_R$, then $E_H = \sum_{e \in E_R} E_{H(e)} = \emptyset$, i.e., $H$ is totally disconnected. $\qquad\square$

For simplifying the technicalities, we restrict our following consideration to the class $\mathcal{C}_{ERG}$ of edge–replacement grammars in the sense of [HK 83+85]. To be more explicit, a typed and well–formed hyperedge–replacement grammar $HRG = (N, T, P, Z)$ is in $\mathcal{C}_{ERG}$ if and only if the right–hand sides of the productions as well as the axiom are (1,1)–graphs. Note that, in this case, each $G \in L(HRG)$ is a (1,1)–graph, i.e., a graph with two distinguished nodes $begin_G$ and $end_G$.

## 3.2 Connectedness

Now we are going to discuss the question "Is $G$ connected?" for a graph $G \in \mathcal{G}_T$ given by a derivation $A^{\bullet} \Longrightarrow R \overset{*}{\Longrightarrow} G$ in an edge–replacement grammar $ERG = (N, T, P, Z)$ in terms of the fibres $e^{\bullet} \overset{*}{\Longrightarrow} G(e)$ induced by the edges of $R$.

Obviously, a graph $G$ is connected whenever $R$ is connected and $G(e)$ is connected for all $e \in E_R$. The converse relationship does not hold as the following example shows.

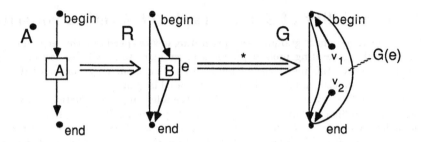

Consider the derivation $A^{\bullet} \Longrightarrow R \stackrel{*}{\Longrightarrow} G$ (in Figure 1) in which $R$ (consisting of a terminal edge as well as a nonterminal edge $e$) is connected and the derived graph $G$ (consisting of three terminal edges) is connected, too. The graph $G(e)$ (marked by the "half-moon") derived from the handle of $e$ is not connected. It is "semi-connected" in the following sense: each node of $G(e)$ is joined with $begin_{G(e)} = begin$ or $end_{G(e)} = end$; addition of an edge between $begin_{G(e)}$ and $end_{G(e)}$ would yield a connected graph. Note that the removal of the edge $e \in E_R$ (the handle of which derives the non-connected graph $G(e)$) from the connected graph $R$ yields a connected graph $R' = R - \{e\}$, again. The connection of the nodes $v_1, v_2 \in V_{G(e)} \subseteq V_G$ is given by three edges, the edge joining the nodes $v_1$ and $begin_{G(e)} = begin$ in $G(e)$, the edge outside from $G(e)$ connecting $begin$ and $end$, and the edge joining $end = end_{G(e)}$ and $v_2$ in $G(e)$.

As the example shows, beside the property "connectedness" the property "semi-connectedness" is needed to characterize "connectedness" of graphs derived from handles. In the following a graph $G \in \mathcal{G}_T$ is said to be *semi-connected* if each node $v \in V_G - \{begin_G, end_G\}$ is joined with $begin_G$ or $end_G$. It is called *connected* if each node $v \in V_G - \{begin_G, end_G\}$ is joined with $begin_G$ and $end_G$.

## Theorem

Let $ERG = (N, T, P, Z)$ be an edge-replacement grammar, $A^{\bullet} \Longrightarrow R \stackrel{*}{\Longrightarrow} G$ a derivation in $ERG$ with $A \in N \cup T$ and $G \in \mathcal{G}_T$, and, for $e \in E_R$, $l_R(e)^{\bullet} \stackrel{*}{\Longrightarrow} G(e)$ the fibre of $R \stackrel{*}{\Longrightarrow} G$ induced by $e$. Then

1. $G$ is connected if and only if $R' = R - \{e \in E_R | G(e) \text{ is not connected}\}$ is connected and $G(e)$ is semi-connected for all $e \in E_R$.
2. $G$ is semi-connected if and only if $R' = R - \{e \in E_R | G(e) \text{ is not connected}\}$ is semi-connected and $G(e)$ is semi-connected for all $e \in E_R$.

**Proof:** Without loss of generality, we may assume that $V_R \subseteq V_G$ and $G = REPLACE(R, repl)$ with $repl(e) = G(e) \subseteq G$ for each $e \in E_R$. 1. Let $G$ be connected. Then $G(e)$ is semi-connected for each $e \in E_R$ which may be seen as follows: If, for some $e \in E_R$, $G(e)$ would not be semi-connected, then there would be a connected component of $G(e)$ neither containing $begin_{G(e)}$ nor $end_{G(e)}$ which would be a connected component in $G$, too. Moreover, $R' = R - \{e \in E_R | G(e) \text{ is not connected}\}$ turns out to be connected: For $v_1, v_2 \in V_R \subseteq V_G$ there exists a path $p$ in $G$, which joins $v_1$ and $v_2$. [4] Running through the path, some nodes of $R$ are visited, say $w_0, \ldots, w_n$. Cutting $p$ in these nodes, one gets a sequence of paths $p_1, \ldots, p_n$ (where $p_i$ connects $w_{i-1}$ and $w_i$ for $i = 1, \ldots, n$). For each $i = 1, \ldots, n$, there exists $e_i \in E_R$ such that $p_i$ belongs to $G(e_i)$. Obviously, $e_i$ connects $w_{i-1}$ and $w_i$ for $i = 1, \ldots, n$. Thus, the edges $e_1, \ldots, e_n$ form a path from $v_1$ to $v_2$ in $R$. Note that $e_1, \ldots, e_n \in E_{R'}$ because the graphs $G(e_1), \ldots, G(e_n)$ are semi-connected as well as begin-end connected and hence connected. Conversely, let $R' = R - \{e \in E_R | G(e) \text{ is not connected}\}$ be connected and $G(e)$ be semi-connected for all $e \in E_R$. Let $v_1, v_2 \in V_G$. Three cases may occur.

---

[4] A *path* is meant to be a sequence $p = e_1, e_2, \ldots, e_n$ of edges such that, for $1 \leq i < n$, $e_i$ and $e_{i+1}$ are adjacent.

Case 1: $v_1, v_2 \in V_R$. Then $R$ has a path from $v_1$ to $v_2$ consisting, say of edges $e_1, \ldots, e_n$, and $G(e_i)$ for $i = 1, \ldots, n$ has a path $p_i$ connecting $begin_{G(e_i)}$ and $end_{G(e_i)}$. Then the sequence of paths $p_1, \ldots, p_n$ forms a path $p$ from $v_1$ to $v_2$ in $G$.

Case 2: $v_1 \in INT_{repl(e)}$ (for some $e \in E_R$) and $v_2 \in V_R$. By semi–connectedness of $repl(e)$ there is a node $v_1' \in EXT_{repl(e)}$ such that $v_1$ and $v_1'$ are joined by a path in $repl(e)$. Moreover, $v_1'$ and $v_2$ are joined by a path in $G$ (cf. case 1). Thus, there is a path from $v_1$ to $v_2$ in $G$.

Case 3: $v_i \in INT_{repl(e_i)}$ for some $e_i \in E_R$ ($i = 1, 2$). Using the same arguments as in case 2, there is a node $v_i' \in EXT_{repl(e_i)}$, such that $v_i$ and $v_i'$ are joined by a path in $repl(e_i)$ ($i = 1, 2$). Moreover, $v_1'$ and $v_2'$ are joined by a path in $G$ (cf. case 1). Thus, there is a path from $v_1$ to $v_2$ in $G$. This completes the proof of the first statement.

2. The second statement can be proved in a similar way. □

For expressing the interrelations between the statements concerning "connectedness" and "semi–connectedness" we make use of the predicates $CONN$ and $CONN'$ defined on pairs $(G, x)$ with $G \in \mathcal{G}_T$ and $x \in \{co, semico\}$ and triples $(R, assign, x)$ with $R \in \mathcal{G}_{NUT}$, $assign : E_R \to \{co, semico\}$, and $x \in \{co, semico\}$, respectively:

- $CONN(G, co)$ if and only if $G$ is connected,
  $CONN(G, semico)$ if and only if $G$ is semi–connected,
- $CONN'(R, assign, co)$ if and only if $R' = R - \{e \in E_R \mid assign(e) \neq co\}$ is connected and $CONN'(R, assign, semico)$ if and only if $R' = R - \{e \in E_R \mid assign(e) \neq co\}$ is semi–connected.

Now the theorem can be reformulated as follows:

## Corollary

Let $ERG = (N, T, P, Z)$ be an edge–replacement grammar, $A^* \Longrightarrow R \stackrel{*}{\Longrightarrow} G$ a derivation in $ERG$ with $A \in N \cup T$ and $G \in \mathcal{G}_T$, and, for $e \in E_R$, $l_R(e)^\bullet \stackrel{*}{\Longrightarrow} G(e)$ the fibre of $R \stackrel{*}{\Longrightarrow} G$ induced by $e$. Then, for $x \in \{co, semico\}$, $CONN(G, x)$ holds if and only if there is a mapping $assign : E_R \to \{co, semico\}$ such that $CONN'(R, assign, x)$ as well as $CONN(G(e), assign(e))$ for all $e \in E_R$.

**Proof:** For $x = co$ (resp. $x = semico$), suppose $CONN(G, x)$ holds. By the theorem, $R' = R - \{e \in E_R \mid G(e) \text{ is not connected}\}$ is connected (resp. semi–connected) and $G(e)$ is semi–connected for all $e \in E_R$. Now choose $assign : E_R \to \{co, semico\}$ by $assign(e) \neq co$ iff $G(e)$ is not connected ($e \in E_R$). Then $CONN'(R, assign, x)$ is satisfied. Moreover, $CONN(G(e), assign(e))$ holds for $e \in E_R$ with $assign(e) = semico$ and, by choice of $assign$, $CONN(G(e), assign(e))$ holds for $e \in E_R$ with $assign(e) = co$, too. Conversely, suppose $assign : E_R \to \{co, semico\}$ is given such that $CONN'(R, assign, co)$ (resp. $CONN'(R, assign, semico)$) and $CONN(G(e), assign(e))$ for $e \in E_R$ hold. Then $R - \{e \in E_R \mid G(e) \text{ is not connected}\} \supseteq R - \{e \in E_R \mid assign(e) \neq co\}$ is connected (resp. semi–connected) and $G(e)$ is semi–connected for all $e \in E_R$ because each connected graph is also semi–connected. □

## 3.3 Eulerian Graphs

We are going to discuss the question "Is $G$ Eulerian?", i.e., "Is it possible to find a path that traverses each edge of the graph exactly once, goes through all nodes, and ends at the starting node?". For this purpose, we use the well–known fact that a graph $G$ is Eulerian if and only if it is connected and every node of $G$ has even degree. The problem whether $G$ is connected (or not) can be handled as in 3.2. Hence it remains to find a criterion for testing whether every node of $G$ has even degree or not.

The "even–degree property" of a graph $G$ with derivation $A^* \Longrightarrow R \stackrel{*}{\Longrightarrow} G$ can be characterized in terms of the fibres $l_R(e)^\bullet \stackrel{*}{\Longrightarrow} G(e)$. The characterization is based on the following observation. Let $R$ be a graph and $v$ be a node in $R$. If an edge $e$ outgoing from $v$ is replaced by some graph $G(e)$, then the degree of $begin_{G(e)}$ has to be taken into account whenever the degree of $v$ in $G$ is considered; accordingly, the degree of $end_{G(e)}$ has to be taken in consideration if $e$ is incoming into the node $v$. Now the degree of $v$ in $G$ can be determined by summing up the degrees of the nodes $begin_{G(e)}$ in $G(e)$ for edges $e$ outgoing from $v \in V_R$ and $end_{G(e)}$ in $G(e)$ for edges $e$ incoming into $v \in V_R$.

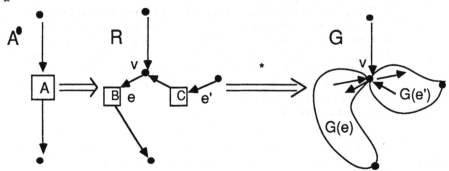

With respect to the question "Does all nodes of a graph $G \in \mathcal{G}_T$ have even degree?" it is useful to know for a graph $G(e) \in \mathcal{G}_T$ whether the degree of all internal nodes is even, whether the degree of the begin–node is even or odd, and whether the degree of the end–node is even or odd. For determing whether all nodes of $G$ have even degree, we have to consider the nodes $v \in V_R \subseteq V_G$, too. Since $R \in \mathcal{G}_{NUT}$ may contain nonterminal edges and nonterminal edges are replaced in the derivation of $G$, we have to make use of the results with respect to the begin–node and end–node of the derived graphs. Thus, the "auxiliary" property concerning $R$ used to compose the results of the components $G(e)$, is more complicated than the "main" properties.

Let $G \in \mathcal{G}_T$, $R \in \mathcal{G}_{NUT}$, $assign : E_R \to \{0,1\}^2$, and $i,j \in \{0,1\}$. Then
- $EUL(G,(i,j))$ expresses the fact that $d_G(begin_G) \bmod 2 = i$, $d_G(end_G) \bmod 2 = j$, and $d_G(v) \bmod 2 = 0$ for all $v \in INT_G$. [5]
- $EUL'(R, assign, (i,j))$ denotes the fact that $ASSIGN(begin_R) \bmod 2 = i$, $ASSIGN(end_R) \bmod 2 = j$, and $ASSIGN(v) \bmod 2 = 0$ for all $v \in INT_R$
  where, for $v \in V_R$, $ASSIGN(v) = \sum_{e \in E_R^+(v)} assign(e)_1 + \sum_{e \in E_R^-(v)} assign(e)_2$ [6]
  with $E_R^+(v) = \{e \in E_R | s_R(e) = v\}$ and $E_R^-(v) = \{e \in E_R | t_R(e) = v\}$.

## Theorem

Let $ERG = (N,T,P,Z)$ be an edge–replacement grammar, $A^\bullet \Longrightarrow R \stackrel{*}{\Longrightarrow} G$ a derivation in $ERG$ with $A \in N \cup T$ and $G \in \mathcal{G}_T$, and, for $e \in E_R$, $l_R(e)^\bullet \stackrel{*}{\Longrightarrow} G(e)$ the fibre of $R \stackrel{*}{\Longrightarrow} G$ induced by $e$. For $i,j \in \{0,1\}$, $EUL(G,(i,j))$ holds if and only if there is a mapping $assign : E_R \to \{0,1\}^2$ such that $EUL(G(e), assign(e))$ for all $e \in E_R$ as well as $EUL'(R, assign, (i,j))$.

**Proof:** Without loss of generality, we may assume that $V_R \subseteq V_G$ and $G(e) \subseteq G$ for all $e \in E_R$. Suppose $EUL(G,(i,j))$ holds. Then, $d_G(begin_G) \bmod 2 = i$, $d_G(end_G) \bmod 2 = j$, and $d_G(v) \bmod 2 = 0$ for $v \in INT_G$. Choosing $assign : E_R \to \{0,1\}^2$ by
$$assign(e) = (d_{G(e)}(begin_{G(e)}) \bmod 2 \, , \, d_{G(e)}(end_{G(e)}) \bmod 2)$$
for all $e \in E_R$, we obtain $EUL(G(e), assign(e))$ for all $e \in E_R$. Moreover, for $v \in V_R$,

$$ASSIGN(v) \bmod 2 = \left( \sum_{e \in E_R^+(v)} assign(e)_1 + \sum_{e \in E_R^-(v)} assign(e)_2 \right) \bmod 2$$

$$= \left( \sum_{e \in E_R^+(v)} d_{G(e)}(begin_{G(e)}) + \sum_{e \in E_R^-(v)} d_{G(e)}(end_{G(e)}) \right) \bmod 2 = d_G(v) \bmod 2.$$

---

[5] For a graph $G$ and a node $v \in V_G$, $d_G(v)$ refers to the degree of $v$ in $G$.
[6] $assign_1$ and $assign_2$ denote the projections of $assign$ to the first and the second component, respectively.

Since, by assumption, $d_G(begin_G) \bmod 2 = i$, $d_G(end_G) \bmod 2 = j$, and $d_G(v) \bmod 2 = 0$ for $v \in INT_R$, we obtain $ASSIGN(begin_R) \bmod 2 = i$, $ASSIGN(end_R) \bmod 2 = j$, and $ASSIGN(v) \bmod 2 = 0$ for $v \in INT_R$, i.e., $EUL'(R, assign, (i, j))$ is satisfied.

Conversely, assume that, for some $assign : E_R \to \{0, 1\}^2$, $EUL(G(e), assign(e))$ for all $e \in E_R$ as well as $EUL'(R, assign, (i, j))$ hold. Then, for $v \in V_G$,

$$d_G(v) \bmod 2 = \left( \sum_{e \in E_R^+(v)} d_{G(e)}(begin_{G(e)}) + \sum_{e \in E_R^-(v)} d_{G(e)}(end_{G(e)}) \right) \bmod 2$$

$$= \left( \sum_{e \in E_R^+(v)} assign(e)_1 + \sum_{e \in E_R^-(v)} assign(e)_2 \right) \bmod 2 = ASSIGN(v) \bmod 2.$$

By assumption, we have $ASSIGN(begin_R) \bmod 2 = i$, $ASSIGN(end_R) \bmod 2 = j$, and $ASSIGN(v) \bmod 2 = 0$ $(v \in INT_R)$. Thus, $d_G(begin_G) \bmod 2 = i$, $d_G(end_G) \bmod 2 = j$, and $d_G(v) \bmod 2 = 0$ for $v \in INT_R$. Consequently, $EUL(G, (i, j))$ is satisfied. $\qquad\square$

**Remarks:** With respect to the questions "Is $G$ Eulerian?" resp. "Has $G$ an Eulerian path from $begin_G$ to $end_G$?" we get the following relations:

(1) $G$ is Eulerian (i.e., $G$ has an Eulerian cycle) if and only if $CONN(G, co)$ and $EUL(G, (0, 0))$.

(2) $G$ has an Eulerian path connecting $begin_G$ and $end_G$ if and only if $CONN(G, co)$ and $EUL(G, (1, 1))$. [7]

## 3.4 Boundedness

In this subsection we will investigate the question "Is $G$ $k$-bounded?". Note that a graph $G$ is said to be $k$-bounded (for some $k \in I\!N$) if the degree of each node is bounded by $k$, i.e., $d_G(v) \leq k$ for all $v \in V_G$.

Similar to the "even-degree property" investigated in 3.3, the "$k$-boundedness property" of a graph $G$ with derivation $A^\bullet \Longrightarrow R \overset{*}{\Longrightarrow} G$ can be characterized in terms of the fibres $l_R(e)^\bullet \overset{*}{\Longrightarrow} G(e)$. The characterization is based on the observation that $k$-boundedness of $G$ is induced by $k$-boundedness of the $G(e)$ provided that, for each $v \in V_R$, the sum of the degrees of the begin-nodes of the $G(e)$ for which $e$ is outgoing from $v \in V_R$ and the end-nodes of the $G(e)$ for which $e$ is incoming into $v \in V_R$ is bounded by k.

Let $G \in \mathcal{G}_T$, $R \in \mathcal{G}_{NUT}$, $assign : E_R \to [k]^2$, and $i, j \in [k]$ (for some $k \in I\!N$) [8]. Then

- $k - BOUNDED(G, (i, j))$ denotes the fact that $\max_{v \in V_G} d_G(v) \leq k$, $d_G(begin_G) = i$, and $d_G(end_G) = j$.
- $k - BOUNDED'(R, assign, (i, j))$ denotes the fact that $\max_{v \in V_R} ASSIGN(v) \leq k$, $ASSIGN(begin_R) = i$, and $ASSIGN(end_R) = j$
  where, for $v \in V_R$, $ASSIGN(v) = \sum_{e \in E_R^+(v)} assign(e)_1 + \sum_{e \in E_R^-(v)} assign(e)_2$
  with $E_R^+(v) = \{e \in E_R | s_R(e) = v\}$ and $E_R^-(v) = \{e \in E_R | t_R(e) = v\}$.

## Theorem

Let $ERG = (N, T, P, Z)$ be an edge-replacement grammar, $A^\bullet \Longrightarrow R \overset{*}{\Longrightarrow} G$ a derivation in $ERG$ with $A \in N \cup T$ and $G \in \mathcal{G}_T$, and, for $e \in E_R$, $l_R(e)^\bullet \overset{*}{\Longrightarrow} G(e)$ the fibre of $R \overset{*}{\Longrightarrow} G$ induced by $e$. For $k \in I\!N$ and $i, j \in [k]$, $k - BOUNDED(G, (i, j))$ holds if and only if there is a mapping $assign : E_R \to [k]^2$ such that $k - BOUNDED(G(e), assign(e))$ for all $e \in E_R$ as well as $k - BOUNDED'(R, assign, (i, j))$.

**Proof:** Without loss of generality, we may assume that $V_R \subseteq V_G$ and $G(e) \subseteq G$ for all $e \in E_R$. Suppose $k - BOUNDED(G, (i, j))$ holds. Then, $\max_{v \in V_G} d_G(e) \leq k$ for $e \in E_R$, i.e., $G(e)$ is $k$-bounded for $e \in E_R$.

---

[7] Notice that a graph $G \in \mathcal{G}_T$ possesses an Eulerian path from $begin_G$ to $end_G$ if and only if it is connected and $begin_G$ and $end_G$ are the only nodes with odd degree.

[8] For $k \in I\!N$, $[k]$ denotes the set $[k] = \{0, \ldots, k\}$.

Choosing $assign : E_R \to [k]^2$ by $assign(e) = (d_{G(e)}(begin_{G(e)}) , d_{G(e)}(end_{G(e)}))$ for $e \in E_R$, $k - BOUNDED(G(e), assign(e))$ is true for all $e \in E_R$. Moreover, for $v \in V_R$,

$$ASSIGN(v) = \sum_{e \in E_R^+(v)} assign(e)_1 + \sum_{e \in E_R^-(v)} assign(e)_2$$

$$= \sum_{e \in E_R^+(v)} d_{G(e)}(begin_{G(e)}) + \sum_{e \in E_R^-(v)} d_{G(e)}(end_{G(e)}) = d_G(v).$$

By assumption, $\max_{v \in V_G} d_G(v) \leq k$, $d_G(begin_G) = i$, and $d_G(end_G) = j$. Thus, we obtain $\max_{v \in V_G} ASSIGN(v) \leq k$, $ASSIGN(begin_R) = i$, and $ASSIGN(end_R) = j$, i.e., $k - BOUNDED'(R, assign, (i,j))$ is satisfied.

Conversely, assume that, for some $assign : E_R \to [k]^2$, $k - BOUNDED(G(e), assign(e))$ for all $e \in E_R$ as well as $k - BOUNDED'(R, assign, (i,j))$ hold. Then, $d_G(v) = d_{G(e)}(v) \leq k$ for $v \in INT_{G(e)}$ and $e \in E_R$ and

$$d_G(v) = \sum_{e \in E_R^+(v)} d_{G(e)}(begin_{G(e)}) + \sum_{e \in E_R^-(v)} d_{G(e)}(end_{G(e)})$$

$$= \sum_{e \in E_R^+(v)} assign(e)_1 + \sum_{e \in E_R^-(v)} assign(e)_2 = ASSIGN(v)$$

for $v \in V_R \subseteq V_G$. Thus,

$$\max_{v \in V_G} d_G(v) = \max(\max_{e \in E_R} \max_{v \in INT_{G(e)}} d_{G(e)}(v) , \max_{v \in V_R} d_G(v)) \leq \max(k, \max_{v \in V_R} ASSIGN(v)).$$

By assumption, $\max_{v \in V_R} ASSIGN(v) \leq k$, $ASSIGN(begin_R) = i$, $ASSIGN(end_R) = j$. Consequently, $\max_{v \in V_G} d_G(v) \leq k$, $d_G(begin_G) = i$, and $d_G(end_G) = j$, i.e., $k - BOUNDED(G, (i,j))$ is satisfied. $\square$

## 3.5 Edge–Colorability

In this subsection we will consider the problem of coloring the edges of a graph $G$, such that no two adjacent edges are uniformly colored. Such a distribution of colors is called a *(proper) edge–coloring* of $G$. Similar to the $k$–boundedness investigated in 3.4, the $k$–edge–colorability of a graph $G$ with derivation $A^\bullet \Longrightarrow R \overset{\bullet}{\Longrightarrow} G$ can be characterized in terms of the fibres $l_R(e)^\bullet \overset{\bullet}{\Longrightarrow} G(e)$. The characterization is based on the observation that the $k$–edge–coloring of $G$ is induced by the $k$–edge–coloring of the $G(e)$ provided that, for each $v \in V_R$, the cardinality of the union of the color sets $COLORS_{G(e)}(begin_{G(e)})$ — the set of colors used for the edges in $G(e)$ incident with $begin_{G(e)}$ — for which $e$ is outgoing from $v \in V_R$ and $COLORS_{G(e)}(end_{G(e)})$ — set of colors used for the edges in $G(e)$ incident with $end_{G(e)}$ — for which $e$ is incoming into $v \in V_R$ is equal to the degree of $v$ in $V_G$ and also is bounded by $k$.

To recall the notion of edge–colorability, let $G \in \mathcal{G}_T$, $k \in \mathbb{N}$, and $K$ be a set with $k$ elements. A mapping $c_G : E_G \to K$ is a $k$–edge–coloring of $G$ if $c_G(e) \neq c_G(e')$ for each pair $e, e' \in E_G$ of adjacent edges. $G$ is said to be $k$–edge–colorable if there is a $k$–edge–coloring of $G$. For $G \in \mathcal{G}_T$, $R \in \mathcal{G}_{NUT}$, $assign : E_R \to \wp(K)^2$, [9] and $X, Y \subseteq K$,

- $k - COLOR(G, (X, Y))$ denotes the fact that there is a $k$–edge–coloring $c_G$ of $G$ such that $COLORS_G(begin_G) = X$ and $COLORS_G(end_G) = Y$ where, for $v \in V_G$, $COLORS_G(v) = \bigcup_{e \in E_G(v)} c_G(e)$ and $E_G(v) = \{e \in E_G | s_G(e) = v \text{ or } t_G(e) = v\}$.

- $k - COLOR'(R, assign, (X, Y))$ refers to the fact that $\#ASSIGN(v) = \#E_R(v) \leq k$ [10] for all $v \in V_R$, $ASSIGN(begin_R) = X$, and $ASSIGN(end_R) = Y$ where, for $v \in V_R$, $ASSIGN(v) = \bigcup_{e \in E_R^+(v)} assign(e)_1 \cup \bigcup_{e \in E_R^-(v)} assign(e)_2$ with $E_R^+(v) = \{e \in E_R | s_R(e) = v\}$ and $E_R^-(v) = \{e \in E_R | t_R(e) = v\}$.

---

[9] For a set $A$, $\wp(A)$ denotes the power set of $A$.

[10] For a set $A$, $\#A$ denotes the number of elements in $A$.

## Theorem

Let $ERG = (N, T, P, Z)$ be an edge–replacement grammar, $A^* \Longrightarrow R \stackrel{*}{\Longrightarrow} G$ a derivation in $ERG$ with $A \in N \cup T$ and $G \in \mathcal{G}_T$, and, for $e \in E_R$, $l_R(e)^* \stackrel{*}{\Longrightarrow} G(e)$ the fibre of $R \stackrel{*}{\Longrightarrow} G$ induced by $e$. For $k \in \mathbb{N}$ and $X, Y \subseteq K$, $k - COLOR(G, (X, Y))$ holds if and only if there is a mapping $assign : E_R \to \wp(K)^2$ such that $k - COLOR(G(e), assign(e))$ for all $e \in E_R$ as well as $k - COLOR'(R, assign, (X, Y))$.

**Proof:** Without loss of generality, we may assume that $V_R \subseteq V_G$ and $G(e) \subseteq G$ for all $e \in E_R$. Suppose $k - COLOR(G, (X, Y))$ holds. Let $c_G : E_G \to K$ be a $k$–edge–coloring of $G$. Because $G(e) \subseteq G$, the restriction of $c_G$ to $E_{G(e)}$ defines an appropriate $k$–edge–coloring $c_{G(e)}$ of each $G(e)$. Choosing $assign : E_R \to \wp(K)^2$ by

$$assign(e) = (COLORS_{G(e)}(begin_{G(e)}) \, , \, COLORS_{G(e)}(end_{G(e)}))$$

for $e \in E_R$, $k - COLOR(G(e), assign(e))$ becomes true for each $e \in E_R$. Furthermore, for $v \in V_R$,

$$
\begin{aligned}
ASSIGN(v) &= \bigcup_{e \in E_R^+(v)} assign(e)_1 + \bigcup_{e \in E_R^-(v)} assign(e)_2 \\
&= \sum_{e \in E_R^+(v)} COLORS_{G(e)}(begin_{G(e)}) + \sum_{e \in E_R^-(v)} COLORS_{G(e)}(end_{G(e)}) \\
&= COLORS_G(v).
\end{aligned}
$$

By assumption, $\#COLORS_G(v) \leq k$ for all $v \in V_R$, $COLORS_G(begin_G) = X$, and $COLORS_G(end_G) = Y$. Thus, we obtain $\#ASSIGN(v) \leq k$ for $v \in V_R$, $ASSIGN(begin_R) = X$, and $ASSIGN(end_R) = Y$. Moreover, $\#ASSIGN(v) = \#COLORS_G(v) = \#E_G(v)$ for $v \in V_R$ because $c_G$ is an edge–coloring of $G$. Hence $k - COLOR(R, assign, (X, Y))$ holds. Conversely, assume that, for some $assign : E_R \to \wp(K)^2$, $k - COLOR(G(e), assign(e))$ for $e \in E_R$ as well as $k - COLOR'(R, assign, (X, Y))$ hold. Then

$$
\begin{aligned}
COLORS_G(v) &= \bigcup_{e \in E_R^+(v)} COLORS_{G(e)}(begin_{G(e)}) + \bigcup_{e \in E_R^-(v)} COLORS_{G(e)}(end_{G(e)}) \\
&= \bigcup_{e \in E_R^+(v)} assign(e)_1 + \bigcup_{e \in E_R^-(v)} assign(e)_2 = ASSIGN(v).
\end{aligned}
$$

By assumption, we have $\#ASSIGN(v) = \#E_G(v) \leq k$ $(v \in V_R)$, $ASSIGN(begin_R) = X$, and $ASSIGN(end_R) = Y$. Consequently, $\#COLORS_G(v) \leq k$ for $v \in V_R$, $COLORS_G(begin_G) = X$, $COLORS_G(end_G) = Y$. Define now $c_G : E_G \to K$ by $c_G(e') = c_{G(e)}(e')$ for $e' \in E_{G(e)}$ and $e \in E_R$. Then $c_G$ becomes an edge–coloring of $G$ because, for $e \in E_R$, $c_{G(e)}$ is an edge–coloring and, for $v \in V_R \subseteq V_G$, $\#COLORS_G(v) = \#ASSIGN(v) = \#E_G(v)$. Hence, $k - COLOR(G, (X, Y))$ is satisfied. $\qquad \square$

# 4. A General View of Compatibility

All the examples given in Section 3 follow a simple scheme. Roughly speaking, a predicate $PR$ holds for $H \in \mathcal{H}_C$ which is derivable from a handle $A^*$ starting with the production $(A, R)$ if and only if $R$ fulfills some auxiliary predicate $PR'$ and $PR$ holds for all $H(e)$, $e \in E_R$. As the example concerning connectedness demonstrates, this view is oversimplified for most applications. To check $PR$ for $H$, one may have to check some other related properties for the $H(e)$ (with respect to the connectedness of $H$, some of the $H(e)$ are allowed to be semi–connected). Therefore, we use families of properties indexed by some finite set $I$ and we need a mapping $assign$ which determines the property out of the family to be checked for each $H(e)$.

## 4.1 Definition (compatibility)

1. Let $C$ be a class of hyperedge–replacement grammars, $I$ a finite index set, $PROP$ a decidable predicate defined on pairs $(H, i)$ with $H \in \mathcal{H}_C$ and $i \in I$, and $PROP'$ a decidable predicate on triples $(H, assign, i)$ with $H \in \mathcal{H}_C$, a mapping $assign : E_H \to I$, and $i \in I$. [11] Then $PROP$ is called $C, PROP'$–compatible if, for all $HRG = (N, T, P, Z) \in C$ and all derivations of the form $A^\bullet \Longrightarrow R \overset{\ast}{\Longrightarrow} H$ with $A \in N$ and $H \in \mathcal{H}_T$, and for all $i \in I$, the following holds:
$PROP(H, i)$ is true if and only if there is a mapping $assign : E_R \to I$ such that
$PROP'(R, assign, i)$ is true and
$PROP(H(e), assign(e))$ is true for all $e \in E_R$.

2. A predicate $PR$ on $\mathcal{H}_C$ is called $C$–compatible if predicates $PROP$ and $PROP'$ and an index $i_0$ exist such that $PR = PROP(-, i_0)$ [12] and $PROP$ is $C, PROP'$–compatible.

**Remark:** Intuitively, a property is compatible if it can be tested for a large hypergraph with a long fibre by checking the smaller components of the corresponding shorter fibres. Such a property must be closed under isomorphisms because the derivability of hypergraphs is independent of the representation of nodes and hyperedges.

Various explicit examples of compatible predicates were discussed in Section 3.

## 4.2 Examples

1. Let, for $H \in \mathcal{H}_C$, $DIS(H)$ denote the fact that $H$ is totally disconnected. Then $DIS$ is $C$–compatible for the class $C$ of all hyperedge–replacement grammars: Take a one–element set $I = \{*\}$ and the predicates $DISCO$ and $DISCO'$ which are given by
$DISCO(H, *)$ if and only if $DIS(H)$ for all $H \in \mathcal{H}_C$,
$DISCO'(H, assign, *) = true$ for all $H \in \mathcal{H}_C$.
By Theorem 3.1, $DISCO$ is $C, DISCO'$–compatible and $DIS = DISCO(-, *)$ is $C$–compatible.

2. By Theorem 3.2, the predicates $CO$ and $SEMICO$ given for all $G \in \mathcal{G}_C$ by
$CO(G)$ if and only if $G$ is connected, i.e. $CONN(G, co)$, and
$SEMICO(G)$ if and only if $G$ is semi–connected, i.e., $CONN(G, semico)$,
are $C$–compatible.

3. By Theorem 3.3, the predicates $EUL_C$ and $EUL_P$ given for all $G \in \mathcal{G}_C$ by
$EUL_C(G)$ if and only if every node of $G$ has even degree, i.e., $EUL(G, (0, 0))$, and
$EUL_P(G)$ if and only if every node of $G$ has even degree, except $begin_G$ and $end_G$, i.e., $EUL(G, (1, 1))$
are $C$–compatible. Moreover, the predicates $EULER_C$ and $EULER_P$ given for all $G \in \mathcal{G}_C$ by
$EULER_C(G)$ if and only if $CO(G) \wedge EUL_C(G)$ and
$EULER_P(G)$ if and only if $CO(G) \wedge EUL_P(G)$
are $C$–compatible provided that the conjunction of $C$–compatible predicates is $C$–compatible.

4. For $k \in \mathbb{N}$, the predicate $k - BOUND$ given for all $G \in \mathcal{G}_C$ by
$k - BOUND(G)$ if and only if $G$ is $k$–bounded, i.e., $\bigvee_{0 < i, j \le k} k - BOUNDED(G, (i, j))$,
is $C$–compatible provided that the disjunction of $C$–compatible predicates is $C$–compatible.

5. For $k \in \mathbb{N}$, the predicate $k - COL$ given for all $G \in \mathcal{G}_C$ by
$k - COL(G)$ if and only if $G$ has a $k$–edge–coloring, i.e., $\bigvee_{X, Y \subseteq K} k - COLOR(G, (X, Y))$,
is $C$–compatible provided that the disjunction of $C$–compatible predicates is $C$–compatible.

The examples in 4.2 show that the question of closureness of $C$–compatibility is very important. In fact, $C$–compatibility is closed under Boolean operations (cf. Theorem 4.3).

---

[11] In the following we assume that all considered predicates are *closed under isomorphisms*, i.e., if a predicate $\Phi$ holds for $H \in \mathcal{H}_C$ and $H \cong H'$, then $\Phi$ holds for $H'$, too.

[12] For $i \in I$, $PROP(-, i)$ denotes the unary predicate defined by $PROP(-, i)(H) = PROP(H, i)$ for all $H \in \mathcal{H}_C$.

## 4.3 Theorem (Closure under Boolean operations)

Let $C$ be a class of hyperedge–replacement grammars and $PR_1$, $PR_2$ be $C$–compatible predicates. Let, for $H \in \mathcal{H}_C$,

$(PR_1 \wedge PR_2)(H)$ if and only if $PR_1(H) \wedge PR_2(H)$,
$(PR_1 \vee PR_2)(H)$ if and only if $PR_1(H) \vee PR_2(H)$,
$(\neg PR_1)(H)$ if and only if $\neg PR_1(H)$.

Then $(PR_1 \wedge PR_2)$, $(PR_1 \vee PR_2)$, and $(\neg PR_1)$ are $C$–compatible.

**Proof:** Let, for $j = 1,2$, $PR_j$ be a $C$–compatible predicate, $PROP_j$, $PROP_j'$ the corresponding predicates, $I_j$ the corresponding index set, and $i_j$ the index such that $PR_j = PROP_j(-, i_j)$.

1. For proving the $C$–compatibility of $(PR_1 \wedge PR_2)$, we define new predicates $PROP$ and $PROP'$ derived from the given ones and show the $C, PROP'$–compatibility of $PROP$ using that of the old ones. Let

$I = I_1 \times I_2$,
$PROP(H, (j_1, j_2)) \iff PROP_1(H, j_1) \wedge PROP_2(H, j_2)$, and
$PROP'(H, assign, (j_1, j_2)) \iff PROP_1'(H, assign_1, j_1) \wedge PROP_2'(H, assign_2, j_2)$

where $assign_1$, $assign_2$ are the projections of $assign$ to the first and the second component, respectively.

Then, for $H \in \mathcal{H}_C$, $(PR_1 \wedge PR_2)(H) \iff PR_1(H) \wedge PR_2(H) \iff PROP_1(H, i_1) \wedge PROP_2(H, i_2) \iff PROP(H, (i_1, i_2))$, i.e., $(PR_1 \wedge PR_2) = PROP(-, (i_1, i_2))$. The $C, PROP'$–compatibility of $PROP$, can be proved by the following reasoning for a derivation $A^* \Longrightarrow R \stackrel{*}{\Longrightarrow} H$ with $A \in N$ and $H \in \mathcal{H}_T$. By definition of $PROP$, $C, PROP_1'$–compatibility of $PROP_1$, and $C, PROP_2'$–compatibility of $PROP_2$, we obtain

$PROP(H, (j_1, j_2)) \iff PROP_1(H, j_1) \wedge PROP_2(H, j_2)$
$\iff$ (A) $\exists assign_1 : E_R \to I_1, \exists assign_2 : E_R \to I_2 :$
$\qquad PROP_1'(R, assign_1, j_1) \wedge PROP_2'(R, assign_2, j_2)$
$\qquad \wedge \bigwedge_{e \in E_R} PROP_1(H(e), assign_1(e)) \wedge \bigwedge_{e \in E_R} PROP_2(H(e), assign_2(e))$.

Choosing $assign : E_R \to I$ as $assign(e) = (assign_1(e), assign_2(e))$ for all $e \in E_R$, $PROP(H(e), assign(e))$ for $e \in E_R$ as well as $PROP'(R, assign, (j_1, j_2))$ become true. Conversely, if, for some mapping $assign : E_R \to I$, $PROP(H(e), assign(e))$ (for $e \in E_R$) as well as $PROP'(R, assign, (j_1, j_2))$ hold, then we obtain (A) choosing $assign_1$ and $assign_2$ as the projections of $assign$ to the first and the second component, respectively. Hence $PROP$ is $C, PROP'$–compatible and $(PR_1 \wedge PR_2) = PROP(-, (i_1, i_2))$ is $C$–compatible.

2. To derive the $C$–compatibility of $(\neg PR_1)$, we define — similar to 1. — new predicates $PROP$ and $PROP'$ and show the $C, PROP'$–compatibility of $PROP$. Let

$I$ be the power set of $I_1$,
$PROP(H, X) \iff \bigwedge_{x \in X} PROP_1(H, x) \wedge \bigwedge_{x \notin X} \neg PROP_1(H, x)$, and
$PROP'(H, assign, X) \iff \bigwedge_{x \in X} \bigvee_{a \in M(assign)} PROP_1'(H, a, x) \wedge \bigwedge_{x \notin X} \bigwedge_{a \in M(assign)} \neg PROP_1'(H, a, x)$

where $M(assign) = \{a : E_R \to I_1 | a(e) \in assign(e) \text{ for all } e \in E_R\}$.

Using the definition of $PROP$ and $C, PROP_1'$–compatibility of $PROP_1$, we get

$PROP(H, X) \iff \bigwedge_{x \in X} PROP_1(H, x) \wedge \bigwedge_{x \notin X} \neg PROP_1(H, x)$
$\iff \bigwedge_{x \in X} [ \bigvee_{a : E_R \to I_1} PROP_1'(R, a, x) \wedge \bigwedge_{e \in E_R} PROP_1(H(e), a(e)) ] \wedge$
$\qquad \bigwedge_{x \notin X} \neg [ \bigvee_{a : E_R \to I_1} PROP_1'(R, a, x) \wedge \bigwedge_{e \in E_R} PROP_1(H(e), a(e)) ]$.

Choosing $assign : E_R \to I$ by $assign(e) = \{i \in I_1 \mid PROP_1(H(e), i)\}$ for all $e \in E_R$, $PROP(H(e), assign(e))$ becomes true for all $e \in E_R$. Moreover, for each $x \in X$, there is a mapping $a : E_R \to I_1$ such that $PROP'(R, a, x)$ and $a(e) \in assign(e)$ for all $e \in E_R$. For $x \notin X$, there is no mapping $a : E_R \to I_1$ such that $PROP'_1(R, a, x)$ and $a(e) \in assign(e)$ for all $e \in E_R$. Thus, $PROP'_1(R, assign, X)$ is satisfied.

Conversely, if $PROP'(R, assign, X)$ and $PROP(H(e), assign(e))$ hold for some mapping $assign : E_R \to I$, then

(B) $\bigwedge\limits_{x \in X} \bigvee\limits_{a \in M(assign)} PROP'_1(R, a, x) \wedge \bigwedge\limits_{e \in E_R} \bigwedge\limits_{x \in assign(e)} PROP_1(H(e), x)$

$\Longrightarrow \bigwedge\limits_{x \in X} [\ \bigvee\limits_{a \in M(assign)} PROP'_1(R, a, x) \wedge \bigwedge\limits_{e \in E_R} PROP_1(H(e), a(e))\ ]$

$\Longrightarrow \bigwedge\limits_{x \in X} PROP_1(H, x)$

(C) $\bigwedge\limits_{x \notin X} \bigwedge\limits_{a \in M(assign)} \neg PROP'_1(R, a, x)$

$\Longrightarrow \bigwedge\limits_{x \notin X} [\ \bigwedge\limits_{a \in M(assign)} \neg PROP'_1(R, a, x) \vee \bigvee\limits_{e \in E_R} \neg PROP_1(H(e), a(e))\ ]$

$\Longrightarrow \bigwedge\limits_{x \notin X} \neg [\ \bigvee\limits_{a \in M(assign)} PROP'_1(R, a, x) \wedge \bigwedge\limits_{e \in E_R} PROP_1(H(e), a(e))\ ]$

$\Longrightarrow \bigwedge\limits_{x \notin X} \neg PROP_1(H, x).$

By (B) and (C), $PROP(H, X)$ becomes satisfied. This completes the proof of the $\mathcal{C}, PROP'$–compatibility of $PROP$.

Finally, we will show how the $\mathcal{C}, PROP'$–compatibility of $PROP$ induces the $\mathcal{C}$–compatibility of $(\neg PR_1)$. Obviously,

$(\neg PR_1)(H) \iff \neg PR_1(H) \iff \neg PROP_1(H, i_1)$

$\iff \neg \bigvee\limits_{i_1 \in S \subseteq I} [\bigwedge\limits_{x \in S} PROP_1(H, x) \wedge \bigwedge\limits_{x \notin S} \neg PROP_1(H, x)\ ]$

$\iff \neg \bigvee\limits_{i_1 \in S \subseteq I} PROP(H, S) \iff \bigwedge\limits_{i_1 \in S \subseteq I} \neg PROP(H, S)$

$\iff \bigwedge\limits_{i_1 \in S \subseteq I} PROP(H, I - S) \iff \bigwedge\limits_{S \subseteq I \text{ with } i_1 \notin S} PROP(H, S).$

The $\mathcal{C}, PROP'$–compatibility of $PROP$ implies the $\mathcal{C}$–compatibility of $PR_S = PROP(-, S)$ for all $S \subseteq I$. Since the conjunction of $\mathcal{C}$–compatible predicates yields a $\mathcal{C}$–compatible predicate, $\neg PR_1$ becomes $\mathcal{C}$–compatible.

3. By $(PR_1 \vee PR_2) = \neg(\neg PR_1 \wedge \neg PR_2)$, the $\mathcal{C}$–compatibility of $(PR_1 \vee PR_2)$ follows from the $\mathcal{C}$–compatibility of $PR_1$, the $\mathcal{C}$–compatibility of $PR_2$, and the fact that the conjunction and negation of $\mathcal{C}$–compatible predicates yields $\mathcal{C}$–compatible predicates (see 1. and 2.). $\qquad\square$

## 4.4 Examples

By Theorem 4.3, the predicates $EULER_C$, $EULER_P$, $k - BOUND$, and $k - COL$ given by

$\quad EULER_C(G)$ if and only if $CO(G) \wedge EUL_C(G)$,
$\quad EULER_P(G)$ if and only if $CO(G) \wedge EUL_P(G)$,
$\quad k - BOUND(G)$ if and only if $\bigvee\limits_{0 \le i, j \le k} k - BOUNDED(G, (i, j))$
$\quad k - COL(G)$ if and only if $\bigvee\limits_{X, Y \subseteq K} k - COLOR(G, (X, Y))$

are $\mathcal{C}$–compatible.

Given a (hyper)graph property $PR$ and a class $\mathcal{C}$ of hyperedge–replacement grammars, we are going to study two types of questions for all $HRG \in \mathcal{C}$: "Does $PR$ hold for some $H \in L(HRG)$?" and "Does $PR$ hold for all $H \in L(HRG)$?". Both questions turn out to be decidable provided that $PR$ is $\mathcal{C}$–compatible. We call this result "metatheorem" because of its generic character: Whenever one can prove the compatibility of a property (and we have given various examples in the previous section), one gets a particular decision result for this property as corollary of the metatheorem.

## 4.5 Theorem

Let $PR$ be $C$-compatible with respect to some class $C$ of hyperedge–replacement grammars. Then, for all $HRG \in C$, it is decidable whether
(1) $PR$ holds for some $H \in L(HRG)$,
(2) $PR$ holds for all $H \in L(HRG)$.

**Sketch of Proof** [13]: 1. Let $PR$ be $C$-compatible for some class $C$ of hyperedge–replacement grammars and $PROP$ and $PROP'$ the corresponding predicates over the index set $I$ such that $PROP$ is $C, PROP'$-compatible and $PR = PROP(-, i_0)$ for some $i_0 \in I$. Let $HRG = (N, T, P, Z) \in C$. Then the set

$$V_{PROP,i} = \{A \in N \cup T | \text{ there is an } H \in \mathcal{H}_C \text{ such that } A^\bullet \overset{*}{\underset{P}{\Longrightarrow}} H \text{ and } PROP(H, i)\} \ (i \in I)$$

can be explicitly constructed: We define recursively sets $V_{PROP,i,k}$ for $i \in I$ and $k \geq 0$. $V_{PROP,i,0}$ contains each terminal symbol $a$ with $PROP(a^\bullet, i)$. $V_{PROP,i,k+1}$ contains $V_{PROP,i,k}$ and each symbol $A$ where $(A, R) \in P$ and a mapping $assign : E_R \to I$ exist such that $PROP'(R, assign, i)$ holds, and $l_R(e) \in V_{PROP,i,k}$ for all $e \in E_R$. These sequences of sets terminate because they are included in the finite set $V = N \cup T$. Hence there is always an $l \geq 0$ with $V_{PROP,i,l} = V_{PROP,i,l+1}$. Finally, it turns out that $V_{PROP,i} = V_{PROP,i,l}$. In other words, the decision problem reduces to checking whether $l(Z) \in V_{PROP,i_0}$ or not.
2. Obviously, $PR$ holds for all $H \in L(HRG)$ if and only if there is no $H \in L(HRG)$ such that $\neg PR$ holds. By Theorem 4.3, $\neg PR$ is $C$-compatible. Thus, the question whether $\neg PR$ holds for some $H \in L(HRG)$ is decidable. Hence the decision problem whether $PR$ holds for all $H \in L(HRG)$ can be reduced to the decision problem whether $\neg PR$ holds for some $H \in L(HRG)$. $\square$

Combining the compatibility results of Section 3 with Theorem 4.5 one obtains a list of decidability results.

## 4.6 Corollary

The question: "Does a given edge–replacement grammar generate some or only graphs with property $PR$?" is decidable for all edge–replacement grammars and all properties in the following list:
(1) A graph is totally disconnected.
(2) A graph is connected (resp. semi–connected).
(3) All nodes of a graph have even degree (resp. only the begin– and the end–node have odd degree).
(4) A graph has an Eulerian cycle (resp. an Eulerian path).
(5) The degree of a graph is bounded by $k$ (for some $k \in I\!N$).
(6) A graph has an edge–coloring with at most $k$ colors (for some $k \in I\!N$).

**Remarks:** 1. Remember that $DIS$ is compatible for arbitrary hyperedge–replacement grammars.
2. Although in most cases we have restricted to the class $C_{ERG}$ of all edge–replacement grammars, we are convinced that our statements hold even if we consider the class of all hyperedge–replacement grammars which generate ordinary graph languages and use nonterminal hyperedges with a bounded number of tentacles. We even guess that the considered properties are compatible for arbitrary hyperedge–replacement grammars (with a bounded number of tentacles) if their definition is properly adapted to hypergraphs (which seems easy with respect to connectedness, degree properties as well as $k$–edge–colorability).
3. As shown in [HKV 87], many other properties — like the existence of Hamiltonian paths and cycles, vertex coloring, and subcontraction — are compatible and the corresponding questions are decidable.

---

[13] For an explicit proof of Theorem 4.5 see [HKV 87]. To give an idea how compatibility and decidability interact we give a sketch of the proof.

# 5. Conclusion

In this paper, a spectrum of graph–theoretic properties has been studied and proved to be *compatible*. Moreover, the set of compatible properties has been shown to be closed under Boolean operations. Using the decidability result for compatible properties saying that if a graph–theoretic property is compatible with the derivation process, it is decidable whether some or all generated graphs and hypergraphs have this property, the graph–theoretic property — mentioned in Section 3 — can be proved as decidable.

Related investigations can be found in the literature.

(1) In [Co 87], Courcelle presents a similar metatheorem for decision problems for a certain type of hyperedge replacement grammars using the quite different framework of algebraic systems. Properties definable by second–order formulas are shown to behave like compatible properties with respect to the decision problems considered in [HKV 87] and in this paper. Although Courcelle's results seem to indicate that definable properties are compatible, the relationship between both concepts has not been revealed completely yet.

(2) Moreover, in a paper by Lengauer and Wanke [LW 88] some decision problems similar to those mentioned here are studied. They use a type of graph grammars, which are variants of hyperedge–replacement grammars, and consider so–called finite properties, which are defined independent of the derivation process. The relationship between finite properties and compatible properties will need a thorough investigation.

(3) The concepts and techniques considered in this paper apply to complexity issues. Compatibility means that a property can be tested for a hypergraph derived by a long fibre (starting from a handle) by testing some hypergraphs derived by shorter fibres. This observation leads to tests for compatible properties, which are very fast under certain conditions (see [Kr 87]). Other authors use similar ideas to come up with efficient solutions of particular graph problems on certain types of graphs. In [Sl 82], Slisenko introduces a class of graph grammars, which can be seen as special hyperedge–replacement grammars, and gives a polynomial–time algorithm testing the existence of Hamiltonian paths in the generated graphs. Lengauer (see, e.g., [Le 85]) and Lautemann (see [La 88]) present efficient tests for a variety of graph problems dealing with hierarchical graphs and decomposition trees respectively.

# Acknowledgment

This paper is very much inspired by the common work with Hans-Jörg Kreowski and Walter Vogler (see [HKV 87]). I wish to thank them. I am also grateful to Detlef Plump and Anne Wilharm for various valuable suggestions.

# Appendix: Hypergraphs and Hyperedge Replacement

This appendix summarizes the basic notions on graphs and hypergraphs as far as needed in the paper. The key construction is the replacement of some hyperedges of a hypergraph by hypergraphs yielding an expanded hypergraph. In our approach, a hyperedge is an atomic item with an ordered set of incoming tentacles and an ordered set of outgoing tentacles where each tentacle grips at a node through the source and target functions. Correspondingly, a hypergraph is equipped with two sequences of distinguished nodes so that it is enabled to replace a hyperedge.

## A.1 Definition (hypergraphs)

1. Let $C$ be an arbitrary, but fixed alphabet, called a set of *labels* (or *colors*).

2. A *hypergraph* over $C$ is a system $(V, E, s, t, l)$ where $V$ is a finite set of *nodes* (or *vertices*), $E$ is a finite set of *hyperedges*, $s : E \to V^*$ and $t : E \to V^*$ [14] are two mappings assigning a sequence

---

[14] For a set $A$, $A^*$ denotes the set of all words over $A$, including the empty word $\lambda$.

of *sources* $s(e)$ and a sequence of *targets* $t(e)$ to each $e \in E$, and $l : E \to C$ is a mapping *labeling* each hyperedge.

3. A hyperedge $e \in E$ of a hypergraph $(V, E, s, t, l)$ is called an $(m, n)$–*edge* for some $m, n \in \mathbb{N}$ if $|s(e)| = m$ and $|t(e)| = n$. [15] The pair $(m, n)$ is the *type* of $e$, denoted by $type(e)$. $e$ is said to be *well–formed* if its sources and targets are pairwise distinct.

4. A *multi–pointed hypergraph* over $C$ is a system $H = (V, E, s, t, l, begin, end)$ where the first five components define a hypergraph over $C$ and $begin, end \in V^*$. Components of $H$ are denoted by $V_H, E_H, s_H, t_H, l_H, begin_H, end_H$, respectively. The set of all multi–pointed hypergraphs over $C$ is denoted by $\mathcal{H}_C$.

5. $H \in \mathcal{H}_C$ is said to be an $(m, n)$–*hypergraph* for some $m, n \in \mathbb{N}$ if $|begin_H| = m$ and $|end_H| = n$. The pair $(m, n)$ is the *type* of $H$, denoted by $type(H)$. $H$ is said to be *well–formed* if all hyperedges are well–formed and the begin–nodes and end–nodes of $H$ are pairwise distinct.

6. Let $H \in \mathcal{H}_C$, $begin_H = begin_1 \ldots begin_m$ and $end_H = end_1 \ldots end_n$ with $begin_i, end_j \in V_H$ for $i = 1, \ldots, m$ and $j = 1, \ldots, n$. Then $EXT_H = \{begin_i | i = 1, \ldots, m\} \cup \{end_j | j = 1, \ldots, n\}$ denotes the set of *external nodes* of $H$. Moreover, $INT_H = V_H - EXT_H$ denotes the set of *internal* nodes of $H$.

**Remarks:** 1. There is a 1–1–correspondence between hypergraphs and $(0,0)$–hypergraphs so that hypergraphs may be seen as special cases of multi–pointed hypergraphs.
2. A hypergraph with $(1,1)$–edges only can be considered as an ordinary directed, edge–labeled graph. The set of $(1,1)$–graphs over $C$ is denoted by $\mathcal{G}_C$.

## A.2 Definition (special hypergraphs)

1. A multi–pointed hypergraph $H$ is said to be a *singleton* if $|E_H| = 1$ and $|V_H - EXT_H| = 0$. $e(H)$ refers to the only hyperedge of $H$ and $l(H)$ refers to its label.

2. A singleton $H$ with $s_H(e) = begin_H$ and $t_H(e) = end_H$ is called a *handle*. If additionally $l_H(e) = A$ and $type(e) = (m, n)$ for some $m, n \in \mathbb{N}$, then $H$ is said to be an $(m, n)$–*handle* induced by $A$ and is denoted by $A(m, n)^{\bullet}$.

## A.3 Definition (subhypergraphs and isomorphic hypergraphs)

1. Let $H, H' \in \mathcal{H}_C$. Then $H$ is called a *(weak) subhypergraph* of $H'$, denoted by $H \subseteq H'$, if $V_H \subseteq V_{H'}$, $E_H \subseteq E_{H'}$, and $s_H(e) = s_{H'}(e)$, $t_H(e) = t_{H'}(e)$, $l_H(e) = l_{H'}(e)$ for all $e \in E_H$. [Note that nothing is assumed on the relation of the distinguished nodes.]

2. Let $H, H' \in \mathcal{H}_C$ and $i_V : V_H \to V_{H'}$, $i_E : E_H \to E_{H'}$ be bijective mappings. Then $i = (i_V, i_E) : H \to H'$ is called an *isomorphism* from $H$ to $H'$ if $i_V^*(s_H(e)) = s_{H'}(i_E(e))$, $i_V^*(t_H(e)) = t_{H'}(i_E(e))$, $l_H(e)) = l_{H'}(i_E(e))$ for all $e \in E_H$ as well as $i_V^*(begin_H) = begin_{H'}$, $i_V^*(end_H) = end_{H'}$. [16] $H$ and $H'$ are said to be *isomorphic*, denoted by $H \cong H'$, if there is an isomorphism from $H$ to $H'$.

Now we are ready to introduce how hypergraphs may substitute hyperedges. An $(m, n)$–edge can be replaced by an $(m, n)$–hypergraph in two steps:
(1) Remove the hyperedge,
(2) add the hypergraph except the external nodes and
hand over each tentacle of a hyperedge (of the replacing hypergraph) which grips to an external node to the corresponding source or target node of the replaced hyperedge.
Moreover, an arbitrary number of hyperedges can be replaced simultaneously in this way.

## A.4 Definition (hyperedge replacement)

Let $H \in \mathcal{H}_C$ be a multi–pointed hypergraph, $B \subseteq E_H$, and $repl : B \to \mathcal{H}_C$ a mapping with $type(repl(b)) = type(b)$ for all $b \in B$. Then the *replacement* of $B$ in $H$ through $repl$ yields the multi–pointed hypergraph $X$ given by

---

[15] For a word $w \in A^*$, $|w|$ denotes its length.
[16] For a mapping $f : A \to B$, the free symbolwise extension $f^* : A^* \to B^*$ is defined by $f^*(a_1 \ldots a_k) = f(a_1) \ldots f(a_k)$ for all $k \in \mathbb{N}$ and $a_i \in A$ ($i = 1, \ldots, k$).

- $V_X = V_H + \sum_{b \in B}(V_{repl(b)} - EXT_{repl(b)})$, [17]
- $E_X = (E_H - B) + \sum_{b \in B} E_{repl(b)}$,
- each hyperedge of $E_H - B$ keeps its sources and targets,
- each hyperedge of $E_{repl(b)}$ (for all $b \in B$) keeps its internal sources and targets and the external ones are handed over to the corresponding sources and targets of $b$, i.e.,

$$s_X(e) = h^*(s_{repl(b)}(e)) \text{ and } t_X(e) = h^*(t_{repl(b)}(e)) \text{ for all } b \in B \text{ and } e \in E_{repl(b)}$$

where $h : V_{repl(b)} \rightarrow V_X$ is defined by $h(v) = v$ for $v \in V_{repl(b)} - EXT_{repl(b)}$,

$h(b_i) = s_i$ ($i = 1, \ldots, m$) for $begin_{repl(b)} = b_1 \ldots b_m$ and $s_H(b) = s_1 \ldots s_m$,

$h(e_j) = t_j$ ($j = 1, \ldots, n$) for $end_{repl(b)} = e_1 \ldots e_n$ and $t_H(b) = t_1 \ldots t_n$.

- each hyperedge keeps its label,
- $begin_X = begin_H$ and $end_X = end_H$.

The resulting multi-pointed hypergraph $X$ is denoted by $REPLACE(H, repl)$.

**Remarks:** 1. The construction above is meaningful and determines (up to isomorphism) a unique hypergraph $X$ if $h$ is a mapping. This is automatically fulfilled whenever the *begin*-nodes and *end*-nodes of each replacing hypergraph are pairwise distinct. If one wants to avoid such a restriction, one has to require that the following application condition is satisfied for each $b \in B$:

If $begin_{repl(b)} = x_1 \ldots x_m$ and $end_{repl(b)} = x_{m+1} \ldots x_{m+n}$ as well as $s_H(b) = y_1 \ldots y_m$ and $t_H(b) = y_{m+1} \ldots y_{m+n}$,

then, for $i, j = 1, \ldots, m+n$, $x_i = x_j$ implies $y_i = y_j$.

2. The replacement of hyperedges is of a simpler nature than it may look like at first sight. The hyperedges of $B$ are removed. For each $b \in B$, the replacing hypergraph $repl(b)$ is added disjointly except for its distinguished nodes. A tentacle of a hyperedge in $repl(e)$ which grips to a begin or end node of $repl(e)$ gets the corresponding source or target node of the replaced hyperedge. All other tentacles of hyperedges retain their nodes. All hyperedges keep their labels. The distinguished nodes of $H$ remain distinguished.

Intuitively, the replacement of some hyperedges means that they are blown up into subhypergraphs, as the figure tells.

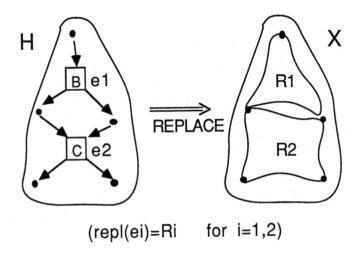

(repl(ei)=Ri    for i=1,2)

---

[17] The sum symbols $+$ and $\sum$ denote the disjoint union of sets; the symbol $-$ denotes the set-theoretic difference.

# References

[CER 79]     V. Claus, H. Ehrig, G. Rozenberg (eds.): Graph–Grammars and Their Application to Computer Science and Biology, Lect. Not. Comp. Sci. 73, 1979

[Co 87]      B.Courcelle: Recognizability and Second Order Definability for Sets of Finite Graphs, Research Report 8634, Bordeaux I University, 1987

[ENRR 87]    H. Ehrig, M. Nagl, A. Rosenfeld, G. Rozenberg (eds.): Graph–Grammars and Their Application to Computer Science, Lect. Not. Comp. Sci. 291, 1987

[ENR 83]     H. Ehrig, M. Nagl, G. Rozenberg (eds.): Graph–Grammars and Their Application to Computer Science, Lect. Not. Comp. Sci. 153, 1983

[HK 83]      A. Habel, H.-J. Kreowski: On Context–Free Graph Languages Generated by Edge Replacement, Lect. Not. Comp. Sci. 153, 143-158, 1983

[HK 85]      A. Habel, H.-J. Kreowski: Characteristics of Graph Languages Generated by Edge Replacement, University of Bremen, Comp. Sci. Report No. 3/85, also in: Theor. Comp. Sci. 51, 81-115, 1987

[HK 87a]     A. Habel, H.-J. Kreowski: May We Introduce to You: Hyperedge Replacement, Lect. Not. Comp. Sci. 291, 15-26, 1987

[HK 87b]     A. Habel, H.-J. Kreowski: Some Structural Aspects of Hypergraph Languages Generated by Hyperedge Replacement, Proc. STACS'87, Lect. Not. Comp. Sci. 247, 207-219, 1987

[HKV 87]     A. Habel, H.-J. Kreowski, W. Vogler: Metatheorems for Decision Problems on Hyperedge Replacement Graph Languages, Techn. Report, short version in: Bull. EATCS 33, 55-62, 1987

[Kr 87]      H.-J. Kreowski: Rule Trees Can Help to Escape Hard Graph Problems, unpublished

[La 88]      C. Lautemann: Decomposition Trees: Structured Graph Representation and Efficient Algorithms, Proc. CAAP'88, Lect. Not. Comp. Sci. 299, 28-39, 1988

[Le 85]      T. Lengauer: Efficient Solution of Biconnectivity Problems on Hierarchically Defined Graphs, Proc. WG'85, Trauner Verlag, Linz, 201-215, 1985

[LW 88]      T. Lengauer, E. Wanke: Efficient Analysis of Graph Properties on Context–Free Graph Languages, Proc. ICALP'88, Lect. Not. Comp. Sci. 317, 379-393, 1988

[RW 86a]     G. Rozenberg, E. Welzl: Boundary NLC Graph Grammars - Basic Definitions, Normal Forms, and Complexity, Inf. Contr. 69, 136-167, 1986

[RW 86b]     G. Rozenberg, E. Welzl: Graph Theoretic Closure Properties of the Family of Boundary NLC Graph Languages, Acta Informatica 23, 289-309, 1986

[Sl 82]      A.O. Slisenko: Context–Free Graph Grammars as a Tool for Describing Polynomial–time Subclasses of Hard Problems, Inf. Proc. Lett. 14, 52-56, 1982

# THE MONADIC SECOND-ORDER LOGIC OF GRAPHS :
## DEFINABLE SETS OF FINITE GRAPHS

**Bruno COURCELLE**

BORDEAUX I UNIVERSITY
Laboratoire d'Informatique (+)
351, COURS DE LA LIBÉRATION
33405 TALENCE, FRANCE

**Abstract** : Every set of finite graphs definable in monadic second-order logic is recognizable in the algebraic sense of Mezei and Wright (no "graph automaton" is provided). We apply this result to the comparison of several definitions of sets of finite graphs , in particular by context-free graph grammars, and by forbidden configurations. It follows that the monadic second order theory of a context-free set of graphs is decidable,and that every graph property expressible in monadic second-order logic is decidable in polynomial time for graphs of a given maximal tree-width.

## Introduction:

Sets of finite graphs can be defined in several ways:

(1) by properties (like k-regularity or connectivity),expressed in certain logical calculi,

(2) by finite sets of forbidden configurations (Kuratowski's characterization of planar graphs is a well-known example),

(3) by graph grammars, generating graphs by means of finitely many rewriting rules (series-parallel graphs can be defined in this way),

(4) by congruences of finite index:this is a new definition yielding the notion of a recognizable set of graphs.

---

**Notes:** (+) Unité de Recherche Associée au CNRS n°726
This work has been supported by the "Programme de Recherches
Coordonnées : Mathématiques et Informatique "
Electronic mail :mcvax!inria!geocub!courcell (on UUCP network)
or : courcell (at) inria.geocub.fr

Monadic second-order logic (MSOL) is an appropriate calculus for expressing graph properties. Forbidden configurations can be expressed in MSOL.So are connectivity, and many other useful properties (but the existence of nontrivial automorphisms cannot be expressed in this way ).

The foundation stone of our theory is the theorem saying that *if a set of finite graphs is definable in monadic second-order logic, then it is recognizable.* This paper presents its consequences.Here are the main results:

1 *Every MSOL graph property is decidable in polynomial time,for graphs the tree-width of which is bounded by a given integer.*

2 *For every context-free set of graphs L,and for every MSOL graph property,one can decide whether all graphs in L (or some graphs in L) satisfy this property; furthermore the graphs of L satisfying it,form a context-free set .*

3 *The four above listed definitions of sets of graphs are compared .*

Applications to algebraic characterizations of sets of finite graphs defined by "forbidden minors" follow.We shall use a few recent results of Robertson and Seymour,and,in particular, their proof of Wagner's conjecture.

The algebraic aspect of these results is based on an algebraic structure on graphs defined and investigated by Bauderon and Courcelle [4,10] .

## BASIC DEFINITIONS

### 1.Graphs

By a graph,we mean a finite oriented graph, the edges of which are labeled by symbols from a fixed set A, and that is equipped with a sequence of vertices called the sequence of **sources**. The length of this sequence is called the **type** of the graph. A graph of type n is also called an **n-graph**. One may have in a graph several edges with the same label, the same origin and the same target. One can also have loops. The sequence of sources of a graph G is denoted by $src_G$.It may have repetitions. Two isomorphic graphs are considered as equal. We denote by $FG_n$ the set of all finite n-graphs.

Formal definitions can be found in Bauderon and Courcelle [4,10] . In these papers, hypergraphs are used.Here ,we only consider graphs with binary edges, i.e.,with edges having two, possibly identical vertices, and sets of such graphs. The context-free graph grammars used to generate sets of graphs may have nonterminal hyperedges.

## 2.Graph operations

If G is an n-graph and G'is an m-graph,then $G \oplus G'$ denotes their disjoint union, equipped with the concatenation of $src_G$ and $src_{G'}$, as sequence of sources. This operation is not commutative.

If i and j are two integers in $[n]$,and if G is an n-graph,then $\theta_{i,j}(G)$ is the n-graph obtained by fusing the i-th source and the j-th source of G.

If $\alpha : [p] \longrightarrow [n]$ is a (total) mapping, if $G \in FG_n$, then $\sigma_\alpha(G)$ is the graph in $FG_p$ consisting of G equipped with $(src_G(\alpha(1)),\ldots,src_G(\alpha(p)))$ as sequence of sources.( We denote by $src_G(i)$ the i-th element of $src_G$).

These operations define thus a **many-sorted algebra of graphs FG**: the set of positive integers $\mathbb{N}$ is the set of sorts, $FG_n$ is the carrier of sort n, the above defined operations define infinitely many operations forming an $\mathbb{N}$-signature H .The symbols of this signature are:

$$\oplus_{n,m} \text{ of profile } (n,m) \longrightarrow n+m$$
$$\theta_{i,j,n} \text{ of profile } n \longrightarrow n$$
$$\sigma_{\alpha,p,n} \text{ of profile } n \longrightarrow p.$$

For every $a \in A$,we also denote by a the 2-graph consisting of one edge labeled by a, with $src_a(1)$ as origin,and $src_a(2)$ as target. We denote by **1** the 1-graph consisting of one vertex that is its unique source.We denote the empty graph by **0**.We let $H_0:=A\cup\{1,0\}$ and $H':=H \cup H_0$.

PROPOSITION [4] : **Every finite graph is denoted by an algebraic expression constructed over H'** .

These algebraic expressions are called graph expressions. We denote by $E_n$ the set of graph expressions of sort n. We denote by val(g) the graph defined by a graph expression g. Hence, $val(g) \in FG_n$ if g belongs to $E_n$.

The width wd(g) of g in $E_n$ is the maximal sort of a symbol of H' occurring in g. Hence $wd(g) \geq n$. The width of a finite n-graph G is $wd(G):= Min\{wd(g)/g\in E_n, val(g)=G\}$. We denote by $FG_n^k$ the set $\{G \in FG_n/wd(G) \leq k\}$, and by $E_n^k$ the set of graph expressions of sort n and of width at most k.

## 3.Context-free graph grammars and equational sets of graphs

Context-free graph grammars using edge- and hyperedge-replacements have been introduced in several papers by Habel and Kreowski,and by Bauderon and Courcelle [4,11,18].They define context-free sets of graphs.

These sets are equational w.r.t.the above introduced algebraic structure .A set is equational iff ,as in Mezei and Wright [22], or Courcelle [7] ,it is a component of the least solution of a system of equations.We denote by Equat(FG)$_n$ the set of equational subsets of FG$_n$. The system associated with a context-free graph grammar is solved in $\mathscr{P}$(FG), the power-set algebra of FG .It is similar to the system classically associated with a context-free grammar, and it is written with U (set union), the operations of H extended to sets in a canonical way , and the constants of H$_0$ .

It is proved in [4] that a set of graphs is context-free iff it is equational.

The elements of a context-free set are all of the same type. They can be constructed with the graph operations that appear in the defining system. Hence L $\subseteq$ FG$_n^k$ for some n and k$\geqslant$n, if L is context-free.The set FG$^k{}_n$ is context-free.

The set of all n-graphs FG$_n$ is not context-free since a complete graph with n vertices is of width larger than n ( see [4]).

## 4.Recognizable sets of graphs.

The recognizable subsets of FG can be defined as in any algebra (Mezei and Wright [22], Courcelle [8,9,12]). A subset L of FG$_n$ is recognizable iff it is a union of classes of a congruence over FG that has finitely many classes of each sort.We let Rec(FG)$_n$ denote the family of recognizable subsets of FG$_n$.See Courcelle [12] for a survey of recognizability in arbitrary algebraic structures.

The family of sets Rec(FG)$_n$ is closed under the Boolean operations, and it contains the sets $\emptyset$ and FG$_n$. Furthermore it is uncountable [8]. This is unexpected but the infiniteness of the signature allows this fact.

This last result helps to establish that Rec(FG)$_n$ and Equat(FG)$_n$ are incomparable (whereas Rec(M)$\subseteq$ Equat(M) if M is an algebra that is generated by a finite signature).But, if the graphs of a recognizable set are of bounded width,this set is equational.Hence the classical inclusion, of the family of recognizable sets in the family of equational ones , holds for sets of graphs of bounded width. The proofs can be found in Courcelle [8].

**5.PROPOSITION** :If L $\in$ Equat(FG)$_n$ and K $\in$ Rec(FG)$_n$ then L$\cap$K $\in$ Equat(FG)$_n$.

See [8,9] for the proof . The construction of a system defining L$\cap$K from one defining L is effective. This result extends the classical one saying that the intersection of a context-free language and a regular one is context-free.

**COROLLARY** : If $L \in Rec(FG)_n$, and L is included in some context-free set, then $L \in Equat(FG)_n$.

For later use, let us mention the following concrete characterization of the family of recognizable subsets of $FG_n{}^k$. We say that a subset K of $E_n{}^k$ is saturated if, for every graph expression e in K , every other graph expression in $E_n{}^k$ denoting the same graph is also in K. Hence, L is a recognizable subset of $FG_n^k$ iff it is the set of values of the graph expressions of a subset K of $E_n{}^k$ that is saturated and recognizable as a set of trees. For a set of trees, recognizable means defined by a finite-state tree-automaton, in the sense of $[14,28]$.

## 6. Graphs as logical structures.

In order to express properties of k-graphs, we define the following symbols :

    **v** : the " vertex" sort,
    **e** : the " edge" sort,
    $s_i$ : a constant of sort **v** for $i=1,\ldots,k$,
    $edg_a$ : a predicate symbol of arity (e,v,v) for all a in A.

With $G \in FG_k$, we associate the logical structure $|G| = \langle V_G, E_G, (s_{iG})_{i \in [k]}, (edg_{aG})_{a \in A} \rangle$ where $V_G$ is the domain of sort **v** (the set of vertices of G), $E_G$ is the domain of sort **e** (the set of edges of G), $s_{iG}$ is the i-th source of G, and $edg_{aG}(e,v_1,v_2) =$ true iff the label of the edge e is a, its origin is $v_1$, and its target is $v_2$.

## 7. Counting monadic second order logic.

We shall build formulas by using object variables $u,x,y,z,u',\ldots$ of sort **v** or **e**, intended to denote vertices or edges, and set variables $U,X,Y,Z,U'$ of sort **v** or **e**, intended to denote sets of vertices or sets of edges.

Let $\mathscr{W}$ be a sorted set of variables $\{u,u',\ldots U,U',\ldots\}$. The sort mapping : $\mathscr{W} \longrightarrow \{v,e\}$ is denoted by $\sigma$. We denote by $\mathscr{W}_s$ the set $\mathscr{W} \cup \{s_1,\ldots,s_k\}$. Uppercase letters will denote set variables, and lowercase letters will denote the remaining elements of $\mathscr{W}_s$, i.e., the object variables and the constants.

The set of atomic formulas consists of :

    $u=u'$ for $u,u' \in \mathscr{W}_s$, $\sigma(u) = \sigma(u')$,
    $u \in U$ for $u,U \in \mathscr{W}_s$, $\sigma(u) = \sigma(U)$,
    $edg_a(u,u_1',u_2')$ for $u,u_1',u_2' \in \mathscr{W}_s$, $\sigma(u) = e, \sigma(u_1') = \sigma(u_2') = v$
    $card_{n,p}(U)$ for $U \in \mathscr{W}$, $0 \leqslant n < p$, $p \geqslant 2$.

In the structure $|G|$, associated with a graph G in $FG_k$, the meaning of these atomic formulas is clear from the above definitions completed as follows. If U denotes a finite set X, then $card_{n,p}(U) =$ true iff $card(X) = n+mp$ for some $m \geqslant 0$.

The language of counting monadic second-order logic ( CMSOL ) is the set of logical formulas formed with the above atomic formulas together with the Boolean connectives $\wedge, \vee, \neg$ , the object quantifications $\forall u$, $\exists u$ (over vertices or edges), and the set quantifications $\forall U$, $\exists U$ (over sets of vertices or sets of edges).

The language of monadic second-order logic (MSOL) is the set of such formulas that are written without the atomic formulas $card_{n,p}(U)$ .

We denote by $\mathscr{CL}_k(\mathscr{W})$ the set of formulas of counting monadic second-order logic with free variables in $\mathscr{W}$. They express properties of graphs of type at most k. The notation $\mathscr{L}_k(\mathscr{W})$ refers similarly to monadic second-order logic.

If $\varphi$ is a closed formula, and if G is a graph, we write $G \models \varphi$ iff $|G| \models \varphi$, and we say that $\varphi$ holds in G, or that G satisfies $\varphi$.

## 8. Definable properties and definable sets of graphs.

A property of graphs is definable if there exists a closed formula in CMSOL that holds in a graph G iff G satisfies this property. A set of graphs L is definable iff the membership in L is a definable property. If the defining formula belongs to MSOL, we say that the property or the set is $\mathscr{L}$-definable.

Properties concerning paths in graphs, like connectivity and existence of cycles, are $\mathscr{L}$-definable .See Courcelle [8,9,11].

It can be proved that the predicates $card_{p,q}(U)$ are not definable in $\mathscr{L}$, hence that $\mathscr{CL}$ is strictly more powerful than $\mathscr{L}$. But $\mathscr{CL}$ and $\mathscr{L}$ are of equal power for expressing properties of graphs having an $\mathscr{L}$-definable linear order :this is proved in Courcelle [8, section 3.11].

## A BASIC RESULT

**9. THEOREM [8]** : Every definable set of graphs is recognizable.

The proof technique for this result is also used in [9]. Since the proof is quite long we cannot reproduce it here.

## APPLICATIONS TO CONTEXT-FREE SETS OF GRAPHS.

**10.**PROPOSITION : Let L be a context-free set of finite graphs $\subseteq FG_n$. Let $\varphi \in \mathscr{CL}_n$. Then : $\{G \in L/G \not\models \varphi\}$ is context-free. One can decide whether $G \not\models \varphi$ for all $G \in L$, and whether $G \not\models \varphi$ for some $G \in L$.

This applies in particular to the context-free sets $L = FG_n^k$, i.e.,to the sets of finite n-graphs of width at most k.

Proof:This result follows from Propositions 5 ,9 and the fact that one can decide whether a context-free set is empty.

REMARK: The time complexity of the algorithms derived from this proposition has a level of exponentiation proportional to the level of nested quantifications in $\varphi$. Efficient algorithms can be investigated for particular formulas that define graph properties of particular interest.The notion of an inductive set of predicates introduced in [9,11] can help to find such algorithms.

### APPLICATIONS

1.The set of dependency graphs of an attribute grammar is context-free.Since the existence of cycles in a graph is expressible in MSOL,the noncircularity test is a special case of Proposition 10.Other interesting properties, like whether the computation of a given attribute uses another given attribute in some (or in all ) dependency graph ,are decidable ,also by Proposition 10.

2. Applications to the analysis of recursive definitions have been given by Courcelle in [13] .

### FORBIDDEN CONFIGURATIONS

### 11.Minors

In this section, we only consider graphs from the reference set $M := FG_o$.

We say that two graphs in M are quasi-isomorphic if they are isomorphic,up to the orientations of the edges and their labels.

Let G and H belong to M. We say that H is contained as a minor in G if G can be transformed into a graph quasi-isomorphic to H, by a finite sequence of edge deletions, of edge contractions, and of deletions of isolated vertices.

Equivalently, H is contained as a minor in G iff there exist two mappings $f : V_H \to \mathcal{P}(E_G)$ and $g : E_H \to E_G$ satisfying the following conditions :

(1) if $v, v' \in V_H$, $v' \neq v$, if $e \in f(v)$, $e' \in f(v')$, then e and e' have no common vertex (hence $e \neq e'$),

(2) for each v, $f(v)$ is the set of edges of a connected subgraph of G,

(3) if $e, e' \in E_H$, and $e' \neq e$, then $g(e') \neq g(e)$,

(4) if $e \in E_H$, if v and v' are the two ends of e, then one end of $g(e)$ belongs to some edge in $f(v)$ and the other to some edge in $f(v')$.

From this alternative definition,(adapted from Robertson and Seymour [24]) it follows that, for every given finite graph H, the property of a graph G reading"H is contained as a minor in G" is $\mathcal{L}$-definable .

We write $H << G$ if H is contained in G as a minor. This binary relation on graphs is transitive.

For every set K of finite graphs , we denote by FORB(K) the set of graphs G in M,such that $H << G$ for no H in K. In order to illustrate this definition,we recall that a graph is planar iff it contains neither $K_5$ nor $K_{3,3}$ as a minor ( Wagner [29] ;see also [16]).

A subset L of M is minor-closed if,for every G in L, every minor of G is in L. If L is minor-closed, then, clearly : L=FORB(M-L). Robertson and Seymour have established a conjecture by Wagner saying that in such a case, there exists a finite subset K of M-L such that L=FORB(K) ([26]).Hence:

**12.THEOREM** : If a set of graphs is minor-closed, then it is $\mathcal{L}$-definable and recognizable.

**13. PROPOSITION** : $FG_0^k$ is minor closed.

Proof : Let G be defined by a graph expression g of width k. Let $H << G$ be quasi-isomorphic to H',where G can be transformed into H' by edge contractions, edge and vertex removals. These transformations can be performed at the level of graph expressions as follows.

Let e be an edge of G labeled by some a in A. There is in g an occurrence of a that corresponds to this edge.

Let G' be the graph obtained from G by deleting this edge. This graph is val(g') where g' is the graph expression resulting from the substitution in g of 1⊕1 for the considered occurrence of a. Clearly wd(g')=wd(g).

Let G" be obtained from G by the contraction of the edge e. We have G"=val(g") where g" is obtained from g by the substitution of $\sigma_\alpha(1)$ for the considered occurrence of a, where $\alpha$ is the unique mapping : $[2] \rightarrow [1]$. Clearly wd(g") =wd(g).

Let finally G"' be obtained from G by the delection of one isolated vertex, then it is not hard to transform g into g"' that defines G"', and is such that wd(g"') $\leqslant$ wd(g). This last transformation is a bit more technical, and we omit the details.

Hence, after repeating these steps several times, one obtains a graph expression defining H', and that is of width at most k. It is easy to see that H and H' have the same width. Hence, H is of width at most k. □

**14.COROLLARY** : (1) The set of 0-graphs of width at most k is $\mathcal{L}$-definable and recognizable.(It is also context-free).

(2) The set of 0-graphs of width (exactly) k has the same properties.

**PROOF** : (1)We know that $FG_0{}^k$ is context-free.The other assertions follow from Propositions 9, 12, and 13.

(2) Since the difference of two recognizable sets is recognizable,the result follows from (1), Proposition 5,and the fact that Rec(P) is a Boolean algebra for every algebra P.The context-freeness follows from Proposition 5. □

Since Theorem 12 is not effective , we do not know the finite set K of forbidden minors that defines $FG_0^k$. Hence ,we know that this set is recognizable,but we do not know the congruence defining it.As a consequence, we do not know the grammar defining the set of graphs of width k, but we know the existence of such a grammar. Hence we know that certain problems are decidable(for example whether all graphs of width k satisfy a formula), without knowing the algorithm. Johnson has already pointed out a similar situation for polynomial decidability in [16].

# 15.Tree-width.

Another notion of width of graph has been introduced by Seymour and Robertson.It is very close to our notion.

A graph is of tree-width at most k if it has a tree-decomposition of width k,that is,if it can be built by gluing in a tree-like fashion "small" graphs each of them having at most k vertices. We denote by twd(G) the tree-width of a graph G. A precise definition of tree-width, and a proof of the following proposition can be found in the appendix.

**16.** Proposition :For every 0-graph G, wd(G)$\leqslant$ 2twd(G)+4 and twd(G)$\leqslant$2wd(G)-1.

Remark:Since the set of graphs of tree-width at most k is minor closed, the results of Proposition 14 hold for the notion of tree-width.In particular,the set TW(k) of graphs of tree-width exactly k is $\mathcal{L}$-definable and context-free.

**17.** PROPOSITION : If K is a finite set of graphs,one of which is planar,then **FORB**(K) is a set of graphs of bounded width. It follows that **FORB**(K) is context-free , $\mathcal{L}$-definable and recognizable. Conversely, if L is context-free and is of the form **FORB**(K) for some finite set of graphs K ,then L= **FORB**(KU{Q})for some planar graph Q.

PROOF : Seymour and Robertson have established in [25] that FORB({H}) is included in the set of graphs of tree-width at most k (an integer computable from H) if H is a planar graph. The result follows from Propositions 16 and 10 .

Let L be context-free and be also equal to FORB(K) for some finite set K. Let k be such that L $\subseteq$ FG$_0$$^k$. We have twd(G)$\leqslant$2k-1 for all G in L (this follows from Proposition 16 ). Let Q be a large grid such that twd(Q)>2k-1. We have L$\subseteq$FORB({Q}), (otherwise Q<<G for some G in L, and twd(Q)$\leqslant$twd(G), contradicting the choices of k and Q ). Hence L=FORB(K) $\cap$ FORB({Q}) = FORB(KU{Q}) . $\square$

## A Comparison Diagram

**18.** Our results can be summarized in the following diagram, where several families of sets of finite 0-graphs are considered. *(On this diagram, the scope of a family name is the largest rectangle, at the upper left corner of which it is written).*

**REC** is the family of recognizable sets of graphs
**CMSOL** is the family of definable sets of graphs
**MSOL** is the family of $\mathscr{L}$-definable sets of graphs
**CF** is the family of context-free (or equational ) sets of graphs
**FORB** is the family of sets defined by a finite set of forbidden minors
**FORB-P** is the family of sets defined by a finite set of forbidden minors one of which is planar
**B** is the family of all subsets of $FG_0^k$ for all k.

Remarks: 1. The families **B** and **REC** are uncountable.

2. Proposition 17 says that **CF ∩ FORB = FORP-P**.

3. The equality of two sets of graphs in **CF∩REC** is decidable, provided each of them is effectively given, as a context-free set (by a grammar), and as a recognizable set ( by a tree-automaton on graph expressions ). This is an immediate consequence of Proposition 5.

4. All inclusions shown on the diagram are proper.All boxes are nonempty, except perhaps the shaded one.Its emptiness raises an open problem.

**OPEN PROBLEM :** Does there exist a recognizable subset of $FG_o^k$, for some k, that is not definable ? (Such a set exists iff the shaded box of the diagram is not empty).

The diagram also locates the following sets of graphs :

$L_G$, the set of square grids,
L, the set of all square grids of size nXn where n is an
     arbitrary integer,belonging to some nonrecursive set B.
P, the set of planar graphs,
T, the set of undirected, unrooted trees,
NC, the set of cyclefree graphs,
SP, the set of series-parallel graphs,
S, the language $\{a^n b^n / n > 0\}$,considered as a set of graphs,
U, the language $\{a^n b^n c^n / n > 0\}$,considered as a set of graphs,
W(k), the set of graphs of width exactly k,
TW(k) ,the set of graphs of tree-width exactly k,
E, the set of discrete graphs (all vertices of which are
     isolated),having an even number of vertices.

It is proved in [8] that $L_G$ is $\mathscr{L}$-definable, and that every subset of $L_G$ is recognizable. The above defined set L is such a set. If it would be defined by a formula $\varphi$, then one would have $G_n \not\models \varphi$ iff $n \in B$, where $G_n$ is the square nxn grid.Since one can decide whether a finite graph satisfies a formula, one would be able to decide whether an integer n belongs to B. This contradicts the choice of B.

It is proved in [8] that E is definable but not $\mathscr{L}$-definable. The reason is that one cannot express in the language $\mathscr{L}$ that the number of elements of a set is even.

The set P is equal to **FORB**$(\{K_{3,3}, K_5\})$ by the above mentionned result by Wagner. It is not in **B** since it contains all grids, and there exist grids of arbitrary large width.

The set NC of cycle-free graphs is **FORB**$(\{G\})$ where G is the graph:

Context-free graph grammars generating SP,T,S, and U are easy to construct. There exist grammars generating W(k) and TW(k), but we do not know them. Note that these six context-free sets are not minor-closed.

It is proved in [8] that S is not recognizable. The same holds for U.

## COMPLEXITY ISSUES

The size of an n-graph G is defined as $n+Card(V_G)+Card(E_G)$.

**19.** PROPOSITION :Let $\varphi$ be a closed formula in CMSOL.For every k and n,one can decide in time $O(\textbf{size}(e))$,whether **val**(e) satisfies $\varphi$, where e is a graph expression in $\textbf{E}_n$ of width at most k.

Proof :The proof of Theorem **9** yields a bottom-up deterministic tree automaton recognizing the set of graph expressions of width at most k, the value of which satisfies $\varphi$ .This automaton can read e, represented as a tree (see [12,14,28]), in time O(size(e)). Hence, **val**(e) satisfies $\varphi$ iff e is accepted by the automaton . □

**20.**THEOREM :Given k,n and $\varphi$ as above ,one can construct an algorithm that says,for every n-graph G and in time $O(\textbf{size}(G) +(\textbf{card}(V_G))^{k+2})$ :
    either that **twd**$(G^0)> k$
    or that **twd**$(G^0) \leqslant k$ and G satisfies $\varphi$
    or that **twd**$(G^0) \leqslant k$ and G does not satisfy $\varphi$.

The proof is given in the appendix. (We denote by $G^0$ the 0-graph obtained by "forgetting" the sources of G).

## RELATED WORKS

*On the properties of the graphs of a context-free set.*

Habel considers in [17] graph properties that are "compatible" with the derivation steps of context-free graph grammars, and establishes a result analogous to our Proposition 10. Her notion of compatibility is essentially equivalent to our notion of an inductive set of predicates, hence to our notion of recognizability.

We establish in [9] another theorem similar to Proposition 10, that holds for a richer class of graph-grammars (the context-free NLC grammars), and a weaker logical language ( the CMSOL where quantifications are restricted to vertices and to sets of vertices ). A result by Seese [27] shows that this result cannot hold if quantifications over sets of edges are allowed.

*On complexity issues.*

In his 1985 column [15], Johnson discusses a few NP-complete problems that become polynomial when restricted to special classes of graphs.

In some cases, polynomial algorithms can be obtained from "structured decompositions" of the graphs. Many results have been obtained along this line. A few references are [2,19,23,30], and a more complete list can be found in [5].

Instead of considering specific problems, we have considered *classes of problems*, described in a *uniform way*. Our metatheorem concerns the following NP-complete problems: Hamiltonian circuit, Partition into triangles, Cubic subgraph, to name a few, that are expressible in MSOL.

Arnborg et al.[3] have introduced the extended monadic second-order logic (EMSOL), where arithmetic is available. The corresponding extension of Theorem 20 holds, and covers a longer list of NP-complete problems. But the set of graphs satisfying an EMSOL formula is **not** recognizable. (The set S shown on diagram 18 is definable in EMSOL, but it is not recognizable).

Bodlaender also considers in [5] classes of problems defined in uniform ways. His definitions seem to be closely related to those of [3].

The aims of Lengauer and Wanke in [21] are similar. Their notion of a "finite" graph property is equivalent to the recognizability of the set of graphs enjoying the property. Some of their results are similar to our results 20 and 10. They extend our Proposition 10 to certain "controlled derivations", hence to non context-free sets of graphs.

The efficiency of the algorithms based on graph decompositions depend on that of finding the appropriate decompositions. There are mainly two cases. If one wishes a tree-decomposition of a given width, then, one can use the algorithm of [1].( There exists an $O(n^3)$-algorithm by the results of Robertson and Seymour, but it is still unknown; see [5,p.106] ). Graph decompositions based on context-free graph grammars offer another possibility. See Lautemann [19,20] for efficient algorithms based on this idea.

Let us finally mention that the technique of Bodlaender [6] can be used to all graph properties definable in CMSOL. It follows that deciding them for graphs of bounded tree-width is in the complexity class NC.

## APPENDIX

This appendix contains a few definitions and proofs that have been omitted in the main text. We give in particular the complete proof of Proposition 16, that is essential in this paper. We need a few new definitions.

### (A.1) Notations

An n-graph is a 5-tuple $G = <V_G, E_G, lab_G, vert_G, src_G>$ where $V_G$ is the set of vertices, $E_G$ is the set of edges, $lab_G : E_G \to A$ associates a label with each edge, $vert_G : E_G \to V_G \times V_G$ associates with each edge its pair of vertices, and $src_G$ is a sequence of n vertices, called the sequence of sources. The type of G, i.e., the length of $src_G$ is denoted by $\tau(G)$. Such a graph is source-separated if $src_G$ has no repetition.

The 0-graph associated with G is the graph $G^0 = <V_G, E_G, lab_G, vert_G, \varepsilon>$, having an empty sequence of sources. We denote by $\emptyset$ the trivial map : $\emptyset \rightarrow [n]$. Hence $G^0 = \sigma_{\emptyset}(G)$. Finally, if H is a 0-graph and s is a sequence of vertices of H, we denote by (H,s) the unique graph G such that $G^0 = H$ and $src_G = s$.

We denote by $S_G$ the set of sources of a graph G.

We say that G is a subgraph of H, and this is written $G \subseteq H$, if $V_G \subseteq V_H$, $E_G \subseteq E_H$, $lab_G = lab_H \upharpoonright E_G$ (i.e., is the restriction of $lab_H$ to the set $E_G$), $vert_G = vert_H \upharpoonright E_G$, and $src_G$ is the subsequence of $src_H$ consisting of the vertices in $V_G$, in their order.

If G and G' are two subgraphs of H, we denote by $G \cup G'$ the unique subgraph K of H such that $V_K = V_G \cup V_{G'}$ and $E_K = E_G \cup E_{G'}$.

If $V \subseteq V_H$ and $E \subseteq E_H$, we denote by $H \upharpoonright (V \cup E)$ the unique subgraph G of H such that $E_G = E$, and $V_G = V \cup V'$ ,where V' is the set of vertices of the edges of E.

In Definition 2 we have introduced a graph operation $\theta_{i,j,n}$ that takes an n-graph, and transforms it by fusing its i-th and j-th sources. For every equivalence relation $\delta$ on $[n]$ (we denote by $[n]$ the set $\{1,\ldots,n\}$, and by $Eq([n])$ the set of equivalence relations over it), we let

$$\theta_{\delta,n}(G) := \theta_{i_1,j_1,n}(\theta_{i_2,j_2,n}(\ldots\theta_{i_k,j_k,n}(G)))\ldots))$$

where $\{(i_1,j_1),\ldots,(i_k,j_k)\}$ is a subset of $[n] \times [n]$ that generates $\delta$. (It is clear that any two sets generating $\delta$ yield the same graph $\theta_{\delta,n}(G)$, and one can choose one of cardinality $\leqslant n-1$).

A derived graph operation is an operation on graphs defined by a finite well-formed term over H (see Definition 2). Hence $\theta_{\delta,n}$ is a derived graph operation.

The size of a graph expression g is defined as the number of occurrences of symbols from H' used to write it. We do not count parentheses as symbols. In defining the size of a graph expression written with the derived operators of the form $\theta_{\delta,n}$ we count for one, each occurrence of such operators. Hence, we consider them as elementary, as in Bauderon and Courcelle [4].

Other derived operations will be introduced below. Each occurrence of one of them will be counted for the number of symbols from H'∪{θ$_{\delta,n}$ / δ ∈ Eq([n])} with which they are written.

Similarly, the width of a graph expression written with derived operators, is defined as the maximum sort of one of symbols of H' occurring there, or in the definition of one of the derived operations occurring there.

**(A.2) Definition : Tree-width.**

Let G be a 0-graph. A tree decomposition of G is a pair (T,f) consisting of an undirected tree T and a mapping f : $V_T$ ⟶ $\mathscr{P}(V_G)$ (the set of subsets of $V_G$) such that :

(1)   $V_G = \cup\{f(i) \; / \; i \in V_T\}$ ,

(2)   every edge of G has its two vertices in f(i) for some i,

(3)   if i,j,k ∈ $V_T$, and if j is on the (unique) path in T from i to k, then f(i) ∩ f(k) ⊆ f(j).

The width of such a decomposition is defined as:

$$\text{Max}\{\text{card}(f(i))/ \; i \in V_T \}-1 \; .$$

The tree-width of G is the minimum width of a tree-decomposition of G. It is denoted by **twd**(G).

It is clear two quasi-isomorphic 0-graphs have the same tree-width. The notion of tree-width is defined in Robertson and Seymour [24,25] for undirected graphs with possibly loops and multiple edges, i.e., for equivalence classes of 0-graphs w.r.t. quasi-isomorphism.

For every k, Arnborg et al. [1] give an $O(n^{k+2})$-algorithm that decides whether a graph G with n vertices has tree-width at most k. When this is the case, the algorithm constructs a tree decomposition (T,f) of G such that card($V_T$) = n-k+1 (or 1 if n≤k-1). (This algorithm actually constructs an embedding of G in a k-tree . Since a k-tree has such a decomposition, the result follows. We refer the reader to [1] for details).

Proposition 16 follows from lemmas (A.3) and (A.5).

**(A.3) Lemma :** For every 0-graph G , **wd**(G) ≤ 2 **twd**(G) + 4.

From a tree-decomposition of G of width k-1, we shall construct a graph expression defining G, of width at most 2k+2. In order to facilitate this construcion, we define a family of derived graph operations.

**(A.4) Definition**

For every one-to-one partial mapping $\pi$ : $[n] \longrightarrow [p]$, we define, for G of type n and G' of type p:

$$G \sqsupset_{\pi,n,p} G' = \sigma_\alpha(\theta_\delta(G \oplus G'))$$

where $\alpha$ is the inclusion map : $[n] \rightarrow [n+p]$, and $\delta$ is the equivalence relation on $[n+p]$ generated by $\{(i,n+\pi(i)) \, / \, i \in dom(\pi) \}$.

The resulting graph is of type n. (The domain of $\pi$, i.e., the set of integers such that $\pi(i)$ is defined is denoted by $dom(\pi)$).

This operator will be used for source-separated graphs. Let G,G',H be source-separated graphs such that :

    (1)  $G \subseteq H$, i.e., G is a subgraph of H,

    (2)  $E_H = E_G \cup E_{G'}$, $E_G \cap E_{G'} = \emptyset$ ,

    (3)  $V_H = V_G \cup V_{G'}$ , $V_G \cap V_{G'} \subseteq S_G \cap S_{G'}$

Then H is isomorphic to $G \sqsupset_{\pi,n,p} G'$ where $n=\tau(G)$, $p=\tau(G')$, and $\pi$ is the partial one-to-one function such that $\pi(i)=j$ iff $src_G(i)=src_{G'}(j)$.

The above conditions (1)-(3) can also be written as follows

    (1')  $H=(\,G^0 \cup G'^{\,0},\, src_G\,)$,

    (2')  $E_G \cap E_{G'} = \emptyset$,

    (3')  $V_G \cap V_{G'} \subseteq S_G \cap S_{G'}$ .

This situation is illustrated on Fig.1. (The thick lines represent the sets of sources of the graphs).

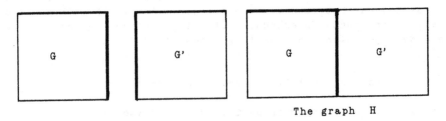

The graph H

Figure 1

**Proof of Lemma (A.3)**

Let $(T,f)$ be a tree-decomposition of $G$ of width $k-1$.

Let us choose a root $rt$ for $T$, and orient its edges in such a way that, for every vertex $i$ of $T$, there is one and only one oriented path from $rt$ to $i$. We write $i' \leqslant i$ iff : $i'=i$ or there exists an oriented path from $i'$ to $i$. Hence $rt \leqslant i$ for all $i$.

We also choose, for each $i \in V_T$, a sequence without repetition $\bar{f}(i)$, the elements of which are those of $f(i)$.

Finally, we choose a partition $\{E(i)/i \in V_T\}$ of $E_G$ such that, the two vertices of an edge in $E(i)$ belong to $f(i)$. (Some of the sets $E(i)$, $i \in V_T$, may be empty).

We let then :

$$H(i) = ( G \upharpoonright (E(i) \cup f(i)) , \bar{f}(i))$$

$$G(i) = ( G \upharpoonright (\cup\{E(i') \cup f(i')/i \leqslant i' , i,i' \in V_T\}) , \bar{f}(i)).$$

Hence $G = G(rt)^0$.

For every $i \in V_T$, we shall construct graph expressions $h_i$ and $g_i$ defining respectively $H(i)$ and $G(i)$.

Since $V_{H(i)} = f(i)$, and has at most $k$ elements, one can find a graph expression $h_i$ of width at most $k+2$, and of size at most $2k+4Card(E(i))+2$ that defines $H(i)$.

If $i$ is a leaf of $T$, then $G(i)=H(i)$ and one can take $g(i):=h(i)$.

Otherwise, let $i$ have successors $i_1,\ldots,i_n$. It is clear that

$$G(i)=( H(i)^0 \cup G(i_1)^0 \cup \ldots \cup G(i_n)^0 , src_{H(i)} ).$$

Let $G(i,j):=( H(i)^0 \cup G(i_1)^0 \cup \ldots \cup G(i_j)^0 , src_{H(i)} )$ for $0 \leqslant j \leqslant n$, with $G(i,0):= H(i)$.

Then, $G(i,j+1) = (G(i,j)^0 \cup G(i_{j+1})^0, src_{G(i,j)})$, the graphs $G(i,j)$ and $G(i_{j+1})$ have no edge in common, and every vertex common to them belongs to $f(i)$, hence, to the intersection of their sets of sources by condition (3) of Definition (A.2). Hence:

$$G(i,j+1) = G(i,j) \sqsupset_{r_{j+1}, m, p_{j+1}} G(i_{j+1})$$

for some appropriate $\pi_j$, where $m=\mathrm{Card}(f(i))$ and $p_{j+1}=\mathrm{Card}(f(i_{j+1}))$, for all $j=0,\ldots,n-1$. It follows that, if we have already constructed $g_{i_1}$, $\ldots,g_{i_n}$, we can take

$$g_i := ((..((h_i \; \beth_{\pi_1,m,p_1} \; g_{i_1} \; )\beth_{\pi_2,m,p_2} \; g_{i_2}) \cdot \cdot \beth_{\pi_n,m,p_n} \; g_{i_n} ).$$

This graph expression defines $G(i)$.

Since $T$ is a finite tree, we can construct,(by bottom-up induction on $T$), such a graph expression for all i. Hence the graph expression $g := \sigma_\emptyset(g_{rt})$ defines $G$.

The width of $g_i$ is

$$\mathrm{Max}\{\mathrm{wd}(h_i),\mathrm{wd}(g_{i_1}),\ldots,\mathrm{wd}(g_{i_n}), \; m+p_1,\ldots,m+p_n\},$$

for all i. Since $m \leqslant k$, $p_1 \leqslant k,\ldots,p_n \leqslant k$, and $\mathrm{wd}(h_i) < 2k+2$ for all i, we obtain $\mathrm{wd}(g_i) \leqslant 2k+2$ for all i, and $\mathrm{wd}(g) < 2k+2$. $\Box$

Remark: The size of $g_i$ is equal to

$$\mathrm{size}(h_i) + 3m + \mathrm{size}(g_{i_1}) + \ldots + \mathrm{size}(g_{i_n}).$$

An elementary computation yields :

$$\mathrm{size}(g) \leqslant 1 + 4 \; \mathrm{Card}(E_G) + t(2k+5).$$

It can be proved that every graph expression g defining a 0-graph G can be transformed in time $O(\mathrm{size}(g))$ into a graph expression g' defining the same graph and such that :

$$\mathrm{wd}(g') \leqslant \mathrm{wd}(g)$$

$$\mathrm{size}(g') \leqslant 6 \; (\mathrm{Card}(E_G)+\mathrm{Card}(\mathrm{Isol}_G))$$

where $\mathrm{Isol}_G$ denotes the set of isolated vertices of G. Since we shall not need this technical result in this paper, we do not prove it.

To complete the proof of Prop.16, we shall prove the following lemma :

(A.5) Lemma : For every 0-graph G, $\mathbf{twd}(G) \leqslant 2\mathbf{wd}(G) - 1$.

We first extend to graphs with sources, the notion of a tree-decomposition.

**(A.6) Definition :**

Let G be a graph of any type. A tree-decomposition of G is a pair $(T,f)$ where T is a rooted tree (its root is denoted by $rt_T$), $(T,f)$ is a tree-decomposition of $G^0$ (as in Definition (A.2)), and $f(rt_T)=S_G$, i.e., is the set of sources of G. Its width is as in Definition (A.2). The definition of the tree- width of G follows immediately.

It is clear that a tree-decomposition of G is also a tree-decomposition of $G_0$. Hence $twd(G^0) \leqslant twd(G)$. But it is easy to find examples where the inequality is strict.

**Proof of Lemma (A.5)**

For every graph expression g of width k, one can construct a tree-decomposition of $val(g)$ of width at most $2k-1$. The construction is by induction on the structure of g.

**Base cases :**

The cases of $g=a$ and $g=1$ are clear. One obtains a tree-decomposition of width at most 1, hence, $\leqslant 2k-1$ ,since we have necessarly $k \geqslant 1$ .

**Inductive cases :**

1 - Let $g=\sigma_\alpha(g')$, where $\alpha : [n] \longrightarrow [p]$, $n,p \leqslant k$. Let $(T',f')$ be a tree decomposition of $G'=val(g')$. One extends it into a tree-decomposition $(T,f)$ of $G=val(g)$, by adding a new root $rt_T$ to $T'$, such that $rt_T$, is its unique successor (in T), $f(rt_T) = S_G$, and $f \upharpoonright T' = f'$. Since $f(rt_T) \subseteq f'(rt_T')$, the width of $(T,f)$ is equal to that of $(T',f')$, hence is $\leqslant 2k-1$ by the induction hypothesis.

2 - Let us now consider the case where $g=g_1 \oplus g_2$, where $\tau(g_1)+\tau(g_2) \leqslant k$, $wd(g_1) \leqslant k$, $wd(g_2) \leqslant k$. Two tree-decompositions $(T_1,f_1)$ and $(T_2,f_2)$, of $G_1=val(g_1)$ and of $G_2=val(g_2)$, of respective widths $k_1'$ and $k_2'$, can be combined into a decomposition $(T,f)$ of $G=val(g)$, as shown below.

$$T =$$

We let f be such that $f \upharpoonright T_1=f_1$, $f \upharpoonright T_2=f_2$ and $f(rt_T)=S_G=S_{G_1} \cup S_{G_2}$ .

The width of T is at most

$$\text{Max}\{k_1', k_2', \text{ Card}(S_{G_1}) + \text{Card}(S_{G_2})-1\} \leqslant 2k-1$$

since $k_i' \leqslant 2k-1$, for i=1,2 (by the induction hypothesis), and since

$$\text{Card}(S_{G_1}) + \text{Card}(S_{G_2}) - 1 \leqslant 2k-1$$

3 - Let us finally consider the case where $g=\theta_{i,j,n}(g')$, with $\text{wd}(g') \leqslant k$, $n \leqslant k$.

It follows from Lemma (A.7) stated and proved below, that $\text{wd}(\text{val}(g)) \leqslant \text{twd}(\text{val}(g'))$. This completes the proof. □

**(A.7) Lemma** : Let $(T,f)$ be a tree decomposition of a 0-graph G. Let v and w be two vertices belonging to $f(\mathbf{rt}_T)$. Let G' be the quotient graph of G defined by the fusion of v and w. Let $h : G \rightarrow G'$ the canonical surjective homomorphism associated with this quotient.Then $(T, hof)$ is a tree-decomposition of G'.

**Proof** : Conditions (1) and (2) of Definition (A.2) are clearly satisfied. Let us verify condition (3).

Let x belong $h(f(i)) \cap h(f(j))$ where i and j belong to $V_T$. Let k belong to the unique path in T from i to j. There are two cases.

**First case** : $x=h(y)$, where $y \in f(i) \cap f(j)$. Then $y \in f(k)$ since $(T,f)$ is a tree-decomposition of G. Hence $x \in h(f(k))$.

**Second case** : The first case does not hold. Hence $x=h(v)=h(w)$ with $v \in f(i)$ and $w \in f(j)$ (or vice-versa). If k belongs to the path from i to $\text{rt}_T$,then $v \in f(k)$ (since $v \in f(\text{rt}_T)$). Hence $x \in h(f(k))$. If k belongs to the path from j to $\text{rt}_T$,then $w \in f(k)$, hence $x \in h(f(k))$. □

**Proof of Theorem 20** : Let G be given. One first applies the algorithm of [1] to $G^0$ and k. In time $O(\text{Card}(V_G)^{k+2}+\text{Card}(E_G))$, one obtains, either the information that $\text{twd}(G^0)>k$, or a tree- decomposition $(T,f)$ of $G^0$, of width at most k, and such that $\text{Card}(V_T) \leqslant \text{Card}(V_G)$.

By Lemma (A.3) and the remark following it, one can construct in time $O(\text{Card}(V_T)+\text{Card}(E_G))$ a graph expression of width at most 2k+4, and of size $O(\text{Card}(E_G)+\text{Card}(V_T))$, that defines $G^0$. It is not hard to transform it into a graph expresssion e of width at most 2k+4+n and of the same size, that defines G.

By applying the algorithm of Proposition 19, one can decide in time $O(\text{size}(e))$, whether G $=\text{val}(e)$ satisfies $\varphi$. The total time complexity is thus $O(\text{size}(G)^{k+2})$. □

**Acknowledgements:**   I   thank   D.Seese   for   useful   informations   and suggestions allowing me in particular to simplify some proofs.

# REFERENCES

[1]     ARNBORG S.,CORNEIL D.,PROSKUROWSKI A.,Complexity     of
        finding   an   embedding   in   a k-tree,SIAM J.of
        Alg. and Discr. Methods 8 (1987)277-284

[2]     ARNBORG S.,PROSKUROWSKI A.,Linear time  algorithms for
        port TRITA NA 8404 (1984),Stockolm.

[3]     ARNBORG S.,LAGERGREN J.,SEESE D.,Which  problems   are
        easy for tree decomposable graphs ? Preprint,
        Nov.1987,and Proc. of ICALP'88,Tampere,Lec.Notes
        in Comput.Sci.317,pp.38-51.

[4]     BAUDERON M.,COURCELLE B.,Graph expressions and  graph
        rewritings, Math Systems Theory 20 (1987)83-127

[5]     BODLAENDER H.,Dynamic   programming  on   graphs with
        bounded tree-width,Proc.ICALP'88,Lec.Notes Comp.
        Sci. 317 ,pp.105-132.

[6]     BODLAENDER H.,NC-algorithms for graphs with small tree-
        width,this volume.

[7]     COURCELLE B., Equivalences and transformations of regu
        lar systems. Applications to recursive program
        schemes and   grammars,   Theor. Comp. Sci. 42
        (1986) 1-122.

[8]     COURCELLE B.,  Recognizability and second order defi
        nability for sets of finite graphs, Report
        I-8634,(revised   version:The  monadic  second
        order logic of graphs,I:Recognizable sets  of
        finite graphs, to appear).

[9]     COURCELLE B.,An axiomatic definition of context-free
        rewriting and its application  to NLC graph
        grammars, Theoret.Comput.Sci.55(1987)141-181.

[10]    COURCELLE B., A representation of graphs by algebraic
        expressions and its use for graph rewriting
        systems, Proceedings of the $3^{rd}$ international
        workshop  on  graph  grammars,  Warrenton,
        Virginia, 1986 in Lec. Notes Comput. Science
        vol.291 ,1987 pp.112-133

[11]     COURCELLE B., On context-free sets of graphs and their
         monadic   $2^{nd}$-order   theory,   same   volume   as
         [10],pp.133-146.

[12]     COURCELLE B.,On recognizable  sets and tree-automata
         Report  8736,  Proc. of the Conference on the
         resolution    of    equations    in    algebraic
         structures (CREAS),M.Nivat and H.Ait-Kaci eds.,
         Academic Press,to appear in 1989.

[13]     COURCELLE B.,On the use of context-free graph grammars
         for   analyzing   recursive   definitions,Second
         French-Japanese   Seminar,L.Kott   ed., North-
         Holland,to appear in 1988.

[14]     DONER J.,Tree acceptors and some of their applications
         J. Comput. System Sci. 4 (1970)406-451.

[15]     JOHNSON D., The NP-completeness column:an ongoing guide
         (16th),J. of Algorithms 6(1985)434-451.

[16]     JOHNSON  D.,The  NP-completeness column: An ongoing gui
         (19 th), J. of Algorithms 8(1987) 285-303.

[17]     HABEL A.,Graph-theoretic properties compatible with gra
         derivations,this volume.

[18]     HABEL A.,KREOWSKI H.J.,Some structural aspects of
         hyperedge  languages  generated  by  hyperedge
         replacements, preprint Oct. 85,and L.N.C.S.
         vol.247,1987,pp.207-219.

[19]     LAUTEMANN C.,Decomposition trees:structured graph repre
         sentations and efficient algorithms,Lec.Notes
         Comp.Sci.299,pp.28-39,1988.

[20]     LAUTEMANN C.,Efficient algorithms on context-free graph
         languages,Proc.ICALP'88,Lec.Notes Comp.Sci.317
         pp.362-378,1988.

[21]     LENGAUER T.,WANKE E.,Efficient analysis of graph proper
         ties on context-free graph languages,Proc.ICALP
         1988,Lec.Notes Comp.Sci.317,pp.379-393.

[22]     MEZEI J., WRIGHT J.,Algebraic automata and context-
         free  sets,  Information and Control 11 (1967)
         3-29.

[23]    MONIEN B.,SUDBOROUGH H.,Bandwidth constrained NP
         complete problems,13th ACM Symp.on Theory of
         computation,1981, pp.207-217.

[24]    ROBERTSON N., SEYMOUR P., Some new results on the well
         quasi-ordering  of  graphs, Annals of Discrete
         Mathematics 23 (1984),343-354,Elsevier Pub.

[25]    ROBERTSON  N.,  SEYMOUR  P.,  Graph  minors  V,
         Excluding  a  planar  graph,  J. Combinatorial
         Theory Ser. B, 41(1986) 92-114.

[26]    ROBERTSON  N.,  SEYMOUR  P.,  Graph  minors  XVI,
         Wagner's conjecture, in preparation, 1988.

[27]    SEESE D., The structure of the models of decidable
         monadic theories of graphs,1987,submitted for
         publication.

[28]    THATCHER J., WRIGHT J., Generalized finite automata
         theory, Math. Systems Theory 2 (1970) 57-81.

[29]    WAGNER K.,Uber eine Eigenschaft der ebenen Komplexe,
         Math. Ann.114(1937)570-590.

[30]    WIMER T.,HEDETNIEMI S.,LASKAR R.,A methodology for
         constructing  linear graph algorithms,DCS,
         Clemson Univ,1985

# On Systems Of Equations Defining Infinite Graphs[1]

*Michel BAUDERON*

*Université de Bordeaux I*
*Département Informatique, I.U.T. 'A'*
*F-33405 TALENCE CEDEX*

**Abstract** : A framework is described in which we can solve equations and systems of equations on oriented edge labelled hypergraphs with a finite sequence of distinguished sources. We show that this cannot be done with the standard order-theoretic methods, but implies the use of some category-theoretic tools and results.

**Résumé** : Nous proposons un cadre pour l'étude d'équations et systèmes d'équations portant sur des hypergraphes orientés, aux arêtes étiquetées et possédant une suite finie de sommets distingués appelés sources. Nous montrons que de telles équations ne peuvent être résolues avec les outils classiques, mais nécessitent des outils et résultats de la théorie des catégories.

## 1. Introduction.

Infinite trees have proved to be a very essential tool for the study of the algebraic semantics of recursive program schemes, where they usually arise as (components of) solutions of systems of equations on trees and numerous works have been devoted to them (a comprehensive treatment of the theory of infinite trees may be found in [Co83]).

However, a program scheme should be more naturally represented as a graph (in fact a directed graph), in order to take into account the natural sharing of some subterms, and the real objects of interest in semantics should then be the infinite graphs arising as solutions of certain systems of

---

[1] This work has been supported by the C.N.R.S PRC "Mathématiques et Informatique"

equations on graphs.

We shall consider here some simple systems of equations which generate some very 'regular' graphs (in a sense that could be made more precise, cf [Ba88]). On the other hand, this very kind of graphs turns out to appear in the generation of regular patterns (cf. the pattern graphs of Caucal [Ca88], the map sequences of Lindenmayer [Li87] or the context-free graphs of Muller and Schupp [MS85]), which is a further reason for the interest we have in their study.

We shall use here the model introduced by Bauderon and Courcelle [BC87] for finite oriented edge labelled hypergraphs with a sequence of distinguished vertices called sources, which we shall readily extend to countable hypergraphs (a fairly similar model for finite graphs has been introduced independently by Habel and Kreowski [HK87]). In section 2, we recall the essential definitions and give a few examples.

Systems of regular equations on graphs are defined in Section 3. In this section we give an intuitive resolution of some sample equations by the usual method of iterated substitution which is commonly used on trees. The validity of this method when used for the resolution of equations on trees is insured by the availability of a metric or order structure which turns the set ot infinite trees either into a complete metric space or an $\omega$-complete many-sorted algebra. For both structures, there exists a well-known fixpoint theorem which one can readily apply.

In the case of infinite graphs, the Dewey enumeration used to define the order (or the metric) structure on trees is no longer available, and looking for an appropriate notion of continuity in a similar way turns out to be very difficult. As a matter of fact we give a counter-example which shows that no reasonable order structure may be defined on the set of infinite hypergraphs.

This difficulty is well-known and has been encountered in a wide range of problems : study of infinite words (Courcelle [Co78] where an *ad hoc* approach is used), theory of domains (Lehmann [Le76], Smyth [Sm78]), abstract data types (Lehmann and Smyth [LS81]), ... and a method has been developped by generalizing the usual concepts and results to the more general setting of $\omega$-categories. Section 4 describes the application of these results to our setting and shows how to give good foundations to the theory of systems of regular equations on hypergraphs. For the reader who is unfamiliar with those concepts, an appendix is provided, which gives a summary of the main concepts and tools that we must use.

This paper is essentially self-contained. Most of the proofs are complete but of a very elementary nature. More involved categorical concepts and proof techniques yield some stronger results. Section 5 is devoted to a very short description of some of those results and to the relationship with some other works (context-free graphs of Muller and Schupp [MS85] and pattern graphs of Caucal [Ca88]) (a complete exposition of these topics could not be included due to space limitations, but may be found in [Ba88]).

## 2. Hypergraphs.

In this paper, we consider oriented edge labelled hypergraphs having a finite sequence of distinguished vertices called sources. More formally, let $A = \bigcup_{n \in \mathbb{N}} A_n$ be a ranked alphabet with rank function $\tau : A \longrightarrow \mathbb{N}$, n be any integer and $[n] = \{p \in \mathbb{N} \ / \ 1 \leq p \leq n\}$. Then :

*(2.1) Definition.* A *(concrete) sourced hypergraph of type* n *(or an n-graph) over* A, is a sextuplet $G = \langle V_G, E_G, vert_G, lab_G, src_G, n \rangle$ where :
- $V_G$ and $E_G$ are countable sets (respectively of *vertices* and *edges*),
- $vert_G : E_G \longrightarrow V_G^*$ assigns to each hyperedge e, a word on $V_G$ representing the ordered sequence of its vertices, $vert_G(k,e)$ denoting the $k^{th}$ element of the sequence,
- $lab_G : E_G \longrightarrow A$ labels every hyperedge e with some letter in A such that $\tau(lab_G(e)) = |vert_G(e)|$,
- $src_G : [n] \longrightarrow V_G$ is the ordered sequence of *sources* of the hypergraph G, namely : $src_G([n]) = src_G(1), \ldots, src_G(n)$.

For the sake of simplicity, we shall hereafter omit the prefix hyper-, and simply talk of graphs. The set of n-graphs over A is denoted by $\mathfrak{C}\mathfrak{G}_n^\infty(A)$, the set of all graphs by $\mathfrak{C}\mathfrak{G}^\infty(A)$, the corresponding sets of finite graphs by omitting the symbol '∞'.

For all graphs of type 0, we shall omit the two last components of the definition and simply talk of graphs. Clearly, with any n-graph G, we can associate a unique graph G° (called the *underlying graph*) by simply discarding these two last components. In this way, we can consider an n-graph G as a triple : $G = \langle G°, src_G, n \rangle$.

A sourced graph is *finite* whenever both sets $V_G$ and $E_G$ are finite. Such graphs have been defined in Bauderon and Courcelle [BC87]. A very similar notion has been introduced independently by Habel and Kreowsky [HK87].

A sourced graph is *bounded* when $V_G$ is finite and has *locally finite degree* when every vertex has a finite degree.

*(2.2) Examples.* Numerous examples of finite sourced graphs may be found in [BC87]. Let us simply recall that with each integer n, we associate the discrete n-graph $\underline{n}$ having n distinct isolated vertices, no edge and a source mapping sending $i \in [n]$ to the $i^{th}$ vertex of $\underline{n}$. This graph will be drawn as :

$$\underset{1}{\circ} \quad \underset{2}{\circ} \quad \underset{3}{\circ} \quad \cdots \quad \underset{n-1}{\circ} \quad \underset{n}{\circ}.$$

Its underlying graph is merely [n].

Similarly, with a letter $a \in A$ with rank n, we associate the n-graph $\underline{a}$ with n distinct vertices, one edge labelled by a, and source mapping sending $i \in [n]$ to the $i^{th}$ vertex of $\underline{a}$ :

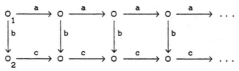

The following 2-graph $G_1$ is infinite, but has locally finite degree :

A formal description of the 2-graph $G_2$ might be :
- $A = \{a, b, c\}$, $V_G = \mathbb{N}$, $E_G = \mathbb{N}$, $n = 2$, $\text{vert}_G(1) = 1.2$,
- $\text{vert}_G(i) = \text{vert}_G(1, i+1) . \text{vert}_G(1, i+2)$,
- $\text{vert}_G(i+1) = \text{vert}_G(1, i) . \text{vert}_G(1, i+3)$,
- $\text{vert}_G(i+2) = \text{vert}_G(2, i) . \text{vert}_G(2, i+3)$,
- $\text{lab}_G(i) = b$, $\text{lab}_G(i+1) = a$, $\text{lab}_G(i+2) = c$,
$\text{src}_G(1) = 1$, $\text{src}_G(2) = 2$.

The following 2-graph $G_2$ is bounded but not finite, hence has not a locally finite degree (a formal definition is left to the reader) :

*(2.3) Definition.*

$x_1, \ldots, x_n, \ldots$ will be called *variables*. The set of *graphs with variables in X over the alphabet A* is the set $\mathfrak{CG}^{\infty}(A \cup X)$.

Let G be an n-graph with one occurrence of a variable x of type m i.e., one edge $e_i$ of which is labelled by x. Let then $\Gamma$ be the (n+m)-graph defined by (where ◄ denotes the restriction of a mapping) :

$$V_\Gamma = V_G, \ E_\Gamma = E_G - \{e_i\}, \ \text{lab}_\Gamma = \text{lab}_G ◄ E_\Gamma, \ \text{vert}_\Gamma = \text{vert}_G ◄ E_\Gamma$$

and $\quad \text{src}_\Gamma : [n+m] \longrightarrow V_\Gamma$

$$k \longrightarrow \text{src}_G(k) \text{ for } 1 \leq k \leq n$$
$$k \longrightarrow \text{vert}_\Gamma(k-n, e_i) \text{ for } n+1 \leq k \leq n+m$$

The graph $\Gamma$ is said to be the *context* of the occurrence of x in G and we write : $G = \Gamma[x]$. The vertices of the edge $e_i$ labelled by x will be called the *occurrence vertices* of x. They are the common vertices of the context and of the occurrence of the variable.

(2.4) *Substitution* [BC87] (or *hyperedge replacement* [HK87]). Let $G = \Gamma[x]$ be an n-graph with one occurrence of a variable x of type m and let H be an m-graph. Then, the result of the substitution of H for x in G is the n-graph $G' = \Gamma[H/x]$ defined by :

$V_{G'} = (V_\Gamma \cup V_H)/_\equiv$ where $\equiv$ is the equivalence relation generated by the pairs $(\text{src}_\Gamma(k+n), \text{src}_H(k))$ for all k, $1 \leq k \leq m$,

$E_{G'} = E_\Gamma \cup E_H$,
$\text{vert}_{G'} = \text{vert}_\Gamma \cup \text{vert}_H$,
$\text{lab}_{G'} = \text{lab}_\Gamma \cup \text{lab}_H$,
$\text{src}_{G'} = \text{src}_G$;

Intuitively, the edge of G corresponding to the variable x has been removed and replaced by the graph H, the sources of H being glued to the corresponding sources of the context of x.

Clearly, this notion extends to the case of multiple occurrences of one variable in a graph, in which case one may define either substitution for one occurrence or simultaneous substitution for several or all occurrences of the variable. In the same way the definition may be extended to sourced graphs with more than one variable.

Formal definition of simultaneous substitution turns out to be fairly awkward. However, since it may be interpreted as iterated substitution for one single occurrence, we will not bother the reader with it and will hereafter consider only this simple case.

*(2.5) Definition* : Let G and G' be two n-graphs. We say that G is a *sub-n-graph* of G' if :

- $V_G \subseteq V_{G'}$,
- $E_G \subseteq E_{G'}$,
- $\text{vert}_G = \text{vert}_{G'} \triangleleft E_G$
- $\text{lab}_G = \text{lab}_{G'} \triangleleft E_G$
- $\text{src}_G = \text{src}_{G'}$.

Note that this notion of sub-n-graphs respects the sources of the n-graphs. However, when there are no sources (i.e., n = 0), this notion is exactly the classical notion of a subgraph.

The *frontier* fr(G',G) of G in G' is the set of vertices of G belonging to an edge of G' which is not in G.

If G = $\Gamma[x]$ is a graph with a variable x and e is an edge labelled with an occurrence of the variable x, the occurrence vertices of (this occurrence) of x are precisely the vertices in the frontier in G of the graph reduced to the edge e.

*(2.6) Proposition* :

(i) Let H be a sub-n-graph of H'. Then there exists an m-graph G such that H' = H[G/x].

(ii) Let H be an n-graph, H' an n'-graph and let us assume that H° is a subgraph of H'° (i.e. with no sources involved) but that H is *not* a sub-n-graph of H'. Then there exists an integer m and an (n'+m)-graph G such that H' = G[H/x].

*(2.7) Proof* :

(i) We let the underlying graph G° of G be defined as follows :

- $V_G = (V_{H'} - V_H) \cup \text{fr}(H',H)$
- $E_G = E_{H'} - E_H$
- $\text{lab}_G = \text{lab}_{H'} \triangleleft E_G$
- $\text{vert}_G = \text{vert}_{H'} \triangleleft E_G$

Let m be the cardinal of the frontier of H in H', $\text{src}_G([m])$ be precisely this frontier and x be a variable x of type m. We can consider H as an (n+m)-graph by adding $\text{src}_G([m])$ to its own sources and set H' = H[G/x].

(ii) Let G° be the same graph as in (i). Then, we let G have as sources $\text{src}_{H'}([n']) \cup \text{fr}(H',H)$. Now, if we let x be a variable of type m = # fr(H',H) we can write H' = G[H/x]. ∎

## 3. Equations and systems of equations on graphs.

*(3.1) Definition.* A *system of regular equations on graphs* is a finite system of the form $\Sigma = \langle x_1 = G_1, \ldots, x_n = G_n \rangle$ where $U = \{x_1, \ldots, x_n\}$ is the set of unknown, and for each integer $i$, $1 \le i \le n$, $G_i \in \mathbb{G}_{\tau(x_i)}(A \cup U)$ (i.e., both $x_i$ and $G_i$ have the same type $\tau(x_i)$).

A *solution* of the system $\Sigma$ is an n-uple of sourced graphs $(\overline{G}_1, \ldots, \overline{G}_n)$ such that :

for each i, $1 \le i \le n$, one has $\overline{G}_i \simeq G_i[\overline{G}_1/x_1, \ldots, \overline{G}_n/x_n]$.

(where $\simeq$ means isomorphic. The intuitive notion is clear, a formal definition will be found in section 4).

*(3.2) Examples of equations and intuitive resolution.*

(i) Let us first consider the following equation :

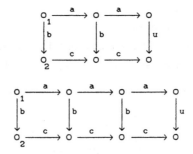

(E1)

An intuitive resolution of this equation by the usual method of iterated substitution should yield successively :

and lead finally to the following solution $G_1$ for the equation (E1) :

(ii) Let us now consider the equation :

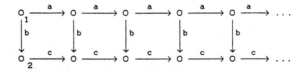

(E2)

The same resolution process clearly leads to the following solution which has not locally finite degree :

However, it is easily checked that the following graph, with an infinity of a-labelled loops is also a solution of the same equation :

This example clearly shows that the solution to an equation isnot necessarily unique.

(iii) As a last example let us consider the following equation :

$$\text{(E3)}$$

The solution is now much less obvious. We shall see later on (section 4) that the 'minimal' solution of this equation is the discrete graph $\underline{2}$. It may be noticed that in this case the solution of the equation is not connected although the original equation was.

*(3.3) Order structure.* The classical approach to the resolution of equations on trees involves the use of an ordered structure on the F-algebra of trees (see e.g., [Co83]. Note that a metric may be used as well).

Indeed, iterated substitution as a means to solve an equation does not directly lead to a solution but to an increasing sequence of trees which are intended to approximate the actual solution. The order structure is necessary to provide a notion of limit and make sure that the sequence actually converges towards a well defined object.

To define this structure on the set of infinite trees, one uses the very essential fact that the nodes of a tree may be given a canonical enumeration (Dewey enumeration) which allows to tell in a unique way that one tree is smaller than an other one (note that this enumeration is used in the definition of the metric structure as well). In the case of hypergraphs, it quickly appears to be impossible to find such an enumeration, and all tries to define an order structure are inefficient. As a matter of fact, it turns out that there is no coherent way to define an order structure on the set of sourced hypergraphs.

Indeed, to be natural, such an order relation should respect graph inclusion. Namely, if G is a sub-n-graph of G', one expects G to be smaller than G' and a reasonable definition of such an order should be something like :

$$G \leq G' \iff G \text{ is a sub-n-graph of } G'$$

Now, to be an order relation, this relation should be asymmetric i.e., one should have :

$$G \leq G' \qquad G' \leq G \quad \Rightarrow \quad G \simeq G'$$

This is satisfied in the finite case, i.e., one can check that the relation we have just defined is an order relation on $\mathfrak{CG}_n(A)$, but this is not necessarily true in the case of infinite graphs as will be shown by the following :

*(3.4) Counter-example* (Courcelle) : Let G and G' be the two following 1-graphs :

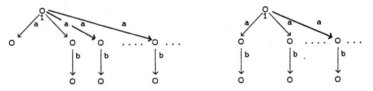

It is fairly clear on this example that $G \leq G'$ and that $G' \leq G$ but that these two graphs are not isomorphic. It must be noted that this counter-example involves two graphs with non locally finite degree. In fact, one has :

*(3.5) Theorem* (Caucal [Ca88]) : If we restrict to locally finite graphs, the relation $\leq$ defined above is an order relation. ∎

However, since graphs with non locally finite degree occur very naturally when solving regular equations on hypergraphs (cf example *3.2.* (ii)) we cannot restrict ourselves to the hypotheses of the previous theorem.

In order to overcome the difficulties that we have just pointed out, we shall have to use some tools and results from elementary category theory.

## 4. The category of graphs

In this section, we shall assume that basic notions from category theory are known. All necessary definitions will be found in the appendix.

*(4.1) Graph morphisms* : Let G and G' be two concrete sourced n-graphs. A *morphism of n-graphs* g : G $\longrightarrow$ G' is a pair g = (Vg, Eg) of arrows in **Set** such that :
$$Vg : V_G \longrightarrow V_{G'}, \qquad Eg : E_G \longrightarrow E_{G'},$$
satisfy the following conditions :
$$lab_G = lab_{G'} \circ Eg$$
$$Vg^* \circ vert_G = vert_{G'} \circ Eg$$
$$Vg \circ src_G = src_{G'}$$
(where $Vg^*$ denotes the canonical extension of Vg to the monoid of words on $V_G$) expressing the commutativity of the following diagrams in the category **Set** :

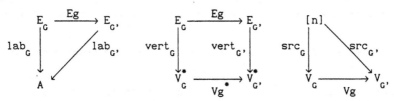

*(4.2) Lemma.* The set of morphisms of concrete n-graphs is closed under composition (defined componentwise). For any sourced graph, the identity morphism is the unique morphism whose components are all set identities. A morphism is an isomorphism if and only if its two components are bijections. ∎

From this lemma (whose proof is straightforward), it follows that, with the morphisms we have just defined, the set $\mathfrak{CG}^\infty_n(A)$ of concrete n-graphs is for any integer n, a category that we shall denote by $\underline{\mathfrak{CG}}^\infty_n(A)$. We shall let $\underline{\mathfrak{CG}}_n(A)$ denote the full subcategory of finite n-graphs.

Some rather straightforward but fairly tedious elementary proofs show that for any integer n, the category $\underline{\mathfrak{CG}}^\infty_n(A)$ satisfies some very essential properties. We shall give the main ideas of the proofs but we shall omit them, not merely because of their tediousness, but mainly because they can be obtained as corollaries of a much stronger and richer construction. For more details, the reader is referred to Bauderon [Ba88]. The definitions of the notions used in the remaining of this section will be found in the appendix.

The main properties of the category $\underline{\mathfrak{CG}}^\infty_n(A)$ are summarized in the following theorem :

     *(4.3) Theorem.* The category $\underline{\mathfrak{CG}}_n^{\infty}(A)$ has the following properties :
(i) it is cocomplete hence is $\omega$-complete and has pushouts,
(ii) the category $\underline{\mathfrak{CG}}_n(A)$ is $\omega$-dense in $\underline{\mathfrak{CG}}_n^{\infty}(A)$
(iii) the discrete n-graph $\underline{n}$ is an initial objet of the category,
(iv) if $\mathfrak{F}$ is an $\omega$-continuous endofunctor of $\underline{\mathfrak{CG}}_n^{\infty}(A)$, its fixpoints form a category which has an initial object.

     *(4.4.) Sketch of an elementary proof :*
(i) this is a consequence of the well known fact that a category has colimits if and only it has coproducts and coequalizers. This fact is easily proved since an n-graph is essentially a pair of sets, and the existence of coproducts and coequalizers follows from their existence in **Set**,

(ii) for an n-graph G, any enumeration of the sets $V_G$ and $E_G$ provides a way to build an $\omega$-sequence of finite n-graphs $G_n$ whose colimit is precisely G,

(iii) this merely follows from the definition of an n-graph morphism,

(iv) this last assertion is a mere rephrasing in the category $\underline{\mathfrak{CG}}_n^{\infty}(A)$ of the standard fixpoint theorem for an $\omega$-endofunctor of an $\omega$-category (cf. appendix, theorem 6.7). ∎

     This properties will allow us to solve equation in an 'almost' classical way. First, we must associate with a regular equation (or a system of regular equations), an endofunctor of the category $\underline{\mathfrak{CG}}_n^{\infty}(A)$ (or an endofunctor of a more complicated category).

     *(4.4) Definition.* Let $x = \Gamma[x]$ be an equation with one variable x of type n. The *functor associated with this equation* is the endofunctor $\mathfrak{F}_{\Gamma}$ of $\underline{\mathfrak{CG}}_n^{\infty}(A)$ defined by substitution, i.e., the endofunctor whose object component sends the n-graph H to the n-graph $\Gamma[H/x]$ :

$$\mathfrak{F}_{\Gamma} : \underline{\mathfrak{CG}}_n^{\infty}(A) \longrightarrow \underline{\mathfrak{CG}}_n^{\infty}(A)$$
$$H \longrightarrow \Gamma[H/x]$$

     More formally, substitution of an n-graph for one occurrence of a variable x of type n may be be defined through the following pushout diagram in the category $\underline{\mathfrak{CG}}_0^{\infty}(A)$ :

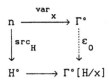

where $\Gamma°$ is the graph underlying the context of the occurrence of $x$, and $\mathrm{var}_x([\underline{n}])$ is the sequence of vertices of the unique $x$-labelled edge. Then, the result of the substitution of $H$ for $x$ is the $n$-graph $\Gamma[H/x] = \langle\Gamma°[H/x], \mathrm{src}_G, n\rangle$.

Morever, the arrow $\varepsilon_0$ (in $\underline{\mathfrak{CG}}_0^\infty(A)$) defined by the right-hand column of the pushout diagram canonically defines an arrow in $\underline{\mathfrak{CG}}_n^\infty(A)$ from $\Gamma = \langle\Gamma°, \mathrm{src}_G, n\rangle$ into $\Gamma[H/x] = \langle\Gamma°[H/x], \mathrm{src}_G, n\rangle$ that we shall call an *augmentation*. This graph morphism is clearly injective on edges and on vertices except perhaps on the occurrence vertices of $x$ (since $H$ may have some non-distinct sources).

This defines the object component of the functor $\mathcal{F}_\Gamma$. The arrow component is defined in a straightforward way :

Let $h : H \longrightarrow H'$ be an arrow in $\underline{\mathfrak{CG}}_n^\infty(A)$. Then $\mathcal{F}h : \mathcal{F}H \longrightarrow \mathcal{F}H'$ is given 'componentwise' : $\mathcal{F}h = \mathrm{Id}_\Gamma$ on $\Gamma$ (except perhaps on the occurrence vertices of $x$), and $\mathcal{F}h = h$ on $H$.

Let now $\Sigma = \langle x_1 = \Gamma_1[x_1,\ldots,x_n], \ldots, x_n = \Gamma_n[x_1,\ldots,x_n]\rangle$ be a system of regular equations. In a similar way which we shall not detail, we can now associate with $\Sigma$, an endofunctor $\mathcal{F}_\Sigma$ of $\underline{\mathfrak{CG}}_{\tau(x_1)}^\infty(A) \times \ldots \times \underline{\mathfrak{CG}}_{\tau(x_n)}^\infty(A)$. The name of this last category will be abbreviated to $C_\Sigma$ (domain of the functor $\mathcal{F}_\Sigma$) or $C_\Gamma$ in the case of a single equation.

*(4.5) Proposition* : The $\omega$-endofunctor associated with an equation or a system of equations is $\omega$-continuous.

*(4.6) Proof* : For the sake of simplicity and without loss of generality, the proof will be carried out in the case of a single equation.

In fact, it merely amounts to proving that substitution of an $n$-graph for a variable $x$ of type $n$ in a graph $\Gamma[x]$ is $\omega$-continuous.

Let us consider an $\omega$-diagram together with its $\omega$-limiting cocone :

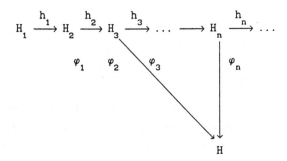

Applying the functor $\mathcal{F}_\Sigma$, we obtain a new $\omega$-diagram together with a cocone over that diagram. We must show that this cocone is an $\omega$-limiting one, i.e., that for any outer cocone with vertex G, there is a unique arrow g : $\mathcal{F}_\Sigma H \longrightarrow G$. Now, since by definition one has $\mathcal{F}_\Sigma H_i = \Gamma[H_i/x]$, there is a canonical augmentation $\varepsilon_0 : \Gamma \longrightarrow \mathcal{F}_\Sigma H_1 = \Gamma[H_1/x]$ :

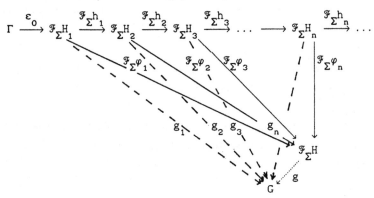

Hence, there is a graph morphism $g_0 = g_1 \circ \varepsilon_0 : \Gamma \longrightarrow G$ which is injective except perhaps on the occurrence vertices of x in $\Gamma$. It follows (proposition 2.6) that there exists a graph G' such that $G \simeq \Gamma[G'/x]$.

Furthermore, it follows from the definition of $\mathcal{F}_\Sigma h_i$, that for each i, i $\in \mathbb{N}$, there is an arrow $g'_i : H_i \longrightarrow G'$.

Now, since H is the $\omega$-limit of the first diagram, there is a unique arrow g' : $H \longrightarrow G'$ which make the diagram commute and $g = \mathcal{F}_\Sigma g'$ is the unique graph morphism from $\mathcal{F}_\Sigma H$ to G that we were looking for. ∎

(4.7). *Resolution of an equation or a system.* From Theorem (4.3) it follows that the set of solutions of a system $\sum$ of regular equations is an initial category. The *initial solution* of the equation (or the system of regular equations) which plays in our context the role of the minimal solution in the case of trees may be built by the classical method of successive approximations, iteratively applying the functor $\mathcal{F}_\Sigma$ starting from the initial object of the category ($\underline{n}$ for an equation of type n, $\underline{[\tau(x_1)]} \times \ldots \times \underline{[\tau(x_n)]}$ for a general system).

Moreover, it follows from the proof of Theorem (6.7) that a more general solution can be built from any object C of the category such that there exists an arrow $C \longrightarrow \mathcal{F}_\Sigma C$, by iteration of the application of the functor $\mathcal{F}_\Sigma$ to the object C.

(4.8) *Examples*. In this paragraph we shall show how the application of the theorem to the examples of section 3.2 actually yields the expected solutions.

(i) Let us consider the first equation :

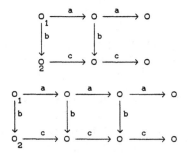

$$\begin{array}{ccc} O_1 & & O_1 \xrightarrow{\ a\ } O \\ \downarrow u & = & \downarrow b \quad\quad \downarrow u \\ O_2 & & O_2 \xrightarrow{\ c\ } O \end{array} \qquad\qquad (E1)$$

The corresponding functor must now be iteratively applied, starting from the initial object $\underline{2}$ in $\mathfrak{C}\mathfrak{G}_2^\infty(A)$, yielding successively :

which finally leads to the following solution $G_1$ for the equation (E1) :

(ii) Let us now consider the equation :

$$\begin{array}{ccc} O_1 & & \left[\begin{array}{c} O_1 \\ \\ \\ O_2 \end{array}\right. \\ \downarrow u & = & \quad a \quad \downarrow u \\ O_2 & & \left. \to O_2 \leftarrow \right. \end{array} \qquad\qquad (E2)$$

The same resolution process yields (starting once again from $\underline{2}$) :

first : 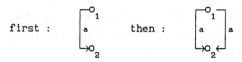 then :

and leads finally to the following expected solution :

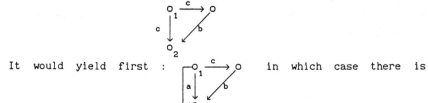

Now, we can start the process from any other 2-graph. Let start for example with the 2-graph $H_0$ :

It would yield first : in which case there is

an(injective) morphism $H_0 \longrightarrow \mathcal{F}_\Sigma H_0$ which allow us to iterate the application of $\mathcal{F}$, yielding then :

and finally :

Last, starting from the graph : , one will obtain through the same process the following solution with an infinity of a-labelled loops :

(iii) As a last example let us consider the following equation :

$$ = \qquad (E3)$$

The initial solution to this equation is now quite trivially found. Starting with $\underline{2}$, one gets the $\omega$-diagram whose terms are constantly equal to the initial 2-graph $\underline{2}$. Starting with any other 2-graph G, one gets as a solution an infinity of copies of G all glued along their sources.

# 5. Further results and related works.

(5.1) *Algebraic expressions for infinite graphs* : In [BC87], Bauderon and Courcelle have defined an algebraic framework which allows them to describe finite graphs by the means of algebraic expressions and to develop a theory of context-free graph grammars in the classical equational way ([MW67], [Co88]).

This algebraic formalism is extended to infinite graphs in [Ba88] where it is shown that :

(i) infinite graphs may be described by algebraic expressions, in such a way that approximation of an infinite graph by an $\omega$-sequence of finite ones is coherent with the approximation of an infinite expression by an increasing sequence of expressions.

(ii) two expressions define the same graph if and only if they are congruent with regard to a certain rewriting system. This result extending the one obtained in [BC87] for finite graphs, is non-trivial since it is not true for infinite words for instance (Courcelle's arrangements, cf [Co78]).

(iii) in this framework, equations on graphs may be interpreted as equations on expressions (or trees) and solved at that level, using the classical theory of regular equations on trees [Co83]. It is then shown that both approaches yield the same results.

(5.2) *Context-free graphs* were defined by Muller and Schupp ([MS85]) to be the graphs of transitions of deterministic push-down automata and shown to be exactly those graphs with a distinguished source, locally finite degree and a finite number of non isomorphic ends. It has been shown independently that these graphs are exactly the graphs that can be generated as pattern graphs ([Ca88]) or obtained as solutions of certain systems of regular equations ([Ba88]).

(5.3) *Unicity of solutions*. As remarked in paragraph 4.7, an equation always has an initial solution but may happen to have some other solutions. For instance, any graph C such that there is an arrow $C \longrightarrow \mathcal{F}_\Gamma C$ generates a new solution. The question then naturally arises to know whether an equation may have a unique solution or several non-isomorphic solutions. It is investigated in [Ba88]. Related results may be found in [Co88].

## 6. Appendix : elements of category theory.

This appendix is merely a brief summary of the basic notions and results from category which are used in the main text. The reader is referred to any standard textbook ([ML71] or [Ma76]) for more details and complete proofs. A more comprehensive study of the category of hypergraphs may be found in [Ba88].

(6.1) A *category* C is a pair $(Ob_C, Hom_C)$ where $Ob_C$ is the class of objects and for any two objects A, B of $Ob_C$, $Hom_C(A,B)$ is the set of arrows from A to B, subject to the following conditions :
- for any objet A, there exist a unique identity arrow $1_A$,
- for any three objects, A, B, C, the composition of arrow is defined : $Hom_C(A,B) \times Hom_C(B,C) \longrightarrow Hom_C(A,C)$

As usual we let **Set** denote the category whose objects are sets and whose arows are mappings between sets. Another category of frequent use in our context is the category $\omega$, which is associated with the first infinite ordinal $\omega$. It is usually represented in the following way (omitting the identities and composed arrows) :

$$0 \longrightarrow 1 \longrightarrow 2 \longrightarrow 3 \ldots \longrightarrow n \longrightarrow \ldots$$

It is well known, that in a similar way, any partially ordered set may be viewed as a category.

An object $\bot$ in a category C is an *initial object* if for any object C of C, there is a unique arrow $\bot_C : \bot \longrightarrow C$.

A category is an *initial category* if it has an initial objet.

(6.2) Let C and D be two categories. A *functor* $\mathcal{F}$ from C to D assigns to each object C of C an objet $\mathcal{F}C$ of D and to each arrow $f : C \longrightarrow D$, an arrow $\mathcal{F}f : \mathcal{F}C \longrightarrow \mathcal{F}D$ such that $\mathcal{F}1_A = 1_{\mathcal{F}A}$ and $\mathcal{F}(f \circ g) = \mathcal{F}f \circ \mathcal{F}g$.

Let J be a category. A *J-diagram* in the category C is a functor $\mathcal{D}$ from J to C. An *$\omega$-diagram* in C is then a functor from $\omega$ into C, or, in other words, a countable sequence of objects and arrows which we shall draw as :

$$C_0 \longrightarrow C_1 \longrightarrow C_2 \longrightarrow C_3 \longrightarrow \ldots \longrightarrow C_n \longrightarrow \ldots$$

A *cocone* over a J-diagram $\mathcal{D}$ in C is an objet K of C (the vertex of the cocone) and a family of arrows $KJ : \mathcal{D}J \longrightarrow K$ for any J in J such that, for any arrow u : $\mathcal{D}I \longrightarrow \mathcal{D}J$ in J one has KI = u ∘ KJ.

A *colimit* for $\mathcal{D}$ is a cocone over $\mathcal{D}$ with vertex C such that, for any cocone with vertex K over $\mathcal{D}$, there is a unique arrow U : C $\longrightarrow$ K which satisfies KJ = U ∘ CJ for every object J in J. The colimit of a J-diagram is unique up to an isomorphism.

The $\omega$-limit of an $\omega$-diagram in C is the colimit of this diagram.

The *coproduct* of a family of objects $C_{I \in I}$ in C indexed by a category I is the colimit of the diagram $\mathcal{G} : I \longrightarrow C_I$, with no arrow component.

Let 2 be the category 1 $\underset{\longrightarrow}{\overset{\longrightarrow}{}}$ 2 (two distinct objets, two distincts arrows and identities). A colimit of a 2-diagram is called a *coequalizer*.

A pushout is a colimit for a diagram over the category V with three objects and only two arrows different from the identities :

(6.3) A category is *cocomplete* if it has all colimits. The standard way to prove this property is to use the following theorem :

*Theorem ([McL71])* : A category has colimits if and only it has coproducts and coequalizers. ∎

*Proposition* If C and D are cocomplete categories, the product category C × D is cocomplete. If C and D are initial with initial objects $\perp_C$ and $\perp_D$, the product category is initial with initial object the pair $(\perp_C, \perp_D)$. ∎

A category is an $\omega$-*category* it is has all $\omega$-limits, i.e., colimits of all $\omega$-diagram. Clearly, a cocomplete category is an $\omega$-category and has coproducts and pushouts as well.

We say that a subcategory C of a category D is $\omega$-*dense* in D if any object D of D is the $\omega$-limit of an $\omega$-diagram in C (in other words, any object in D may be approximated by a countable sequence of objects in C).

(6.5) A functor $\mathcal{F}$ from C to D is $\omega$-*continuous* if it preserves the $\omega$-limits and the related $\omega$-limiting cocones ($\omega$-cones).

*(6.6) Definition* Let $\mathcal{F}$ be an endofunctor of a category **C**. A *fixpoint* of $\mathcal{F}$ is a pair $(C,u)$ where $C$ is an object of **C** and $u : \mathcal{F}C \longrightarrow C$ is an isomorphism.

A detailled study of the following theorem will be found in Adamek and Koubek [AK79] where it is dated back to Lambek [La68]. Related works and results are those of Lehmann [Le76], Lehmann and Smyth [LS81].

*(6.7) Theorem : existence of a fixpoint.* Let **C** be an initial $\omega$-category and $\mathcal{F}$ be an $\omega$-continuous endofunctor of **C**. Then the fixpoints of $\mathcal{F}$ form an initial category, whose initial object is the colimit of the following $\omega$-diagram :

$$\bot \longrightarrow \mathcal{F}\bot \longrightarrow \mathcal{F}^2\bot \longrightarrow \mathcal{F}^3\bot \longrightarrow \ldots \longrightarrow \mathcal{F}^n\bot \longrightarrow \ldots$$

*Proof* : We sketch the proof of this theorem not because of its difficulty or of its intrinsic interest, but merely because it is a constructive proof which shows how a fixpoint may be built and thereby yields a process to solve our equations on graphs. As usual, the construction of a fixpoint is made by successive approximations of the solution. Indeed, from any object C such that there is an arrow $u : C \longrightarrow \mathcal{F}C$ in **C** we can generate an $\omega$-diagram :

$$C \xrightarrow{u} \mathcal{F}C \xrightarrow{\mathcal{F}u} \mathcal{F}^2C \xrightarrow{\mathcal{F}^2u} \mathcal{F}^3C \xrightarrow{\mathcal{F}^3u} \ldots \longrightarrow \mathcal{F}^nC \xrightarrow{\mathcal{F}^nu} \ldots$$

Since **C** is an $\omega$-category, this $\omega$-diagram has an $\omega$-limit $\Xi$ corresponding to the inner $\omega$-limiting cocone in the following commutative diagram :

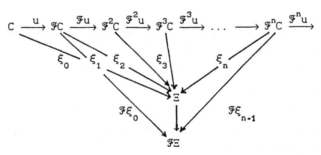

. . .

Now since $\mathcal{F}$ is an $\omega$-continuous functor, when applied to the inner cocone, it generates the outer cocone (dotted lines) of the same diagram. It follows that these two cocones are $\omega$-limiting for the following diagram :

$$\mathcal{F}C \xrightarrow{\mathcal{F}u} \mathcal{F}^2C \xrightarrow{\mathcal{F}^2u} \mathcal{F}^3C \xrightarrow{\mathcal{F}^3u} \ldots \longrightarrow \mathcal{F}^nC \xrightarrow{\mathcal{F}^nu} \ldots$$

hence that the arrow $\Xi \longrightarrow \mathcal{F}\Xi$ is an isomorphism.

Of course, the previous construction applies to the initial object $\perp_C$ of $C$, since there is always (by initiality of $\perp_C$ ) a unique arrow $\perp_C \longrightarrow \mathcal{F}\perp_C$. Hence, the initial object in a category generates in all cases a fixpoint $\Omega_{\mathcal{F}}$ of the functor. It is then easily checked that $\Omega_{\mathcal{F}}$ is initial in the category of fixpoints of $\mathcal{F}$, or, in other words, that for any other fixpoint $\Xi$ of $\mathcal{F}$ there exists a unique arrow $\Omega_{\Xi} : \Omega \longrightarrow \Xi$ such that the following diagram is commutative :

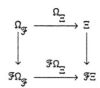

This concludes the proof.■

# 7. References

[AK79] ADAMEK J., KOUBEK V., Least fixed points of a functor, J. Comput. System Sci. 19 (1979) 163-178

[Ba88] BAUDERON M., Infinite hypergraphs, Research Report, Bordeaux, 1988

[BC87] BAUDERON M., COURCELLE M., Graph expressions and graph rewritings, Math. Systems Theory 20, 83-127 (1987)

[Ca88] CAUCAL D., Pattern graphs, Research report, Rennes 1988

[Co78] COURCELLE B., Frontiers of infinite trees, RAIRO, Informatique théorique, vol 12, n°4, 1978, 319-337

[Co83] COURCELLE B., Fundamental properties of infinite trees, Theor. Com. Sci. 25 (1983), 95-169.

[Co85] COURCELLE B., Equivalences and transformations of regular systems. Applications to recursive program schemes and grammars, Theor. Comp. Sci.42 (1986) 1-122.

[Co88] COURCELLE B. The monadic 2nd order theory of graphs IV : every equational graph is definable, Research report 8830, Bordeaux 1988

[HK87] HABEL A, KREOWSKY H-J., Some structural aspects of hypergraph languages generated by hyperedge replacement, Lect. Not. Comp. Sci; 247, 207-219 (1987) and May we introduce to you : hyperedge replacement, Lect Not. Comp. Sci.,291, 15-2-, 1987

[La68] LAMBEK, J.,A fixpoint theorem for complete category, Math. Z. 103 (1968), 151-161

[Le76] LEHMANN D.J., Categories for fixpoint semantics, 17th symposium on Foundations of Computer Science, Houston 1976, 122-126

[LS81] LEHMANN D.J., SMYTH M.B., Algebraic specification of data types, Math. Systems Theory 14 (1981) 97-139

[Li87] LINDENMAYER A, Models for multicellular development : characterization, inference and complexity of L-systems, in Trends, techniques and problems in Theoretical Computer Science, Lect. Not.Comp. Sci. 281, 1987, 138-168

[Ma76] MANES E., Algebraic theories, Springer-Verlag 1976

[Mc71] McLANE S., Category for the working mathematician, Springer-Verlag 1971

[MS85] MULLER D., SCHUPP P, The theory of ends, pushdown automata and second-order logic, Theor. Comp. Sci. 37 (1985) 51-75

[PS82] PLOTKIN G., SMYTH M.B., The category theoretic solution to recursive domain equations, SIAM J. Comput. Vol. 11, No 4 (1982)

[Sm78] SMYTH M.B., Powerdomains, Journ of Comp. and Syst. Sci., 16, 1978, 23-36

# Fault Tolerant Networks of Specified Diameter[1]

H. Meijer & R. Dawes
Department of Computing and Information Science
Queen's University
Kingston, Ontario, Canada

## ABSTRACT

We construct directed graphs of specified diameter. We are
especially interested in the minimal number of arcs required to
construct 2-connected digraphs. Classes of graphs considered are
unconstrained digraphs, digraphs without cycles of length 2, and
digraphs with fixed indegree and outdegree.

## Definitions, Notation, and Assumptions

All graphs in this paper are directed graphs with no loops or multiple arcs.
The vertex set of a graph will be denoted by **V**. We make the following
definitions:

A graph is antisymmetric if it contains no cycles of length 2.

A graph is regular if all vertices have the same indegree and outdegree.

The diameter $\Delta(G)$ of graph G $= \max_{x,y \in V} \{distance(x,y)\}$

---

[1]This work was supported by the Natural Sciences and Engineering
Council of Canada.

The k-th order diameter $\Delta^k(G)$ of graph G = $\max_{\substack{H \subseteq V \\ |H| \leq k}} \{\Delta(G - H)\}$

ie. the largest diameter that may result from the deletion of k or fewer vertices from G.

A graph G is k-connected if $\Delta^{k-1}(G) \neq \infty$, for $k \geq 1$. A k-connected graph is said to be (k-1)-fault tolerant, since the deletion of any k-1 vertices will not disconnect the graph.

The function a(n,d), introduced in [1], is the minimum number of arcs needed to construct a graph G on n vertices with $\Delta(G) = d$. Types of graphs discussed in this paper were graphs which are regular, graphs which are antisymmetric, graphs which are both regular and antisymmetric, and unconstrained graphs. For each of these classes of graphs, [1] gives construction algorithms to justify the bounds on a(n,d). We now give constructions for 1-fault tolerant graphs (ie. $\Delta^1(G) \neq \infty$) for these classes, and determine $\Delta^k(G)$, k > 1, for some of these graphs as well. We define the function $a^1(n,d)$ to be the minimum number of arcs needed to construct a graph on vertices with $\Delta(G) = d$, such that the relation

$$\Delta^1(G) \leq 2 \Delta(G)$$

is satisfied. We give lower and upper bounds on $a^1(n,d)$.


Arc-Minimal Graphs: the function a(n,d)

Upper and lower bounds on a(n,d) are given in [1]. These results are summarized in Table 1.

The upper bounds are demonstrated by constructive methods. Some of the non-regular graphs constructed by these algorithms have vertices of very high indegree and/or outdegree. These graphs have the disappointing property that deletion of these "bottleneck" vertices may disconnect the graph, that is to say these graphs are not 1-fault tolerant. In the next sections we shall construct 1-fault tolerant graphs. As we shall see, the regular graphs constructed in [1] already have this property, so for these graphs a(n,d) and $a^1(n,d)$ have the same upper and lower bounds. For non-regular graphs, however, our constructions yield lower and upper bounds on $a^1(n,d)$ different from a(n,d).

|  | Lower Bound | Upper Bound | Restrictions |
|---|---|---|---|
| Unconstrained Graphs | $2n - d$ | $2n - d$ | $d = 2,3$ |
|  | $n + 1$ | $\left[ \dfrac{1}{\lfloor \frac{d}{2} \rfloor} + 1 \right] (n-1) + O(1)$ | $d \geq 4$ |
| Antisymmetric Graphs | $\dfrac{n}{2} \log \dfrac{n}{2}$ | ? | $d = 2$ |
|  | $2n - 3$ | $2n - 3$ | $d = 3$ |
|  | $n + 1$ | $\left[ \dfrac{1}{\lfloor \frac{d}{2} \rfloor} + 1 \right] (n-1) + O(1)$ | $d \geq 4$ |
| Regular Graphs | $n \sqrt[d]{\dfrac{n}{2}}$ | $dn(\sqrt[d]{n} - 1) + n$ | $d \geq 2$ |
| Antisymmetric Regular Graphs | $n \sqrt[d]{\dfrac{n}{2}}$ | $dn(\sqrt[d]{n} - 1)$ | $d \geq 2,$ $n = k^d$ for some k |

TABLE 1 : $a(n,d)$

## Unconstrained Graphs

All of the unconstrained graphs constructed in [1] have $\Delta^1(G) = \infty$.

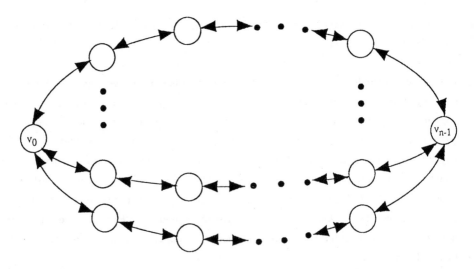

FIGURE 1

Figure 1 illustrates a 2-connected unconstrained graph of diameter d.    The paths from $v_0$ to $v_{n-1}$ are all of length d, with the exception of one path which may have length < d.    It is clear that this construction gives graphs with $\Delta(G) = d$.    Moreover, the graphs are 2-connected, with

$$\Delta^1(G) = 2d - 1$$
$$\Delta^i(G) = \infty \quad \forall\; i > 1$$

The number of disjoint paths from $v_0$ to $v_{n-1}$ is $\left\lceil \dfrac{n - 2}{d - 1} \right\rceil$ from which we derive

$$a^1(n,d) \leq \left\lceil \frac{n - 2}{d - 1} \right\rceil * 2d$$

To derive a lower bound on $a^1(n,d)$, we observe that each vertex must have indegree $\geq 2$ and outdegree $\geq 2$, else removing a vertex's sole predecessor or successor would disconnect the graph.    Thus $a^1(n,d) \geq 2n$.


## Antisymmetric Graphs


Figure 2 illustrates the construction of antisymmetric graphs with diameter $d \geq 4$, as given in [1].    These graphs contain approximately $n + \dfrac{2n}{d}$ arcs.

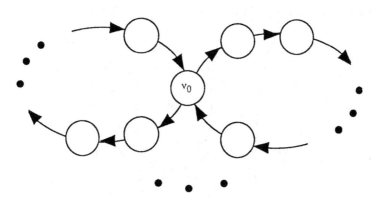

FIGURE 2

When $n \equiv 1 \bmod \left\lfloor \dfrac{d}{2} \right\rfloor$, these graphs consist of a set of cycles each of length $1 + \left\lfloor \dfrac{d}{2} \right\rfloor$.    For other values of n, trivial additions to the graph are made to accomodate the excess vertices.    These graphs are not 2-connected, since the removal of any vertex disconnects the graph.    However, 2-connected antisymmetric graphs with diameter $\geq 3$ may be constructed as follows.

Let $DC_m$ denote a directed cycle of length m on vertices labelled 0 to m-1, with all arcs $(i, i+\lceil\sqrt{m}\rceil)$ added, where addition is performed modulo m. $DC_m$ thus has 2m arcs, and can easily be shown to have diameter no more than

$$\lceil\sqrt{m}\rceil + \lfloor\sqrt{m}\rfloor - 2$$

We shall use r(m) to denote this quantity.

Observe that $DC_4$ and $DC_6$ are not antisymmetric. For $DC_4$ this is unavoidable, but for $DC_6$ this problem is solved by using arcs of the form $(i, i+\lfloor\sqrt{m}\rfloor)$ instead of $(i, i+\lceil\sqrt{m}\rceil)$. This does not affect the discussion which follows.

Consider the mimimum distance in $DC_m$ from any vertex to either member of the set $\{0, \lfloor\frac{m}{2}\rfloor\}$. Again, it can be shown without difficulty that this distance is not more than

$$\left\lfloor \frac{\left\lceil \frac{m-2}{2} \right\rceil}{\lceil\sqrt{m}\rceil} \right\rfloor + \lceil\sqrt{m}\rceil - 1$$

We shall refer to this as j(m).

Now consider the graph formed by taking t copies of $DC_m$ (t $\geq$ 2) and identifying vertices 0 and $\lfloor\frac{m}{2}\rfloor$ across all the copies. An example of such a graph, with t = 2 and m = 6, is illustrated in Figure 3.

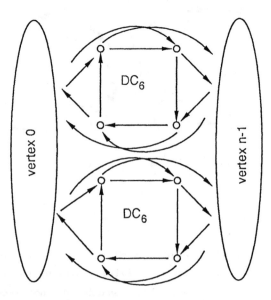

FIGURE 3

The diameter of such a graph is clearly no more than $r(m) + j(m)$, regardless of the number of copies of $DC_m$ used in the construction. Furthermore, $\Delta^1(DC_m)$ can be shown to be $\leq r(m) + 1$. Observe that smaller cycles may also be attached, without increasing the diameter.

For a given desired diameter d, therefore, we must find the maximum value of m such that $r(m) + j(m) \leq d$. Straightforward manipulation determines that for $d \geq 5$, $m \geq \left\lceil \frac{2(d+1)}{7} \right\rceil^2$. In point of fact, for $d = 3$ and $d = 4$, we achieve a similar bound by constructing the $DC_m$ cycles with $m = 5$ and $m = 6$, respectively, using arcs of the form $(i, i+\lfloor \sqrt{m} \rfloor)$, rather than $(i, i+\lceil \sqrt{m} \rceil)$, as mentioned above.

Using this lower bound on m, we see that a graph on n vertices with diameter d may be constructed in this fashion from a set of t copies of $DC_m$, where $m = \left\lceil \frac{4(d+1)^2}{49} \right\rceil$, and $t = 1 + \lceil \frac{n-m}{m-2} \rceil$. As observed above, one of the cycles may in fact contain fewer than m vertices. In this graph, all but two vertices have outdegree 2, and the two remaining vertices each have outdegree 2t. Thus the total number of arcs used is $2(n-2) + 4t$.

It is clear that these graphs are 1-fault tolerant. Furthermore, we observe that after the deletion of any vertex, the diameter cannot exceed $2(r(m) + 1)$, which cannot exceed $2(r(m) + j(m))$, twice the original diameter. Thus we arrive at the following upper bounds on $a^1(n,d)$ for antisymmetric graphs:

$$a^1(n,3) \leq \frac{10n - 6}{3}$$

$$a^1(n,4) \leq 3n - 6,$$

and, for larger values of d,

$$a^1(n,d) \leq 2n + \frac{49n - 4(d+1)^2 + 4}{(d+1)^2 - 25}$$

Observe that for $d = 2$, a graph constructed in this manner would not be antisymmetric, since it would require the incorporation of copies of $DC_4$. We treat this case separately.

## Antisymmetric Graphs of Diameter 2

Let $H_n$ and $H_m$ be any antisymmetric graphs on n and m vertices respectively, with $\Delta(H_n) = \Delta(H_m) = 2$.    Without loss of generality, assume n ≥ m > 2.    We construct an antisymmetric graph of diameter 2 on     n + m + 2     vertices as follows:

Let **v** and **w** be vertices not in $H_n$ or $H_m$.

For each vertex **x** in $H_n$, create the arcs (**xv**) and (**wx**).

For each vertex **y** in $H_m$, create the arcs (**yw**) and (**vy**).

For each vertex **x** in $H_n$, create an arc (**xy**) where **y** is a vertex in $H_m$.    This must be done in such a way that each vertex in $H_m$ is used at least once.

For each vertex **x** in $H_n$, create an arc (**yx**) where **y** is a vertex in $H_m$ such that (**xy**) is not an arc of the graph.    Again, this must be done in such a way that each vertex in $H_m$ is used at least once.

There are numerous ways of satisfying this condition - any will serve the purpose of our argument.

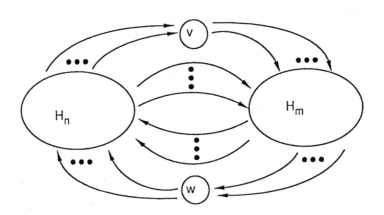

FIGURE 4

Figure 4 shows the general structure of these graphs.    Figure 5 illustrates an example of the construction when n = m = 3.

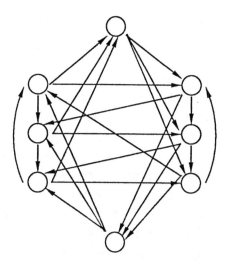

FIGURE 5

These graphs are clearly antisymmetric, with $\Delta(G) = 2$. Further, if we let $T(n) = \min\limits_{H_n} \{|\text{set of arcs in } H_n|\}$, then for $n = 2 + 2k$, we have

$$T(n) \le 4k + 2k + 2*T(k) \le 3n + 2*T(\frac{n}{2})$$

from which it is clear that $T(n) \le 3n*\log_2 n + n*T(i)$, for some small value of i. Thus $a(n,2) \le 3n*\log_2 n + O(n)$. This resolves an outstanding problem suggested by [2], and replaces the "?" in Table 1.

Regarding the fault tolerance of these graphs, we observe that if G is constructed in the indicated manner from $H_n$ and $H_m$, then $\Delta^1(G) = 3$. Thus it follows immediately that $a^1(n,2) \le 3n*\log_2 n + O(n)$ as well.

Regular Graphs of Diameter d

Regular graphs of diameter d were constructed in [1] as follows:

*i)* Let the set of vertices be $\{0, 1, \ldots, n-1\}$

*ii)* Let S be a set of integers $\{S_i\}$ such that

$$\{x \mid x = \sum_{S_i \in T} S_i, \; |T| \le d, \; T \subset S\} \ge \{1, 2, \ldots, n-1\}$$

where summation is performed mod n.

*iii*) For each vertex *i*, create an arc from vertex *i* to vertex
$i+S_j$, for all $S_j \in S$

This construction produces graphs with diameter $\leq$ d. Antisymmetric graphs will be constructed if $S$ is determined such that $\forall S_i, S_j \in S$, $S_i + S_j \not\equiv 0$ (mod n). (Note: from this point, all equivalences are mod n.)

We will now consider the fault tolerance of these graphs. Without loss of generality, we will discuss only paths from vertex 0 to all other vertices. Let $\mathbb{P}$ be a path from vertex 0 to any other vertex **v**. If $\mathbb{P}$ contains arcs generated by two or more elements of $S$, we call $\mathbb{P}$ a <u>mixed</u> path. We may characterize $\mathbb{P}$ by the elements of $S$ that generate the arcs of $\mathbb{P}$. Formally, if $\mathbb{P} = \{e_1, .., e_t\}$ where $e_i$ is generated by $S_{j_1}$, then we will write $\mathbb{P} = \{S_{j_1}, S_{j_2}, \ldots, S_{j_t}\}$.

<u>Lemma 1</u>: Let $\mathbb{P}$ be a path from vertex 0 to vertex **w**. Then if $\mathbb{Q}$ is defined by permuting the $S_{j_1}$ in $\mathbb{P}$, $\mathbb{Q}$ is also a path from 0 to **w**.
<u>Proof</u>: Trivial. ■

We call $\mathbb{P}$ <u>degenerate</u> if there exists a non-empty subset of the $S_{j_1}$ values in $\mathbb{P}$ that sums to 0. This is equivalent to some permutation of $\mathbb{P}$ containing a cycle. Henceforth we shall consider only paths which are not degenerate. Observe that any shortest path must be non-degenerate.

<u>Lemma 2</u>: Let **v** and **w** be vertices in a regular graph generated as described. If there exists a non-degenerate mixed path from 0 to **w**, then there exists a non-degenerate mixed path from 0 to **w** that does <u>not</u> traverse vertex **v**.
<u>Proof</u>: Let $\mathbb{Q} = \{S_{j_1}, \ldots, S_{j_t}\}$ be a mixed 0w-path of minimal length. Since every permutation of $\mathbb{Q}$ is a mixed 0w-path, if the lemma does not hold, then all permutations of $\mathbb{Q}$ must traverse **v**. Let $\mathbb{P}$ be a permutation of $\mathbb{Q}$ such that the length of the subpath from 0 to **v** is maximized over all permutations of $\mathbb{Q}$. Let $\mathbb{P}_1$ be the subpath of $\mathbb{P}$ joining vertex 0 to **v**, and let $\mathbb{P}_2$ be the subpath of $\mathbb{P}$ from **v** to **w**. Since $\mathbb{P}$ is mixed, $\exists$ an $S_k$ in $\mathbb{P}_1$ which is different from some $S_1$ in $\mathbb{P}_2$. Let $S_k$ be the last such element of $\mathbb{P}_1$. Let $\mathbb{P}'$ be the path produced by exchanging $S_k$ and $S_1$. We now show that $\mathbb{P}'$ is a path joining 0 to **w** in G\v. It suffices to show that $\mathbb{P}'$ does not contain **v**. If $\mathbb{P}'$ contains **v**, it must be reached by a subpath no longer than $\mathbb{P}_1$, by our choice of $\mathbb{P}$. If $S_k$ is the last value in $\mathbb{P}_1$, then $\mathbb{P}'$ cannot contain **v**, since $S_k \neq S_1$. Thus, since

all elements in $\mathbb{P}_1$ after $S_k$ are equal to $S_1$, we deduce that $S_k + t*S_1 \equiv r*S_1$ for some integers t and r, $r \leq t+1$. However, since $S_k \neq S_1$, we have $r \neq t+1$, so $r \leq t$. Thus $S_k + (t-r)*S_1 \equiv 0$. This leads immediately to the conclusion that $\mathbb{P}$ is degenerate. ∎

<u>Lemma 3</u>: Let $\mathbb{P}$ be a non-empty path consisting of i arcs generated by $S_j$, followed by k arcs generated by $S_1$, for any i,j,k,1. Then $\mathbb{P}$ is degenerate if and only if $\mathbb{P}$ contains a cycle.
<u>Proof</u>: By definition, if $\mathbb{P}$ contains a cycle, it is degenerate. Conversely, if $\mathbb{P}$ is degenerate, $s*S_j + t*S_1 \equiv 0$ for some $s \leq i$, $t \leq k$. $\mathbb{P}$ contains a sequence of s $S_j$ arcs followed by t $S_1$ arcs, which form a cycle. ∎

<u>Theorem</u>: Let G be a regular graph generated by $S = \{S_1,\ldots,S_k\}$, $k \geq 2$, with $\Delta(G) = d < \infty$. Then $\Delta^1(G) < \infty$.
<u>Proof</u>: Suppose that $\Delta^1(G) = \infty$. Then ∃ two vertices **v** and **w** such that in G\\**v**, no path joins vertex 0 to vertex **w**. Let $\mathcal{P}$ be the set of all non-degenerate paths joining vertex 0 to vertex **w** in G. We consider two cases:

*i)* $\mathcal{P}$ contains a mixed path. Then by Lemma 2, G\\**v** contains a path from 0 to **w**.

*ii)* $\mathcal{P}$ contains no mixed path. Let $\mathbb{P}$ be any path from 0 to **w**. Suppose $\mathbb{P}$ consists entirely of arcs generated by $S_1$. Let $a = n / \gcd(S_j,n)$, for some arbitrary $j \neq 1$. Now consider the simple directed cycle
$$\mathbf{w} \to \mathbf{w} + S_j \to \mathbf{w} + 2S_j \to \ldots \to \mathbf{w} + aS_j = \mathbf{w}.$$
Call this cycle $\mathbb{C}$.

If $0 \in \mathbb{C}$, then the path in $\mathbb{C}$ from 0 to **w** either intersects $\mathbb{P}$ (at some $i*S_1$ other than 0 or **w**) or it doesn't. In the first case, we may construct a mixed 0w-path $\mathbb{Q}$, consisting of some number of $S_1$ arcs followed by a number of $S_j$ arcs. Since $\mathcal{P}$ contains no mixed path, $\mathbb{Q}$ must be degenerate. By Lemma 3, $\mathbb{Q}$ contains a cycle. Deleting all cycles from $\mathbb{Q}$ leaves a non-degenerate 0w-path $\mathbb{R}$. $\mathbb{R}$ must be mixed, since otherwise **w** would be an internal vertex of the path $\mathbb{P}$ or the cycle $\mathbb{C}$. This contradicts the assumption regarding $\mathcal{P}$.

If $\mathbb{P}$ and the path in $\mathbb{C}$ from 0 to **w** intersect only at 0 and **w**, then 0 and **w** are connected by two disjoint paths in G, at least one of which must exist in G\\**v**.

On the other hand, suppose $0 \notin \mathbb{C}$. Let the vertices of $\mathbb{C}$ be $\{\mathbf{w}, \mathbf{x}_1, \mathbf{x}_2, \ldots, \mathbf{x}_{a-1}\}$. Either $\mathbb{C} \cap \mathbb{P} = \{\mathbf{w}\}$, or $\mathbf{x}_i \in \mathbb{C} \cap \mathbb{P}$, for some i.

Suppose $\mathbb{C} \cap \mathbb{P} = \{w\}$. Clearly $v \notin \mathbb{C}$. But now $\mathbb{P} \cup \mathbb{C}$ contains a mixed path $\mathbb{Q}$ from $0$ to $x_{a-1}$ (ie. $\mathbb{P}$ followed by the first $a-1$ arcs of $\mathbb{C}$). By Lemma 3, if $\mathbb{Q}$ is degenerate, it contains a cycle. Deleting all cycles from $\mathbb{Q}$ must leave a mixed non-degenerate path from $0$ to $x_{a-1}$, since $\mathbb{P} \cap \mathbb{C} = \{w\}$. Thus $\exists$ a mixed non-degenerate $0x_{a-1}$-path in $G\backslash v$. This path, plus the arc $(x_{a-1}w)$ forms a path from $0$ to $w$ in $G\backslash v$.

Finally, suppose $0 \notin \mathbb{C}$, and $x_i \in \mathbb{C} \cap \mathbb{P}$. Clearly there exists a mixed path from $0$ to $w$, consisting of some number of $S_1$ arcs followed by a number of $S_j$ arcs. By the same argument as is used above, this yields a mixed non-degenerate path from $0$ to $w$, contradicting the assumption that $\mathcal{P}$ contains no mixed paths. ∎

We will now demonstrate that under certain conditions on $S$, we can predict $\Delta^1$ for these graphs.

Theorem: Let $G$ be a regular graph generated by $S = \{S_1, S_2, \ldots, S_k\}$, with $\Delta(G) = d$, such that $\forall S_i$, $\exists S_j$, $i \neq j$ and $S_j \neq a * S_i$, for all $a$ where $-d < a < 0$. Then $\Delta^1(G) \leq d + 1$.

Proof: Let $w$ be any vertex. There is a shortest path from $0$ to $w$ of length $\leq d$. That is, $\sum_{i=1}^{k} x_i S_i \equiv w$ for some vector $\underline{x}$ with $0 \leq x_i$, $1 \leq i \leq k$, and $\sum_{i=1}^{k} x_i \leq d$. We will use the notation $[\underline{x}]$ to denote the following path:

$$0 \to S_1 \to 2S_1 \to \cdots \to x_1 S_1 \to x_1 S_1 + S_2 \to \cdots \to x_1 S_1 + x_2 S_2 \to \cdots \to \sum_{i=1}^{k} x_i S_i = w$$

Since this is also a shortest path from $0$ to $w$ and is non-degenerate, it contains no vertex more than once.

First, suppose that $[\underline{x}]$ is a mixed path. By the argument used above, we observe that for any vertex $v$ in $[\underline{x}]$, $0$ is connected to $w$ by an alternate path of equal length which does not contain $v$.

Second, suppose that $[\underline{x}]$ is not mixed. Without loss of generality assume that $x_1 > 0$, so we have $x_1 * S_1 \equiv w$. Also, let $S_2 \in S$ be such that $S_2 \neq a * S_1$, for any $a$ where $-d < a < 0$. Define vertex $z$ by $z \equiv w - S_2$. Let $[\underline{y}]$ be a shortest path from $0$ to $z$, where $[\underline{y}] = (y_1, y_2, \ldots, y_k)$. Now suppose that vertex $v \equiv t * S_1$ is deleted from path $[\underline{x}]$. If $[\underline{y}]$ is mixed or does not contain $v$, then there exists a path of the same length as $[\underline{y}]$ from $0$ to $z$ in $G\backslash v$. Thus the distance from $0$ to $w$ is $\leq d+1$. On the other hand, suppose $[\underline{y}]$ is not mixed, and does contain the deleted vertex $v \equiv t * S_1$. Since

$S_2 \not\equiv a * S_1$ for any a where $-d < a < 0$, we know that $y_1 = 0$. Let $y_i$ be the single non-zero element of $\underline{y}$. Since $[\underline{x}]$ and $[\underline{y}]$ intersect at $\mathbf{v}$, we have $t * S_1 \equiv r * S_i$, for some r. However, since $[\underline{x}]$ and $[\underline{y}]$ are shortest paths, we must have $t = r$.

Then $[\underline{u}]$ with $u_1 = x_1 - t$

$\qquad\qquad u_i = t$

$\qquad\qquad u_j = 0 \qquad j \neq i, \; j \neq 1$

is a mixed shortest path from 0 to $\mathbf{w}$. $\qquad\qquad$ ∎

The following examples demonstrate that there exist graphs G with $\Delta^1(G) = \Delta(G) + 1$, and that the diameter can increase by more than 1 if the specified condition on S is not satisfied.

Example 1:     $n = 15$, $S = \{3,5\}$, and $\Delta(G) = 6$.  We have $3 \not\equiv a*5$ and $5 \not\equiv a*3$ for all a where $-6 < a < 0$.  Examination reveals that $\Delta^1(G) = 7$.

Example 2:     $n = 8$, $S = \{1,7\}$ and $\Delta(G) = 4$.   The condition on S is not satisfied, since $7 \equiv -1*1$.   $\Delta^1(G) = 6$.

Example 3:     $n = 30$, $S = \{9,17,26\}$, and $\Delta(G) = 8$.   Again the condition is violated since $26 \equiv -2*17$ and $9 \equiv -3*17$.   $\Delta^1(G) = 10$.

Summary

We summarize the bounds on $a^1(n,d)$ in Table 2.

| | Lower Bound | Upper Bound | Restrictions |
|---|---|---|---|
| Unconstrained Graphs | $2n$ | $\left\lceil \dfrac{n-2}{d-1} \right\rceil * 2d$ | $d \geq 2$ |
| Antisymmetric Graphs | $\dfrac{n}{2} \log \dfrac{n}{2}$ | $3n * \log_2 n + O(n)$ | $d = 2$ |
| | $2n$ | $\dfrac{10n - 6}{3}$ | $d = 3$ |
| | $2n$ | $3n - 6$ | $d = 4$ |
| | $2n$ | $2n + \dfrac{49n - 4(d+1)^2 + 4}{(d+1)^2 - 25}$ | $d \geq 5$ |
| Regular Graphs | $n \sqrt[d]{\dfrac{n}{2}}$ | $dn(\sqrt[d]{n} - 1) + n$ | $d \geq 2$ |
| Antisymmetric Regular Graphs | $n \sqrt[d]{\dfrac{n}{2}}$ | $dn(\sqrt[d]{n} - 1)$ | $d \geq 2$, $n = k^d$ for some $k$ |

TABLE 2 : $a^1(n,d)$

References

[1] R. Dawes and H. Meijer, Arc-Minimal Digraphs of Specified Diameter, J. of Comb. Math. and Comb. Computing 1(1987), 85-96.

[2] G. Katona and E. Szemerédi, On a Problem of Graph Theory, Studia Sci. Math. Hungar. 2(1967), 23-28.

# DFS TREE CONSTRUCTION: ALGORITHMS and CHARACTERIZATIONS

(Preliminary version)

Ephraim Korach  and  Zvi Ostfeld

Computer Science Department
Technion - Israel Institute of Technology
Haifa 32000, Israel

## ABSTRACT

The Depth First Search ($DFS$) algorithm is one of the basic techniques which is used in a very large variety of graph algorithms. Every application of the $DFS$ involves, beside traversing the graph, constructing a special structured tree, called a $DFS$ tree. In this paper, we give a complete characterization of all the graphs in which every spanning tree is a $DFS$ tree. These graphs are called $Total-DFS-Graphs$. The characterization we present shows that a large variety of graphs are not $Total-DFS-Graphs$, and therefore the following question is naturally raised: *Given an undirected graph $G=(V,E)$ and an undirected spanning tree T, is T a DFS tree of G?* We give an algorithm to answer this question in linear ($O(|E|)$) time.

## 1. INTRODUCTION

The Depth First Search ($DFS$) algorithm is one of the basic techniques which is used in

a very large variety of graph algorithms. The history of this algorithm (in a different form)

goes back to 1882 when Tremaux' algorithm for the maze problem was first published (see

[BLW, page 18]).

The impact of *DFS* grew rapidly since the Hopcroft and Tarjan version of it was published (see [Ta], [HT a], [HT b], and [HT c] ).

In many areas of computer science, this algorithm is used, and lately it also has penetrated the field of parallel and distributed algorithms (e.g. [AA], [Aw], [KO], [LMT], [Re], and [Ti]).

Every use of the *DFS*, beside traversing the graph, constructs a special structured directed rooted tree, called a *DFS tree*, that may be used subsequently.

In this paper, we raise two important questions, regarding the structure of the *DFS* tree that is obtained, and discuss their solutions.

(i)  First, we are interested in determining in which graphs, every spanning tree can be obtained as a *DFS* tree. These graphs are called *Total–DFS–Graphs* (*T–DFS–G*). In section 2 we give a complete characterization of these graphs.

It turns out that a large variety of graphs are not *Total–DFS–Graphs*, and therefore the following question is naturally raised:

(ii)  *Given an undirected graph $G=(V,E)$ and an undirected spanning tree T of G, is T a DFS tree (T-DFS) in G?*

Question (ii) is answered by a linear ($O(|E|)$) time algorithm in section 3. The algorithm gives as an output all the vertices $S = \{ s \in V : T_s$ - the directed tree rooted at $s$ - is a *DFS* tree in $G$ }. If $T$ is not a *DFS* tree in $G$ (i.e. $S$ is the empty set) then the algorithm can supply an $O(|V|)$ proof for that fact. Details about that feature of the algorithm will appear in the final version of this paper.

One can think about many applications of the results presented here. For example, we can constructively solve the following problem: Let $G$ be an undirected graph with a unique minimum undirected spanning tree $T$ (e.g. where no two edges have the same weight). Is it possible to run a *DFS* in such a way that $T$ will be the *DFS* tree ? (This question might be important in a distributed network).

Another application is where we consider the vertices of the graph $G=(V,E)$ as a collection of tasks where every task has to be executed once and for every two tasks $v_i$, $v_j \in V$, $(v_i, v_j) \in E$ if and only if $v_i$ and $v_j$ can not be executed simultaneously (for example they use a common resource). Let us consider the following computation model: (i) We choose a task $s \in V$ which is the first task to be executed. (ii) For every other task $v \in V- \{ s \}$ we choose a unique predecessor $f(v) \in V$ such that $(f(v),v) \in E$. (iii) Every task $v \in V- \{ s \}$ can start execution at any time after $f(v)$ is completed. Clearly, a necessary condition for a possible execution of all tasks is that the set of edges $\{ f(v) \to v \}$ is circuit free and hence this edges are the edges of a directed spanning tree of $G$ which is rooted at $s$. Let $T$ be an undirected spanning tree of $G$ and let $s \in V$. By Proposition 2.3 below it follows that $T_s$, the directed tree rooted at $s$, represents a safe execution of all these tasks (no overlap in the execution of tasks $v_i$, $v_j \in V$, may occur if $(v_i, v_j) \in E$), starting from $s$ if and only if $T_s$ is a *DFS* tree in $G$.

Question (ii) was independently answered in [HN] in a different approach. However, we think that our algorithm has some advantages over the other algorithm. For example, we can not see how one can extend the algorithm of [HN] to supply a short proof for a negative answer to question (ii) above.

Another result that concerns a *DFS* tree in a 2-isomorphic copy of a graph $G$ appears in [Sy].

Extensions to the directed and parallel cases appear in [KO].


## 2. TOTAL DFS GRAPHS

**Definition:** An undirected connected graph $G$ is called *Total-DFS-Graph* $(T-DFS-G)$ if every spanning tree of $G$ is a *DFS* tree.

**Definition:** A graph $G$ is called $k$-parallel-path graph if the edges of $G$ can be partitioned into $k$ internally vertex disjoint paths between two vertices.

A $k$-parallel-path graph $G$ is completely defined by a positive integral vector $l = (l_1, \ldots, l_k)$ of length $k$, where $l_i$ is the length of the $i$-th path in $G$ (the number of edges in it). W.l.o.g. we assume that $l_i \geq l_j$ for $i < j$.

**Notation:** A $k$-parallel-path graph with a vector $l = (l_1, \ldots, l_k)$ is denoted by $PPG(l_1, \ldots, l_k)$.

**Theorem 2.1 (Characterization of Total-DFS-Graphs):** A connected simple graph $G$ is a $T-DFS-G$ if and only if $G$ has at most one non-separable component $C$ with at least three vertices and $C$ is one of the following $k$-parallel-path graphs:

$$PPG(x, y) \qquad x \geq 2, y \geq 1$$
$$PPG(x, 2, 1), PPG(x, 2, 2) \qquad x \geq 2$$
$$PPG(2, 2, 2, 1), PPG(2, 2, 2, 2)$$

**Remark:** Let $G$ be a simple graph and let $G'$ be the result of adding loops and parallel edges to $G$. Then $G'$ is a $T-DFS-G$ if and and only if $G$ is a $T-DFS-G$. Therefore, in the rest of the section we assume w.l.o.g. that all graphs are simple.

Next we give the proof of Theorem 2.1. The reader who is not interested in the proof can move immediately to Corollary 2.11.

In order to prove the theorem we need the lemmas and propositions stated and proved below.

**Lemma 2.2:** Let $G$ be a *Total–DFS–Graph* ($T-DFS-G$). Then there are two vertices which are in all circuits of $G$.

The proof of the lemma appears after Proposition 2.9.

**Definitions:**   Let $T$ be an undirected spanning tree of an undirected graph $G = (V, E)$ and let $s \in V$. $T_s$ is the tree $T$ together with the orientation that makes $s$ the root of $T$. $T_s$ is called a *DFS* tree ($T\text{-}DFS$) in $G$ if it can be constructed by a *DFS* run in $G$. We use the same terminology also for the undirected tree $T$: $T$ is also called a *DFS* tree ($T\text{-}DFS$) in $G$ if there exists a vertex $s \in V$ such that $T_s$ is a $T\text{-}DFS$ in $G$.

Let $T$ be a directed spanning tree and let $a$ and $b$ be two vertices in $T$. If there is a directed path in $T$ from $a$ to $b$, we say that $a$ is an *ancestor* of $b$ and $b$ is a *descendant* of $a$. A vertex is a descendant and an ancestor of itself.

**Property PDFS:**   $T_s$ has the property *PDFS* if for every edge $(a, b) \in E$ $a$ is either an ancestor of $b$ or a descendant of $b$.

**DFS orientation:**   Let $T$ be a spanning tree of $G = (V, E)$ and let $s \in V$. If $T_s$ has the property *PDFS*, then the orientation given to T is called *DFS orientation of T*.

**Proposition 2.3:**   Let $T$ be a spanning tree of $G$ and let $s \in V$, then $T_s$ is a *DFS* tree in $G$ if and only if $T_s$ has the property *PDFS*.

**Proof:** See [Ta, Theorem 1].  □

(The symbol: " □ " stands for "end of the proof" or "end of the statement and a proof is not provided").

**Proposition 2.4:**   Let $T$ be the underlying undirected tree of a *DFS* tree in a graph $G$ and let $(u, v) = e \in E(G) - E(T)$ be an edge between two leaves of $T$. Then the root of the *DFS* tree must be either $u$ or $v$.

**Proof:** Clearly, in the *DFS* tree, the in-degree of every vertex except for the root is equal to one and the in-degree of the root is zero. $e$ is an edge between a vertex and one of its descendants, and w.l.o.g. assume that $v$ is a descendant of $u$. Hence there is a directed path from $u$ to $v$, and since $u$ is a leaf in $T$ it has only one edge directed out of it. Therefore $u$ is a root. □

**Proposition 2.5:** If $G$ is a $T$–$DFS$–$G$ then every connected subgraph of $G$ is a $Total$–$DFS$–$Graph$.

**Proof:** Let $G'$ be a connected subgraph of $G$ and let $T'$ be any spanning tree of $G'$. Clearly we can extend $T'$ to a spanning tree $T$ of $G$. Since $G$ is a $T$–$DFS$–$G$ then $T$ is a $DFS$ tree and hence, by Proposition 2.3, $T$ together with the orientation induced by a $DFS$ run has the property $PDFS$ in $G$.

Let $OT$ be the tree $T$ together with the orientation induced by the $DFS$ run. Let $OT'$ be the tree $T'$ with the orientation induced by $OT$. Since $OT$ has the property $PDFS$ in $G$ then $OT'$ also has the property $PDFS$ in $G'$ and hence, by Proposition 2.3, $T'$ is a $DFS$ tree in $G'$. $\square$

**Proposition 2.6:** Every two circuits in a $Total$–$DFS$–$Graph$ have at least two vertices in common.

**Proof:** Let $C_1 = (v_1 v_2 ... v_k)$ and $C_2 = (u_1 u_2 ... u_m)$ be any two circuits in $G$. Since $G$ is a simple graph then $m, k \geq 3$. Let $\pi$ be the shortest path connecting $C_1$ and $C_2$ and w.l.o.g. assume that the ends of $\pi$ are $v_1$ and $u_1$ (possibly $v_1 = u_1$).

Assume, in contradiction, that $|V(C_1) \cap V(C_2)| \leq 1$ (equality holds if and only if $v_1 = u_1$) and consider the subgraph $G' = C_1 \cup C_2 \cup \pi$.

Let $C'_1, C'_2$ be the paths obtained from $C_1, C_2$ by deleting the edges $(v_2, v_3)$, $(u_2, u_3)$, respectively.

Let $T' = \pi \cup C'_1 \cup C'_2$. Clearly $T'$ is a spanning tree of $G'$ and has four leaves $v_2, v_3, u_2, u_3$. By proposition 2.4 one of $v_2$ and $v_3$ must be the root of the $DFS$ tree and also one of $u_2$ and $u_3$ must be the root of the $DFS$ tree which is impossible. Therefore $T'$ is not a $DFS$ tree in $G'$ which implies that $G'$ is not a $T$–$DFS$–$G$ and therefore, by Proposition 2.5, $G$ is not a $T$–$DFS$–$G$ in contradiction to the assumption. $\square$

**Proposition 2.7:** If $G$ is a homeomorph of $K_4$ then $G$ is not a $Total$–$DFS$–$Graph$.

**Proof:** Let $v_1, v_2, v_3$ and $v_4$ be the vertices of degree three in $G$. Let $\pi_{ij}$, $1 \leq i, j \leq 4$ be the six paths between $v_i$ and $v_j$ homeomorphic to the edges of the $K_4$. Let $l_{12}, l_{23}$, and $l_{13}$ be edges on $\pi_{12}, \pi_{23}$ and $\pi_{13}$, respectively. One can see that $G-\{l_{12}, l_{13}, l_{23}\}$ is a spanning tree of $G$ which is not a $T-DFS$. $\square$ (A proof of this proposition appears also in [Sy].)

**Corollary 2.8:**    A *Total–DFS–Graph* does not contain a subgraph homeomorphic to $K_4$.

$\square$

**Proposition 2.9:**    Let $G$ be a $T-DFS-G$ with at least 2 circuits, and assume that $a, b \in V$ are contained in all circuits of $G$. Then there are three paths, pairwise internally disjoint between $a$ and $b$ in $G$.

**Proof:** Let $C_1$ and $C_2$ be any two different circuits in $G$ with edge sets $EC_1$ and $EC_2$. Then $EC_1 \Delta EC_2$ is the edge set of a collection of circuits $CY$ in $G$, where $\Delta$ denotes the set symmetric difference (i.e. $EC_1 \Delta EC_2 = (EC_1 - EC_2) \cup (EC_2 - EC_1)$). If $CY$ is a single circuit $C$ then since $\{a, b\} \subseteq C$, it contains two disjoint paths between $a$ and $b$. Clearly $C_1 \cap C_2$ is another path between $a$ and $b$ and these three paths are pairwise internally disjoint. If $CY$ is more than one circuit then since all the circuits contain both $a$ and $b$, there are at least three paths between $a$ and $b$ which are pairwise internally disjoint. $\square$

**Proof of Lemma 2.2:** By induction on the dimension of the cycle space (see [Tu]) of $G$.
Let $G$ be a $T-DFS-G$. The lemma is trivially true for graphs of dimension $\leq 1$. Clearly if the cycle space of $G$ is of dimension 2 then by Proposition 2.6, one can see that the lemma is true. Assume it is true for all $T-DFS-G$ with cycle space of dimension less or equal $k$ for a fixed $k \geq 2$ and for contradiction let us assume that it is not true for a given $T-DFS-G$ with cycle space of dimension $k+1$. Let $T$ be any spanning tree of $G$ and let $e \in E(G)-E(T)$. Then $G-e$ is also a $T-DFS-G$ (by Proposition 2.5) and has cycle space of dimension $k$. By the induction hypothesis, there are two vertices $a$ and $b$ which are common to all circuits of $G-e$. By Proposition 2.9 there are three paths between $a$ and $b$ which are pairwise internally disjoint, say, $\pi_1, \pi_2$ and $\pi_3$. Since we assumed that the claim of the lemma is not true

or $G$, there is a circuit $C_e$ containing $e$ that does not contain both $a$ and $b$, w.l.o.g. assume that $C_e$ does not contain $a$.

Let $C_1 = \pi_1 \cup \pi_2$, then by proposition 2.6, $C_e$ and $C_1$ has at least two vertices in common and w.l.o.g. we may assume that $C_e$ has a vertex $x$ which is internal to $\pi_1$.

Let $C_2 = \pi_2 \cup \pi_3$, then by similar arguments we have that $C_e$ has a vertex $y$ internal to $\pi_2$ or to $\pi_3$.

Let $\pi_c$ be the part of $C_e$ between $x$ and $y$ that does not contain $b$, and let $\pi''_c$ be a minimal subpath of $\pi_c$ that has two ends in two out of $\{\pi_1, \pi_2, \pi_3\}$. Clearly, $\pi''_c$ is internally disjoint from all the three paths $\pi_1$, $\pi_2$, and $\pi_3$, and its ends are on the interior of two of these three paths, w.l.o.g. assume that it starts with $s$ on $\pi_1$, ends in $t$ in $\pi_2$ and avoids $\pi_3$. Then $\pi_1 \cup \pi_2 \cup \pi_3 \cup \pi''_c$ form an homeomorph of $K_4$ (where $a, b, s, t$ are the four points of the $K_4$) contradicting Corollary 2.8. This completes the proof of Lemma 2.2. $\square$

**Lemma 2.10:**    The complete set of (simple) $k$-parallel-path graphs which are $T-DFS-G$ is the following:

     1. all k-parallel-path graphs for $k = 1, 2$;

     2. for k = 3: $PPG(x, 2, 1)$, $PPG(x, 2, 2)$;       $x \geq 2$

     3. for k = 4: $PPG(2, 2, 2, 1)$, $PPG(2, 2, 2, 2)$.

**Proof:** It is easy to check that all graphs in the above list are $T-DFS-G$. We show that no other graph can be added to the list.

Claim a:  For $k \geq 5$ there is no $T-DFS-G$.

     *Proof:* By Proposition 2.5 it is enough to prove the claim only for $k = 5$. Since there are no parallel edges then there are four paths, say $\pi_1$, $\pi_2$, $\pi_3$ and $\pi_4$ of length $\geq 2$. Let $a, b \in V$ be the end vertices of the paths in $G$. Let $l_1 \in \pi_1$ and $l_2 \in \pi_2$ be the edges with end in $a$ and let $l_3 \in \pi_3$, $l_4 \in \pi_4$ be the edges with end in $b$. Clearly, the tree $G - \{l_1, l_2, l_3, l_4\}$ is not a *DFS* tree in $G$.

Claim b: For $k = 3$ every PPG$(x,y,z)$ where $x,y \geq 3$, $z \geq 1$ is not a $T-DFS-G$.

> *Proof:* Since there are two paths say $\pi_1$ and $\pi_2$ of length $\geq 3$, there are two edges $l_1 \in \pi_1$ and $l_2 \in \pi_2$ that are both between vertices of degree two and by Proposition 2.4 the tree $G - \{l_1, l_2\}$ is not a $T-DFS$.

Claim c: For $k = 4$ there is no $T-DFS-G$ in the following form:

> PPG$(x,y,z,t)$  $x \geq 3$, $y,z \geq 2$, $t \geq 1$.

> *Proof:* Let $\pi_x, \pi_y, \pi_z, \pi_t$ be the corresponding paths of $G$ between the two vertices $a,b \in V$ and let $l_1 \in \pi_x$ be an edge between two vertices of degree two in $\pi_x$. Let $l_2 \in \pi_y$ have $a$ as an end and let $l_3 \in \pi_z$ have $b$ as an end.

> One can show that the tree $G \setminus \{l_1, l_2, l_3\}$ is not a $T-DFS$.

It is easy to see that a $k$-parallel-path graph that does not belong to the set in the statement of the lemma is eliminated from being a $T-DFS-G$ by one of the above claims. $\square$

**Proof of Theorem 2.1:**    Clearly every graph which is a tree is a $T-DFS-G$.

By Proposition 2.6 if $G$ has more than one nonseparable component with at least three vertices in each then $G$ is not a $T-DFS-G$.

If $G$ is a $T-DFS-G$ that has only one nonseparable component $C$ with at least three vertices then by Lemma 2.2 - $C$ is a $k$-parallel-path graph. By Lemma 2.10 and Proposition 2.5 if $C$ is not one of the list in the statement of the theorem then $G$ is not a $T-DFS-G$.

One can see that if $G'$ is a $T-DFS-G$ and if $G''$ is obtained from $G'$ by adding an edge $(u,v)$ such that $u \in V(G')$, $v \notin V(G')$ then $G''$ is also a $T-DFS-G$. Therefore if $C$ is one of the list in the statement of the theorem then $G$ is obtained from it by adding edges as it is described above and therefore $G$ is a $T-DFS-G$. $\square$

**Corollary 2.11:** There is a sequential algorithm with $O(|E|)$ time complexity to decide whether a given graph $G = (V,E)$ is a $T-DFS-G$. A parallel implementation of this algorithm runs in $O(\log |V|)$ time complexity and uses $O(|E|)$ processors, on a concurrent-read, concurrent-write parallel RAM (CRCW PRAM).

**Proof:** The following algorithm is based on the characterization given in Theorem 2.1. The algorithm has four steps:

(1) Decomposition of $G$ into its 2-connected components.

(2) Checking that there is at most one 2-connected component $C$ with at least three vertices.

(3) Checking that in $C$ there are at most two vertices where the degree (in $C$) is greater than two.

(4) Computing all the paths of length less than three between those two vertices.

Step (1) is done using the algorithm of Tarjan and Vishkin [TV] which has a linear $(O(|E|))$ sequential implementation and a parallel implementation which runs in $O(log\ |V|)$ time and uses $O(|E|)$ processors on a CRCW PRAM. One can see that steps (2)-(4) are easily done without affecting the complexity stated above. $\square$


## 3. DFS ORIENTATION of a TREE

In the previous section we have seen the characterization of *Total−DFS−Graphs* (in Theorem 2.1). It follows that a large variety of graphs are not *Total−DFS−Graphs* and therefore the question: *"Given an undirected graph $G=(V,E)$ and an undirected spanning tree T of G, is T a DFS tree in G?"* becomes important. In this section we solve this question by presenting a linear $(O(|E|))$ time algorithm which gets as an input the pair $(G,T)$ and gives as an output the set of all the vertices $s \in V$ such that $T_s$ is a $T−DFS$ in $G$ if any exists. If none exists then the algorithm answers that $T$ is not a *DFS* tree in $G$.

Before presenting the algorithm we need Proposition 3.1 and Observation 3.2.

**Proposition 3.1:** Let $G$ be an undirected graph with a spanning tree $T$. Let $G_i$ be a subgraph of $G$ with a spanning tree $T_i$ which is the restriction of $T$ to $G_i$. If $T$ is a *DFS* tree in

$G$ then $T_i$ is a *DFS* tree in $G_i$.

**Proof:** By Proposition 2.3 every *DFS* orientation of $T$ is a *DFS* orientation of $T_i$. $\square$

It is well known that the root of a *DFS* tree in a 2-connected graph has degree one (in the tree). Hence we have the following observation.

**Observation 3.2:** Let $G=(V,E)$ be a 2-connected graph and let $T$ be an undirected spanning tree of $G$. Then one of the following holds:

(1) $T$ is a path and there are two *DFS* orientations of $T$ (where the roots of the oriented trees are the two ends of the path).

(2) $T$ is not a path and there is at most one *DFS* orientation of $T$ (there is at most one leaf $s$ in $T$ where $T_s$ has the property *PDFS*).

**Proof:** If $T$ is a path then (1) is true. Assume in contradiction that $T$ is not a path and has two *DFS* orientations $T_a$ and $T_b$ where $a$ and $b$ are two leaves. Since $T$ is not a path there must be a vertex $v \in V$ in $\pi_{a,b}$ such that $d_T(v) \geq 3$ ( $d_T(v)$ is the degree of $v$ in the tree $T$ ). $v$ is the root of $d_T(v)-1 \geq 2$ subtrees $T_1, T_2, \ldots, T_{d_T(v)-1}$ of $T_a$ that do not contain $a$ and w.l.o.g. we assume that $T_1$ does not contain $b$. Since $T_a$ is a *DFS* tree and since $v$ is not a separating vertex in $G$, there must be a back edge in $T_a$, $(u,w) \in E$ such that: (i) $w \neq v$ ; (ii) $w \in T_1$; (iii) $u \neq v$ ; (iv) $u$ is an ancestor of $v$ in $T_a$. Clearly, $(u,w)$ is a cross edge in $T_b$ and by Proposition 2.3, $T_b$ is not a $T-DFS$ in $G$, a contradiction. $\square$

## Algorithm DECIDE

**input:** An undirected graph $G=(V,E)$ and an undirected spanning tree $T$ of $G$.

**output:** A decision whether $T$ is a *DFS* tree in $G$.

If the decision is positive then the set of all the vertices $s \in V$ such that $T_s$ is a *DFS* tree in $G$ is given.

The algorithm consists of the following four main phases:

*Phase* 1: Decomposition of $G$ into its 2-connected components using Tarjan's algorithm [Ta] and constructing the related super-structure $S(G) = (A \cup C, F)$ (see [Ev], [LP]). The vertices of $S(G)$ are partitioned into two groups $A$ and $C$ where $A$ is the set of all the separating vertices (called also articulation points) of $G$ and $C$ is the set of all vertices of $S(G)$ that represent the 2-connected components of $G$.

*Phase* 2: For each 2-connected component $G_i = (V_i, E_i)$ of $G$ with $T_i$ - the induced subtree of $T$ in $G_i$ - we solve algorithmically whether $T_i$ is a *DFS* tree in $G_i$.

If the answer is positive, we get all (one or two) the roots of a *DFS* run that gives $T_i$ in $G_i$.

If the answer is negative for at least one component, then by Lemma 3.1, $T$ is not a *DFS* tree in $G$.

This phase consists of the following sub-phases which are applied for every $(G_i, T_i)$.

1) If $T_i$ is a path then its ends $(s_1, s_2 \in V_i)$ are the two roots. After that, Phase 2 is terminated for $G_i$.

If $T_i$ is not a path then there is at most one vertex which is a candidate for being the root, and this vertex must be a leaf. In the following sub-phases we look for this candidate.

2) Choose an arbitrary leaf and denote it by $s$.

Check whether $T_i$ rooted at $s$ has the property *PDFS*.

If yes then $s$ is the *DFS* root of $T_i$ in $G_i$ and Phase 2 is terminated for $G_i$.

If not then by Proposition 2.3, there is at least one edge $e \in E$ that is a cross edge in $T_i$ rooted at $s$ (and is found during the checking process performed in this sub-phase).

3) We range label $T_i$ from $s$ (see algorithm *Range –Labeling* below).

4) Starting from $s$ we search for another leaf, say $r$, to be a candidate for a root (see algorithm *Search* below).

5) Check whether $T_i$ rooted at $r$ has the property *PDFS*.

If the answer is negative, then $T_i$ is not a *DFS* tree in $G_i$ (and hence by Proposition 3.1, $T$ is not a *DFS* tree in $G$). In this case the algorithm stops.

If the answer is positive then $r$ is the *DFS* root of $T_i$ in $G_i$.

*Phase* 3:  Compute $\vec{S}(G)=(A \cup C, \vec{F})$, the result of the orientation of $S(G)=(A \cup C, F)$ according to Phase 2 as follows:

Let $G_i$ be a 2-connected component of $G$ represented by the vertex $c_i \in C$. Then every edge $(c_i, a_j) \in F$ (where $a_j \in G_i$) is directed from $c_i$ to $a_j$. If $T_i$ rooted at $a_m$ is a $T$–*DFS* in $G_i$ we direct $(c_i, a_m)$ also from $a_m$ to $c_i$ (i.e. $(c_i, a_m)$ is bidirected).

*Phase* 4:  The decision is made according to Lemma 3.3 below.

We find all the roots of the oriented super-structure (if any exists) (see algorithm *Find –Root* below). If $c_i \in C$ is a root of $\vec{S}(G)$ then all (one or two) the roots of a *DFS* run that gives $T_i$ in $G_i$ are output (i.e. they are the roots of a *DFS* run in $G$ that gives $T$).

**Lemma 3.3:** The following statements are equivalent:

(1)  $T$ is a *DFS* tree in $G=(V, E)$ rooted at $s \in V$.

(2)  The oriented super-structure of $G$: $\vec{S}(G)=(A \cup C, \vec{F})$ has a root $c_i \in C$ (that represents the 2-connected componnent $G_i$ in $G$) such that $s$ is a root of a *DFS* run in $G_i$ that gives $T_i$ - the induced sub tree of $T$ in $G_i$.

**Proof:** Assume $T_s$ is a $T$–*DFS* in $G$ and let $G_i$ be any 2-connected component in $G$ that contains $s$. Consider a *DFS* run in $G$ that gives $T_s$. Let us denote by $s_j$ the first vertex to be discovered during the search in the 2-connected componnent $G_j$ where $j \neq i$. It is clear that

(i) $s_j$ is a separating vertex in $G$. (ii) $T_j$ - the induced sub tree of $T$ in $G_j$- rooted at $s_j$ is a

$T$–$DFS$ in $G_j$. It is easily checked (by induction) that there is a feasible path from $c_i$ to $c_j$

in $\vec{S}(G)$ (i.e. no edge in the path is directed only in the opposite direction of the path).

Conversely, let $c_i \in C$ be a root of $\vec{S}(G)$ and let $s \in G_i$ be such that $T_i$ rooted at $s$ is a $DFS$

tree in $G_i$, and let us assume for contradiction that $T_s$ is not a $DFS$ tree in $G$. By Proposi-

tion 2.3 there is a cross edge $e \in E$ in $T_s$ and clearly, $e$ is in $G_j$ ($\neq G_j$). Let $s'_j$ be the first

vertex in $G_j$ that is discovered during a $DFS$ run in $T_s$. It is clear that: (i) $s'_j \in A$. (ii) $T_j$

rooted at $s'_j$, is not a $T$–$DFS$ in $G_j$. Hence the edge $(c_j, s'_j)$ is not bidirected in $\vec{S}(G)$

which implies that there is no feasible path in $\vec{S}(G)$ from $c_i$ to $c_j$, a contradiction. $\square$

**DFS-labeling:** A tree $T = (V, E)$ is $DFS$–$labeled$ by a bijection $k : V \rightarrow \{1, 2, ..., |V|\}$ if

there is a $DFS$ run in $T$ such that for every vertex $v \in V$, $k(v) = i$ if and only if $v$ is the $i$-th

vertex to be discovered during that $DFS$ run.

**Range-Labeling** of the edges of a $DFS$–$labeled$ tree: every edge $e = (f(v), v)$ (where $f(v)$

is the father of $v$ in the tree) is labeled by $R(e) = (min(e), max(e))$. This is the range of the

labels of all vertices that are descendants of $v$ in the tree (recall that $v$ is a descendant of

itself).

**Observation 3.4:** Let $T$ be a $DFS$–$labeled$ tree with range labeling and let

$e = (f(v), v) \in T$. Then

(i) For every vertex $u \in T$, if $min(e) \leq k(u) \leq max(e)$ then $u$ is a descendant of $v$.

(ii) If $min(e) \leq j \leq max(e)$ then there is a vertex $w \in T$ such that $k(w) = j$ and $w$ is a descen-

dant of $v$. $\square$

**Algorithm Range-Labeling**

**input:** A tree $T$ and a vertex $s \in T$.

**output:** $DFS$ labeling of $T$, where $s$ is labeled by 1, together with range labeling of the edges.

(1) Mark all the edges "unused"; $i := 0$; $v := s$;

(2) $i := i+1$; $k(v) := i$;

(3) if $v$ has no unused incident edges, go to Step (6);

(4) Choose an unused incident edge $e = (v, u)$; Mark $e$ "used";

   $f(u) := v$; $v := u$; $min(e) := i+1$;

(5) go to Step (2);

(6) if $k(v) = 1$ then Stop;

(7) $max(e = (f(v), v)) := i$; $v := f(v)$; go to Step (3);

{The above algorithm is taken from [Ev page 56] endowed with edges labeling by $min$ and $max$.}

**Algorithm Search**    (The algorithm searches for a root candidate).

**input:** A 2-connected graph $G = (V, E)$ where $|V| = n$; a spanning tree $T$ of $G$ that is

   $DFS$-$labeled$ by a bijection $k : V \rightarrow \{1, 2, ..., n\}$ and range-labeled; a leaf $s \in T$

   where $k(s) = 1$ and $T_s$ is a directed tree which does not have the property $PDFS$.

**output:** $r \in V$ (the only candidate for being the root of the $DFS$ tree).

(1) Mark all the edges "unused";

(2) $lower\_candidate := 2$; $upper\_candidate := n$; $v := s$;

   {$lower\_candidate$ and $upper\_candidate$ are the lower and

   upper bounds on the values of the possible candidate for a root.}

(3) while there is an unused non tree edge $e = (v, u)$ do

    begin

      mark $e$ "used";

      if $lower\_candidate < k(u) \leq upper\_candidate$ then update the following:

        begin

            $lower\_candidate := min(f(u), u)$ ; $upper\_candidate := max(f(u), u)$;

        end;

      {Clearly, in this case $e$ is a cross edge in $T_x$ for every vertex $x \in V$ such that before the

        update $lower\_candidate \leq k(x) \leq upper\_candidate$ and $x$ is not a descendant of

        $u$ in $T_s$.}

    end;

(4) let $e = (v, u)$ be the unused tree edge adjacent to $v$ {in this case, only one is left}.

    $v := u$; mark $e$ "used";

(5) if $d_T(v) = 2$ go to Step (3); { $d_T(v)$ is the degree of $v$ in the tree $T$.}

(6) if $d_T(v) = 1$ then output $v$ and Stop;

    {The leaf $v$ is the only candidate to become the root.}

(7) {in this case $d_T(v) > 2$} find an unused tree edge $e'$ emanating from $v$ in $T_s$ such that

    $min(e') \leq lower\_candidate$ and $max(e') \geq upper\_candidate$;

    {there is exactly one such edge.}

(8) while there is an unused tree edge $e = (v, u) \neq e'$ do

    begin

      mark $e$ "used";

      perform $SideSearch(u)$

    end;

(9) go to Step (3);

**SideSearch** $(s)$

(It is a subroutine of *Search*.)

(1) $v := s$;

(2) if $v$ has no unused incident edges, go to Step (6);

(3) Choose an unused incident edge $e = (v, u)$. Mark $e$ "used";

    if $e$ is a tree edge then { in this case $f(u) = v$ } $v := u$ and go to Step (2);

(4) { $e = (v, u)$ is a non tree edge}

    if *lower_candidate* $< k(u) \le$ *upper_candidate* then update the following:

        begin

            *lower_candidate* $:= min(f(u), u)$;   *upper_candidate* $:= max(f(u), u)$;

        end;

      {Clearly, in this case $e$ is a cross edge in $T_x$ for every vertex $x \in V$ such that before the

      update *lower_candidate* $\le k(x) \le$ *upper_candidate* and $x$ is not a descendant of

      $u$ in $T_s$.}

(5) go to Step (2);

(6) if $v = s$ then RETURN;

(7) $v := f(v)$ and go to Step (2);

**Lemma 3.5:** Algorithm *Search* is correct.

**Proof:** Since we have as an input a 2-connected graph it is obvious that $T$ is a *DFS* tree in $G$ only if there is a leaf $s'$ ($\ne s$) in $T$ such that $T_{s'}$ is a $T$–*DFS* in $G$. A vertex $x \in V$ is eliminated from being a root of $T$ only when there is a non tree edge $e \in E$ which is a cross edge in $T_x$ (and hence, by Proposition 2.3, $T_x$ is not a *DFS* tree in $G$).

In order to conclude the proof we have to justify the observation that in step (7) there is exactly one tree edge $e'$ emanating from $v$ in $T_s$ such that $min(e') \le$ *lower_candidate* and $max(e') \ge$ *upper_candidate*. At that moment we have scanned all the vertices in $G$ that are not descendants of $v$ in $T_s$. Since $v$ is not a separating vertex in $G$ there is at least one non tree edge $(a,b)$, $a \neq v$, such that $a$ is a descendant of $v$ in $T_s$ and $b$ is not a descendant of $v$ in $T_s$. The first such edge that was used (in Step (3) of the algorithm *Search* or in Step (4) of the subroutine *SideSearch* ) fixed a range between *lower_candidate* and *upper_candidate* that is contained in the range of the tree edge $v \to w \in T_s$ only if $w$ is an ancestor of $a$ in $T_s$. Since every new range between *lower_candidate* and *upper_candidate* is contained in the previous one, the existence and the uniqueness of $e'$ follows. $\square$

**Algorithm Find-Root**

The algorithm finds all the roots in a mixed tree. (The algorithm is used for finding all the roots in the oriented super structure.)

**input:** A tree $T$ where every edge in $T$ is either directed (denoted by $u \to v$ ) or bidirected

(undirected, denoted by $(u,v)$, represents two anti parallel edges).

**output:** All the roots of the tree (if any exists) or an answer that $T$ has no root.

(1) Mark all the edges "unused"; Choose a vertex $s \in V$; $i := 0$; $v := s$;

(2) $i := i+1$; $k(v) := i$;

(3) if $v$ has no unused incident edges, go to Step (6);

(4) Choose an unused incident edge $e = v \to u$ or $e = u \to v$ or $e = (v,u)$;

   Mark $e$ "used"; $f(u) := v$; $v := u$;

(5) go to Step (2);

(6) if $k(v) = 1$ $\{s$ is a root of $T\}$ then go to (9);

(7) if $f(v) \to v \in T$ or $(f(v), v) \in T$ then $v := f(v)$ and go to Step (3);

(8) $\{v \to f(v) \in T$ , $T$ has a root only if $v$ is a root of $T.\}$

   Check whether $v$ is a root of $T$ $\{$using *DFS* from $v$ $\}$;

   If not output " $T$ has no root " and Stop;

(9) $\{v$ is a root of $T\}$ Output all the vertices $u \in V$ such that there is a bidirected path from $u$

   to $v$ in $T$.

**Lemma 3.6:** Algorithm *Find–Root* is correct.

**Proof:** It is obvious that if $k(v) = 1$ in Step (6) then $s$ is a root because every directed edge in $T$ has the same orientation as in $T_s$. If the algorithm finds an edge $v \rightarrow f(v) \in T$ in Step (8) then it is clear that every root must be a descendant of $v$ in $T_s$. Being in this stage of the algorithm means that we have backtracked from all the descendants of $v$ in $T_s$ according to Step (7), which means that there is a feasible path in $T$ from $v$ to all of its descendants in $T_s$. Therefore, $T$ has a root only if $v$ is a root of $T$.

The correctness of Step (9) is obvious (there is a bidirected path between every two roots in the tree). $\square$

**Theorem 3.7:** Algorithm *DECIDE* is correct and has linear time complexity. $\square$

## 4. CONCLUSIONS

In this paper a complete characterization of *Total–DFS–Graphs* is presented. This characterization implies efficient recognition of *Total–DFS–Graphs* both by sequential and parallel algorithms.

Moreover, it implies that a large variety of graphs are not *Total–DFS–Graphs*. Therefore a linear time algorithm for deciding whether a given undirected spanning tree is a *DFS* tree in a given undirected graph $G = (V, E)$ is provided.

A linear algorithm for this latter decision problem in digraphs is given in [KO]. However, solving the analogous problem for mixed graphs by a linear algorithm is still open.

# ACKNOWLEDGMENT

THIS RESEARCH WAS PARTIALLY SUPPORTED BY TECHNION V.P.R. FUND No. 120-0683

# REFERENCES

[AA]   A. Aggarwal and R.J. Anderson, A random NC algorithm for depth-first search, *In Proc. of the 19th ACM Symposium on Theory of Computation*, 325-334, (1987). To appear in *Combinatorica*.

[Aw]   B. Awerbuch, A new distributed depth-first-search algorithm, *Information Processing Letters* , 20 (3), 147-150, (1985).

[BLW] N.L. Biggs , E.K. Lyod and R.J. Wilson, *Graph Theory 1736-1936*, Clarendon Press, Oxford, (1977).

[Ev]   S. Even, *Graph Algorithms*, Computer Science Press, Potomac, MD , (1979).

[HN]   T. Hagerup and M. Nowak, Recognition of Spanning Trees Defined by Graph Searches, Technical Report A 85/08, Universität des Saarlandes, Saarbrücken , West Germany, June 1985.

[HT a] J.E. Hopcroft and R.E. Tarjan, Dividing a graph into triconnected components, *SIAM Journal on Computing*, 2(3), (1973).

[HT b] J.E. Hopcroft and R.E. Tarjan, Efficient algorithms for graph manipulation, *Comm. ACM,* 16(6), 372-378, (1973).

[HT c] J.E. Hopcroft and R.E. Tarjan, Efficient planarity testing, *J. ACM,* 21, 549-568, (1974).

[KO]   E. Korach and Z. Ostfeld, On the Possibilities of DFS Tree Constructions: Sequential and Parallel Algorithms, Technical Report no. 508, CS Dept. Technion, May 1988.

[LMT] K.B. Lakshmanan, N. Meenakshi and K. Thulasiraman, A time-optimal message-efficient distributed algorithm for Depth-first-Search, *Information Processing Letters,* 25, 103-109, (1987).

[LP]   A. LaPaugh and C.H. Papadimitriou, The even-path problem for graphs and digraphs, *Networks*, 14, 507-513, (1984).

[Re]   J.H. Reif, Depth-first search is inherently sequential, *Information Processing Letters*, 20, 229-234, (1985).

[Sy]   M.M. Syslo, Series-parallel graphs and depth-first search trees, *IEEE Transactions on Circuits and Systems*, 31(12), 1029-1033, (1984).

[TV]   R.E. Tarjan and U. Vishkin, An efficient parallel biconnectivity algorithm, *SIAM Journal on Computing*, 14(4), 862-874, (1985).

[Ta]   R.E. Tarjan, Depth-first search and linear graph algorithms, *SIAM Journal on Computing*, 1, 146-160, (1972).

[Ti]   P. Tiwari An Efficient Parallel Algorithm for Shifting the Root of a Depth First Spanning Tree, *Journal of Algorithms*, 7, 105-119, (1986).

[Tu]   W.T. Tutte, Lectures on matroids, *J. Res. Nat. Bur. Stand.* 69B, 1-48, (1965).

# SERIALIZABLE GRAPHS

K. Vidyasankar
Department of Computer Science
Memorial University of Newfoundland
St. John's, Newfoundland
Canada A1C 5S7

## Abstract

A database system is a collection of data items, read or written by transactions in a possibly
rleaved fashion. An interleaved execution is assumed to be correct if the sequence of transaction
s, called *history*, is *serializable*, that is, the effect of the execution is equivalent to that of some
al execution of the transactions. We give a graph-theoretic analogue of serializable histories. We
le a new class of graphs, called *serializable graphs*, whose properties are such that (i) a serializable
·h can be associated with each serializable history, and this can be done for various notions of seri-
bility of histories and for serializability under various sets of constraints, and (ii) a serializable
ry, in fact a serial one, can be associated with each serializable graph. We use serializable graphs
haracterize in an intuitive manner serializable histories involving general multi-step transactions,
e some data items may be accessed by several read and write steps in an arbitrary fashion, and
e involving nested transactions. The main graph-theoretic properties used in these characteriza-
s are a directed cutset matching property and graph contraction.

## 1. Introduction

A database system consists of a set of data items, and a set of transactions each comprising of
ral partially ordered atomic steps that either read or write the value of a data item. The transac-
s execute their steps in a possibly interleaved fashion. The sequence of steps in an execution is
d a *history*. A *serial* history corresponds to a serial, that is, noninterleaving execution of the
sactions. A *serializable* history corresponds to any concurrent execution whose effect is equivalent
hat of some serial execution of the transactions. Only serializable executions are normally accepted
orrect executions.

The theory of serializability of histories has been extensively studied in the literature [2-5, 8-12].
 main direction in the study involves different notions of serializability which correspond to
rent interpretations of "equivalence" to serial histories. For example, in *view-serializability* [12]
values read by all the transactions and the final values of all the data items are the same as in
e serial history; in *state-serializability* [12] only the final values of the data items need be the same
in *S-serializability* [11] only the values read by the specified subset $S$ of the transactions need be
same.

Another main direction is the study of serializability under some constraints. Here the transac-
order in an equivalent serial history must satisfy certain constraints, which are usually derived
n some syntactic properties of the histories. For example, in *strict serializable class SSR* [8], if a
·saction $T_i$ finishes execution before a transaction $T_j$ starts, in the given history, then $T_i$ must

appear before $T_j$ in the equivalent serial order. Different constraints give rise to different subclass of serializable histories. Examples are the classes DSR, Q, 2PL and SSR of [8], and the classes W WR, RW, and WRW of [5, 10]. This study helps to identify subclasses of serializable histories th have polynomial membership test. The membership problem is NP-complete for the general class.

A third direction deals with different transaction models. Normally it is assumed that each da item is accessed by at most one read step and at most one write step in a transaction, and if bot these steps do occur then the read step precedes the write step in the transaction partial order. W call this the *simple* transaction model. In a *general multi-step* model, several read steps and likewi several write steps of a transaction may access the same data item, and these steps may be partial ordered arbitrarily. A *nested transaction* consists of several subtransactions; that is, instead of grou ing all the atomic steps into a single transaction, we group them into several subtransactions, and t subtransactions are grouped into higher level subtransactions, and so on. Most of the studies in t literature are confined to the simple model. Extensions to the other models are being discusse recently [1,3,6,7,9].

Several characterizations of serializability (usually view- or state-serializability, with no co straints and for the simple model) exist in the literature. Most of them construct a directed grap from the given history, for example, polygraph [8], TIO-graph [5] and transaction dag [9]. All the graphs contain vertices corresponding to transactions, and edges representing, in some form, t *reads-from relation* (an edge from $T_i$ to $T_j$ if transaction $T_j$ reads a data item $X$ written by tra saction $T_i$ in the given history) and perhaps a few more edges. In all these cases, serializable histori are characterized in terms of acyclicity and an additional property of the graph; this property different for each of the above graphs, depending on how the graph is defined.

A similar approach is followed in [10]. From a given history $h$, a 1-source 1-sink directed grap $H(h)$, called the *history graph* of $h$, is obtained. Here the reads-from relation is represented labelled edges, the label referring to the data item read. (An edge with label $X$ will be called an $X$ edge.) It is shown [10] that $h$ is serializable iff by adding a few unlabelled edges to $H(h)$ we ca obtain another graph, called *transaction precedence graph* $TP(h)$, such that (i) it is acyclic and (ii) satisfies the *exclusion property*, namely, no directed cutset has $X$-edges corresponding to two differe values of $X$ (written by two different transactions) at least one of which is useful. A value is *useful* it is read by at least one transaction in the history; otherwise it is *useless*. Thus the additional pr perty (along with acyclicity) in this characterization is the exclusion property which is a *a form matching property for directed cutsets.*

In this paper, we go a step further. The history graphs constructed above have several chara teristics that are not central to the serializability issue. We isolate the central properties and u them to define a new class of graphs, called *serializable graphs*. This definition uses only graph term nology, with no reference to database terms (except in the name of the graphs where the reference intentional). Then a history will be serializable iff its history graph is a serializable graph. In add tion, we show that with any serializable graph we can associate a serializable, in fact a serial, history

The serializable graph definition covers many common notions of serializability, each notion gi ing rise to different reads-from relation and hence different history graph, and allows for any set constraints, each adding a different set of unlabelled edges to the history graph. Thus only the hi tory graphs will be different in each case, but they all must be serializable graphs for the histories be serializable. The serializable graph definition allows also natural characterization of serializab histories involving general multi-step transactions, and those involving nested transactions. Here a s

graphs obtained from the history graph by graph contraction must be serializable for the history to
e serializable. Thus the serializable graph definition is helpful in characterizing a very large class of
rializable histories in an intuitive manner.

In section 2, we give the basic terminology and definitions. In section 3, we define history graphs
nd characterize serializable histories. Serializable graphs are defined in section 4. A property called
onsecutive L-serializability is discussed in section 5. This property is used in characterizing serializ-
ble histories involving general multi-step transactions in section 6, and those involving nested tran-
actions in section 7. Section 8 concludes the paper.

## 2. Basic Terminology and Definitions

A database system consists of a set $D$ of data items and a set $T = \{T_0, T_1, ..., T_n, T_f\}$ of tran-
actions. A *transaction* is a finite partially ordered set of steps. Each *step* is either a *read step* read-
g (exactly) one data item, or a *write step* writing (exactly) one data item. First we assume that each
ata item is accessed by at most one read step and at most one write step in a transaction, and if
oth these steps do occur, then the read step precedes the write step. We call this the *simple* transac-
on model. We consider other, general, models in sections 6 and 7. A transaction is a *write-only*
ansaction if it does not have any read steps. It is a *read-only* transaction if it does not have any
rite steps. The transaction $T_0$ is a fictitious write-only *initial transaction* which writes the initial
alues of all the data items, and $T_f$ is a fictitious read-only *final transaction* that reads the values of
l data items after all transactions have completed.

A read step (write step) of transaction $T_i$ reading (writing) the data item $X$ is denoted
$R_i[X](W_i[X])$. A set of read steps of $T_i$, unrelated by the partial order, reading a subset $C$ of $D$ and
curring together, is denoted $R_i[C]$. For example, $R_i[\{X,Y\}]$ denotes the unrelated read steps
$R_i[X]$ and $R_i[Y]$ occurring together in any order; for simplicity, we write this as $R_i[X,Y]$. Similar
otation is followed for the write steps. A *history* $h$ of $T$ is a sequence of the steps of $T$ representing
e execution of the transactions in a possibly interleaved fashion, starting with $W_0[D]$ and ending
ith $R_f[D]$. Note that the steps of each transaction in $h$ must satisfy the partial order. A history is
rial if there is no interleaving, that is once a transaction starts executing, it finishes without any
her transaction executing some step in between.

A transaction $T_j$ *reads X from* transaction $T_i$ in a history $h$ if $W_i[X]$ is the last write $X$ before
$_j[X]$ in $h$. The *reads-from* relation $rf$ of a history $h$ is defined as follows:

$$rf(h) = \{(T_i, X, T_j) : T_j \text{ reads } X \text{ from } T_i\}.$$

wo histories $h$ and $h'$ (on the same set of transactions $T$) are *equivalent* if $rf(h) = rf(h')$. A his-
ry $h$ is *serializable* if there exists a serial history $h'$ equivalent to $h$. This is the *view-serializability*
tion of [12].

## 3. Serializable Histories

Let $h$ be a history of $T = \{T_0, T_1, T_2, ..., T_n, T_f\}$. A write $X$ by transaction $T_i$ (also the value
at is written) is *useless* in $h$ if no transaction reads this value, that is, there is no $T_j$ for which
$_i, X, T_j)$ is in $rf(h)$; otherwise, it is *useful*.

*Definition 3.1.* The *history graph* of $h$, denoted $H(h)$, is a directed graph constructed as follow

(a) The vertex set of $H(h)$ is $T \cup T'$ where $T' = \{T'_{iX}:T_i$ has a useless write $X$ in $h\}$. T set $T'$ is a set of dummy transactions, one for each useless write in $h$.

(b) The edge set of $H(h)$ has the following:

    (i)   an edge labelled $X$ from $T_i$ to $T_j$, for each $(T_i, X, T_j)$ in $rf(h)$;

    (ii)  an edge labelled $X$ from $T_i$ to $T'_{iX}$ for each useless write $X$ of $T_i$, for all $T_i$;

    (iii) an unlabelled edge from each vertex in $T'$ to $T_f$;

    (iv) an unlabelled edge from each read-only transaction other than $T_f$ to $T_f$; and

    (v)  an unlabelled edge from $T_0$ to each write-only transaction other than $T_0$. $\square$

We use $T_i$ to refer to transaction $T_i$ and also the vertex $T_i$ in the graph. An edge $\alpha$ from to $T_j$ is denoted $(T_i, T_j)$. Here, $T_i$ is the *positive end* $p\alpha$ of $\alpha$, and $T_j$ is the *negative end* $n\alpha$ of We also refer to $\alpha$ as an *outdirected* edge of $T_i$ and an *indirected* edge of $T_j$. A *source* is a vert with no indirected edges, and a *sink* is a vertex with no outdirected edges. An *X-edge* refers to edge labelled $X$. The labelled edges incident to $T'$ are *useless*; all other labelled edges are *useful*.

*Example 3.1.* Let $h$ be the following history.
$$W_0[D]R_1[X]R_2[Y]W_1[Z]R_5[Z]W_5[X,Y]R_2[X,Z]$$
$$W_2[X,Z]R_4[X]W_3[X,Y]R_f[D]$$

The history graph $H(h)$ is given in Figure 1. The stars ($*$) indicate the useful edges. $\square$

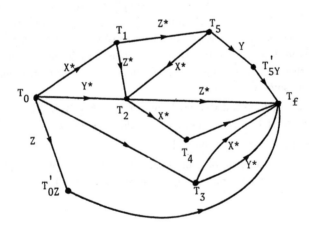

Figure 1

**Proposition 3.1**. *A history graph has the following properties.*

(a) *It is a 1-source 1-sink graph.*

(b) *All edges between vertices in* $(T-\{T_0, T_f\}) \cup T'$ *are labelled.*

(c) *All unlabelled edges either start from* $T_0$ *or end in* $T_f$ .

(d) *No two indirected edges of any vertex have the same label.*

(e) *If* $\alpha$ *is a useful X-edge for some X in D, then* $p\alpha$ *is not the positive end of any useless X-edge.*

(f) *A vertex that is the negative end of a useless edge is not the (negative or positive) end of any other (useful or useless) labelled edge.*

(g) *For every X in D, there is at least one X-edge starting from* $T_0$.

(h) *For every X in D, there is exactly one X-edge ending in* $T_f$ .

(i) *No useless edge is incident to* $T_f$ . □

It is clear that two histories $h$ and $h'$ of $T$ are equivalent iff $H(h) = H(h')$.

We now introduce some more graph terminology. The *coboundary* $\delta S$ of a subset $S$ of the vertex set of a graph $G$ is the set of edges each with one end in $S$ and the other not in $S$. By a *coboundary* of $G$, we refer to a set of edges that is the coboundary of some subset of the vertex set of $G$. A *cutset* is a minimal nonnull coboundary. (A set is *minimal* with a given property if it has that property but none of its proper subsets has the property.) An edge $\alpha$ of a coboundary $\delta S$ is *outdirected* if $p\alpha$ is in $S$; it is *indirected* if $n\alpha$ is in $S$. The set of outdirected edges of $\delta S$ is denoted $\delta_{out} S$, and the set of indirected edges $\delta_{in} S$. Now $\delta S$ is *outdirected* if $\delta_{in} S$ is null, it is *indirected* if $\delta_{out} S$ is null; in either case, it is *directed*. A coboundary is *directed* if it can be expressed as a directed coboundary of some subset of the vertex set. A *directed cutset* is a directed coboundary that is a cutset.

*Definition 3.2.* A *Transaction Precedence graph* of $h$, denoted $TP(h)$, is a graph having the following properties.

(a) The vertex set of $TP(h)$ is the same as the vertex set of $H(h)$.

(b) The edge set of $TP(h)$ includes the edge set of $H(h)$, and perhaps a few additional unlabelled edges. (These unlabelled edges may start and end at any vertices.)

(c) *Exclusion property.* No directed cutset in $TP(h)$ that contains a useful X-edge $\alpha$ contains any other (useful or useless) X-edge $\beta$ with $p\beta \neq p\alpha$, for any X in D. □

Note that $TP(h)$ can be obtained from $H(h)$ by repeatedly applying the following construction: there is a directed cutset with a useful X-edge $\alpha$ and another X-edge $\beta$, with $p\beta$ different from $p\alpha$, then add an unlabelled edge either from $n\alpha$ to $p\beta$, or from $n\beta$ to $p\alpha$, thus making that cutset indirected.

*Example 3.2.* For the history $h$ in Example 3.1, Figure 2 shows a $TP(h)$ graph obtained by adding the broken lines to $H(h)$. □

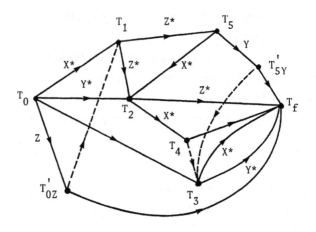

Figure 2

The following theorem has been proved in [10].

**Theorem 3.2.** *A history h is serializable iff there exists an acyclic TP(h).* □

## 4. Serializable Graphs

In this section, we identify the graph properties underlying the problem of serializability of h
tories, and use them to define a new class of graphs called *serializable graphs*. Essentially, we wa
the definition such that a history $h$ is serializable iff the history graph $H(h)$ is a serializable grap
Several properties of history graphs are stated in Proposition 3.1. These and the crucial property th
there must exist an acyclic $TP(h)$ include all the essential properties of serializable graphs. But th
also contain some nonessential properties that unnecessarily restrict the domain of serializable grapl
These nonessential properties fall into two classes: those whose removal enables covering many oth
notions of serializability of histories; and those whose removal does not affect the serializability of h
tories.

The serializability notion considered in section 3 is *view-serializability* [12]. Here a history $h$
equivalent to a serial history $h'$ in the sense that the values read by all the transactions in $h$ are t
same as in $h'$, and the final values in the database after the execution of $h$ are the same as those aft
$h'$. The latter condition can be stated as the values read by $T_f$ are the same in both $h$ and $h'$. 
weakening this equivalence criterion, several other notions of serializability can be defined. In *stat*
*serializability* of [12], the equivalence is only with respect to the values read by $T_f$. In
*serializability* of [3], the equivalence is with respect to the values read by the transactions, excludi
$T_f$. These notions are generalized in [11] to $S$-serializability: for a specified subset $S$
$T = \{T_0, T_1, ..., T_n, T_f\}$, a history $h$ is $S$-*serializable* if all the transactions in $S$ read the same valu
in $h$ as in some serial history. Given a history $h$ and a subset $S$ of $T$, a graph $H(h,S)$ is defined

.1]. This graph differs from $H(h)$ in that the reads-from-relation is with respect to only those reads that are useful to $S$. Then, for $S$ equal to $T$, $H(h,T)$ is the same as $H(h)$. The graphs $H(h,S)$ also satisfy all the properties of Proposition 3.1 except (h) which we will show later as nonessential. a that paper, a graph $TP(h,S)$ is also defined the same way as $TP(h)$, and it is shown that $h$ is $S$-erializable iff there exists an acyclic $TP(h,S)$. Thus several notions of serializability are already overed in the properties of Proposition 3.1. We show below that some further notions can be overed.

It is known that testing whether a history is serializable is NP-complete. With the intention of etting subclasses of serializable histories having polynomial membership test, histories which are seri-lizable under some constraints $c$ have been studied -- for example, the class $DSR$, $Q$, $2PL$ and $SSR$ in ] and the classes $WW$, $WR$, $RW$, $WRW$ and $RR$ in [5]. The constraints are usually derived from ome syntactic properties of histories. (For example, in the *strict serializable class SSR* a transaction $;$ that finishes execution before a transaction $T_j$ starts must appear before $T_j$ in the equivalent erial order). These constraints give rise to some precedence relation among transactions. This rela-on is added, in the form of unlabelled edges between vertices in $T$, to $H(h)$ to get $H[c](h)$, and we ook for the existence of an acyclic $TP[c](h)$, which is defined the same way as $TP(h)$.

Now we want serializable graphs to include $H[c](h)$ graphs for which there exist acyclic $TP[c](h)$ graphs. Hence we allow unlabelled edges between any vertices of $T$. In fact, we will allow hem between any vertices of $T \cup T'$, and discard properties (b) and (c) of Proposition 3.1.

In another notion of serializability discussed in [11] (which is $((T - \{T_f\})+C)$-serializability in hat paper), the equivalence of histories is with respect to the values read by all the transactions and he final values for just the subset $C$ of data items, instead of the entire set of data items $D$. This mounts to substituting the read steps $R_f[C]$ in place of $R_f[D]$ for $T_f$, and considering the history raph for the resulting history. Hence $T_f$ will be incident to $X$-edges only for some elements $X$ in . We want these graphs also to be covered in the serializable graph definition. Hence we discard he property (h) of Proposition 3.1. Note that property (d) ensures that no two indirected edges of $'_f$ have the same label.

Now there are two properties whose removal does not affect the serializability of histories. First e note that the useless writes of $T_0$ need not be represented in history graphs. Thus we need not ave $X$-edges starting from $T_0$ for all $X$ in $D$, and the property (g) can be discarded. That is, $T_0$ eed only write the data items whose values are read by some transactions. Second, by the definition f the history graphs, for each data item $X$ written by transaction $T_i$ there is at least one $X$-edge out-irected from $T_i$. We can show easily that an extra useless $X$-edge starting from $T_i$ can be added ithout affecting serializability. In fact we can add any number of useless $X$-edges starting from $T_i$ or each write $X$ of $T_i$. This can be done whether the write $X$ of $T_i$ is useful or useless. Hence, pro-erty (e) can be discarded.

We are now ready to define serializable graphs. Here the marked edge set would correspond to e useful edge set.

*Definition 4.1.* A *serializable graph* $G(V;E,L,M,I)$ is a finite directed graph with vertex set $V$, dge set $E$, a subset $L$ of $E$ called *labelled edge set* whose labels are from $I$, and a subset $M$ of $L$ called *arked edge set*, satisfying the following properties.

(a) It is a 1-source 1-sink graph.

(b) Each edge is assigned at most one label. If the label is $X$, it is called $X$-edge.

(c) A vertex that is the negative end of an unmarked labelled edge is not incident to any other (marked or unmarked) labelled edge.

(d) No unmarked labelled edge is incident to the sink vertex.

(e) No two indirected edges of any vertex have the same label.

(f) There exists a (possibly empty) set of unlabelled edges whose addition to $G$ yields a graph $G'$ which is acyclic and which satisfies the following property for each $X$ in $I$: no direct cutset in $G'$ that contains a marked $X$-edge $\alpha$ contains any other (marked or unmarked) edge $\beta$ with $p\,\beta \neq p\,\alpha$. (The graph $G'$ is called a *full serializable graph* of $G$.) □

Now a history $h$ is serializable iff $H(h)$ is a serializable graph. This property holds even for serializability satisfying some constraints. We now show that each serializable graph "corresponds" some serializable history. In fact, we will show the correspondence to a serial history itself. First allow $T_f$ to read an arbitrary subset of $D$. Now we define a generalized history graph.

For a history $h$, an *expanded history graph*, denoted $EH(h)$, is obtained the same way as $H(h)$ except that (i) any (finite) number of useless $X$-edges starting from $T_i$ can be added for each write of $T_i$, for each $i$, and (ii) the $X$-edges for the useless writes of $T_0$ need not be represented. An $ETP(h)$ is a graph that contains $EH(h)$ and satisfies the exclusion property.

Note that whereas $H(h)$ is unique for each $h$, there exist several $EH(h)$ for the same $h$.

**Theorem 4.1.** *Let $G(V;E,L,M,I)$ be a serializable graph. Then there exists a set $D$ of data items, a set $T = \{T_0, T_1, ..., T_n, T_f\}$ of transactions and a serial history $h$ of $T$ such that the transitive closure of a full serializable graph of $G$ is an acyclic $ETP(h)$.*

*Proof.* First let $D$ equal $I$, $T_0$ the source and $T_f$ the sink of $G$. By property (d) of the definition of serializable graphs, no unmarked labelled edges are incident to $T_f$. By (e), no two edges incident to $T_f$ have the same label. For each $X$-edge incident to $T_f$, include $R_f[X]$ in transaction $T_f$. The vertices which are not the negative ends of unmarked labelled edges are $T_i$ for different. Let transaction $T_i$ include a read step $R_i[X]$ if there is an $X$-edge indirected at $T_i$, and a write step $W_i[X]$ if there is an $X$-edge outdirected at $T_i$. Transaction $T_0$ has a write step $W_0[X]$ for each read or written by some transaction $T_i$. Let the set $T$ contain all the transactions defined so far.

Suppose $\beta$ is an unmarked labelled edge, say, $X$-edge. Then $p\beta$ is not the negative end of an unmarked labelled edge, by Definition 4.1(c). Therefore $p\beta$ is $T_i$ for some $i$ (including 0). Call it $T'_{iX}$. If there are several unmarked $X$-edges starting from $T_i$, call their negative ends by appropriate different names.

Let $G'$ be a full serializable graph of $G$. Take any topological sort of $G'$ and consider the order which the transactions in $T$ appear in the sort. Let $h$ be the corresponding serial history. Construct an $EH(h)$ $H'$ such that for each $X$ written by $T_i$, $H'$ contains exactly as many useless $X$-edges as does for each $i$ between 0 and $n$. It is clear that the marked edges of $G$ correspond to the useful edges of $H'$. Now $G$ and $H'$ have the same labelled edges. They may differ in unlabelled edges. Clearly, the transitive closure of the full serializable graph $G'$ is an acyclic $ETP(h)$ (except for differing labels dummy transactions). □

## 5. Consecutive L-serializability

Let $T$ be $\{T_0, T_1, ..., T_n, T_f\}$ where $T_f$ may read some subset of $D$, and $h$ a history of $T$. For a subset $L$ of $T$, we say that $h$ is *consecutive L-serializable*, denoted *[L]-serializable*, if there is a serial history $h'$ equivalent to $h$ such that the steps of $L$ occur consecutively in $h'$.

*Example 5.1.* The history in Example 3.1 is not $[\{T_1, T_4\}]$-serializable, whereas it is $[\{T_2, T_5\}]$-serializable. □

In this section, we derive necessary and sufficient conditions for $[L]$-serializability.

*Definition 5.1.* Let $h$ be a history of $T$. For a subset $L$ of $T$, the *history graph of $h$ for $L$*, denoted $H_L(h)$, is constructed as follows.

(a) The vertex set of $H_L(h)$ contains $L$, a special vertex $L_0$ if $T_0$ is not in $L$ (otherwise, $T_0$ itself is called $L_0$), and a special vertex $L_f$ if $T_f$ is not in $L$ (otherwise, $T_f$ itself is called $L_f$).

(b) The edge set of $H_L(h)$ contains all the edges of $H(h)$ at least one end of which is in $L$. Their incidence relationship with the vertices in $L$, labels and useful or useless status are the same as in $H(h)$.

(c) The edges of $\delta_{in} L$ in $H(h)$ have $L_0$ as their positive end in $H_L(h)$. The unlabelled, and useful labelled edges of $\delta_{out} L$ in $H(h)$ have $L_f$ as their negative end. If multiple edges with same label occur, they are replaced by a single edge with that label; note that all these edges will be useful.

(d) Each useless labelled edge of $\delta_{out} L$ in $H(h)$ is indirected at a new vertex in $H_L(h)$, and an unlabelled edge is added from that vertex to $L_f$.

(e) There are no other edges or vertices in $H_L(h)$. □

*Definition 5.2.* Let $h$ be a history of $T$. For a subset $L$ of $T$, the *L-contracted history graph of $h$*, denoted $H_{[L]}(h)$ is constructed as follows.

(a) It contains a new vertex $T_L$ in place of the vertex set $L$, and all the remaining vertices of $H(h)$.

(b) It contains all the edges of $H(h)$ which have at least one end not in $L$. All the edges of $\delta_{in} L$ end in $T_L$, and the edges in $\delta_{out} L$ start from $T_L$. All other incidence relationships of the edges are the same as in $H(h)$; their labels and useful or useless status are also the same. If multiple edges with same label occur, they are replaced by a single edge with that label; note that all these edges will be useful. □

*Example 5.2.* For the history $h$ in Example 5.1, for $L = \{T_2, T_5\}$, $H_L(h)$ is given in Figure 3, and $H_{[L]}(h)$ in Figure 4. □

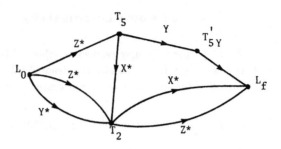

Figure 3

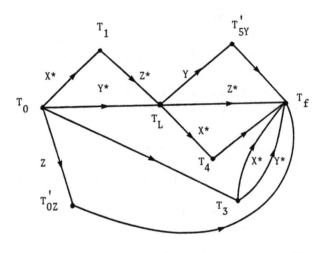

Figure 4

**Theorem 5.1.** *A history h is [L]-serializable iff both $H_L(h)$ and $H_{[L]}(h)$ are serializable grap* *(treating useful edges as marked edges).*

*Proof.* First suppose $h$ is [L]-serializable. Then there is a serial history $h'$ equivalent to $h$ su that the steps of $L$ occur consecutively in $h'$. Clearly $H(h')$ equals $H(h)$. Let $t = ((PT),(L),(ST$ correspond to the transaction order in $h'$. That is, the transactions in $PT$ occur before $L$, and the in $ST$ occur after $L$. For each set $A$, $(A)$ denotes the sequence in which the members of $A$ occur in We note that if $T_0$ is in $L$, it occurs first in $(L)$ and $PT$ is empty, and if $T_f$ is in $L$, it occurs last $(L)$ and $ST$ is empty.

First we show that the set $L$ can be considered as one "super-transaction", say $T_L$, which is a a simple transaction. The read steps of $T_L$ will be the read steps of transactions in $L$ that read fr transactions in $PT$. If several read steps read the same data item $X$, they must read $X$ from the sa transaction in $PT$ since $h'$ is serial; hence $T_L$ will contain just one read $X$ step. The write steps $T_L$ will consist of one step for each data item written by transactions in $L$, even if several of the write the same data item. Let all the read steps precede all the write steps. Now consider the transaction order $((PT), T_L, (ST))$ and let $h'_{[L]}$ be the corresponding serial history. It is straightforward to verify that $H(h'_{[L]})$ equals $H_{[L]}(h)$, thus the latter is a serializable graph.

Now let $t_L$ be the order $(L_0, (L), L_f)$, where $L_0$ contains a write $X$ if some $L_j$ reads $X$ from some transaction in $PT$ in $h'$, and $L_f$ contains a read $X$ if some transaction in $ST$ reads $X$ from some transaction in $L$ in $h'$. (Again, if $PT$ is empty then $L_0$ is omitted, and if $ST$ is empty then $L_f$ is omitted from $t_L$.) Let $h'_L$ be the serial history corresponding to $t_L$. It is easy to verify that $H(h'_L)$ equals $H_L(h)$, and hence the latter is serializable.

For the converse, assume both $H_L(h)$ and $H_{[L]}(h)$ are serializable graphs. Let $((PT), T_L, (ST))$ denote the transaction order in a topological sort of $H_{[L]}(h)$ omitting the dummy transactions, and $h'_{[L]}$ the corresponding serial history. Clearly $H(h'_{[L]})$ equals $H_{[L]}(h)$. Let $(L_0, (L), L_f)$ be the transaction order in a topological sort of $H_L(h)$, omitting the dummy transactions again, and let $h'_L$ be the corresponding serial history. Then $H(h'_L)$ equals $H_L(h)$. Now let $t$ be $((PT), (L), (ST))$ and let $h'$ be the serial history corresponding to $t$. We will show that $h'$ is equivalent to $h$, that is $H(h')$ is the same as $H(h)$.

Let us consider an $X$-edge $(T_a, T_b)$ for transactions $T_a$ and $T_b$. If $T_a$ and $T_b$ are both in $PT$ or both in $ST$, $(T_a, T_b)$ is in $H(h')$ iff it is in $H(h'_{[L]})$ which is $H_{[L]}(h)$, iff it is in $H(h)$. If $T_a$ and $T_b$ are both in $L$, then $(T_a, T_b)$ is in $H(h')$ iff it is in $H(h'_L)$, which is $H_L(h)$ iff it is in $H(h)$. If $T_a$ is in $PT$ and $T_b$ is in $L$, then $(T_a, T_b)$ is in $H(h')$ iff $(T_a, T_L)$ is in $H(h'_{[L]})$, which is $H_{[L]}(h)$, and $(L_0, T_b)$ is in $H(h'_L)$, which is $H_L(h)$. Since $H_{[L]}(h)$ is serializable, $(T_a, T_L)$ is the only $X$-edge incident to $T_L$. Hence the above is possible iff $(T_a, T_b)$ is in $H(h)$. The only remaining case is where $T_a$ is in $L$ and $T_b$ is in $ST$. Then $(T_a, T_b)$ is in $H(h')$ iff $(T_a, L_f)$ is in $H(h'_L)$ which equals $H_L(h)$ and $(T_L, T_b)$ is in $H(h'_{[L]})$ which equals $H_{[L]}(h)$. This is possible iff $(T_a, T_b)$ is in $H(h)$, since $(T_a, L_f)$ is the only $X$-edge incident to $L_f$ in $H_L(h)$. □

Now we consider a straightforward generalization of the above result. Let $T$ be $T_0, T_1, ..., T_n, T_f\}$ and $h$ a history of $T$. For subsets $L_1, L_2, ..., L_k$ of $T$, $k \geq 1$, we say that $h$ is $\{L_1, L_2, ..., L_k\}$-*serializable* if there is a serial history $h'$ equivalent to $h$ such that the steps of $L_i$, for each $i$ between 1 and $k$, occur consecutively in $h'$. For simplicity, we drop the curly brackets in the set notation, in the following.

When the subsets $L_i$ are mutually disjoint, we can define $H_{L_i}(h)$ for each set $L_i$ as before, and define $H_{[L_1, L_2, ..., L_k]}(h)$ as the graph obtained from $h$ by contracting each set $L_i$ to a single vertex $T_{L_i}$, as before. We now have the following theorem.

**Theorem 5.2.** *Let $h$ be a history of $T$ and $L_1, L_2, ..., L_k$ mutually disjoint subsets of $T$. Then $h$ is $[L_1, L_2, ..., L_k]$-serializable iff $H_{L_i}(h)$ for each $i$ between 1 and $k$, and $H_{[L_1, L_2, ..., L_k]}(h)$ are serializable graphs.* □

We now consider another generalization which is useful in section 7.

**Theorem 5.3.** *Let $h$ be a history of $T$, and $A, B_1, B_2, ..., B_k$, for $k \geq 1$, be such that $A \subseteq T, B_i \subseteq A$ for each $i$, and $B_i \cap B_j = \phi$ for $i \neq j$. Then $h$ is $[A, B_1, B_2, ..., B_k]$- serializable iff $H_{[A]}(h)$, $H_{B_i}(h)$ for each $i$, and $H_{A[B_1, ..., B_k]}(h)$ are all serializable graphs, where the last graph is obtained by contracting each set $B_i$ to a single vertex in $H_A(h)$.* □

## 6. General Multi-step Transactions

In this section, we consider general multi-step transactions. Formally, a transaction $T_i$ is a fin[i] partially ordered set of steps, where each step $t_{ij}$ is either a read step or a write step accessing so[me] data item $X$. Several read steps and likewise several write steps may access the same data item. T[he] partial order among these steps may be quite arbitrary. The value written in a write step is an un[in]terpreted function of the "most recent" values read in the read steps that precede this write step [in] the partial order. We assume that for each data item $X$ read by $T_i$, a *private copy* $X_i$ is created (f[ol]lowing the idea of [9]), and a read $X$ step $t_{ij}$ consists of two "substeps" $R_{ij}[X]$ and $W_{ij}[X_i]$ occur[ing] together in that order. That is, each read step updates $X_i$ to the value just read. A write $X$ step [i] following the read steps that read data items $X$, $Y$ and $Z$ in the partial order is assumed to consist [of] the substeps $R_{ik}[X_i,Y_i,Z_i]$ and $W_{ik}[X]$ in that order. Thus each step of $T_i$ is considered as a simp[le] "subtransaction".

Now an execution of a transaction is any serial execution of its subtransactions, preserving t[he] partial order among them. Two different executions of the same transaction may give differe[nt] results, even when no other transaction is executing concurrently. For example, suppose the steps a[re] $R[X]$, $R[Y]$ and $W[X]$, and the partial order specifies only that $R[Y]$ precedes $W[X]$, leavi[ng] $R[X]$ and $W[X]$ unordered. Then $R[X]$ may be executed either before or after $W[X]$, and the tw[o] executions will read different values for $X$. Such transactions have been called *ambiguous* in [3]. [We] formalize this notion as below.

With each transaction $T_i$, we will associate a fictitious initial subtransaction $t_{i0}$ that writes the data items read by the steps of $T_i$, and a fictitious final subtransaction $t_{if}$ that reads all the da[ta] items written by the steps of $T_i$. We define a *history* of $T_i$, called $h_i$, as a sequence of the steps [of] the subtransactions of $T_i$ preserving the partial order among them, starting with all the steps of [i] and ending with all the steps of $t_{if}$. A *serial* history is one in which the steps of subtransactio[ns] occur serially. Since we insist that all the steps of all the subtransactions must occur together in a[n] execution of $T_i$, we will be interested only in serial histories of $T_i$.

*Definition 6.1* [3]. A transaction $T_i$ is *unambiguous* if in the execution of any serial history [of] $T_i$, the values read by all the subtransactions including $t_{if}$ are identical; otherwise, $T_i$ is *ambiguo[us]*. □

We do not assume that the subtransactions, even when executed alone, preserve the consisten[cy] of the database. On the other hand, a transaction when executed alone is supposed to preserve co[n]sistency [4]. We formalize the consistency notion by the following correctness criterion.

*Definition 6.2.* Any serial execution of the subtransactions of $T_i$ starting with $t_{i0}$ and endi[ng] with $t_{if}$ is *correct*. □

Therefore a transaction need not be unambiguous to be consistent. We now address the serializability of histories involving general multi-step transactions.

Let $h$ be a history of a set $T = \{T_0, T_1,..., T_n, T_f\}$ of transactions. Let $ST_i$ denote the set of subtransactions of $T_i$, for $i$ between 1 and $n$, excluding the fictitious subtransactions. Let $ST$ denote $\{T_0, T_f\} \cup ST_1 \cup \cdots \cup ST_n$. Now we can treat $h$ of $T$ as a history $Sh$ of $ST$, and a history graph $(Sh)$ can be constructed representing each member of $ST$ by a different vertex. The graph $H(Sh)$ will also contain unlabelled edges corresponding to the partial orders among the subtransactions of each $T_i$. From Theorem 5.2, we have the following.

**Theorem 6.1**. *A history $h$ of $T$ is serializable iff the history $Sh$ of $ST$ is $[ST_1, ST_2,..., ST_n]$-serializable, that is, iff $H_{ST_i}(Sh)$, for each $i$ between 1 and $n$, and $H_{[ST_1, ST_2, \ldots, ST_n]}(Sh)$ are serializable graphs.* □

We note that since we insist that all the steps of subtransactions be executed consecutively, $H_{ST_i}(Sh)$ will be a serializable graph iff the vertex $L_f$ (where $L = ST_i$) is incident to at most one $X$-edge for any $X$.

## 7. Nested Transactions

Our model for general multi-step transactions in the previous section and Theorem 5.1 on consecutive $L$-serializability allow us to characterize serializable histories involving nested transactions also, in a natural way. Nested transactions have been dealt with by several authors [1,3,6,7]. We define nested transactions as $m$-transactions, as in [3].

*Definition 7.1.* For any $m \geq 1$ an *m-transaction* $T = (A, \leq)$ is defined inductively as follows.

*Basis: $m = 1$.* A *1-transaction* $T = (A, \leq)$ is a simple transaction with a finite set $A$ of atomic steps (that is, read and write steps) and partial order $\leq$ on $A$.

A *T-history* is any sequence of the steps of $T$ obeying the partial order. Any $T$-history of a 1-transaction is *serial*.

*Induction Step:* $m \geq 1$. An *(m+1)-transaction* $T = (A, \leq)$ is a finite set $A$ of $m$-transactions $\{T_1,..., T_n\}$ together with an arbitrary partial order $\leq$ on $A$.

A *T-history* $h$ is any sequence of the atomic steps of $T$ such that

(a)  the restriction of $h$ to the steps of $T_i$ is a $T_i$-history, for each $i$ between 1 and $n$; and

(b)  if $T_i \leq T_j$ then all the steps of $T_i$ occur before all the steps of $T_j$ in $h$.

A *T-history* is *serial* if it is a concatenation of serial $T_i$-histories in some order. □

A *history* of $T = \{T_0, T_1,..., T_n, T_f\}$ is a sequence of the steps of $T_0$ (in any order), followed by a $T'$-history, where $T' = (\{T_1,..., T_n\}, \leq)$, followed by the steps of $T_f$ (in any order). Again, a history is *serializable* if it is equivalent to a serial history as before.

We now characterize the serializability of histories. First we develop some notation. Let $T_i$ an $m$-transaction. Each $k$-transaction, for $k$ between 1 and $m$, that is used in the definition of $T_i$ is *subtransaction* of $T_i$. The set of subtransactions of $T_i$ is denoted $\mathrm{sub}\,T_i$. The set of 1-transactio used in the definition of $T_i$ is denoted $ST_i$. We use $\mathrm{sub}ST_i$ to denote the collection of the sets $S$ for $T_a$ in $\mathrm{sub}\,T_i$.

Let $h$ be a history of a set $T = \{T_0, T_1, ..., T_n, T_f\}$ of transactions. Let $ST$ den $\{T_0, T_f\} \cup ST_1 \cup \cdots \cup ST_n$. (Note that $T_0$ and $T_f$ are 1-transactions.) Now we can treat $h$ $T$ as a history $Sh$ of $ST$, and a history graph $H(Sh)$ can be constructed representing each member $ST$ by a different vertex. The graph $H(Sh)$ will also contain unlabelled edges corresponding to t partial orders among the subtransactions. Now the following theorem can be proved in a straightfo ward manner using Theorem 5.3.

**Theorem 7.1.** *A history $h$ of $T$ is serializable iff the history $Sh$ of $ST$ is $[\mathrm{sub}ST]$-serializable, t is, iff for any sets $A, B_1, B_2, ..., B_k$, for $k \geq 1$, in $\mathrm{sub}ST$ such that $B_i \subseteq A$ for each $i$ and $B_i \cap B_j = \phi$ $i \neq j$, $Sh$ is $[A, B_1, B_2, ..., B_k]$ -serializable.* $\square$

## 8. Discussion

We have defined a new class of graphs, called serializable graphs, with properties relating those of serializable histories. The database problem of whether a history is serializable is n transformed into a graph-theoretic problem of whether the corresponding history graph is serializab This holds for various notions of serializability of histories, and for serializability under various sets constraints -- only the history graph will differ in each case. With general multi-step and nested tra saction models, the graph-theoretic problem is whether a set of graphs obtained from the histo graph by graph contraction are serializable. Thus the serializable graph definition is helpful characterizing a very large class of serializable histories in a natural and intuitive manner. We co sider this as the main advantage over the other characterizations of serializable histories in the lite ture. We have shown also that with each serializable graph, we can associate a serializable history, fact, a serial history itself.

Various characterizations of serializable histories exist in the literature. Those of Sethi [9] a Papadimitriou [8] are very closely related to ours. Sethi constructs a graph, called transaction d $D(h)$, from a history $h$, and asserts that $h$ is serializable iff $D(h)$ has a one-pebbling. The gra $D(h)$ is vertex-oriented, that is, it contains "a vertex for each instantaneous value taken by an it in the database", whereas the history graph is edge-oriented, containing an edge for each value. T one-pebbling of $D(h)$ corresponds to $H(h)$ being serializable. Sethi also deals with general multi-st transactions. Our treatment is very much the same as his. Our model is the "edge-version" of model, and our characterization Theorem 6.1 is the "serializable graph version" of his Theorem [9].

We have defined serializable graphs using only graph terminology, with no reference to datab terms. Papadimitriou [8] defines polygraphs the same way. A history $h$ is serializable iff the po graph $P(h)$ that corresponds to $h$ is acyclic. Thus $P(h)$ being acyclic corresponds to $H(h)$ bei serializable. One difference between these two classes of graphs is that polygraphs have unlabell edges, whereas serializable graphs have labelled edges. The edge labels help to identify the exclusi property (in the definition of $TP(h)$) which is fundamental to the definition of serializable graphs.

**Acknowledgement**: This research is supported in part by the Natural Sciences and Engineering Research Council of Canada Individual Operating Grant A-3182.

# References

. Beeri, C., Bernstein, P.A., Goodman, N., Lai, M.Y., and Shasha, D.E. A concurrency control theory for nested transactions. *Proc. 2nd ACM SIGACT-SIGOPS Symp. on Principles of Distributed Computing*, August 1983.

. Bernstein, P.A., Shipman, D.W., and Wong, W.S. Formal aspects of serializability in database concurrency control, *IEEE Trans. Software Eng. SE-5*, 3 (May 1979), 203-215.

. Brzozowski, J.A. On models of transactions. Technical Report #84001, Department of Applied Mathematics and Physics, Kyoto University, Japan, April 1984.

. Eswaran, K.P., Gray, J.N., Lorie, R.A. and Traiger, I.L. The notions of consistency and predicate locks in a database system. *Comm. ACM 19*, 11 (Nov. 1976), 624-633.

. Ibaraki, T., Kameda, T., and Minoura, T. Serializability with constraints, *ACM TODS 12*, 3(Sept. 1987), 429-452.

. Lynch, N.A. Concurrency control for resilient nested transactions. *Proc. 2nd ACM SIGACT-SIGMOD Symp. on Principles of Database Systems*, March 1983, 166-181.

. Moss, T.E.B. Nested transactions: an approach to reliable distributed computing. Ph.D. Thesis, Technical Report MIT/LCS/TR-260, MIT Laboratory for Computer Science, Cambridge, MA. 1981.

. Papadimitriou, C.H. The serializability of concurrent database updates. *J. ACM 26*, 4(Oct. 1979), 631-653.

. Sethi, R. A model of concurrent database transactions. *Proc. 22nd IEEE Symp. Foundation of Comp. Sci.*, Oct. 1981, 175-184.

0. Vidyasankar, K. A simple characterization of database serializability. *Proc. 5th Conf. on Foundations of Software Technology and Theoretical Computer Science, India*, Lecture Notes in Computer Science 206, Berlin, Heidelberg, New York: Springer 1985, 329-345.

1. Vidyasankar, K. Generalized theory of serializability. *Acta Informatica 24*, 1(Feb. 1987), 105-119.

2. Yannakakis, M. Serializability by locking. *J. ACM 31*, 2(April 1984), 227-244.

# TRANSITIVE CLOSURE ALGORITHMS FOR
# VERY LARGE DATABASES*

Joachim Biskup, Holger Stiefeling
Institut fuer Informatik
Hochschule Hildesheim
Samelsonplatz 1
D - 3200 Hildesheim
West-Germany

*Abstract:* Nontraditional applications of database systems require the efficient evaluation of recursive queries. The transitive closure of a binary relation has been identified as an important and frequently occurring special case. Traditional algorithms for computing the transitive closure, as developed in the field of algorithmic graph theory, hold both the operand relation and the result relation within directly addressable main memory. The newly anticipated applications, however, deal with very large relations that do not fit into main memory and therefore must be blockwise paged to and from secondary storage. Thus we have to design algorithms and optimization methods for computing the transitive closure of very large relations. We survey and compare various such algorithms and methods in a unifying manner. In particular we identify eight basic strategies to generate and to refine transitive closure algorithms: algebraic manipulation, implementation of the join operator, reusage of newly generated tuples, enforcement of some ordering of tuples, blocking of adjacency lists, tuning and preprocessing, taking advantage of topological order, and selection of an access structure for adjacency lists. The analysis demonstrates the great variety of options on the different description levels and how they are compatible. Based on experiments some specific algorithms are recommended.

*) Research was done as a part of the ESPRIT-Project 311 "Advanced Data and Knowledge Management System", supported by the European Community under a subcontract with Nixdorf Computer Corporation.

# 1 Introduction

Recently it has become obvious that the facilities of todays database
systems are too restricted in many cases. Non-traditional application
areas as PROLOG, expert systems, and knowledgebases require the
efficient evaluation of recursive queries ([Ag 87], [BiRaSt 87], [Da
87], [JaAgNe 87], [Ro 86], [VaBo 86]). In this context a special
recursive query, the transitive closure of a binary relation, has
attracted a lot of attention ([AgJa 87], [Io 86], [Lu 87], [LuMiRi 87],
[VaBo 86]). It is easy to define, and many general recursive queries
can be expressed in a language containing the operators of the
relational algebra plus an operator for the transitive closure.

On the one side evaluation procedures have been developed on the level
of the relational algebra. These include e.g. the NAIV, SEMINAIV, and
LOGARITHMIC methods. On the other side there exist promising algorithms
for computing the transitive closure in the field of graph theory.
Because they have been developed as main memory algorithms and work
over boolean matrices a reformulation is necessary before utilizing
them in the context of database systems. This group contains e.g. the
algorithms of Warshall, Warren, Bloniarz/Fischer/Meyer, and
Goralcikova/Koubek ([Wars 62], [Warr 75], [BlFiMe 76], [GoKo 79]). In
the course of reformulating these methods the level of the relational
algebra turns out to be too high. Instead the algorithms must be
tailored to the data structures which store the relations.

Transitive closure algorithms can be categorized into iterative methods
(depending on the length of the underlying graph) and direct methods
([AgJa 87]). The algorithms of the first group are all iterative,
whereas those of the second group are all direct.

The main goals of our paper are:

- The description of all algorithms on the same level.
- The categorization of the algorithms with respect to some basic
  generation strategies. We start from the most simple algorithm and
  develop all the other algorithms by applying the generation
  strategies.
- The simplification of the comparison of algorithms developed in
  different contexts. Especially the fact is revealed that in
  general direct algorithms are superior to iterative ones.
- The summary of further optimization ideas collected from
  different fields.

We have identified the following set of basic generation strategies.

GEN 1) **Selection of a method on the algebraic level** (for
       computing the transitive closure).

Alternatives are the NAIV method, the SEMINAIV method, and the LOGARITHMIC method.

GEN 2) **Selection of a join operator.**

All algorithms use a join operator in a more or less hidden way. We consider two different methods for performing a join.

GEN 3) **Selection of the kind of insertions.**

Simple iterative algorithms insert newly generated tuples into a separate copy. Thus these tuples are used only in the next iteration step to generate further tuples. More sophisticated algorithms reuse generated tuples in the same iteration step if possible by dynamic insertions into the current data structure.

GEN 4) **Selection of the insertion order.**

In general insertions (into the adjacency lists) are performed according to some ordering, which depends on the access structure. However, in some cases the maintenance of this ordering is discontinued, and the new values are inserted according to a strategy which enforces the immediate reusage of generated tuples.

GEN 5) **Application of the method of blocking.**

The method is described in detail in section 3.7. The first blocked algorithms have been presented in [Ki 80] for the join and in [AgJa 87] resp. [Lu 87] for the transitive closure.

GEN 6) **Application of tuning and preprocessing techniques.**

By tuning we denote an acceleration technique for iterative blocked algorithms, which reduces the number of iteration steps. Preprocessing is a technique which has to be applied to some direct blocked algorithms in order to guarantee their correctness. The two techniques resemble each other in their way of acting.

GEN 7) **Utilization of a topological order.**

If the relation is acyclic and a toplogical order of the values is available, this order can be utilized to compute the transitive closure in a single pass of the underlying graph.

GEN 8) **Selection of an access structure for adjacency lists.**

Alternatives are balanced trees and dynamic hashing methods.

The paper is organized as follows. In section 2 we give some definitions. Most of the generation strategies are described in section

3 (GEN 1-5, 7). In section 4 we start with the description of the most simple NAIV algorithm in our framework and develop all other algorithms by applying the generation strategies. In section 4.2 we also describe the tuning and preprocessing techniques for blocked algorithms (GEN 6) guided by some examples. The access structures for adjacency lists (GEN 8) are examined in section 5. Section 6 addresses the problem of storage allocation for blocked algorithms. Finally a summary is given in section 7.


## 2 Definitions

Let $r$ be a relation with the attributes A and B, let $r^+$ be the transitive closure of $r$, and let $V_A(r)$ resp. $V_B(r)$ be the set of values occurring for attribute A resp. B.

We assume the relation to be given in the form of adjacency lists. The lists can be a set of successor lists $succ[x_1]$, ..., $succ[x_k]$ for the values in $V_A(r)$ or a set of predecessor lists $pred[y_1]$, ..., $pred[y_1]$ for the values in $V_B(r)$.

$y \in succ[x]$   :<=>   $(x,y) \in r$,

$x \in pred[y]$   :<=>   $(x,y) \in r$

We define the type "SADLIST" to be a set of adjacency lists.

Furthermore we assume the lists of the input to be sorted according to a total ordering $f$, which depends on the access structure (see section 5), and in general we maintain the sorting under insertions. This allows us to perform the union of two lists in linear time. However, in some special cases, which are stated explicitly, the maintenance of the sorting is discontinued. Then the new values are inserted according to some strategy which enforces the reusage of generated tuples in the same iteration step. Define

$x<<y$ :<=> $f(x)<f(y)$,

$y>>x$ :<=> $f(x)<f(y)$.

The access to the lists can be realized by a dynamic hashing method or by a balanced tree. We neglect the access structure for the next two sections and return to this issue in section 5.

The assumption of having adjacency lists as input to the algorithms seems very reasonable to us. For the additional effort of generating the lists from other data structures in advance pays off during the main parts of the algorithms. With the exception of pathological cases, this holds always when computing the whole transitive closure. It might not be true for more special recursive queries like computing the successors of a single value. But this topic lies beyond the scope of this paper.

Often we will argue in terms of graphs. Graphically, a relation r can be represented as a directed graph $G(V,E)$, where a node $x \in V$ represents a domain value of $V_A(r) \cup V_B(r)$ and an edge $(x,y) \in E$ represents a tuple $(x,y) \in r$. Cardinalities are denoted by $| \ |$.


# 3 Basic concepts

## 3.1 Methods on the algebraic level (GEN 1)

First, we shortly present the basic algorithms on the level of the relational algebra.

Let $r_1 \cdot r_2 := \pi[r_1.A, r_2.B](r_1 \ x \ r_2)$, where $r_1$ and $r_2$ are relations possessing the same attributes A and B, and x denotes the equijoin with respect to the join attributes $r_1.B$ and $r_2.A$.

*NAIV* method:

```
old_r:=Ø ;
r⁺:=r ;
δ:=r⁺-old_r ;
WHILE δ <> Ø DO
   old_r:=r⁺ ;
   r⁺:=r⁺ ∪ r⁺ ∘ r ;
   δ:=r⁺-old_r ;
END ;
```

*SEMINAIV* method:

```
old_r:=Ø ;
r⁺:=r ;
δ:=r⁺-old_r ;
WHILE δ <> Ø DO
   old_r:=r⁺ ;
   r⁺:=r⁺ ∪ δ ∘ r ;
   δ:=r⁺-old_r ;
END ;
```

*LOGARITHMIC* method:

```
r⁺:=r ;
δ1:=r ;
δ2:=r ;
WHILE δ1 <> Ø DO
   δ2:=δ2 ∘ δ2 ;
   δ1:=r⁺ ∘ δ2 ;
   r⁺:=r⁺ ∪ δ1 ∪ δ2 ;
END ;
```


## 3.2 Join methods (GEN 2)

The most frequently used operation in the algorithms is a compound join and union operation. It performs an equijoin of two relations $r_1$ and $r_2$ ($r_1 \cdot r_2$) and inserts the generated tuples into a relation $r_3$.

We define two different operators of this kind over adjacency lists. They differ in the sequence of processing the lists of $r_1$ and $r_2$, and they differ in the form of the adjacency lists for $r_1$. One operator requires successor lists, whereas the other requires predecessor lists.

The *nested-loop* operator:

```
PROCEDURE nl_join (VAR succ₁, succ₂, succ₃: SADLIST) ;
BEGIN
  FOR EACH x ∈ Vₐ(r₁) DO
    FOR EACH y ∈ succ₁[x] DO
      succ₃[x]:=succ₃[x] ∪ succ₂[y];
    END;
  END;
END nl_join.
```

The *merging-scan* operator:

```
PROCEDURE ms_join (VAR pred₁, succ₂, succ₃: SADLIST);
BEGIN
  FOR EACH y ∈ Vₐ(r₂)∩Vᵦ(r₁) DO
    FOR EACH x ∈ pred₁[y] DO
      succ₃[x]:=succ₃[x] ∪ succ₂[y];
    END;
  END;
END ms_join.
```

nl_join is based on the "nested-loop join algorithm" ([ScTi 83]), which is used e.g. by System R and INGRES.

ms_join is based on the "merging-scan join algorithm" ([ScTi 83]), which is used e.g. by System R as an alternative to the nested-loop join algorithm.

The elements of the innermost FOR-loops are selected in the sequence of occurance in the list. succ₁ and pred₁ can be updated dynamically by the identification of parameters or by triggered insertions (see below). Then the innermost FOR-loop is performed like a WHILE-loop.

## 3.3 Utilization of a topological order (GEN 7)

GEN 7 corresponds to the sequence in which the elements of the outmost FOR-loop are selected. In general they are selected by a scan, and hence their sequence depends on the chosen access structure. However, in one special case we will select the elements in topological order.

## 3.4 Order of insertions (GEN 4)

GEN 4 corresponds to the way of how new values are inserted into lists. We consider three insertion strategies. The first performs the insertions according to an ordering f, which depends on the access structure (see section 5). For now it is sufficient to think of f as the logical order. The second strategy appends the new values to the tail of the list, and the third inserts them just behind the current value of y. We denote the three versions by the suffices _sorted, _tail, and _current. The second and third strategy have deficiencies in performing the union of lists. Hence there must exist good reasons for applying them. In our case this will be a drastical reduction of the number of iteration steps. In conclusion we observe the first strategy to be sufficient for ms_join. However, for nl_join we apply all three strategies.


## 3.5 Triggering of actions

Later the basic join operators are slightly modified for each transitive closure algorithm. The modifications are actions which are triggered immediately after a new tuple has been inserted into $succ_3$. They can be of two kinds:
- the setting of the halting variable for iterative methods
- the insertion of the new tuple into the lists of a further variable.

Syntactically the triggered actions are written in braces behind the responsible operator.


## 3.6 Kind of insertions (GEN 3)

GEN 3 is concerned with the problem of how to reuse appropriate generated tuples in the same iteration step. This goal can be achieved in two ways.


## 3.6.1 Identification of parameters (GEN 3a)

First the reusage is possible by the identification of parameters.
If the actual parameters for $r_1$, $r_2$, and $r_3$ are pairwise distinct and no insertions into $r_1$ and $r_2$ are triggered, then both operators compute $r_3 \cup r_1 \cdot r_2$. However, we allow the actual parameters to be identical. If the third parameter is identified with the first or second parameter, appropriate generated tuples can be reused in the same iteration step. Then the following condition holds for the result r:
-if the first and third parameter are identical:

$$r_1 \cup r_1 \cdot r_2 \subseteq r \subseteq r_1 \cup (r_1 \cdot r_2^+)$$

-if the second and third parameter are identical:

$$r_2 \cup r_1 \cdot r_2 \subseteq r \subseteq r_2 \cup (r_1^+ \cdot r_2)$$

-if all three parameters are identical:

$r_1 \cup r_1 \cdot r_1 \; c \; r \; c \; r_1^+$

The actual size of the result depends on the insertion strategy and on the structure of the underlying graph.

In the extreme case this means that an operator computes the whole transitive closure.

## 3.6.2 Triggering insertions into the joined relations (GEN 3b)

A second possibility to reuse tuples is given by triggering insertions into $r_1$ or $r_2$.

This method is more flexible because it allows to insert only those tuples into the lists of the first two parameters which really can be reused.

Similar conditions as in section 3.6.1 hold now for the result r in the three corresponding cases

-where insertions into $r_1$ are triggered,

-where insertions into $r_2$ are triggered,

-where insertions into $r_1$ and $r_2$ are triggered.

Now the application of GEN 3 corresponds to the selection of parameters for identification and the specification of triggered insertions into $r_1$ and $r_2$.

## 3.7 Blocking (GEN 5)

Paging issues become important when processing a large amount of data, and the transitive closure can be of enormous size even for a moderate small relation. Hence we assume to deal with data not fitting into the main memory, and it seems worthwhile to focus more on reducing the amount of paging.

An approach in this direction is the method of blocking. The basic idea can already be described for both join-union operators. Define

$$succ_j(S) := \bigcup_{x \in S} succ_j[x] \quad \text{for } S \; c \; V_A(r_j), \text{ and}$$

$$pred_j(S) := \bigcup_{y \in S} pred_j[y] \quad \text{for } S \; c \; V_B(r_j).$$

The partitioning has to satisfy the following condition.

$x \in S_1, \; y \in S_j, \; i<j \Rightarrow x<<y$

For the blocked algorithms we only consider the "sorted" insertion strategy (GEN 4). Therefore we omit the suffix _sorted.

```
PROCEDURE blocked_nl_join (VAR succ₁, succ₂, succ₃: SADLIST);
BEGIN
  Partition Vₐ(r₁) into sets S₁, ..., Sₖ;
  FOR i:=1 TO k DO
    FOR EACH y ∈ succ₁(S₁) DO
      FOR EACH x ∈ S₁ DO
        IF y ∈ succ₁[x] THEN
          succ₃[x]:=succ₃[x] ∪ succ₂[y];
        END;
      END;
    END;
  END;
END blocked_nl_join.


PROCEDURE blocked_ms_join (VAR pred₁, succ₂, succ₃: SADLIST);
BEGIN
  Partition Vₐ(r₂)∩Vв(r₁) into sets S₁, ..., Sₖ;
  FOR i:=1 TO k DO
    FOR EACH x ∈ pred₁(S₁) DO
      FOR EACH y ∈ S₁ DO
        IF x ∈ pred₁[x] THEN
          succ₃[x]:=succ₃[x] ∪ succ₂[y];
        END;
      END;
    END;
  END;
END blocked_ms_join.
```

For blocked_nl_join (blocked_ms_join) we call the set containing the lists $succ_1[x]$ and $succ_3[x]$ for $x \in S_1$ ($pred_1[y]$ and $succ_2[y]$ for $y \in S_1$) a block.

The algorithms become attractive if the sets $S_1$ are chosen in a way such that the blocks can be held in the main memory while processing them. For then the lists $succ_2[y]$ (in blocked_nl_join) respectively $succ_3[x]$ (in blocked_ms_join) have to be fetched from the secondary storage only once for each block. In contrast these lists are fetched once for each $x \in V_A(r_1)$ respectively for each $y \in V_A(r_2) \cap V_B(r_1)$ in the worst case of the non blocked versions. Hence, with respect to the amount of paging the innermost FOR-loop is not a real loop.

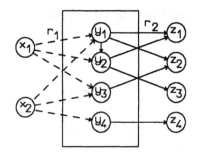

Figure 1.

The basis of blocked_nl_join, the "nested-block method", has already been presented in [Ki 80].

Figure 1 illustrates a situation occurring for blocked_ms_join.

The lists $succ_3[x_1]$ and $succ_3[x_2]$ are loaded once for the set $\{y_1,...,y_4\}$ to insert the values $z_1$, $z_2$, $z_3$, and $y_2$ into $succ_3[x_1]$ and $z_1$, $z_2$, $z_4$, and $y_2$ into $succ_3[x_2]$.

Figure 2 illustrates a situation occurring for blocked_nl_join.

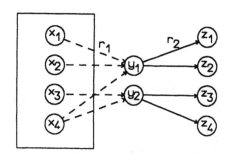

Figure 2.

The lists $succ_2[y_1]$ and $succ_2[y_2]$ are loaded once for the set $\{x_1,...,x_4\}$1. The problem of how to find appropriate sets $S_1$ and how to organize the paging in detail is postponed to section 6. The generation strategies of tuning and preprocessing are described in section 4.2 guided by some examples. Of course they can only be applied to blocked algorithms.


# 4 Categorization of transitive closure algorithms

## 4.1 Non blocked algorithms

In this section we start with the description of the NAIV method in our framework and then develop the other algorithms by applying the generation strategies described in sections 1 and 3.

We assume the relation, over which the transitive closure should be computed, to be available in two versions, one in form of successor lists and the other in form of predecessor lists. These two sets of original lists are denoted by $succ_0$ and $pred_0$. They are not allowed to be modified, and each algorithm copies the required lists in advance by calling the procedure "copy". The renaming of a set of lists is expressed by a simple assignment. The capital letter I indicates an iterative algorithm, and the capital letter D indicates a direct algorithm.

The most simple algorithm is the NAIV method. With respect to the two join operators we obtain two versions (GEN 2).

```
MODULE naiv_1_nl;
INPUT succ₀ (* successor lists for r *): SADLIST;
OUTPUT succ₃ (* successor lists for r⁺ *): SADLIST;
LOCAL succ₁, succ₂: SADLIST;
      stop        : BOOLEAN;
BEGIN
  succ₁:=succ₀;
  succ₂:=succ₀;
  copy(succ₀,succ₃);
  stop:=TRUE;
  nl_join_sorted(succ₁,succ₂,succ₃);
    {IF a new tuple is inserted into succ₃ THEN stop:=FALSE END}
  WHILE NOT stop DO
    stop:=TRUE;
    copy(succ₃,succ₁);
    nl_join_sorted(succ₁,succ₂,succ₃);
      {IF a new tuple is inserted into succ₃
        THEN stop:=FALSE END}
  END;
END naiv_1_nl.    (I)
```

```
MODULE naiv_1_ms;
INPUT succ₀ (* successor lists for r *),
      pred₀ (* predecessor lists for r *): SADLIST;
OUTPUT succ₃ (* successor lists for r⁺ *): SADLIST;
LOCAL pred₁, succ₂: SADLIST;
      stop        : BOOLEAN;
BEGIN
  pred₁:=pred₀;
  succ₂:=succ₀;
  copy(succ₀,succ₃);
  stop:=TRUE;
  ms_join_sorted(pred₁,succ₂,succ₃);
    {IF a new tuple is inserted into succ₃ THEN stop:=FALSE END}
  WHILE NOT stop DO
    stop:=TRUE;
    copy(succ₃,succ₂);
    ms_join_sorted(pred₁,succ₂,succ₃);
      {IF a new tuple is inserted into succ₃
        THEN stop:=FALSE END}
  END;
END naiv_1_ms.    (I)
```

Obviously the difference of the two algorithms can not be described on the level of the relational algebra. Instead it can only be described on the level of adjacency lists.

The same holds for the algorithms naiv_2_nl and naiv_2_ms, which improve naiv_1_nl by identifying succ₁ and succ₃ (yielding succ₁) respectively naiv_1_ms by identifying succ₂ and succ₃ (yielding succ₂)

(GEN 3a).

The improvement results in reducing the number of iteration steps because some tuples generated in one iteration step are reused in the same step to generate further tuples. The syntax of the two algorithms is straightforward, and therefore it is omitted here.

Now, by triggering an insertion into $succ_2$ respectively $pred_1$ after a new tuple has been inserted into $succ_1$ respectively $succ_2$ (GEN 3b) we get remarkable results.

For naiv_2_ms we get the well known direct Warshall algorithm ([Wars 62]).

```
MODULE Warshall;
INPUT succ₀ (* successor lists for r *),
      pred₀ (* predecessor lists for r *): SADLIST;
OUTPUT succ₂ (* successor lists for r⁺ *): SADLIST;
LOCAL pred₁: SADLIST;
BEGIN
  copy(pred₀,pred₁);
  copy(succ₀,succ₂);
  ms_join_sorted(pred₁,succ₂,succ₂);
    {IF a new tuple is inserted into succ₂
     THEN insert this tuple into pred₁ END}
END Warshall.     (D)
```

While the concept of triggering actions is a nice means for categorizing the algorithms, here it has some deficiencies in performance. It is not always efficient to trigger the insertion of a new tuple into $pred_1$ immediately after it has been inserted into $succ_2$. In many cases it is better to defer these insertions until the innermost FOR-loop is totally processed for some $y \in V_A(r) \cap V_B(r)$. This version of the Warshall algorithm can be obtained by deleting the triggered action and augmenting ms_join by another FOR-loop, which performs the update of the lists in $pred_1$.

```
FOR EACH x ∈ pred₁[y] DO
  succ₂[x]:=succ₂[x] ∪ succ₂[y];
END;
FOR EACH z ∈ succ₂[y] DO
  pred₁[z]:=pred₁[z] ∪ pred₁[y];
END;
```

For naiv_2_nl we get a direct algorithm which comes very near to the Warren algorithm ([Warr 75]). Now even $succ_1$ and $succ_2$ can be identified (yielding $succ_1$). Otherwise the triggering of an insertion into $succ_2$ would result in two identical sets of lists.

```
MODULE naiv_3_nl;
INPUT succ₀ (* successor lists for r *): SADLIST;
OUTPUT succ₁ (* successor lists for r⁺ *): SADLIST;
BEGIN
   copy(succ₀,succ₁);
   nl_join_sorted(succ₁,succ₁,succ₁);
   nl_join_sorted(succ₁,succ₁,succ₁);
END naiv_3_nl.    (D)
```

For each $(x,y) \in r^+$ the first call generates a path $x=x_1, \ldots, x_k=y$ in succ$_1$ such that $x_i \ll x_{i+1}$ for $1 \leq i < k-1$, and the second call generates the tuple $(x,y)$ in succ$_1$. Hence two iterations are sufficient.

The *Warren* algorithm can be obtained from naiv_3_nl by a little modification of the join operator. The innermost FOR-loop has to be modified as follows.

FOR EACH $y \in$ succ$_1[x]$: $y \ll x$ DO (for the first call), and

FOR EACH $y \in$ succ$_1[x]$: $y \gg x$ DO (for the second call)

One of the improvements of naiv_2_nl over naiv_1_nl has been the reduction of the number of iteration steps. This has been achieved by a processing of just generated tuples in the same iteration step. Unfortunately only those tuples satisfying some condition are reused in the same step.

It is now possible to enforce that each generated tuple is reused. We only have to discontinue the maintenance of the sorting, and the new values have to be inserted behind the value y which just is in processing. This can be achieved by replacing the "sorted" insertion strategy by one of the other two strategies (GEN 4), thus yielding a direct algorithm.

We obtain two well known methods.

```
MODULE repetitive_breadth_first_search;
INPUT succ₀ (* successor lists for r *): SADLIST;
OUTPUT succ₁ (* successor lists for r⁺ *): SADLIST;
LOCAL succ₂: SADLIST;
BEGIN
   copy(succ₀,succ₁);
   succ₂:=succ₀;
   nl_join_tail(succ₁,succ₂,succ₁);
END repetitive_breadth_first_search.    (D)
```

*repetitive_depth_first_search* can be obtained by replacing nl_join_tail by nl_join_current.

The attempt to further improve this method by identifying succ$_2$ and succ$_1$ is not attractive. The advantage of having only one set of lists is outperformed by the disadvantage that the lists which have to be examined during the search become longer. However, a more sophisticated attempt is possible for the case where the relation is acyclic and a topological order of the values is available. Beside the identification of $succ_2$ and $succ_1$ (yielding $succ_1$), we now select the values $x \in V_A(r)$ of the outmost FOR-loop in topological order (GEN 7), and we modify the innermost FOR-loop of nl_join_tail as follows.

```
l:=|succ₁[x]|;
FOR the first l values in succ₁[x] DO
```

We call this algorithm *one_pass_traversal*. For the case where it can be applied, it is highly efficient. In the field of graph theory it has been presented in [GoKo 79].

An algorithm which improves the NAIV method on the relational level is the SEMINAIV method (GEN 1).

```
MODULE seminaiv_4_nl;
INPUT succ₀ (* successor lists for r *): SADLIST;
OUTPUT succ₄ (* successor lists for r⁺ *): SADLIST;
LOCAL succ₁, succ₂, succ₃: SADLIST;
      stop              : BOOLEAN;
BEGIN
   succ₁:=succ₀;
   succ₂:=succ₀;
   succ₃:=∅;
   copy(succ₀,succ₄);
   stop:=TRUE;
   nl_join_sorted(succ₁,succ₂,succ₄);
      {IF a new tuple is inserted into succ₄
       THEN insert this tuple into succ₃; stop:=FALSE END}
   WHILE NOT stop DO
      stop:=TRUE;
      succ₁:=succ₃;
      succ₃:=∅;
      nl_join_sorted(succ₁,succ₂,succ₄);
         {IF a new tuple is inserted into succ₄
          THEN insert this tuple into succ₃; stop:=FALSE END}
   END;
END seminaiv_4_nl.    (I)
```

The corresponding algorithm *seminaiv_4_ms* can be obtained in a straightforward manner (GEN 2).

On the description level of adjacency lists we can further improve the algorithms by inserting appropriate new tuples into $succ_1$ respectively $succ_2$ instead of inserting them into $succ_3$. Tuples are appropriate if they can still be processed in the same iteration step. The improved algorithm *seminaiv_5_nl* (*seminaiv_5_ms*) differs from seminaiv_4_nl (seminaiv_4_ms) only in the form of the initialization for $succ_1$ ($succ_2$) and the form of the triggered action (GEN 3b), which now are:

```
copy(succ₀,succ₁) (copy(succ₀,succ₂)), and
```

```
{IF a new tuple (x,z) is inserted into succ₄ THEN
   IF z>>y (x>>y) THEN
      insert this tuple into succ₁ (succ₂);
   ELSE
      insert this tuple into succ₃;
      stop:=FALSE;
   END;
 END}
```

Here $y$ corresponds to the variable of nl_join (ms_join).

We observe that seminaiv_5_ms is a non blocked version of the Lu algorithm ([Lu 87]).

Obviously the last improvement reduces the number of iteration steps for the SEMINAIV method.

For completeness purposes we give also a description of the LOGARITHMIC method in our framework (GEN 1). Let insert(succ₁,succ₃) perform the insertion of all tuples from succ₁ into succ₃.

```
MODULE logarithmic_nl;
INPUT succ₀ (* successor lists for r *): SADLIST;
OUTPUT succ₄ (* successor lists for r⁺ *): SADLIST;
LOCAL succ₁, succ₂, succ₃: SADLIST;
      stop             : BOOLEAN;
BEGIN
  succ₁:=succ₀;
  succ₂:=∅;
  succ₃:=succ₀;
  copy(succ₀,succ₄);
  stop:=TRUE;
  nl_join_sorted(succ₁,succ₁,succ₂);
  insert(succ₂,succ₄);
  nl_join_sorted(succ₂,succ₃,succ₄);
    {IF a new tuple is inserted into succ₄
     THEN stop:=FALSE END}
  WHILE NOT stop DO
    stop:=TRUE;
    succ₁:=succ₂;
    succ₂:=∅;
    nl_join_sorted(succ₁,succ₁,succ₂);
    copy(succ₄,succ₃);
    insert(succ₂,succ₄);
    nl_join_sorted(succ₂,succ₃,succ₄);
      {IF a new tuple is inserted into succ₄
       THEN stop:=FALSE END}
  END;
END logarithmic_nl.    (I)
```

## 4.2 Blocked algorithms

For the blocked algorithms we will focus on the most important ones and on some special features related to this concept.

The blocked versions of naiv_1_nl, naiv_1_ms, naiv_2_nl, naiv_2_ms, seminaiv_4_nl, seminaiv_4_ms, seminaiv_5_nl, and seminaiv_5_ms can be obtained form their non blocked versions by simply replacing nl_join_sorted (ms_join_sorted) by blocked_nl_join (blocked_ms_join). Observe that the block of blocked_naiv_2_nl contains only one set of lists. This fact obviously increases the sizes of the sets $S_i$ and thus is important in reducing the amount of paging. On the other hand, for blocked_seminaiv_4_nl and blocked_seminaiv_5_nl it seems reasonable to include the set of lists $succ_3[x]$ for $x \in S_i$ into the block, too.

Now we will focus on an acceleration method which we call *tuning* (GEN 6). It can be applied to all those iterative algorithms which allow insertions into the adjacency lists of the joined relations. Of the above mentioned algorithms these are blocked_naiv_2_nl, blocked_naiv_2_ms, blocked_seminaiv_5_nl, and blocked_seminaiv_5_ms.

An aim of all iterative methods is to generate as much tuples as possible during one iteration step. Each of the four algorithms offers the possibility to generate some tuples requiring only a very little amount of paging. For blocked_naiv_2_nl (blocked_naiv_2_ms) this is the case when tuples $(x,y)$, $x,y \in S_i$, are detected in $succ_1$ ($succ_2$). For then the lists $succ_1[x]$ and $succ_1[y]$ ($succ_2[x]$ and $succ_2[y]$) are both in the main memory and the tuning operation

$$succ_1[x]:=succ_1[x] \cup succ_1[y] \quad (succ_2[x]:=succ_2[x] \cup succ_2[y])$$

can easily be performed without doing any paging. The tuples in question can be detected in two situations:
- when starting the processing of a block $S_i$, the lists $succ_1[x]$ ($succ_2[x]$) for $x \in S_i$ are scanned for such tuples
- when a new tuple $(x,y)$, $x,y \in S_i$, is inserted during the processing of a block $S_i$.

The tuning for blocked_seminaiv_5_nl is more complex and is omitted here for lack of space. However, we present a tuned and blocked version of seminaiv_5_ms because we see it as a possible implementation of the Lu algorithm ([Lu 87]). Define

$$S(S_i,succ_3):=\{(x_1,x_k) \mid x_1 \in S_i, \text{ there exist } x_2,\ldots,x_{k-1} \in S_i:$$
$$x_1 \in succ_3[x_{l-1}] \text{ for } 1 < l \leq k\}, \text{ and}$$

$$P(S_i,pred_3):=\{(x_k,x_1) \mid x_1 \in S_i, \text{ there exist } x_2,\ldots,x_{k-1} \in S_i:$$
$$x_1 \in pred_3[x_{l-1}] \text{ for } 1 < l \leq k\}, \text{ and}$$

$$V_A(succ):=\{x \in V_A(r) \mid succ[x] <> \emptyset\}.$$

```
MODULE tuned_blocked_seminaiv_5_ms;
INPUT succ₀ (* successor lists for r *),
      pred₀ (* predecessor lists for r *): SADLIST;
OUTPUT succ₄ (* successor lists for r⁺ *): SADLIST;
LOCAL pred₁, succ₂, succ₃: SADLIST;
      stop              : BOOLEAN;
BEGIN
  pred₁:=pred₀;
  copy(succ₀,succ₃);
  copy(succ₀,succ₄);
  stop:=FALSE;
  WHILE NOT stop DO
    stop:=TRUE;
    succ₂:=succ₃;
    succ₃:=∅;
    Partition Vₐ(succ₂)∩Vв(r) into sets S₁,...,Sₖ;
    FOR i:=1 TO k DO
      Compute the tuples of S(S₁,succ₄) and insert them into
      succ₄, and insert those tuples which are new in succ₄ also
      into succ₂;
      FOR EACH x ∈ pred₁(S₁)-S₁ DO
        FOR EACH y ∈ S₁ DO
          IF x ∈ pred₁[y] THEN
            succ₄[x]:=succ₄[x] ∪ succ₂[y];
              {IF a new tuple (x,z) is inserted into succ₄ THEN
                  IF x>>y THEN insert this tuple into succ₂;
                          ELSE insert this tuple into succ₃;
                              stop:=FALSE;
                  END;
              END}
          END;
        END;
      END;
    END;
  END;
END tuned_blocked_seminaiv_5_ms.    (I)
```

The tuning of blocked_seminaiv_5_ms is performed as follows. When the processing of a set $S_1$ starts, the lists $succ_2[x]$ and $succ_4[x]$ for $x \in S_1$ are loaded. Then for the tuples $(x,y)$, $x,y \in S_1$, in $succ_4$ the tuning operation

```
succ₄[x]:=succ₄[x] ∪ succ₄[y];
  {IF a new tuple is inserted into succ₄
   THEN insert this tuple into succ₂ END}
```

is performed. Finally the lists $succ_4[x]$ for $x \in S_1$ are replaced by the lists $pred_1[x]$ for $x \in S_1$, and the ordinary processing of the block is continued.

Let us now briefly summarize the general procedure for tuning. The ordinary sequence of processing is interrupted. Some tuning operation is performed to generate new tuples requiring only a little amount of paging. Finally the ordinary processing is continued. The main goal of tuning is the reduction of the number of iteration steps.

Next we will present some direct blocked algorithms. Here tuning is not reasonable because the number of iteration steps can't be reduced. Instead we have to deal with another problem. The correctness of the direct algorithms strictly depends on the sequence of processing. Now the blocking modifies the sequence in order to reduce the amount of paging. Whereas the straightforward blocked version of the Warren algorithm still satisfies the necessary conditions on the sequence, the straightforward blocked version of the Warshall algorithm violates them (see below). Hence, it is necessary to divise means to correct this. First let us consider the blocked Warren algorithm.

```
MODULE blocked_naiv_3_nl;
INPUT succ₀ (* successor lists for r *): SADLIST;
OUTPUT succ₁ (* successor lists for r⁺ *): SADLIST;
BEGIN
  copy(succ₀,succ₁);
  blocked_nl_join(succ₁,succ₁,succ₁);
  blocked_nl_join(succ₁,succ₁,succ₁);
END blocked_naiv_3_nl.    (D)
```

The *blocked_Warren* algorithm can be obtained from blocked_naiv_3_nl in the same way as we have obtained the Warren algorithm from naiv_3_nl. The first presentation of blocked_Warren has been given in [AgJa 87].
Now, let us return to the Warshall algorithm. A straightforward attempt of blocking results in just replacing ms_join_sorted by blocked_ms_join. Unfortunately the new algorithm does not correctly compute the transitive closure. Consider the following situation for $x \ll z \ll y$, $y, z \in S_1$.

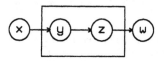

Figure 3.

For the non blocked algorithm, the operation $succ_1[y]:=succ_1[y] \cup succ_1[z]$ is performed before the operation $succ_1[x]:=succ_1[x] \cup succ_1[y]$. However the blocked algorithm performs the two operations the other way round, which results in missing the tuple $(x,w)$ for the transitive closure. This problem can easily be solved by applying a *preprocessing* step (GEN 6) to the block $S_1$ just before the ordinary processing of this block is started. We present two different alternatives for the preprocessing step. The first one computes the tuples of $S(S_1,succ_2)$ and $P(S_1,pred_1)$ and inserts them into $succ_2$ respectively $pred_1$.

During the preprocessing step of the second alternative only the tuples of $S(S_1, succ_2)$ and $P(S_1, pred_1) \frown S_1 x S_1$ are computed and inserted into $succ_2$ respectively $pred_1$. The remaining necessary updates of the lists $pred_1[x]$ for $x \in S_1$ are deferred to a later moment. This algorithm has been presented in [AgJa 87] as blocked_Warshall_with_Predecessors.
Taking into account the remark of how to efficiently perform insertions into $pred_1$, we obtain the following algorithms.

```
MODULE blocked_Warshall_with_exhaustive_preprocessing;
INPUT  succo (* successor lists for r *),
       predo (* predecessor lists for r *): SADLIST;
OUTPUT succ2 (* successor lists for r⁺ *): SADLIST;
LOCAL pred1: SADLIST;
BEGIN
  copy(predo,pred1);
  copy(succo,succ2);
  Partition VA(r)∩VB(r) into sets S1, ..., Sk;
  FOR i:=1 TO k DO
    Compute the tuples of S(S1,succ2) and insert them into succ2;
    Compute the tuples of P(S1,pred1) and insert them into pred1;
    FOR EACH x ∈ pred1(S1)-S1 DO
      FOR EACH y ∈ S1 DO
        IF x ∈ pred1[y] THEN
          succ2[x]:=succ2[x] ∪ succ2[y];
        END;
      END;
    END;
    FOR EACH z ∈ succ2(S1)-S1 DO
      FOR EACH y ∈ S1 DO
        IF z ∈ succ2[y] THEN
          pred1[z]:=pred1[z] ∪ pred1[y];
        END;
      END;
    END;
  END;
END blocked_Warshall_with_exhaustive_preprocessing.    (D)


MODULE blocked_Warshall_with_reduced_preprocessing;
INPUT  succo (* successor lists for r *),
       predo (* predecessor lists for r *): SADLIST;
OUTPUT succ2 (* successor lists for r⁺ *): SADLIST;
LOCAL pred1: SADLIST;
BEGIN
  copy(predo,pred1);
  copy(succo,succ2);
  Partition VA(r)∩VB(r) into sets S1, ..., Sk;
  FOR i:=1 TO k DO
    Compute the tuples of S(S1,succ2) and insert them into succ2;
    Compute the tuples of P(S1,pred1)∩S1xS1 and insert them  into
                                                            pred1;
    FOR EACH x ∈ pred1(S1)-S1 DO
      FOR EACH y ∈ S1 DO
        IF x ∈ pred1[y] THEN
          succ2[x]:=succ2[x] ∪ succ2[y];
```

```
            FOR EACH z ∈ (succ₂[y]⌐S₁) DO
               pred₁[z]:=pred₁[z] ∪ {x};
            END;
         END;
      END;
   END;
   FOR EACH z ∈ succ₂(S₁)-S₁ DO
      FOR EACH y ∈ S₁ DO
         IF z ∈ succ₂[y] THEN
            pred₁[z]:=pred₁[z] ∪ pred₁[y];
         END;
      END;
   END;
END;
END blocked_Warshall_with_reduced_preprocessing.    (D)
```

## 5 Access structures for adjacency lists (GEN 8)

In this section we examine which access structures are suitable for the adjacency lists (GEN 8).

Several algorithms require an ordering for the values of $V:=V_A(r) \cup V_B(r)$. At least the applicability of the Warren algorithm essentially depends on such an ordering.

Let $f:V\to\{1,\ldots,n\}$ (bijective) denote a total ordering. For an adjacency list succ[x] respectively pred[x] we call x the "key" of the list. Let page(x) be that page of the adjacency lists which contains the key x, and let pos(x) be the position of x in page(x) (page(x) = ∞, iff the key x does not occur in the data structure). page(x) and pos(x) vary over the lifetime of the data structure.

Now we have elaborated that the following conditions have to be claimed for f.

i) f must be efficiently computable.

ii) for all $x,y \in V$: $f(x)>f(y) \land$ page(x)=page(y)<> ∞

$\Rightarrow$ pos(x)>pos(y) $\land$ (for all $z \in V$: $f(x)>f(z)>f(y)$

$\land$ page(z)<> ∞ $\Rightarrow$ page(x)=page(z) $\land$

pos(x)>pos(z)>pos(y))

The last condition guarantees that the pages of the adjacency lists can be arranged into a sequence so that a scan of this arrangement yields the keys in ascending order with respect to f. Furthermore we want to ensure that the scan can be performed efficiently. Therefore we claim that

iii) the page where the first key (with respect to f) is located can be read requiring at most 1 page access.

iv) given a key x and page(x), the page where the next key is located can be read requiring at most 1 page access.

Finally we observe that an often used operation is the random access to a key. Thus we claim that

v) given a key $x$, page($x$) can be read requiring at most $\lceil \log |V| \rceil$ page accesses.

Now the comparison operators ($<<,>>$) and the FOR-loops over sets have to be evaluated with respect to $f$, and the "sorted" insertions have to be performed according to $f$.

(In many cases the condition for the insertion strategy can be relaxed again.)

Examples for appropriate access structures are $B^+$-trees ([Co 79]) and extendible hashing ([Fa 79]).


# 6 Storage allocation

For the topic of main memory storage allocation there arise several questions. We can only give a summary of these questions and of some ideas for solution.

The first question is how to maintain the storage for the blocked algorithms. In section 3 we have already mentioned that the method of blocking is only reasonable if the block can be held in the main memory while processing it. First we assume a block to be of fixed size. One approach is the total division of the main memory into areas with their own paging procedures. In general we identify at least five areas:

i) for the lists of the block
ii) for the access structures to the lists of the block
iii) for the other adjacency lists
iv) for the access structures to the other adjacency lists
v) for performing operations like splitting

Another approach is the division of the main memory into a first part containing areas with their own paging procedures and a second part, which is controlled by the paging procedure of the underlying storage management system. Details are omitted for lack of space.

A third approach allows pages to be locked in the main memory. The whole main memory is controlled by the paging procedure of the underlying storage management system, which proceeds as usual. However, it is not allowed to page out a locked page. With this mechanism we can lock a block in the main memory while processing it, and we can unlock it when the processing is finished.

For the first and second approach a question stays open. Shall the division be static or dynamic? The third approach implies that there is no part of the main memory where we can utilize our own paging procedure. We can only keep pages in the main memory, and we can release them again.

The second question under the topic of storage allocation is how to find the partition for the sets $S_i$. If the size of a block remains static while processing it, then the solution is straightforward. The area for the block is filled up successively with adjacency lists until it is full. Thus, the sizes are determined for the blocks when they have their turn. However, if the size of a block grows dynamically while processing it, the strategy has to be augmented or modified.
A first approach augments the strategy. Each time when the size of a block grows beyond the limit, the lists for the greatest value have to be discarded from the block. Thus the partition is altered dynamically if necessary. In general there exist two alternatives how to proceed with the discarded lists. We can undo the effect of the partial updates, which have already been performed, by simply not writing them back. Alternatively, we can write them back saving some computation for the next block. Whether both or any of these alternatives can be applied, has to be decided for each special case.

# 7 Summary

Due to interrelations there exists no ranking list for the significance of the generation strategies. Instead we observed that the most significant goals are
- the reduction of the number of data structures and
- the reduction of the number of iterations (leading to direct algorithms).
Thus, in each special case, the significance of a strategy can be measured by what it contributes to achieve these goals.

## 7.1 Tabular summary of the presented algorithms

i=j indicates that the ith and jth data structures have been identified, Ti indicates that insertions into the ith data structure are triggered.

| | GEN 1 | GEN 2 | GEN 3 | GEN 4 | GEN 5 | GEN 6 | GEN 7 |
|---|---|---|---|---|---|---|---|
| naiv_1_nl | NAIV | nl | - | sorted | - | | scan |
| naiv_1_ms | NAIV | ms | - | sorted | - | | scan |
| naiv_2_nl | NAIV | nl | 1=3 | sorted | - | | scan |
| naiv_2_ms | NAIV | ms | 2=3 | sorted | - | | scan |
| warren | NAIV | nl | 1=2=3 | sorted | - | | scan |
| warshall | NAIV | ms | 2=3,T1 | sorted | - | | scan |
| repetitive_bfs | NAIV | nl | 1=3 | tail | - | | scan |
| repetitive_dfs | NAIV | nl | 1=3 | current | - | | scan |
| one_pass_traversal | NAIV | nl | 1=2=3 | tail | - | | top. |
| seminaiv_4_nl | SEMI | nl | T3 | sorted | - | | scan |
| seminaiv_4_ms | SEMI | ms | T3 | sorted | - | | scan |
| seminaiv_5_nl | SEMI | nl | T1,T3 | sorted | - | | scan |
| seminaiv_5_ms | SEMI | ms | T2,T3 | sorted | - | | scan |
| logarithmic_nl | LOG | nl | - | sorted | - | | scan |
| blocked_semi_5_ms | SEMI | ms | T2,T3 | sorted | + | tuned | scan |
| blocked_warren | NAIV | nl | 1=2=3 | sorted | + | - | scan |
| blocked_warshall | NAIV | ms | 2=3,T1 | sorted | + | prepr | scan |

## 7.2 Summary of further optimization ideas

i) The Warshall algorithm not only computes a set of successor lists
(succ$_2$) for r$^+$, additionally it computes a set of predecessor lists
(pred$_1$) for r$^+$. If the predecessor lists are not of further interest,
then some triggered insertions into pred$_1$ can be omitted. This is the
case when for the newly generated tuple (x,z) it holds z<<y. For then
the list pred$_1$[z] is not referenced in the further processing.
Syntactically the new version of the Warshall algorithm can be obtained
by modifying the triggered action as follows.

```
{IF a new tuple (x,z) is inserted into succ₂ THEN
   IF z>>y THEN
     insert this tuple into pred₁;
   END;
 END;}
```

ii) For each iterative algorithm there exists a variable (for adjacency
lists) which is never updated. This is succ$_2$ for naiv_1_nl, naiv_2_nl,
seminaiv_4_nl, and seminaiv_5_nl, and it is pred$_1$ for naiv_1_ms,
naiv_2_ms, seminaiv_4_ms, and seminaiv_5_ms.
Now Lu ([Lu 87]) suggests to delete dynamically those tuples of these

adjacency lists which will not be used in the remaining iterations to generate new tuples. While performing nl_join (ms_join) the tuples of $succ_2$ ($pred_1$) which don't take part in the join can be removed. For the naiv_... algorithms the size of $succ_2$ ($pred_1$) can only be reduced during the first iteration step. However, for the seminaiv_... algorithms the size can be reduced during each iteration step.

iii) For seminaiv_5_ms we can identify $succ_2$ and $succ_3$ if we allow the deletion of the list $succ_2[y]$ after the processing of the key y is finished.

iv) For logarithmic_nl the halting condition can be improved. If one of the calls nl_join_sorted($succ_1,succ_1,succ_2$) or insert($succ_2,succ_4$) generate no tuples, the algorithm can be stopped. Furthermore it is possible to identify $succ_3$ and $succ_4$. The identification results in a reduction of the number of iteration steps, but it increases the size of the second operand for the call nl_join_sorted($succ_2,succ_3,succ_4$).

v) If the size of $V_B(r)$ is known initially, this fact can be utilized by most of the algorithms. For example the processing of the innermost FOR-loop of nl_join can be stopped when $succ_3[x]$ reaches the maximal length $|V_B(r)|$.

vi) It seems attractive to design a blocked repetitive_breadth_first_search algorithm. However, here the blocking is not straightforward because the insertion strategy is not "sorted".

## 7.3 Experiments

The algorithms naiv_1_nl, naiv_2_nl, naiv_2_ms, seminaiv_4_nl, seminaiv_5_nl, seminaiv_5_ms, logarithmic_nl, warshall, warren, repetitive_bfs, and one_pass_traversal have been implemented in MODULA 2. Simulations have been performed for differently structured input relations measuring the amount of paging. We observe that repetitive_bfs, warren, and one_pass_traversal (for acyclic relations) clearly outperform the other algorithms. The most superior algorithm was determined in each case by the size and structure of the input relation and the buffer size.

# References

Ag 87     Agrawal, R., "Alpha: An Extension of Relational Algebra to Express a Class of Recursive Queries," Proc. IEEE 3rd Int. Conf. Data Engineering, Los Angeles, Feb. 1987, pp. 580-590.

AgDaJa 88  Agrawal, R., Dar, S., Jagadish, H.V., "Transitive Closure Algorithms Revisited: The Case of Path Computations," Submitted for publication.

AgJa 87   Agrawal, R., Jagadish, H.V., "Direct Algorithms for Computing the Transitive Closure of Database Relations," Proc. of the 13th VLDB Conf., Brighton 1987, pp. 255-266.

BiRaSt 87  Biskup, J., Raesch, U., Stiefeling, H., "An Extended Relational Query Language for Knowledgebase Support," Hildesheimer Informatik-Berichte, 1987, Hochschule Hildesheim.

BiRaSt 87a Biskup, J., Raesch, U., Stiefeling, H., "Report on Structure and Complexity of Several Augmented Relational Algebras," ESPRIT Project 311: ADKMS, Deliverable D12.

BiRaSt 87b Biskup, J., Raesch, U., Stiefeling, H., "Survey on Algorithms for Computing the Transitive Closure and Related Operators," ESPRIT Project 311: ADKMS, Deliverable D13.

BlFiMe 76  Bloniarz, P.A., Fischer, M.J., Meyer, A.R., "A Note on the Average Time to Compute the Transitive Closure," Proc. of the Int. Colloquium on Automata, Languages and Programming, 1976, pp. 425-434.

Co 79     Comer, D., "The Ubiquitous B-Tree," ACM Computing Surveys 11, 2, 1979, pp. 121-138.

Da 87     Dayal, U. et al, "PROBE - a Research Project in Knowledge-Oriented Database Systems: Preliminary Analysis," Technical Report CCA-85-03, 1985, Computer Corporation of America.

Fa 79     Fagin, R. et al, "Extendible Hashing - A Fast Access Method for Dynamic Files," ACM Transactions on Database Systems 4, 3, 1979, pp. 315-344.

GoKo 79   Goralcikova, A., Koubek, V., "A Reduct-and-Closure Algorithm for Graphs," Proc. Mathematical Foundations of Computer Science, 1979, pp. 301-307.

Io 86     Ioannidis, Y.F., "On the Computation of the Transitive Closure of Relational Operators," Proc. of the 12th VLDB Conf., Kyoto 1986, pp. 403-411.

JaAgNe 87  Jagadish, H.V., Agrawal, R., Ness, L., "A Study of Transitive Closure as a Recursion Mechanism," Proc. ACM SIGMOD Int. Conf. on Management of Data, San Francisco, May 1987, pp. 331-344.

Ki 80     Kim, W., "A New Way to Compute the Product and Join of Relations," Proc. ACM SIGMOD Int. Conf. on Management of Data, Los Angeles, May 1980, pp. 179-187.

Lu 87     Lu, H., "New Strategies for Computing the Transitive Closure of Database Relations," Proc. of the 13th VLDB Conf., Brighton 1987, pp. 267-274.

LuMiRi 87 Lu, H., Mikkilineni, K., Richardson, J.P., "Design and Evaluation of Algorithms to Compute the Transitive Closure of a
Database Relation," Proc. IEEE 3rd Int. Conf. Data Engineering, Los Angeles, Feb. 1987, pp. 112-119.

Ro 86     Rosenthal, A. et al, "Traversal Recursion: A Practical
Approach to Supporting Recursive Applications," Proc. ACM
SIGMOD Int. Conf. on Management of Data, Washington, May
1986, pp. 166-176.

ScTi 83   Scalas, M.R., Tiberio, P., "The Use of the Nested-Block
Method for Computing Joins," IEEE Int. Computer and Applications Conf. (COMPSAC), November 1983, pp. 455-463.

VaBo 86   Valduriez, P., Boral, H., "Evaluation of Recursive Queries
Using Join Indices," Proc. of the 1st Int. Conf. on Expert
Database Systems, pp. 197-208.

Warr 75   Warren, H.S., "A Modification of Warshall's Algorithm for
Transitive Closure of Binary Relations," Comm. ACM 18, 4,
1975, pp. 218-220.

Wars 62   Warshall, S., "A Theorem on Boolean Matrices," Journal of the
ACM 9, 1962, pp. 11-12.

# A Graph-Based Decomposition Approach for
# Recursive Query Processing

Dietmar Seipel
Lehrstuhl für Informatik I, University at Wuerzburg
Am Hubland, D - 8700 Wuerzburg, W.-Germany

## Abstract

In practice most recursive logic queries to a deductive database are linear. According to /JaAgNe 87/ every linearly recursive query can be expressed by one so-called **transitive closure query** and several non-recursive queries.

A transitive closure query TC refers to two binary relations R and Q. It corresponds to a so-called **database digraph** DG = $(V_R, R)$ and a set $V_{goal}$ of special nodes in DG. Given an arbitrary $c \in V_R$. The major step in the bottom-up evaluation of TC is to efficiently compute the set of all $(c,Y) \in R^*$, $Y \in V_{goal}$.

The first technique clusters the database digraph, which yields a **connection hypergraph** CH, a **clustergraph** CG and a skeleton graph SG. CG is the homomorphic image of DG w.r.t. the clustering, SG is a small special subgraph of the transitive closure $DG^* = (V_R, R^*)$ of DG. Based on CG and CH the query TC is decomposed into auxiliary queries that refer only to the relevant parts of R. This reduces the amount of intermediate tuples generated during query evaluation. Augmenting the database by the edges of the skeleton graph speeds up the evaluation of the auxiliary queries for arbitrary constants 'c'.
The second technique refines the decomposition approach by a further clustering technique using some special knowledge about the structure of R. Every cluster contains one **prototypical element**, such that it suffices to solve auxiliary queries for the prototypes, since the results for other elements can then be derived easily from the prototypical results.

The combination of both clustering techniques is implemented in a Prolog-based advanced **deductive database system**, that solves traversal recursion problems on very large relations R, e.g. computes connecting paths in DG.

# 1. INTRODUCTION

An overview on deductive databases and recursive query processing is given by /GaMiNi 84/ and /BaRa 86/, respectively.

Recursion is used as an augmentation of relational query languages that extends the power of the interfaces to relational database systems to that of general purpose languages. This benefits the application of database technology to artificial intelligence and automated engineering design and yields knowledge base systems.

Thus, efficient processing of recursive queries in large database systems is highly desirable. Performance is measured in CPU, I/O and storage costs, cf. /HaLu 86/.

The compiled approach transforms a logic query into a **relational algebra program ( RAP )**, cf. /VaBo 86/, which contains iterative loops for recursion. The RAP resembles a set oriented bottom-up evaluation of the query instead of PROLOG's tuple-wise interaction with the database. This allows to use efficient methods for doing massive joins. In addition RAP's using few iterations handling large joins are desirable for the same reason.

For reducing CPU costs efficient recursive query processing must cut down the amount of irrelevant tuples generated during bottom-up evaluation and avoid duplicate computation of tuples. Tuples deriving answers are called **relevant** to a query. The **magic sets method** and the **counting method**, cf. /BaMaSaUl 86/, rewrite linearly recursive rules, such that bottom-up evaluation simulates the top-down passing of bindings given by the query goal. This reduces the amount of generated potentially relevant tuples. For more general recursion there are several generalisations of the magic set and counting method, cf. /BeRa 87/, /SaZa 86a/, /SaZa 86b/. The **wavefront methods**, cf. /HaLu 86/, among which the methods of /HeNa 84/ and the naive evaluation are well-known, also pass bindings. They group those joins, which reduce the size of intermediate relations. The methods of /AhUl 79/ and /KiLo 86/ transform a query equivalently by **pushing selections** into the least fixpoint operator. For linear recursion duplicate computations are avoided by the **delta-iteration** methods, cf. /BaRa 86/, /GüKieBay 87/.

For reducing I/O costs it is necessary to design algorithms that minimize the amount of paging, cf. /AgJa 87/, /HaHe 87/, /HaQaCh 88/, /Lu 87/.

Linear recursion frequently occur in deductive databases. According to /JaAgNe 87/ every linearly recursive query can be expressed by one so-called **transitive closure query** - computing the transitive closure $R^*$ of a binary database relation R - and several non-recursive queries. Thus, in most cases transitive closure is a sufficiently powerful primitive recursion operator for deductive databases.

We introduce methods for evaluating transitive closure queries. The relation R is mapped to a directed graph, the so-called **database digraph**, cf. /HaQaCh 88/. The database digraph is decomposed suitably into smaller parts, which are investigated first. The results for the decomposed parts support an easy generation of the overall query result. Optionally the database can be augmen-

ted by a certain subset of $R^*$, that is 'not too large' and 'query independent'.

This yields the following improvements to classical transitive closure algorithms:
(i) The number of iterations in bottom-up evaluation is reduced.
(ii) The total amount of tuples in intermediate relations is reduced, if
firstly for every query the relation R is 'pruned' by deleting irrelevant
tuples, and secondly the decomposition algorithms locally operate on the
decomposed parts of the database and on the augmentation, which is much smaller than $R^*$.
If R is very large, then intermediate relations may considerably exceed the
size of main memory. In a virtual memory environment our methods reduce the
amount of paging.
(iii) When considering the bottom-up evaluation of several arbitrary transitive closure queries to the database, there may be large intermediate relations that are computed several times. Although it is impossible to store all
intermediate relations for later use, we can preprocess some essential common
intermediate relations and augment the database by them in order to avoid
duplicate computation.
(iv) Our algorithm decomposes a query into several independent queries. Thus
it can easily be extended to become a distributed or parallel algorithm.

This paper is organized as follows. Section 2 gives some important definitions
and notations. In section 3 the first query processing technique is introduced. By applying a certain **clustering** technique to the database digraph the
query TC is **decomposed** into a set of auxiliary queries. Section 4 contains the
second query processing technique. It is a refinement of the first approach in
the context of a very large, but specially structured relation R. A further
**clustering** technique is used, where every cluster contains one special element, the **prototype**. Subqueries are solved only for the prototypes. Based on
the special structure of R the results for all the other elements of a cluster
can be derived from the prototypical results.


## 2. DEFINITIONS AND NOTATIONS


A **deductive database** DDB is given by a conventional relational database EDB,
the extensional database, and a logic program LP. An intensional database IDB
can be derived from EDB and LP.

A **logic program** LP consists of a set of horn clauses of the form 'head :
body.', so-called rules. Rules have a non-empty body, and they can contain
function symbols.

For example r1, r2 and r3 denote rules:

```
r1:    p(X,Y) :- q(X,Y).
r2:    p(X,Y) :- p(X,Z), r(Z,Y).
r3:    p(X,Y) :- s(X,X1), p(X1,Y1), r(Y1,Y).
```

There is a one-to-one correspondence between the relations $S \in EDB \cup IDB$ and the predicate symbols occurring in LP: an n-ary predicate symbol s corresponds to an n-ary relation $S \subset D^n$ ( D is the universe of discourse of LP ).

A **query goal** is an atom $G = p(X1,...,Xn)$, where Xi, $1 \leq i \leq n$, is a term over function symbols, constants and variables. A **logic query** is a triple $Q = < G,LP,EDB >$. Q is evaluated bottom-up by the well-known **fixpoint iteration**, which derives an intensional database IDB. The result Answer(Q) consists of all tuples that can be derived for the intensional relation P, provided that they match the query goal.

For example consider $LP = \{r1,r3\}$, where $EDB = \{Q,R,S\}$ and $IDB = \{P\}$. Assume $Q = \{(b,c)\}$, $S = \{(a,b)\}$ and $R = \{(c,d)\}$. Then the tuples (b,c) and (a,d) can be derived. If $G = p(a,Y)$, then Answer(Q) = $\{(a,d)\}$.

A rule is called **linearly recursive**, if exactly one predicate symbol s in its body is recursive and if s is the predicate symbol in the head of the rule, cf. /BaRa 86/. E.g. the rules r2 and r3 are linearly recursive. A logic program or logic query containing only non-recursive and linearly recursive rules is called linearly recursive.
According to /JaAgNe 87/ every linearly recursive logic query can be expressed by the help of **transitive closure queries**. A transitive closure query corresponds to a so-called **database digraph**.

## Definition 1:
(i)  A <u>transitive closure query</u> $TC = < G_0,LP_0,EDB_0 >$ consists of
- an arbitrary goal $G_0$,
- the logic program $LP_0 = \{ r1,r2 \}$:
```
      r1 :    p(X,Y) :- q(X,Y).
      r2 :    p(X,Y) :- p(X,Z), r(Z,Y).
```
- and an arbitrary extensional database $EDB_0 = \{ R,Q \}$, where R and Q are binary relations.

(ii) The <u>database digraph</u> corresponding to TC is defined by $DG = (V_R,R)$,
$V_R = \{ x,y \mid \exists (x,y) \in R \}$.
The set of goal nodes in DG is given by $V_{goal} = \{ x \in V_R \mid \exists (x,y) \in Q \}$.

## Definition 2:
Assume a binary relation S.
(i)  The relations $S^n$, $n \in N_0$, are given by $S^1 = S$ and
$S^{n+1} = \{ (x,z) \mid \exists (x,y) \in S, (y,z) \in S^n \}$, $n \in N_+$.

(ii) The <u>transitive closure</u> of the relation S is denoted by $S^* = \bigcup_{n \in N} S^n$.

**Remark 3:**

Assume a transitive closure query TC = < p(c,Y),$LP_0$,$EDB_0$ >. Then
  answer(TC) = { (c,y) | ( $\exists$ z : (c,z)$\in R^*$, (z,y)$\in$Q ) <u>or</u> ( (c,y)$\in$Q ) }.

Thus, the transitive closure query TC = < p(c,Y),$LP_0$,$EDB_0$ > is evaluated in two steps.
- Find all goal nodes 'g' that can be reached from 'c' by a path in DG.
- Find all 'a', such that (g,a)$\in$Q, for a goal node 'g' reachable from 'c'.

A <u>relational algebra program</u> ( RAP ) for evaluating logic queries consists of several relational algebra expressions with associated control structures - while, for, if -, cf. /VaBo 86/.

Given two binary relations R and S. Triple(R,S) := $Proj_{1,4}$( $Join_{2=1}$(R,S) ) := { (a,c) | $\exists$ (a,b)$\in$R, (b,c)$\in$S }. The triple-operator, that is known from transitive closure algorithms for binary relations R, can be expressed as Triple(R,R).

Consider the transitive closure query TC=<p(X,Y),$LP_0$,}R,Q}>, where Q={ (x,x) | x$\in V_R$ }. Classical relational algebra programs for computing the transitive closure Answer(TC)=$R^*$ are given by /VaBo 86/:

```
Delta-Iteration-TC ( R,R* )          Logarithmic-TC ( R,R* )
  P := R;                              P := R; T := R;
  Delta_P := R;                        Delta_P := R;
  while Delta_P ≠ ∅ do                 while Delta_P ≠ ∅ do
  begin                                begin
    Delta_P := Triple(Delta_P,R);        T := Triple(T,T);
    Delta_P := Delta_P \ P;              Delta_P := Triple(P,T);
    P := P ∪ Delta_P;                    P := P ∪ T ∪ Delta_P;
  end;                                 end;
  R* := P.                             R* := P.
```

Delta-Iteration-TC executes O(n) iterations, Logarithmic-TC executes only O(log(n)) iterations for a relation R of depth n, where the depth of a binary relation R is the length of the longest path in DG=($V_R$,R), cf. /VaBo 86/, /Io 86/.
But Delta-Iteration-TC handles smaller relations during each iteration and allows for a much better binding passing than Logarithmic-TC for handling general transitive closure queries.

# 3. THE DECOMPOSITION APPROACH

## 3.1. Clustering of Database Digraphs

We consider <u>partitions</u> of the node set of a database digraph.

**Definition 4:**
Assume a database digraph DG = $(V_R,R)$.
(i)   A ( disjoint ) partition P = { $V_1,...,V_k$ } of $V_R$ is called <u>type-1-</u>
      <u>clustering</u> ( shortly clustering ). A set $V_i \in P$ is called a <u>cluster</u>.
(ii)  $R_{conn}$ = { $(v,v') \in R$ | v and v' are elements of different clusters }
      denotes the set of <u>connecting edges</u>.
      $V_{conn}$ = { $v,v' \in V_R$ | $(v,v') \in R_{conn}$ } denotes the set of <u>connecting nodes</u>.
(iii) $DG_i = (V_i,R_i)$ denotes the <u>subgraph</u> of DG induced by $V_i$, where
      $R_i$ = { $(v,v') \in R$ | $v,v' \in V_i$ }.

Coarsening a database digraph according to a clustering of its node set yields
the <u>connection hypergraph</u>.

**Definition 5:**
Given a database digraph DG = $(V_R,R)$ and a type-1-clustering P of $V_R$.
(i)   The <u>connection hypergraph</u> for DG and P is given by
      CH = $(V_{conn},R_{conn} \cup P_{conn})$, where $P_{conn}$ = { $V_i \cap V_{conn}$ | $1 \leq i \leq k$ }.
(ii)  A sequence $(w_1,w_2,...,w_m)$ of nodes $w_i \in V_h$, $1 \leq i \leq m$, is called a <u>hyperpath</u>,
      iff ( $\forall 1 \leq i < m$ : ( $(w_i,w_{i+1}) \in R_{conn}$ ) <u>or</u> ( $\exists V_j \in P : w_i,w_{i+1} \in V_j$ ) ) ).

A partition of the set of nodes of a database digraph induces a <u>clustergraph</u>.
The clustergraph is the homomorphic image of the database digraph, if a node
of the database digraph is mapped to the cluster containing it.
A heuristic, that will benefit the query evaluation algorithms, is to choose
partitions of database digraphs, such that the clustergraph has more than a
certain number of nodes and as few as possible edges.

**Definition 6:**
Given a database digraph DG = $(V_R,R)$ and a type-1-clustering P of $V_R$.
(i)   The <u>clustergraph</u> CG = $(V_c,E_c)$ for DG and P is given by $V_c$ = P and
      $E_c$ = { $(V_i,V_j)$ | ( $\exists (v_i,v_j) \in R_{conn}$ ) : ( $v_i \in V_i$, $v_j \in V_j$ ) }.
(ii)  $P_{goal}$ = { $V_g \in P$ | $V_g$ contains a goal node } denotes the set of
      <u>goal clusters</u>.

A <u>path</u> $(v_1,v_2,...,v_n)$ in DG corresponds to a certain hyperpath in CH and a
certain path in CG, respectively:
- the <u>hyperpath</u> $(w_1,w_2,...,w_m)$ in CH is a coarsening of the original path,
  that is constructed by omitting all nodes $v_i$, $1<i<n$, such that neither
  $(v_i,v_{i+1})$ nor $(v_{i-1},v_i)$ is in $R_{conn}$.

154

- the <u>path</u> $(c_1,c_2,...,c_1)$ in CG is obtained from the hyperpath: firstly, every node $w_i$, $1\leq i\leq m$, is replaced by the unique cluster $s_i$ containing $w_i$, and secondly, in the resulting sequence $(s_1,s_2,...,s_m)$ every subsequence $(s_i,s_{i+1})$, $1\leq i<m$, of equal clusters is replaced by one cluster $s_i$.

Then the path in DG and the hyperpath in CH is called a <u>refinement</u> of the hyperpath in CH and the path in CG, respectively.

## Example 7:

The following figures show a <u>database digraph</u> DG, a <u>clustering</u> P = {$V_1,V_2,V_3$}, where $V_1$ = {a,b,c,d,p}, $V_2$ = {e,f,g,h,i,j}, $V_3$ = {k,l,m,n,o}, and the corresponding <u>clustergraph</u> CG.

DG:

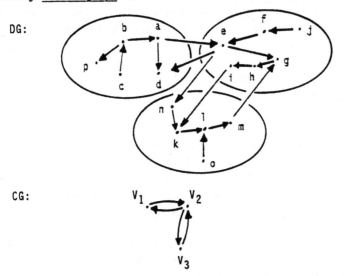

CG:

The path ( c,b,a,e,g,h,i,k,l ) in the database digraph DG can be coarsened to a hyperpath ( a,e,i,k ). The corresponding clusterpath is ( $V_1,V_2,V_3$ ).

## 3.2. The Three-Phase Decomposition Algorithm

TC = <p(c,Y),LP$_0$,{R,Q}> is evaluated in three phases:
- Firstly, all <u>clusterpaths</u> from the start cluster, i.e. the unique cluster containing 'c', to a goal cluster are generated.
  The clusterpaths will reduce the amount of generated irrelevant tuples, in the following phases, since clusters in DG that lie on none of the cluster-paths can be excluded from further observation.
- Secondly, these clusterpaths are refined to <u>hyperpaths</u>, which yield a <u>decomposition</u> of the query TC.
- Thirdly, it is tested which hyperpaths can be refined to paths in DG. Thus, a set of goal nodes is constructed, that yields <u>solutions</u> to TC.

**Definition 8:**
Consider the subgraphs $DG_i=(V_i,R_i)$ of a database digraph DG induced by a clustering $P = \{ V_1,...,V_k \}$.
(i)   $cc\_nr(V_i)$ denotes the number of strongly connected components within $DG_i$, that intersect $V_{conn}$.
(ii) $cc\_nr = cc\_nr(V_1) + ... + cc\_nr(V_k)$.

For every goal node $g \in V_{goal}$, that is reachable from c in DG, usually there are several clusterpaths $(c_1,c_2,...,c_l)$ that can be refined to a path from c to g in DG. But for every goal g there is a clusterpath of at most cc_nr clusters, that can be refined to a path from c to g in DG.

**Lemma 9:**
Assume a database digraph $DG = (V_R,R)$ and $(v,v') \in R^*$.
Then there exists a clusterpath $CP=(c_1,c_2,...,c_l)$, $1 \leq l \leq cc\_nr$, such that
(i)   CP can be refined to a path from v to v' in DG, and
(ii) every cluster $V_i$ occurs at most $cc\_nr(V_i)$ times in the path.

According to the list-notation of PROLOG a list L=[H1,...,Hn,T1,...,Tm] can be denoted by [H1,...,Hn|T], where H1,...,Hn form the head and T=[T1,...,Tm] is called tail. We additionally introduce an analogous list-notation. Assume a list S=[S1,...,Sm]. Then the list L=[S1,...,Sm,E1,...,En] can be denoted by <S|E1,...,Em> or by <S1,...,Sn,E1,...,Em>. E.g. [a,b,c,d] = <<a,b,c>|d> = <<a,b>|c,d>.

We need a built-in function count, where
count( X,L ) = I, iff X occurs exactly I times in the list L.

The first query is given by   $Q_1 = < G_1,LP_1,EDB_1 >$.

$G_1$     = clusterpath( c,CP ),
$LP_1$    = { cr1,cr2,cr3,cr4 },
$EDB_1$ = { $V_c$, $P_{goal}$, $E_c$ }.

cr1   :    clusterpath( X,[CL|T] ) :- CL$\in V_c$, X$\in$CL, cll( [CL|T] ).
cr2   :    cll( [CL] ) :- CL$\in P_{goal}$.
cr3   :    cll( [CL1,CL2|T] ) :- (CL1,CL2)$\in E_c$,
                             count( CL1,[CL2|T] ) < cc_nr( CL1 ),
                             cll( [CL2|T] ).

Answer($Q_1$) consists of all pairs (c,CP), such that CP is a clusterpath beginning with the start cluster and ending with a goal cluster.

The second query is given by $Q_2 = < G_2, LP_2, EDB_2 >$.

$G_2$ = hyperpath( c,HP ),
$LP_2$ = { hr1,hr2,hr3,hr4 },
$EDB_2$ = { $R_{conn}$, $P_{goal}$, CLUSTERPATH }, CLUSTERPATH = answer($Q_1$).

hr1 :    hyperpath( X,HP ) :- clusterpath( X,CP ), hp1( CP,HP ).
hr2 :    hp1( [CL],[] ) :- $CL \in P_{goal}$.
hr3 :    hp1( [CL1,CL2|CP],[X1,X2|HP] ) :
                ($X1,X2) \in R_{conn}$, $X1 \in CL1$, $X2 \in CL2$, hp1( [CL2|CP],HP ).

Answer($Q_2$) consists of all pairs (c,HP), such that HP is a hyperpath beginning in the start cluster and ending with a connecting node in a goal cluster.

Note, that the queries $Q_1$ and $Q_2$ are safe, i.e. their answers are finite. This holds, since all paths given by answer($Q_1$) have length at most cc_nr and all hyperpaths given by answer($Q_2$) have length at most $2 \cdot cc\_nr-2$.

The third query $Q_3 = < G_3, LP_3, EDB_3 >$ is equivalent to the original transitive closure query TC.

$G_3$ = p( c,Y ),
$LP_3$ = { sr1,sr2,sr3,sr4,sr5,sr6,sr7,sr8 },
$EDB_3$ = { R, Q, $V_{conn}$, $R_{conn}$, $V_c$, $P_{goal}$, HYPERPATH },
            HYPERPATH = answer($Q_2$).

sr1 :    p( X,Y ) :- hyperpath( X,HP ),
                    HP = [X1|_], p3( CL1,X,X1 ), test( HP ),
                    HP = <_|Z1>, p3( CL2,Z1,Z ), q( Z,Y ).

sr2 :    test( [X1] )           :- $X1 \in V_{conn}$, $CL \in P_{goal}$, $X1 \in CL$.
sr3 :    test( [X1,X2|HP] ) :- $e_s$( X1,X2 ), test( [X2|HP] ).

sr4 :    $e_s$( X1,X2 ) :- $X1 \in V_{conn}$, $X2 \in V_{conn}$, p3( CL,X1,X2 ).
sr5 :    $e_s$( X1,X2 ) :- ($X1,X2) \in R_{conn}$.

sr6 :    p3( CL,X1,X1 ) :- $CL \in V_c$, $X1 \in CL$.
sr7 :    p3( CL,X1,X3 ) :- r3( CL,X1,X2 ), p3( CL,X2,X3 ).

sr8 :    r3( CL,X1,X2 ) :- r( X1,X2 ), $CL \in V_c$, $X1 \in CL$, $X2 \in CL$.

Thus, the following theorem holds.

### Theorem 10:
The logic queries TC=<p(c,Y),$LP_0$,{R,Q}> and $Q_3$ are equivalent, i.e. answer(TC)=answer($Q_3$).

For every $(c, HP) \in answer(Q_2)$ the hyperpath $HP = (w_1, w_2, \ldots, w_m)$ defines $m+1$
<u>auxiliary queries</u> $SQ_i = \langle G_i, \{sr4, sr5, sr6, sr7, sr8\}, \{R, Q, V_{conn}, R_{conn}, V_c\} \rangle$,
$0 \leq i \leq m$, where $G_0 = p3(CL1, w_0, w_1)$, $w_0 = c$, $G_i = e_s(w_i, w_{i+1})$, $1 \leq i \leq m-1$, and $G_m = p3(CL2, w_m, w_{m+1})$, $w_{m+1} = Z$.
These queries can be transformed to transitive closure queries that are less
complex than TC, since they refer only to parts of R:
- either to the set of edges $R_j$ of a <u>subgraph</u> $DG_j = (V_j, R_j)$ of DG, if $V_j$ contains $w_i$ and $w_{i+1}$,
- or to the set of <u>connecting edges</u> $R_{conn}$, if $w_i$ and $w_{i+1}$ are connecting nodes
  in different clusters.

The <u>skeleton graph</u> SG corresponding to a database digraph $DG = (V_R, R)$ and a
clustering P contains an edge $(v, v')$ between two connecting nodes $v, v' \in V_{conn}$,
iff $(v, v') \in R_{conn}$ or there is a path from v to v' in some subgraph $DG_i$ of DG.

<u>Definition 11:</u>
Given a database digraph $DG = (V_R, R)$ and a type-1-clustering P of $V_R$.
The <u>skeleton graph</u> for DG and P is given by $SG = (V_{conn}, E_s)$, where
$E_s = R_{conn} \cup ( ( R_1^* \cap V_{conn}^2 ) \cup \ldots \cup ( R_k^* \cap V_{conn}^2 ) )$.

Using the skeleton graph SG speeds up the computation of $R^*$. In most cases SG
is much smaller than the transitive closure $DG^* = (V_R, R^*)$.

<u>Lemma 12:</u>
Assume a database digraph $DG = (V_R, R)$ and a clustering $P = \{ V_1, \ldots, V_k \}$.
Then $(v, v') \in R^*$, iff there is some $i \in \{1, \ldots, k\}$, such that $(v, v') \in R_i^*$, or there
are $v_1, v_2 \in V_{conn}$, $i, j \in \{1, \ldots, k\}$, such that $(v, v_1) \in R_i$, $(v_2, v') \in R_j$, $(v_1, v_2) \in E_s^*$.

<u>Example 13:</u>  ( continued )
The following figure shows the skeleton graph $SG = (V, E_s)$.
In our example $|E_s| = 12$, $|R^*| = 117$. Thus, $E_s$ it is much smaller than $R^*$.

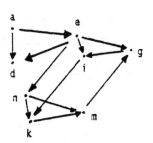

A <u>modification</u> $Q_3'$ of the third query $Q_3$ is more efficient, if the set $E_s$ of
edges of the skeleton graph SG is precomputed by the query $Q_4$.

The <u>modified third query</u> is given by $Q_3' = < G_3', LP_3', EDB_3' >$.

$G_3'$ = p( c,Y ),
$LP_3'$ = { sr1,sr2,sr3,sr6,sr7,sr8 },
$EDB_3'$ = { R3, Q, $R_{conn}$, $V_c$, $E_s$, HYPERPATH },
        HYPERPATH = answer($Q_2$), $E_s$ = answer($Q_4$), R3 = answer($Q_5$).

The <u>fourth query</u> is given by $Q_4 = < G_4, LP_4, EDB_4 >$.

$G_4$ = $e_s$( X1,X2 ),
$LP_4$ = { sr4,sr5,sr6,sr7 },
$EDB_4$ = { R3, $V_{conn}$, $R_{conn}$, $V_c$ }, R3 = answer($Q_5$).

It holds that answer($Q_4$) = $E_s$.

The <u>fifth query</u> is given by $Q_5 = < G_5, LP_5, EDB_5 >$.

$G_5$ = r3( CL,X1,X2 ),
$LP_5$ = { sr8 },
$EDB_5$ = { R, $V_c$ }.

Answer($Q_5$) contains all triples (CL,X1,X2), such that $C \in V_c$ is a cluster, and (X1,X2) is an edge within that cluster, i.e. (X1,X2)$\in$R, X1,X2$\in$C.

The skeleton graph and the relation R3 depend on the database digraph DG=($V_R$,R) and on the type-1-clustering P, but they are independent of the query goal G. Thus, they are suitable for evaluating arbitrary transitive closure queries TC = < p(c,Y),$LP_0$,{R,Q} >.


## 3.3. The Simplified Decomposition Algorithm

The simplified query $Q_6 = < G_6, LP_6, EDB_6 >$ does not deal with paths in the different graphs. It is based on the <u>skeleton graph</u> SG=($V_{conn}$,$E_s$) and assumes, that the relations $E_s$ and R3 are preprocessed. The main step is the computation of the intensional relation $TC\_E_s \subseteq E_s^*$.

$G_6$ = p( c,Y ),
$LP_6$ = { tr1,tr2,tr3,sr6,sr7,sr8,sr9 },
$EDB_6$ = { Q, $V_{conn}$, $V_c$, $P_{goal}$, $E_s$, R3 }, $E_s$ = answer($Q_4$), R3 = answer($Q_5$).

tr1 :   p( X1,X5 ) :- p3( CL,X1,X2 ), tc_$e_s$( X2,X3 ), p3( CL,X3,X4 ),
                q( X4,X5 ).

tr2 :   tc_$e_s$( X1,X1 ) :- X1$\in V_{conn}$, CL$\in P_{goal}$, X1$\in$CL.
tr3 :   tc_$e_s$( X1,X3 ) :- $e_s$( X1,X2 ), tc_$e_s$( X2,X3 ).

Thus, as a consequence of lemma 5, the following theorem holds.

**Theorem 14:**
(i)  The logic queries TC=<p(c,Y),$LP_0$,{R,Q}> and $Q_6$ are equivalent,
     i.e. answer(TC)=answer($Q_6$).
(ii) The intensional relation TC_$E_s$ can be derived in less or equal to
     2·scc_nr-2 iterations.

Thus, <u>bottom-up evaluation</u> of $Q_6$ improves algorithm Delta-Iteration-TC:
- the number of iterations is reduced by grouping large joins referring to $E_s$,
  cf. rule tr3,
- the total amount of tuples in intermediate relations is reduced, since
  TC_$E_s \subset E_s^*$ and $E_s^*$ is much smaller than $R^*$,
- the amount of paging is reduced.


## 4. CLUSTERING BASED ON PROTOTYPES


Consider the transitive closure query TC = < p(c,Y),$LP_0$,{R,Q} >. Assume that
the database relation R is very large, but from the special <u>semantic</u> structure
of R a clustering of the database digraph DG=($V_R$,R) can be deduced with the
following property:
every cluster contains one special element, the prototype, and it suffices to
solve the auxiliary queries ( cf. the query $Q_3$ ) for the prototypes, based on
the special structure of R the results for all the other elements of a cluster
are derived from the prototypical results.

Then, the decomposition approach can be refined yielding a more efficient
query processing for a class of logic queries including traversal recursion.


### 4.1. Traversal Recursion

In /RoHeiDaMa 86/ recursive logic queries that model traversals of a directed
graph are called traversal recursions. The MVV-Expert, cf. /BayGüKieStOb 87/,
is an already existing expert system dealing with traversal recursion. It
computes shortest connections between two stations in the subway system of
Munich. In the previous section of this paper reachability in a directed graph
was decided by logic queries.

The logic query $Q_t$ = < path( [a],[b],Y ),{ pr1,pr2,pr3,pr4,pr5 },{ E } >
determines all paths from node 'a' to node 'b' in a graph G=(V,E):

```
pr1 :    path( A,B,Y ) :- q( A,B,Y ).
pr2 :    path( A,B,Y ) :- r( A,B,A1,B1 ), path( A1,B1,Y ).

pr3 :    q( A,B,Y ) :- A = <S|X>, B = [X|T], Y = concat( A,T ).

pr4 :    r( A,B,C,B ) :- A = <S|X>, e( X,Z ), C = <S|X,Z>.
pr5 :    r( A,B,A,D ) :- B = [Z|T], e( X,Z ), D = [X,Z|T].
```

The variables $A,B,C,D,A1,B1$ denote paths in G. A path $(v_1,...,v_n)$ is represented by a list $[v_1,...,v_n]$ of nodes $v_i \in V$. The function concat yields the concatenation of two lists.

Then the intensional relations R and Q corresponding to r and q, respectively, are very large, since there are a lot of paths in a graph.

The rules pr3, pr4 and pr5 resemble a <u>bidirectional search</u>. Consider the following pair $( A,B ) = ( (v_1,...,v_n),(w_1,...,w_m) )$, $n,m \in N_+$, of paths in G:
(i)   ( A,B ) can be expanded to
      a new pair $( (v_1,...,v_n,v_{n+1}),(w_1,...,w_m) )$, if $(v_n,v_{n+1}) \in E$, or
      a new pair $( (v_1,...,v_n),(w_0,w_1,...,w_m) )$, if $(w_0,w_1) \in E$.
(ii)  ( A,B ) yields a path $Y = A \cdot B$ in G from $v_1$ to $w_m$, if $v_n = w_1$,
      where $p_1 \cdot p_2 := (v_1,...,v_n,w_2,...,w_m)$ denotes the concatenation of tow
      paths $p_1 = (v_1,...,v_n)$ and $p_2 = (w_1,...,w_m)$ in G, if $v_n = w_1$.

We have implemented a refinement of the decomposition approach for query processing, that yields an efficient evaluation of $Q_t$. This refinement is based on the following <u>observation</u>.
- Assume two nodes $a,b \in V$.
- Assume, that we know $S = answer( < path([a],[b],Y),LP_t,DB_t > )$,
  i.e. the set { $([a],[b],Y)$ | Y is a path from 'a' to 'b' in G }.

Then for all paths p,q in G, such that p ends with 'a' and q starts with 'b', $answer( < path(p,q,Y), LP_t,DB_t > )$ can be computed easily from the 'prototypical result' S, since it is equal to { ( $p,q,p \cdot u \cdot q$ | $([a],[b],u) \in S$ ).


## 4.2. Clusterings, Prototypes, and Transformation Functions

### Definition 15:
Assume a database digraph $DG=(V_R,R)$, a <u>partition</u> $P' = \{W_1,...,W_l\}$ of $V_R$ and a set $D' = \{d_1,...,d_l\} \subset V_R$, such that $W_i \cap D' = \{d_i\}$, $1 \leq i \leq l$, $l \in N_+$.
(i)    P' is called a <u>type-2-clustering</u>.
(ii)   W(x) denotes the unique set $W_i \in P'$ containing $x \in V_R$.
(iii)  The unique element $d_i \in W_i$ is called the <u>prototype</u> of $W_i$.

Obviously, there is a one-to-one relationship between a type-2-clustering P' = $\{W_1,...,W_1\}$ and the corresponding set D' = $\{d_1,...,d_1\}$ of prototypes. Further it holds $W_i = W(d_i)$, $1\leq i\leq 1$, and P' = $\{ W(d_i) \mid 1\leq i\leq 1 \}$.

## Definition 16:

Assume a database digraph DG=$(V_R,R)$, a type-2-clustering P' and a set D' of prototypes.
A function tf: $V_R \times D' \times V_R$ --> $V_R$ is a <u>transformation function</u> for R, P' and D', iff for all $1\leq i\leq 1$, $x\in W_i$, $n\in N_0$ it holds $R_n(x)$ = $\{ tf(x,d_i,y) \mid y\in R_n(d_i) \}$, where $R_n(x) := \{ y\in V_R \mid$ there is a path from x to y in DG of length n $\}$.

Thus, a transformation function for R, P' and D' is useful for deriving $R_n(x)$ for all $x\in W_i$ from the <u>prototype</u> $d_i$ of W(x) and the 'prototypical result' $R_n(d_i)$.

Type-2-clusterings can be applied to the logic query $Q_t$, which solves the <u>path finding problem</u> in a graph G = (V,E).
Consider the database digraph DG=$(V_R,R)$ for the intensional relation R = answer( < r(A,B,X,Y), {pr4,pr5}, {E} > ).

Consider the following type-2-clustering P' of $V_R$ and the corresponding set D' of representatives:
- D' := { ( (v),(v') ) | v,v'$\in$V },
  W( ( (v),(v') ) ) := { ( p,q )$\in V_R$ | p ends with v and q starts with v' },
- P' := { W( $d_i$ ) | $d_i \in$ D' }.

A transformation function tf for R, P' and D' is given by
- tf( (p,q), $d_i$, (p',q') ) := (p·p',q'·q),
  for $d_i\in$D', (p,q)$\in$W($d_i$), $n\in N_0$, (p',q')$\in R_n(d_i)$.

## Example 17: ( continued )
Consider the graph of example 7 as G=(V,E). Given the prototype ( (e),(k) ), and the element ( (b,a,e),(k,1) ) $\in$ class( ( (e),(k) ) ).
- $R_1$( ( (e),(k) ) ) = { ( (e,g),(k) ), ( (e,n),(k) ), ( (e,d),(k) ),
                              ( (e),(i,k) ), ( (e),(n,k) ) },
- $R_1$( ( (b,a,e),(k,1) ) ) = { ( (b,a,e,g),(k,1) ), ( (b,a,e,n),(k,1) ),
                              ( (b,a,e,d),(k,1) ), ( (b,a,e),(i,k,1) ),
                              ( (b,a,e),(n,k,1) ) },
Thus $R_1$( ( (b,a,e),(k,1) ) ) = { tf( ( (b,a,e),(k,1) ),( (e),(k) ), y ) |
                              y $\in R_1$( ( (e),(k) ) ) }.


## 4.3. Combination of Type-1- and Type-2-Clusterings

Assume the <u>database digraph</u> DG=$(V_R,R)$ for a transitive closure query TC=$<p(c,Y),LP_0,\{R,Q\}>$, and

- a <u>type-1-clustering</u> P = { $V_1,...,V_k$ } of $V_R$,
  such that there is
- a <u>type-2-clustering</u> P' = { $W_1,...,W_1$ } of $V_R$, that is a refinement of P',
  a corresponding set of <u>prototypes</u> D' = { $d_1,...,d_1$ }, where $W_i=W(d_i)$,
  $1\leq i\leq 1$, and
  a <u>transformation function</u> tf for R, P' and D'.

The type-1-clustering yields a <u>connection hypergraph</u> CH = $(V_{conn}, R_{conn} \cup P_{conn})$,
a <u>clustergraph</u> CG=$(V_c,E_c)$, where $V_c$=P, and a set $P_{goal}$ of goal clusters.

Two queries $Q_1''$ and $Q_2''$ that are similar to $Q_1$ and $Q_2$, respectively, compute
relation HYPERPATH, that consists of all pairs ( c,($w_1,w_2,...,w_m$) ), such that
($w_1,w_2,...,w_m$) is a hyperpath in CH, $w_1$ is in the start cluster, $w_m$ is in a
goal cluster and all nodes $w_1,...,w_m$ are prototypes:

The <u>modified third query</u> $Q_3'' = < G_3'',LP_3'',EDB_3'' >$ is a refinement of $Q_3$. The
incorporation of a transformation function makes it more efficient than $Q_3$.

- $G_3''$  = $G_3$ = p( c,Y ),
- $LP_3''$ = { mr1,mr2,mr3,sr4,sr5,sr6,sr7,sr8 },
- $EDB_3''$ = { R, Q, $V_{conn}$, $R_{conn}$, $V_c$, $P_{goal}$, HYPERPATH },
         HYPERPATH = answer($Q_2''$).

```
    mr1 :    p( X,Y ) :- hyperpath( X,HP ),
                         transform( X,HP,Z ),
                         q( Z,Y ).

    mr2 :    transform( X,[X1],Z ) :-
                  X ∈ W(X1),
                  p3( CL,X1,Z1 ), Z ∈ tf( X,X1,Z1 ).
    mr3 :    transform( X,[X1,X2|HP],Z ) :-
                  X ∈ W(X1),
                  e_s( X1,Y1 ), Y ∈ tf( X,X1,Y1 ),
                  Y ∈ W(X2),
                  transform( Y,[X2|HP],Z ).
```

The properties of the transformation function imply the following theorem.

**Theorem 18:**
The logic queries TC and $Q_3''$ are equivalent, i.e. answer(TC)=answer($Q_3''$).

## 5. FINAL REMARKS


We introduced two techniques for processing **transitive closure queries** TC. Following the classification of /HaHe 87/ we used the following **query processing techniques** to handle redundancy on the different levels of query evaluation.

(i) On the **precompilation** level the query given by a set of horn clauses is transformed according to the decomposition approach.

(ii) On the **iteration** level the number of iterations is reduced by grouping large joins. The preprocessing iterations quickly evaluate the skeleton graph SG resembling a coarsening of the original database digraph. Based on SG a few further iteration steps compute the result to TC. The well-known binding propagation techniques for evaluating transitive closure queries should confine the bottom-up evaluation to the relevant facts.

(iii) On the **tuple processing** level the decomposition approach is level relaxed, cf. /HaQaCh 88/. I.e. there is no restriction on the distance from 'c' to 'Y' in DG for tuples (c,Y) derived in the i-th iteration.

For the bottom-up evaluation delta-iteration techniques should be used, since we are dealing with linear recursion.

(iv) On the **file accessing** level the amount of paging is reduced, since the auxialiary queries in the decomposition approach can be evaluated without extensive paging. The simplified decomposition approach generates fewer intermediate tuples, since it operates on the skeleton graph.

Note, that the simplified decomposition approach itself can be transformed to a transitive closure query. Then our method could be applied recursively. Note also, that the simplified decomposition approach specified for TC can be easily modified to work for arbitrary transitive closure queries.


## REFERENCES

/AhUl 79/            **A. Aho, J. Ullman:**
                     'Universality of Data Retrieval Languages',
                     Proc. POPL 1979.


/AgJa 87/            **R. Agrawal, H. Jagadish:**
                     'Direct Algorithms for Computing the Transitive Closure of
                     Database Relations', Proc. VLDB 1987.


/BaMaSaUl 86/        **F. Bancilhon, D. Maier, Y. Sagiv, J. Ullman:**
                     'Magic Sets and Other Strage Ways to Implement Logic
                     Programs', Proc. PODS 1986.

/BaRa 86/        **F. Bancilhon, R. Ramakrishnan:**
'An Amateur's Introduction to Recursive Query Processing
Strategies', Proc. ACM SIGMOD 1986.

/BayGüKieStOb 87/   **R. Bayer, U. Güntzer, W. Kiessling, W. Strauß,
J. Obermaier:**
'Deduktions- und Datenbankunterstützung für Experten-
systeme',
GI-Fachtagung, 'Datenbanksysteme für Büro, Technik und
Wissenschaft', Informatik-Fachberichte 136, 1987.

/GaMi 78/        **H. Gallaire, J. Minker:**
'Logic and Databases', Plenum Press, New York, 1978.

/GaMiNi 78/      **H. Gallaire, J. Minker, J. Nicolas:**
'An Overview and Introduction to Logic and Data Bases',
in /GaMi 78/.

/GaMiNi 81/      **H. Gallaire, J. Minker, J. Nicolas:**
'Advances in Data Base Theory, Vol. 1',
Plenum Press, New York, 1981.

/GaMiNi 84/      **H. Gallaire, J. Minker, J. Nicolas:**
'Advances in Data Base Theory, Vol. 2',
Plenum Press, New York, 1984.

/GaMiNi 84/      **H. Gallaire, J. Minker, J. Nicolas:**
'Logic and Databases: A Deductive Approach',
ACM Computing Surveys, vol. 16(2), 1984.

/GüKieBay 87/     **U. Güntzer, W. Kiessling, R. Bayer:**
'On the Evaluation of Recursion in ( Deductive ) Database
Systems by Efficient Differential Fixpoint Iteration',
Proc. Int. Conf. on Data Engineering 1987.

/HaQaCh 88/      **J. Han, G. Qadah, C. Chaou:**
'The Processing and Evaluation of Transitive Closure
Queries', Proc. Intl. Conf. Extending Database
Technology 1988, LNCS 303.

/HaHe 87/        **J. Han, L. Henschen:**
'Handling Redundancy in the Processing of Recursive
Database Queries', Proc. ACM SIGMOD 1987.

/HeNa 84/        **L. Henschen, S. Naqvi:**
'On Compiling Queries in Recursive First-Order Data
Bases', JACM, vol. 31(1), 1984.

/HaLu 86/        **J. Han, H. Lu:**
'Some Performance Results on Recursive Query Processing
in Relational Database Systems',
Proc. Int. Conf. on Data Engineering 1986.

/Io 86/        **Y. Ioanidis:**
'On the Computation of the Transitive Closure of
Relational Operations', Proc. VLDB 1986.

/JaAgNe 87/        **H. Jagadish, R. Agrawal, L. Ness:**
'A Study of Transitive Closure as a Recursion Mechanism',
Proc. ACM SIGMOD 1987.

/KiLo 86/        **M. Kifer, E. Lozinskii:**
'Filtering Data Flow in Deductive Databases',
Proc. ICDT 1986.

/KiLo 87/        **M. Kifer, E. Lozinskii:**
'Implementing Logic Programs as a Database System',
Proc. Int. Conf. on Data Engineering, 1987.

/Lu 87/        **H. Lu:**
'New Strategies for Computing the Transitive Closure of
a Database Relation', Proc. VLDB 1987.

/RoHeiDaMa 86/        **A. Rosenthal, S. Heiler, U. Dayal, F. Manola:**
'Traversal Recursion: A Practical Approach to Supporting
Recursive Apllications', Proc. ACM SIGMOD 1986.

/SaZa 86a/        **D. Sacca, C. Zaniolo:**
'On the Implementation of a Simple Class of Logic Queries
for Databases', Proc. PODS 1986.

/SaZa 86b/        **D. Sacca, C. Zaniolo:**
'The Generalized Counting Method for Recursive Logic
Queries', Proc. ICDT 1986.

/Ul 85/        **J. Ullman:**
'Implementation of Logical Query Languages for
Databases', ACM TODS, vol. 10(3), 1985.

# Construction of Deterministic Transition Graphs from Dynamic Integrity Constraints

Udo W. Lipeck
Informatik, Abteilung Datenbanken
Technische Universität Braunschweig
Postfach 3329, D-3300 Braunschweig
Fed. Rep. Germany

Dasu Feng
Computer Centre
Institute of Architecture
and Engineering, Chongqing
P. Rep. China

**Abstract** – Here systems, in particular database systems, are considered whose dynamic behaviour is characterized by state sequences that evolve stepwise, and whose integrity constraints are specified by means of temporal logic. Monitoring temporal formulae in state sequences can be reduced to following paths in transition graphs by only checking nontemporal edge labels in each state. This paper presents an algorithm how to construct deterministic transition graphs from temporal formulae in a bottom-up way corresponding to the formula structure. These graphs ensure at least provisional admissibility of system behaviour up to a present state and at most potential admissibility of future behaviour. Moreover, deterministic graphs have considerable advantages over general transition graphs.

## 1. Introduction

This paper deals with the problem how to ensure at system runtime that a stepwise evolving sequence of system states satisfies long-term correctness conditions. Dynamic system behaviour from initialization up to a present state should be monitored without knowing its future continuation and without remembering complete past states.

In particular, we think of integrity monitoring in database systems. In addition to *static* integrity constraints describing correct database contents at each moment, *dynamic* constraints describing correct database behaviour over the course of time have to be considered. The latter determine which sequences of database states are "admissible". As suggested in [e.g., Se80, CaCF82, Ku84, LiEG85, FiS86], they are formally specified by means of temporal logic: It allows to express conditions on state sequences, since predicate logic is extended by temporal operators like "**from... always ... until ...**" or "**sometime ... before ...**". Thus, arbitrary past and future states within a sequence can be related by one formula.

Traditionally, temporal logic has mainly been applied to verify concurrent programs by proving properties of execution paths [MaP81, Ma82, ClES86, Kr87] or to synthesize programs from given temporal specifications [Wo83, MaW84]. The latter work utilizes decision procedures for the satisfiability of formulae in propositional-temporal logic, i.e. the extension of propositional (!) logic by temporal operators: From temporal formulae, so-called "model graphs" are constructed whose paths correspond to admissible sequences of (program) states. Then, a formula is satisfiable iff the resulting graph is nonempty. Complexity issues have been investigated in [SiC85].

Our monitoring problem differs from verification, since the set of executable state sequences is not fully prescribed at definition time: The system must be able to react to every user action chosen from some basic commands, so that a runtime monitor is needed that accepts only those sequences which satisfy the constraints. Techniques for satisfiability analysis, however, have proven to be helpful here. In our former papers on monitoring dynamic integrity [LiS87, SaL87], we have adapted and generalized the construction of model graphs from simple programs to database systems, and we have introduced their application as monitoring schemes. Since the graphs are used to examine state transitions, we call them "transition graphs".

A transition graph can be constructed from a dynamic integrity constraint during system definition time. The initial node is labelled with the original constraint, whereas the other nodes contain such constraint parts or modifications that remain to be monitored in the future. The edges, however, are labelled with nontemporal conditions that must be checked after each state transition. During runtime, a monitor has to interpret the graph by following paths that correspond to the given state sequence. In each new state, it has to check the outgoing edge labels of all reached nodes. As long as one edge applies, the state sequence is accepted up to the present state. Thus, evaluation of temporal formulae in state sequences can be reduced to tests of nontemporal formulae in states. Details of the monitor algorithm are given in [LiS87].

Originally, transition graphs were constructed from pure propositional-temporal formulae (with boolean variables as the only basic formulae) by a tableau-driven method as in [Wo83, MaW84]. Our higher level algorithm in [LiS87, SaL87] works with arbitrary nontemporal basic formulae and utilizes an iterative conversion of temporal formulae into a disjunctive normalform. The resulting graphs often are nondeterministic, i.e., more than one edge can apply in some states and more than one path has to be followed for one state sequence. On principle, deterministic graphs could be derived afterwards similar to the powerset construction of automata theory.

In this paper we present a totally different construction algorithm, which directly delivers deterministic graphs. It composes subgraphs in a bottom-up manner corresponding to the formula structure. These so-called "standard" transition graphs show considerable advantages over general transition graphs, since they support incremental constructions and they simplify monitoring.

The rest of the paper is structured as follows. The next two sections briefly introduce definitions of temporal formulae and general transition graphs. Then sections 4 and 5 present and verify the construction of standard transition graphs. This work is based on detailed elaborations in [FeL87, Li87].

## 2. Temporal Formulae

Properties of single states can be expressed by formulae of (many-sorted) first order predicate logic using functions and predicates on objects as nonlogical symbols. States are structures where the sort, function, and predicate symbols are appropriately interpreted. Apart from these symbols, variables, logical connectives (∧,∨,¬,...), and quantifications (∀,∃) over objects appear in formulae. Properties of state *sequences*, however, are expressed by *temporal formulae* that may be built from nontemporal (i.e. first-order) basic formulae by using logical connectives, object quantification, and

- temporal quantification by **always, sometime,** or **next,**
- bounded temporal quantification by **from** ... **always/sometime** ... **before/until.**

Integrity constraints are expressed only in **propositional-temporal logic** avoiding object quantifications outside basic parts. Formulae are evaluated in infinite sequences of states $\underline{\sigma}$ = $\langle \sigma_0, \sigma_1, ... \rangle$ for substitutions $\theta$ of their free variables by objects occurring in the states.

**Notation 2.1:** If a formula $\varphi$ is **valid** in $\underline{\sigma}$ for $\theta$, we write "$[\underline{\sigma},\theta] \vDash \varphi$". "$\underline{\sigma} \vDash \varphi$" means that $\varphi$ is valid in $\underline{\sigma}$ for arbitrary substitutions; in this case, the sequence $\underline{\sigma}$ is called **admissible** wrt $\varphi$. ☐

Validity of propositional-temporal formulae is defined inductively:

**Definition 2.2:** Let $\alpha, \psi, \tau$ be temporal formulae, $\rho$ a nontemporal formula, $\underline{\sigma}$ a state sequence, and $\theta$ a substitution. $\underline{\sigma}_i$ denotes the i-th tail sequence $\langle \sigma_i, \sigma_{i+1}, ... \rangle$, and the index $\mu\tau$ stands for the first occurrence of $\tau$, i.e. $\mu\tau = \min (\{j | [\underline{\sigma}_j, \theta] \vDash \tau\} \cup \{\infty\})$.

(0)   $[\underline{\sigma},\theta] \vDash \rho$   iff $\rho$ is valid in the state $\sigma_0$ for $\theta$ (in the classic sense).

(1)   Logical connectives $\wedge, \vee, \neg, ...$ are interpreted as usual.

(2a)   $[\underline{\sigma},\theta] \vDash$ **always** $\psi$      iff for all $i \geq 0$: $[\underline{\sigma}_i, \theta] \vDash \psi$

(2b)   $[\underline{\sigma},\theta] \vDash$ **sometime** $\psi$     iff there exists $i \geq 0$, such that $[\underline{\sigma}_i, \theta] \vDash \psi$

(3)   $[\underline{\sigma},\theta] \vDash$ **next** $\psi$         iff $[\underline{\sigma}_1, \theta] \vDash \psi$

(4a)   $[\underline{\sigma},\theta] \vDash$ **always** $\psi$ **before** $\tau$   iff for all $i \geq 0$: $i < \mu\tau$ implies $[\underline{\sigma}_i, \theta] \vDash \psi$

(4b)   $[\underline{\sigma},\theta] \vDash$ **sometime** $\psi$ **before** $\tau$   iff there exists $i \geq 0$, such that $i < \mu\tau$ and $[\underline{\sigma}_i, \theta] \vDash \psi$

(5)   $[\underline{\sigma},\theta] \vDash$ **always/sometime** $\psi$ **until** $\tau$   like (4a/b), with $i \leq \mu\tau$ instead of $i < \mu\tau$

(6)   $[\underline{\sigma},\theta] \vDash$ **from** $\alpha$ [**holds**] $\psi$    iff $[\underline{\sigma}_{\mu\alpha}, \theta] \vDash \psi$ (if $\mu\alpha < \infty$) / **true** (otherwise) ☐

Thus, nontemporal formulae are evaluated in the first state $\sigma_0$ of a given state sequence $\underline{\sigma}$. Temporal quantification by **always**, **sometime**, or **next** requires the argument formula $\psi$ to be valid in every, some, or the first tail sequence of $\underline{\sigma}$. This requirement can be restricted by a **before** or **until** clause to those tail sequences which start before the first occurrence of an "end" condition $\tau$ (excluding or including it). **from** instead involves a "start" condition $\alpha$. If $\psi$ is a nontemporal formula, the conditions simply refer to states $\sigma_i$ instead of tail sequences $\underline{\sigma}_i$.

**Laws 2.3:** For arbitrary temporal formulae $\alpha, \psi, \tau$, the following equivalences hold:

| | | |
|---|---|---|
| $\neg$ **sometime** $\psi$ [**before** $\tau$] | $\Leftrightarrow$   **always** $\neg \psi$ [**before** $\tau$] | (dualities) |
| **always** $\psi$ | $\Leftrightarrow$   $\psi \wedge$ **next** (**always** $\psi$) | (temporal recursions) |
| **sometime** $\psi$ | $\Leftrightarrow$   $\psi \vee$ **next** (**sometime** $\psi$) | |
| **always** $\psi$ **before** $\tau$ | $\Leftrightarrow$   $\tau \vee (\psi \wedge$ **next** (**always** $\psi$ **before** $\tau$)) | |
| **sometime** $\psi$ **before** $\tau$ | $\Leftrightarrow$   $\neg\tau \wedge (\psi \vee$ **next** (**sometime** $\psi$ **before** $\tau$)) | |
| **from** $\alpha$ **holds** $\psi$ | $\Leftrightarrow$   $(\alpha \wedge \psi) \vee (\neg \alpha \wedge$ **next** (**from** $\alpha$ **holds** $\psi$)) | |

(Analogous laws hold for "**until**"-clauses.) ☐

When specifying dynamic integrity, we assume that the complete behaviour of a system is characterized by an infinite sequence consisting of an initial state, the past states, the present state, and a possible or imaginary future. Thus constraints are meant as restrictions on infinite sequences. According to the definition of admissibility, the formulae have to be evaluated in these sequences for arbitrary objects.

**Example 2.4:** An "automobile registration" database might involve the following object sorts, functions, predicates (i.e. Bool-valued functions), and constraints:

**sorts**     CAR, CAR-OWNER

**functions**

produced, registered,
   destroyed:           CAR $\longrightarrow$ Bool            } /* status of cars */

manuf:                CAR $\longrightarrow$ CAR-OWNER    /* manufacturer of a car */

owner:                CAR $\longrightarrow$ CAR-OWNER    /* current owner      */

is-MANUF, is-GARAGE,
   is-PERSON:          CAR-OWNER $\longrightarrow$ Bool  } /* partition of car-owners */

**variables**   c: CAR, co: CAR-OWNER

**constraints**

(C1)    **from** produced(c) **holds** (**sometime** registered(c) **before** destroyed(c))

     ∧ **from** registered(c) **holds** (**always** registered(c) **before** destroyed(c))

/* Each car c must be "registered" sometime after its production and before its destruction. (A certain date or a sale by its manufacturer would be a more realistic condition.) Then it must remain registered as long as it is not destroyed. */

(C2)    **from** owner(c)≠manuf(c) **always** ¬is-MANUF(owner(c))

     ∧ **always** ( (owner(c)=co ∧ is-GARAGE(co)) ⇒

            **always** owner(c)=co **before** is-PERSON (owner(c)) )

/* When c is no longer owned by its manufacturer, it may never be passed to any manufacturer again. Whenever it belongs to a garage co, it must remain in this property before it is sold to a person. */      □

During runtime, only a finite prefix of system behaviour is known, namely the sequence from an initial state up to the present state, whereas future continuations depend on the users' choice of state transitions. Integrity monitoring, however, has to judge correctness at each moment from such a finite sequence. For this purpose, the following degrees of partial validity/ admissibility seem to be reasonable:

**Definitions 2.5:** Let a finite state sequence $\underline{\sigma} = \langle \sigma_0, \sigma_1, ..., \sigma_{n-1} \rangle$, n≥0, a substitution θ and a temporal formula φ be given.

(a) φ is **stationarily valid** in $\underline{\sigma}$ (for θ) iff it is valid in the constantly continued sequence $(\underline{\sigma} \circ \langle \sigma_{n-1}, \sigma_{n-1}, ... \rangle)$.

(b) φ is **potentially valid** in $\underline{\sigma}$ (for θ) iff there exists some infinite continuation $\underline{\sigma}' = \langle \sigma_1', \sigma_2', .... \rangle$, such that $[\underline{\sigma} \circ \underline{\sigma}', \theta] \models \varphi$.

(c) φ is **provisionally valid** in $\underline{\sigma}$ iff it is valid in $(\underline{\sigma} \circ \underline{\lambda})$ according to def. 2.2, where $\underline{\lambda}$ is a distinguished tail sequence assumed to satisfy arbitrary formulae.     □

Stationary admissibility (a) of a finite sequence says that any state manipulation can be finished after the present state without invalidating the constraint. In many situations, only the weaker requirement of potential admissibility (b) can be satisfied, since future manipulations are expected to handle pending constraint parts like **sometime**-arguments. Thus this notion is most desirable as the goal of monitoring; it guarantees at least one correct continuation. Provisional admissibility (c), however, is sufficient, if one optimistically assumes that every constraint can become valid at some future time. It excludes all definitive constraint violations that have occurred up to the present state, e.g. by exceeding an end condition.

# 3. Transition Graphs

The essential means of reducing constraints on entire state sequences to conditions on state transitions are the following graphs.

**Definition 3.1:** Let $\varphi$ be a propositional-temporal formula over nontemporal basic formulae BF. A **transition graph** $T=\langle V,E,v_0,\nu,\eta\rangle$ **for** $\varphi$ consists of

- a rooted directed graph $\langle V,E,v_0\rangle$ with a finite set V of nodes, edges $E\subseteq V\times V$, and a root $v_0\in V$,
- a node labelling $\nu$ by propositional-temporal formulae over BF with $\nu(v_0)=\varphi$,
- an edge labelling $\eta$ by propositional formulae (i.e. logical combinations) over BF.  $\square$

Transition graphs shall be used to analyse state sequences by searching corresponding paths, whose edge labels successively hold in the states of a sequence. Such computations require to "mark" the nodes that have been reached at each moment for each substitution:

**Definition 3.2:** The **marking** $M_T(\varrho,\theta)$ of T for a finite state sequence $\varrho=\langle\sigma_0,...,\sigma_{n-1}\rangle$ with length n and for a substitution $\theta$ of the free variables in the labels (i.e. in $\varphi$) is defined as the following sets of nodes:

for n=0: $M_T(\varrho,\theta)=\{v_0\}$    for n>0: $M_T(\varrho,\theta)=\mathrm{tr}(M_T(\langle\sigma_0,...,\sigma_{n-2}\rangle,\theta),\sigma_{n-1})$

where    $\mathrm{tr}(M,\sigma)=\{\,v'\in V\mid \exists v\in M\ \exists e=(v,v')\in E:\ [\sigma,\theta]\models\eta(e)\,\}$    for any $M\subseteq V$  $\square$

The definition of tr is called **transition rule** and formalizes the stepwise processing of the graph: A node v' is marked after passing a state $\sigma$ iff there exists an edge e from a marked node v to v' such that the label $\eta(e)$ is valid in $\sigma$. If no edge starting from the old marking applies, tr yields the empty set, and the inspected state sequence is no longer "accepted".

**Definition 3.3:** A finite state sequence $\varrho$ is **accepted** by T iff the marking $M_T(\varrho,\theta)$ is nonempty for every substitution $\theta$.  $\square$

If the transition graph is to be used for monitoring the temporal formula $\varphi$ in state sequences, it has to obey a semantic restriction on its labels:

**Definition 3.4:** T is **correct** iff the following equivalence holds in infinite state sequences for each node v with outgoing edges $e_k=(v,v_k)$, k=1,...,d (d$\geq$0):

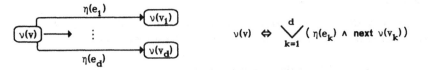

$$\nu(v) \Leftrightarrow \bigvee_{k=1}^{d}(\,\eta(e_k)\wedge\mathrm{next}\ \nu(v_k))$$

$\square$

A node label is valid in a state sequence iff at least one outgoing (nontemporal!) edge label holds in the first state and the label of the corresponding target node is valid in the first tail sequence. Thus node labels in a graph must be constraint parts or modifications that remain to be monitored in the future.

**Example 3.5:** The following (nondeterministic) transition graphs for non-nested temporal operators ($\psi,\tau$ nontemporal) are correct because of laws 2.3:

By induction on length of state sequences, a theorem can be concluded from correctness. It states the basic relationship between the validity of a formula $\varphi$ in an infinite sequence $\sigma$ and the acceptance of $\sigma$ by a correct transition graph T for $\varphi$.

**Theorem 3.6:** For each substitution $\theta$ and each $n \in \mathbb{N}$ holds:

$$[\sigma,\theta] \models \varphi \quad \text{iff} \quad T \text{ accepts } \sigma^{(n)} := \langle \sigma_0,...,\sigma_{n-1} \rangle \quad \text{and} \quad [\sigma_n,\theta] \models \bigvee_{v \in M_T(\sigma^{(n)},\theta)} \nu(v) \quad (*)$$

**Corollary 3.7:** $\varphi$ is potentially valid in a finite sequence $\sigma^{(n)}$ (for $\theta$) iff T accepts $\sigma^{(n)}$ and the formula $(*)$ is satisfiable.

Thus acceptance up to a present state is only a necessary condition for potential admissibility. In addition, the formula represented by the marked nodes must be satisfiable in some future continuation. If, e.g., all nodes are labelled with satisfiable formulae only, the notions are equivalent; in this case, the graph is called **reduced.**

Particularly useful for monitoring are special transition graphs:

- In *deterministic* graphs, at most one outgoing edge applies in each state, so that markings consist of single nodes only.

- In *iteration-invariant* graphs, markings do not change on a state iteration; then monitoring can be restricted to true state changes only.

**Definition 3.8:** A transition graph is called **deterministic** iff for each node $v$ with outgoing edges $e_k = (v,v_k)$, $k=1,...,d$ holds:

$$\eta(e_k) \wedge \eta(e_{k'}) \Leftrightarrow \textbf{false} \quad \text{for all } k,k' \in \{1,...,d\},\ k \neq k'$$

i.e., labels of different outgoing edges exclude each other.

Within our construction, we will use **completed** transition graphs where an additional node **error** (with label **false**) exists, and where each node has an outgoing edge to **error** labelled with the logical complement of the other outgoing edge labels. The definition of correctness is not touched by that completion, whereas markings must never include the node **error**.

**Lemma 3.9:** A completed deterministic transition graph T is correct (def. 3.4) iff for each node $v$ and each outgoing edge $e_k = (v,v_k)$, $k \in \{1,...,d\}$, holds:

$$\nu(v) \wedge \eta(e_k) \Leftrightarrow \eta(e_k) \wedge \textbf{next } \nu(v_k)$$

**Definition 3.10:** A transition graph is called **iteration-invariant** iff there exists a loop $e'=(v',v')$ for each edge $e=(v,v')$, such that $\eta(e) \Rightarrow \eta(e')$.

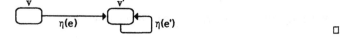

# 4. Graph Construction

Given a temporal formula $\varphi$, the following algorithm constructs a deterministic transition graph for φ in a bottom-up manner corresponding to the composition of φ from subformulae: The graph is built by stepwise composition or manipulation of smaller graphs for subformulae. Essentially, the algorithm defines such operators on transition graphs that correspond to the logical and temporal operators.

**Algorithm 4.1:** The *standard transition graph* for φ, denoted by $ST_φ$ or by $[[φ]]$ in the diagrams, is constructed inductively according to the structure of φ:

**(1)** For φ ≡ **true** or φ ≡ **false** we simply define:

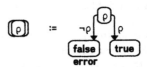

(Loops at nodes labelled with "**true**" or "**false**" will be omitted in the sequel.)

**(2)** For other nontemporal formulae ρ:

**(3ff.)** Now let φ be composed of a logical or temporal operator and formulae $φ_1/φ_2$. The following system of (recursive) rules defines corresponding operators on standard transition graphs. We inductively assume that such graphs have already been constructed for $φ_1$ and $φ_2$; let the outgoing edges of their initial nodes and the target nodes be labelled as follows:

(Note that a target node may coincide with the source node.)

In particular, we have standard transition graphs for the γ-formulae, too: the subgraphs reachable from the correspondingly labelled nodes.

**(3)** In the case φ ≡ $φ_1 ∧ φ_2$, a node with label "$φ_1 ∧ φ_2$" is generated which has an outgoing edge labelled "$β_{1i} ∧ β_{2j}$" and leading into the standard transition graph for $γ_{1i} ∧ γ_{2j}$ for each pair (i,j), i=1,..,d / j=1,...,e. That graph can in turn be constructed by applying this rule recursively. (For φ ≡ $φ_1 ∨ φ_2$ by analogy.) We abbreviate these rules to following diagrams:

Rules for all other cases are given by diagrams only:

**(4)**  **(5)**

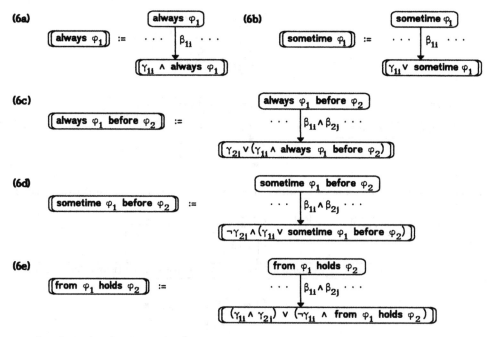

(6a)

$$\llbracket \text{always } \varphi_1 \rrbracket \ :=$$

always $\varphi_1$
$$\cdots \ \Big\downarrow \beta_{1i} \ \cdots$$
$$\llbracket \gamma_{1i} \wedge \text{always } \varphi_1 \rrbracket$$

(6b)

$$\llbracket \text{sometime } \varphi_1 \rrbracket \ :=$$

sometime $\varphi_1$
$$\cdots \ \Big\downarrow \beta_{1i} \ \cdots$$
$$\llbracket \gamma_{1i} \vee \text{sometime } \varphi_1 \rrbracket$$

(6c)

$$\llbracket \text{always } \varphi_1 \text{ before } \varphi_2 \rrbracket \ :=$$

always $\varphi_1$ before $\varphi_2$
$$\cdots \ \Big\downarrow \beta_{1i} \wedge \beta_{2j} \ \cdots$$
$$\llbracket \gamma_{2j} \vee (\gamma_{1i} \wedge \text{always } \varphi_1 \text{ before } \varphi_2) \rrbracket$$

(6d)

$$\llbracket \text{sometime } \varphi_1 \text{ before } \varphi_2 \rrbracket \ :=$$

sometime $\varphi_1$ before $\varphi_2$
$$\cdots \ \Big\downarrow \beta_{1i} \wedge \beta_{2j} \ \cdots$$
$$\llbracket \neg\gamma_{2j} \wedge (\gamma_{1i} \vee \text{sometime } \varphi_1 \text{ before } \varphi_2) \rrbracket$$

(6e)

$$\llbracket \text{from } \varphi_1 \text{ holds } \varphi_2 \rrbracket \ :=$$

from $\varphi_1$ holds $\varphi_2$
$$\cdots \ \Big\downarrow \beta_{1i} \wedge \beta_{2j} \ \cdots$$
$$\llbracket (\gamma_{1i} \wedge \gamma_{2j}) \vee (\neg\gamma_{1i} \wedge \text{from } \varphi_1 \text{ holds } \varphi_2) \rrbracket$$

etc. (similar rules for "... until ...")

Even the direct recursions above, i.e. applications of operators to the same arguments like in "$\gamma_{1i} \wedge$ **always** $\varphi_1$" are well-defined, since any construction of one new "level" of the graph needs to know only the outermost level of the argument graph, which is just defined by the respective rule. During the construction steps, following meta-rules must always be observed: Only such edges are included whose labels cannot be transformed into **false** by means of propositional laws. Each new node or edge label is simplified according to propositional equivalences as far as possible; then outgoing edges (of the same node) and nodes with propositionally equivalent labels are merged. (Node merging includes appropriate edge changes.) For these manipulations, the entire subgraph constructed so far has to be taken into consideration as a global context. – Finally, the **error** node can be deleted.  □

**Example 4.2:** The standard transition graph for the formula (C1) of example 2.4 (c:CAR)

> from produced(c)  **holds** (**sometime** registered(c) **before** destroyed(c))
> ∧ **from** registered(c) **holds** (**always** registered(c) **before** destroyed(c))

is constructed in following steps, using the abbreviations

$$\pi \equiv \text{produced(c)} , \quad \rho \equiv \text{registered(c)} , \quad \text{and} \quad \delta \equiv \text{destroyed(c)} .$$

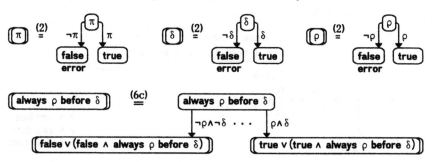

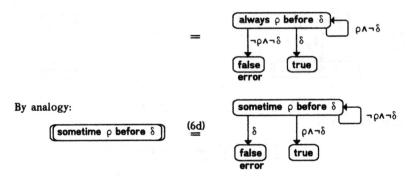

By analogy:

$$\boxed{\boxed{\text{sometime } \rho \text{ before } \delta}} \quad \overset{\text{(6d)}}{=}$$

Graphs for formulae **from π holds φ₂** with a nontemporal condition π can be derived generally:

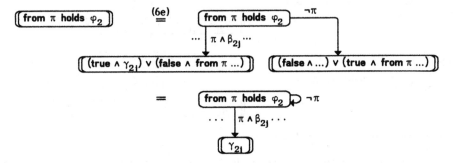

Then the following graphs result for the two subformulae of φ: (Edges into **error** are omitted.)

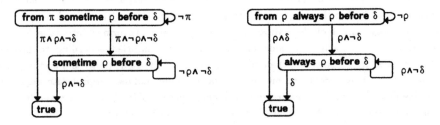

Now, recursive application of the conjunction rule (3) yields the final graph below, if we additionally assume the static constraint: registered(c) ⇒ produced(c), i.e. ρ ⇒ π .

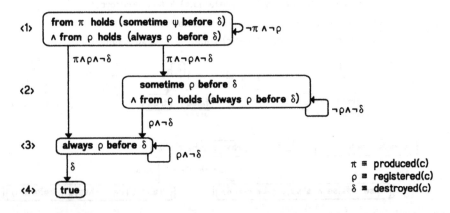

$$\pi \equiv \text{produced(c)}$$
$$\rho \equiv \text{registered(c)}$$
$$\delta \equiv \text{destroyed(c)}$$

Intuitively the graph represents the lifecycle of each object c wrt (C1); the nodes correspond to situations like ⟨1⟩ ≙ "in production", ⟨2⟩ ≙ "to be registered", ⟨3⟩ ≙ "normal (registered) life", and ⟨4⟩ ≙ "after destruction". Missing edges indicate possible constraint violations: The graph forbids, e.g., any deregistration ($\neg\rho$) after the first registration ($\rho$) and any destruction ($\delta$) without preceding registration ! In this sense, transition graphs help a system designer to analyse the local impacts of the global integrity specification on state transitions.  □

**Example 4.3:** For the other constraint (C2), the main bottom-up construction steps are listed below. Simplifications of edge labels make use of the mutual exclusion between the predicates is-MANUF, is-GARAGE, is-PERSON and of the implication:

$$\text{owner}(c) = \text{manuf}(c) \Rightarrow \text{is-MANUF}(\text{owner}(c)).$$

- $\varphi_{21} \equiv$ **from** owner(c)≠manuf(c) **always** ¬is-MANUF(owner(c))

$[\![\varphi_{21}]\!]$ = **from** owner(c)≠manuf(c) **always** ¬is-MANUF(owner(c)) ⟲ owner(c) = manuf(c)

[owner(c)≠manuf(c) ∧] ¬is-MANUF(owner(c))

**always** ¬is-MANUF(owner(c)) ⟲ ¬is-MANUF(owner(c))

- $\varphi_{221} \equiv \gamma \Rightarrow$ **always** owner(c)=co **before** is-PERSON(owner(c))
  where $\gamma \equiv$ owner(c)=co ∧ is-GARAGE(co)

$[\![\varphi_{221}]\!]$ = ¬γ ∨ **always** owner(c)=co **before** is-PERSON(owner(c))

¬γ        γ[∧ owner(c)=co ∧ ¬is-PERSON(co)]

**true** ◄        **always** owner(c)=co **before** is-PERSON(owner(c)) ⟲

is-PERSON(owner(c))        owner(c)=co ∧ ¬is-PERSON(owner(c))

- $\varphi_{22} \equiv$ **always** $\varphi_{221}$

$[\![\varphi_{22}]\!]$ $\overset{(6c)}{=\!=}$

**always** $\varphi_{221}$

¬γ        γ

**true ∧ always** $\varphi_{221}$        **(always ... before ...) ∧ always** $\varphi_{221}$

= 

¬γ ⟲ $\varphi_{22}$

[¬γ ∧] is-PERSON(owner(c))        γ        owner(c)=co ∧ ¬is-PERSON(owner(c))

**always** owner(c)=co **before** is-PERSON(owner(c)) ∧ $\varphi_{22}$ ⟲

- $\varphi_2 \equiv \varphi_{21} \wedge \varphi_{22}$ (constraint C2): see next page

Here, the nodes can be connected with following situations for a substitution of (c,co):

⟨1⟩ ≙ car c owned by its manufacturer
⟨2⟩ ≙ car c owned by the garage co
⟨3⟩ ≙ car c owned by a person or by a garage ≠ co

Sales to manufacturers are correctly excluded by the graph.  □

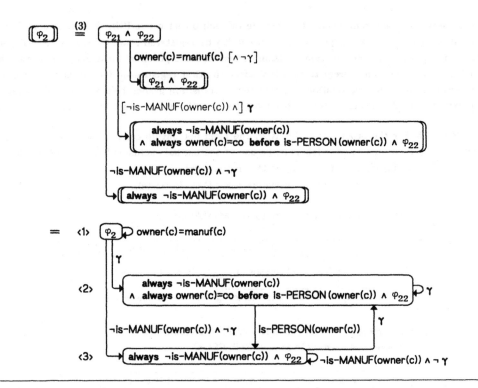

**Remark:** Obviously, the rules given in the algorithm above may be read as a graph replacement system, provided one considers (restricted) logical simplification of labels as an elementary means to identify nodes and edges. □

## 5. Verification

**Theorem 5.1:** Algorithm 4.1 terminates and constructs a correct and deterministic transition graph $ST_\varphi$ for each propositional-temporal formula $\varphi$. □

**Proof:** All properties are shown by induction on formula structure and thus on application of rules in the algorithm:

*Termination* is guaranteed by the principle of earliest simplification, since only finitely many different node labels can be generated for every operator. Let standard transition graphs for $\varphi_1/\varphi_2$ be given. Case (4) $\varphi \equiv \neg\varphi_1$ does not change the number of nodes, case (5) $\varphi \equiv$ **next** $\varphi_1$ increases it by 1, and case (3) $\varphi \equiv \varphi_1 \wedge \varphi_2$ only introduces pairwise conjunctions of given labels. Nodes in $ST_\varphi$ for $\varphi \equiv$ **always** $\varphi_1$ **before** $\varphi_2$ (6c) are labelled by formulae propositionally equivalent to combinations

$$\xi_1 \vee (\zeta_1 \wedge \xi_2) \vee \ldots \vee (\zeta_1 \wedge \ldots \wedge \zeta_p \wedge \xi_q) \vee (\zeta_1 \wedge \ldots \wedge \zeta_p \wedge \varphi)$$

with $\zeta_i$ and $\xi_j$ node labels in $ST_{\varphi_1}$ and $ST_{\varphi_2}$, respectively. Again, only finitely many propositionally inequivalent labels can be formed. The other cases are analogous.

*Determinism* is obviously propagated from basic graphs by using either original edge labels or conjunctions; thus, given partitions are retained or refined.

The *correctness* criterion 3.9 can be concluded from laws of propositional logic, from commutability of **next** and logical operators $\wedge, \vee, \neg$, and from laws 2.3 (temporal recursions) for the temporal operators. E.g.:

(3)   ⟨node label⟩ $\wedge$ ⟨(outgoing) edge label⟩ ≡

$$(\varphi_1 \wedge \varphi_2) \wedge (\beta_{1i} \wedge \beta_{2j}) \Leftrightarrow (\varphi_1 \wedge \beta_{1i}) \wedge (\varphi_2 \wedge \beta_{2j})$$

$$\overset{1}{\Leftrightarrow} (\beta_{1i} \wedge \textbf{next } \gamma_{1i}) \wedge (\beta_{2j} \wedge \textbf{next } \gamma_{2j})$$

$$\Leftrightarrow (\beta_{1i} \wedge \beta_{2j}) \wedge \textbf{next } (\gamma_{1i} \wedge \gamma_{2j})$$

$$\equiv \text{⟨edge label⟩} \wedge \textbf{next} \text{ ⟨target node label⟩}$$

(6c)   (**always** $\varphi_1$ **before** $\varphi_2$) $\wedge$ $(\beta_{1i} \wedge \beta_{2j})$

$$\overset{2}{\Leftrightarrow} (\varphi_2 \vee (\varphi_1 \wedge \textbf{next} (\textbf{always } \varphi_1 \textbf{ before } \varphi_2)) \wedge (\beta_{1i} \wedge \beta_{2j})$$

$$\Leftrightarrow ((\varphi_2 \wedge \beta_{2j}) \wedge \beta_{1i}) \vee ((\varphi_1 \wedge \beta_{1i}) \wedge \beta_{2j} \wedge \textbf{next} (\textbf{always } \varphi_1 \textbf{ before } \varphi_2))$$

$$\overset{1}{\Leftrightarrow} ((\beta_{2j} \wedge \textbf{next } \gamma_{2j}) \wedge \beta_{1i}) \vee ((\beta_{1i} \wedge \textbf{next } \gamma_{1i}) \wedge \beta_{2j} \wedge \textbf{next} (\textbf{always } \varphi_1 \textbf{ before } \varphi_2))$$

$$\Leftrightarrow (\beta_{1i} \wedge \beta_{2j}) \wedge \textbf{next} (\gamma_{2j} \vee (\gamma_{1i} \wedge \textbf{always } \varphi_1 \textbf{ before } \varphi_2))$$

1) by induction hypothesis   2) by temporal recursion

(4)   Only the negation operator crucially needs a completed and deterministic graph:

$$\neg \varphi_1 \wedge \beta_{1i} \overset{1}{\Leftrightarrow} \neg \left( \bigvee_{k=1}^{d} (\beta_{1k} \wedge \textbf{next } \gamma_{1k}) \right) \wedge \beta_{1i}$$

$$\Leftrightarrow \beta_{1i} \wedge (\neg \beta_{1i} \vee \textbf{next } \neg \gamma_{1i}) \wedge \bigwedge_{k \neq i} (\neg \beta_{1k} \vee \textbf{next } \neg \gamma_{1k})$$

$$\Leftrightarrow (\beta_{1i} \wedge \textbf{next } \neg \gamma_{1i}) \wedge \left( \bigwedge_{k \neq i} (\neg \beta_{1k}) \vee \bigwedge_{k \neq i} (\textbf{next } \neg \gamma_{1k}) \right)$$

$$\overset{(*)}{\Leftrightarrow} \beta_{1i} \wedge \textbf{next } \neg \gamma_{1i}$$

1) by induction hypothesis (using definition 3.4 of correctness)

(*) holds, since the edge labels $\beta_{1k}$ form a partition, so that $\beta_{1i} \Leftrightarrow \bigwedge_{k \neq i} (\neg \beta_{1k})$.   □

**Theorem 5.2:** If $\varphi$ does not contain the **next**-operator, $ST_\varphi$ is iteration-invariant.   □

**Proof:** Graphs for nontemporal formulae (cases 1,2) trivially contain loops as required by definition 3.10, namely the (implicit) loops labelled "**true**" at the "**true**"/"**false**"-nodes. Iteration-invariance is inductively obtained for general formulae, since all graph operators except **next** (5) transform given loops into loops again: Let us assume that appropriate loops labelled with $\beta'_{1i}$ and $\beta'_{2j}$ exist for the node/edge labels $\varphi_1/\beta_{1i}$ and $\varphi_2/\beta_{2j}$, respectively:

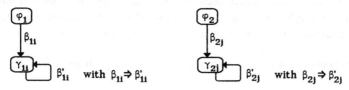

Along these edges, our construction yields the following first two levels of transition graphs for composite formulae:

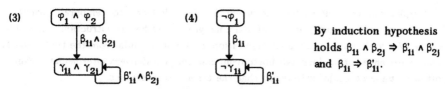

By induction hypothesis holds $\beta_{1i} \wedge \beta_{2j} \Rightarrow \beta'_{1i} \wedge \beta'_{2j}$ and $\beta_{1i} \Rightarrow \beta'_{1i}$.

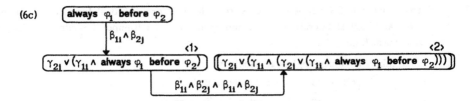

(6c)

Due to idempotency laws, nodes ‹1› and ‹2› are labelled equivalently:

$$\gamma_{2j} \vee (\gamma_{11} \wedge (\gamma_{2j} \vee (\gamma_{11} \wedge \text{ always } \varphi_1 \text{ before } \varphi_2)))$$
$$\Leftrightarrow \gamma_{2j} \vee (\gamma_{11} \wedge \gamma_{2j}) \vee (\gamma_{11} \wedge \gamma_{11} \wedge \text{ always } \varphi_1 \text{ before } \varphi_2)$$
$$\Leftrightarrow \gamma_{2j} \vee \quad\quad (\gamma_{11} \wedge \text{ always } \varphi_1 \text{ before } \varphi_2)$$

So they concide, and the loop satisfies $\beta_{11} \wedge \beta_{2j} \Rightarrow \beta'_{11} \wedge \beta'_{2j} \wedge \beta_{11} \wedge \beta_{2j}$ because of the induction hypothesis. Similar derivations work for all cases under (6). □

**Corollary and Remark 5.3:** Every finite state sequence $\underline{\sigma}$ which is potentially admissible wrt $\varphi$ is accepted by $ST_\varphi$ (because of corollary 3.7 and correctness of $ST_\varphi$). Every accepted sequence is provisionally admissible wrt $\varphi$ [Li87]. □

Thus, the constructed graphs guarantee a degree of acceptance that lies between provisional and potential admissibility (def. 2.5). In other words, these graphs detect all errors in system behaviour up to the present state (violations of provisional admissibility), as well as some (not all) inevitable errors in future behaviour at the earliest possible moment (violations of potential admissibility). In addition to algorithm 4.1, further *reductions* may be applied to delete nodes and paths with labels unsatisfiable in finite or infinite continuations [Wo83, LiS87, FeL87]. If the basic formulae of $\varphi$ can be interpreted independently (like boolean variables), the resulting graph will be reduced, so that exactly all potentially admissible sequences are accepted.

# 6. Conclusions

Monitoring dynamic integrity constraints given as temporal formulae has been reduced to passing through transition graphs, so that only changing sets of static (nontemporal) conditions need to be checked in each single state. We have presented an algorithm how to construct correct transition graphs from temporal formulae. The graphs guarantee a satisfactory degree of partial admissibility for each state sequence up to a present state.

Our standard transition graphs in particular have following advantages over general transition graphs:

- The bottom-up construction principle allows to *reuse* graphs that have already been constructed for subformulae with multiple occurrences or for typical formula patterns.

- Graphs can be constructed *incrementally*, i.e. additions to temporal formulae only lead to corresponding modifications of existing graphs. Other constructions like those in [Wo83, LiS87] would require to start computation for a new formula from the beginning. Moreover, our experience is that (manual) bottom-up construction delivers intermediate results that can often be understood intuitively, so that the error rate decreases.

- Generally, an integrity monitor that interprets transition graphs has to store all presently marked nodes in order to provide the historic information necessary to check long-term constraints. Although no complete past states need to be remembered, the amount of such extra information might be quite large in a database, since graphs have to be inspected for all occurring objects (substituted for variables). Here, deterministic graphs like ours help to reduce storage and computation costs: They require to store *only one node per substitution* and to check disjoint outgoing edges of only one node. Even if the graphs have more nodes than corresponding nondeterministic graphs, storage for the graphs themselves can be neglected, since each graph is stored only once.

Future theoretical work will tackle the problem how to integrate graph reductions into the bottom-up construction, so that the advantages from above apply to reduced graphs, too. On the practical side, the construction algorithm will become part of an interactive support system for database design, since transition graphs help to analyse integrity specification and to prepare integrity monitoring.

# References

[CaCF82] Castilho,J.M.V.de/ Casanova,M.A./ Furtado,A.L.: A Temporal Framework for Database Specifications. Proc. Int. Conf. on Very Large Data Bases 1982, 280-291

[ClES86] Clarke,E.M./ Emerson,E.A./ Sistla,A.P.: Automatic Verification of Finite-State Concurrent Systems Using Temporal Logic Specifications. ACM Trans. on Progr. Lang. and Sys. 8 (1986), 244-263

[FeL87] Feng,D.S./ Lipeck,U.W.: Monitoring Temporal Formulae Deterministically (in German). Informatik-Bericht Nr. 87-06, Techn. Univ. Braunschweig 1987

[FiS86] Fiadeiro,J./ Sernadas,A.: The INFOLOG Linear Tense Propositional Logic of Events and Transactions. Information Systems 11 (1986), 61-85

[Kr87] Kröger,F.: Temporal Logics of Programs. Springer-Verlag, Berlin 1987

[Ku84] Kung,C.H.: A Temporal Framework for Database Specification and Verification. Proc. Int. Conf. on Very Large Data Bases 1984, 91-99

[Li87] Lipeck,U.W.: On Dynamic Integrity of Databases: Fundamentals of Specification and Monitoring (in German). Habilitation Thesis, Informatics, TU Braunschweig, 1987

[LiEG85] Lipeck,U.W./ Ehrich,H.-D./ Gogolla,M.: Specifying Admissibility of Dynamic Database Behaviour Using Temporal Logic. Proc. IFIP Work. Conf. on Theoretical and Formal Aspects of Information Systems (A.Sernadas/ J.Bubenko/ A.Olive, eds.), North-Holland, Amsterdam 1985, 145-157

[LiS87] Lipeck,U.W./ Saake,G.: Monitoring Dynamic Integrity Constraints Based on Temporal Logic. Information Systems 12 (1987), 255-269

[Ma82] Manna,Z.: Verification of Sequential Programs: Temporal Axiomatization. In: Theoretical Foundations of Programming Methodology (M.Broy/ G.Schmidt, eds.) Reidel Publ. Co., Dordrecht 1982, 53-101

[MaP81] Manna,Z./ Pnueli,A.: Verification of Concurrent Programs: The Temporal Framework. In: The Correctness Problem in Computer Science (R.S.Boyer/ J.S.Moore, eds.), Academic Press, 1981, 215-273

[MaW84] Manna,Z./ Wolper,P.: Synthesis of Communicating Processes from Temporal Logic Specifications. ACM Trans. on Progr. Lang. and Sys. 6 (1984), 68-93

[SaL87] Saake,G./ Lipeck,U.W.: Foundations of Temporal Integrity Monitoring. Proc. IFIP Work. Conf. on Temporal Aspects in Information Systems 1987 (C.Rolland et al., eds.), North-Holland, Amsterdam 1988, 235-249

[Se80] Sernadas,A.: Temporal Aspects of Logical Procedure Definition. Information Systems 5 (1980), 167-187

[SiC85] Sistla,A.P./ Clarke,E.M.: The Complexity of Propositional Linear Temporal Logic. Journal of the ACM 32 (1985), 733-749

[Wo83] Wolper,P.: Temporal Logic Can Be More Expressive. Information and Control 56 (1983), 72-99

# (TIME x SPACE)-EFFICIENT IMPLEMENTATIONS
# OF HIERARCHICAL CONCEPTUAL MODELS

Nicola Santoro

School of Computer Science, Carleton University, Ottawa, Canada
and
Istituto di Scienze dell'Informazione, Universitá di Bari, Bari, Italy

## INTRODUCTION

Given an application on a data object ("the real world"), a model of the data object where all and only the relationships relevant to the application are contained is called a *relational conceptual model* To1,To2]; if the relationships are (or can be) expressed as binary relations (e.g., in [A, BFP, M, Sa, Se], the model will be called *binary*.

An *implementation* of a conceptual model is a refinement of the model where only some relationships are made explicit, the others derivable indirectly through computation. For any conceptual model, there usually exists a large number of implementations; associated with each implementation there is a *space* cost (e.g., the number of explicitly stored relationships) and a *time* cost (e.g., the amount of computation needed to derive an implicit relationship). The overall cost, or *complexity*, of an implementation will be a composite measure of both its time and space costs. In turn, the complexity of a conceptual model can be defined as the smallest of the complexities of its implementations.

In this paper, we consider a (large) class of binary conceptual models, the *hierarchical conceptual models*. This class palys an important role in that any conceptual model can be decomposed in a set of independent hierarchical models; hence, many of the results for hierarchical models can be extended to non-hierarchical ones. We investigate for this class the *time x space* complexity which is known to be bounded (above) by $O(n^2)$, where n is the number of domains in the model. In this paper, we show that there always exist an implementation of a hierarchical conceptual model whose *time x space* complexity is at most $O(n \log n)$, and present a technique for constructing such an implementation. These results are obtained by rephrasing the problem in terms of *supporting graphs* and developing a novel partitioning technique for trees.

The paper is organized as follows. In the next section, the problem is formally introduced and reformulated in terms of supporting graphs. In section 3, a tree-partitioning technique is presented and employed to obtain the claimed bound; finally, in section 4, the results are extended to a class of non-hierarchical conceptual models.

## 2. CONCEPTUAL MODELS AND SUPPORTING GRAPHS

### 2.1 Conceptual Models and Implementations

In this section, the basic terminology and definitions used thoughout the paper is briefly introduced; a complete characterization of the framework can be found in [Sa, ST].

Given two sets X and Y, a *total binary relation* r between X and Y is a subset of the Cartesian product $(X \cup \{\mu\}) \times (Y \cup \{\mu\})$ where $\mu \notin X \cup Y$ is a distinguished element called *null value* and

1. $\forall x \in X \quad \exists y \in Y \cup \{\mu\} : (x,y) \in r$;
2. $\forall y \in Y \quad \exists x \in X \cup \{\mu\} : (x,y) \in r$.

The *transpose* of a total binary relation r, denoted by $\overline{r}$, is the total binary relation defined by $(x,y) \in \overline{r}$ iff $(y,x) \in r$. In the following all relations are assumed to be total; where no ambiguity arises, the notation [X,Y] and $[\overline{X,Y}]$ will also be used to refer to a relation r between X and Y and its transpose, respectively.

Given two relations $r_1 = [X,Y]$ and $r_2 = [Y,Z]$ the *composition* of $r_1$ and $r_2$, denoted by $r_1 \cdot r_2$, is the relation constructively defined as follows:

1. $\forall x \in X, z \in Z,$     if $\exists y \in Y : (x,y) \in r_1$ and $(y,z) \in r_2$, then $(x,z) \in r_1 \cdot r_2$;
2. $\forall x \in X,$     if $\forall z \in Z \ (x,z) \notin r_1 \cdot r_2$ then $(x,\mu) \in r_1 \cdot r_2$;
3. $\forall z \in Z,$     if $\forall x \in X \ (x,z) \notin r_1 \cdot r_2$ then $(\mu,z) \in r_1 \cdot r_2$.

Given a set $R = \{[X_i, Y_i], 1 \leq i \leq n\}$ of binary relations, a *simple sequence* on R is any finite sequence $S = < [X_{i1}, Y_{i1}], ..., [X_{ik}, Y_{ik}]>$ of relations in R such that

1. $Y_{ij} = X_{i(j+1)}$     $(1 \leq j < k)$;
2. $[\overline{X_{ij}, Y_{ij}}] \notin S$     $(1 \leq j \leq k)$;
3. $[X_{ij}, Y_{ij}] \neq [X_{il}, Y_{il}]$ for $l \neq j$.

A simple sequence $S = < [X_{i1}, Y_{i1}], ..., [X_{ik}, Y_{ik}]>$ is said to be a *cycle* if $Y_{ik} = X_{i1}$. Let $S(R)$ denote the set of all simple sequences on R. In the following, only simple sequences will be considered.

A binary relational *conceptual model* on a set D of data elements is a triple $I = (D,R,C)$ where

1. $D \subseteq 2^{|D|}$ is a finite non-empty set of *domains*;
2. R is a set of total binary *relations* between domains such that if $r \in R$ then $\overline{r} \in R$;
3. $C \subseteq R \times S(R)$ is a set of time-invariant *constraints* between relations such that $(r, <r>) \in C$ for all $r \in R$.

The meaning of $(r, <r_1, ..., r_k>) \in C$ is that non accidentally $r = r_1 \cdot r_2 \cdot ... \cdot r_k$, where '·' denotes the operation of composition of relations. A conceptual model $I = (D,R,C)$ is *complete* if for every $X, Y \in D$ there exists in R at least one relation between A and B. Since every conceptual model can easily be made complete by appropriate use of the null value, only complete conceptual models will be considered here.

A conceptual model $I = (D,R,C)$ is *hierarchical* if for every cycle $S = <r_1, ..., r_k>$ there exists a relation

$_j \in S$ such that $(r_j, <r_{j-1}, r_{j-2},..., r_1, r_k, r_{k-1},...,r_{j+1}> \in C$; see figure 1.

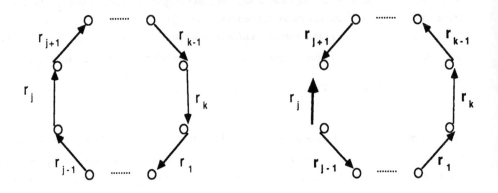

Figure 1. Meaning of $(r_j ,<r_{j-1}, r_{j-2},..., r_1, r_k, r_{k-1},...,r_{j+1}> \in C$: in the cycle on the left, relation $r_j$ is equivalent to the composition $r_{j-1} \cdot r_{j-2} \cdot ... \cdot r_1 \cdot r_k \cdot r_{k-1} \cdot ... \cdot r_{j+1}$

An *implementation* of a conceptual model $I=(D,R,C)$ is a triple $I_i=(D,R_i,f_i)$ where

1. $R_i \subseteq R$;
2. $f_i: R \rightarrow S(R_i)$ is such that $(r,f_i(r)) \in C$ for all $r \in R$.

Informally, the mapping $f_i$ specifies the computation needed to derive a relation of model $I$ which is not explicitly available in implementation $I_i$; namely, $f_i(r)$ is the sequence of relations which must be composed to compute r. Note the similarity between the mapping $f_i$ and the *search path* in independent accessing models [Se]: to the users, all relations appear to be equally accessible (as in the conceptual model) although in the implementation there exists a retrieval mechanism (the mapping), hidden to the users, to access the relations.

## 2.2 SpacexTime Complexity

Let $I$ denote the set of all implementations of $I$. The *space complexity* of $I_i \in I$, denoted by $s(I_i)$, is the number of relations explicitly stored in $I_i$; i.e., $s(I_i)=|R_i|$. The *time complexity* of $I_i \in I$, denoted by $t(I_i)$, is the lenght of the longest sequence of relations in $R_i$ which need to be composed to determine a relation in R; i.e., $t(I_i)= \max\{ \| f_i(r) \| : r \in R\}$, where $\|S\|$ denotes the number of elements in sequence S. The *time x space complexity* of a conceptual model $I$, denoted by $c(I)$ is

$$c(I) = \max \{ s(I_i) \times t(I_i) : I_i \in I \}.$$

For a hierarchical conceptual model $I$, any implementation $I_i$ with $G_i$ acyclic is space optimal since $s(I_i) = 2\,(|D|-1)$. On the other hand, the time complexity $t(I_i)$ of such an implementation is equal to the diameter of $G_i$; i.e., in the worst case $t(I_i)=O(|D|)$. Together, these bounds imply a worst-case quadratic *space x time* complexity: $c(I_i)=O(|D|^2)$.

The model itself can be considered an implementation $I_0=(D,R,\iota)$, called the *complete* implementation, where $\iota(r)=\langle r\rangle$ for all $r\in R$. This implementation is obviously time optimal since $t(I_0)=1$; on the other hand, the space complexity is $s(I_0)= 2\,|D|\,(|D|-1)$. Thus, also in this case, a quadratic *space x time* complexity $c(I_i)=O(|D|^2)$ is implied. For a summary of these bounds, see Table 1.

| implementation | space | time | space  x  time |
|---|---|---|---|
| $I_i$ with $G_i$ acyclic | $O(n)$ | $O(n)$ | $O(n^2)$ |
| complete implementation $I_0$ | $O(n^2)$ | $O(1)$ | $O(n^2)$ |
| implementation proposed here | $O(n \log n)$ | $O(1)$ | $O(n \log n)$ |

Table 1. Summary of existing and new bounds, where $n=|D|$.

In the following section, it is shown how to construct an implementation of a hierarchical conceptual model whose *space x time* complexity is $O(|D| \log |D|)$, improving on the existing quadratic bounds for the *space x time* complexity. Furthermore, the proposed implementation is time optimal; this shows that it is possible to reduce the order of magnitude of space of the complete implementation without increasing the order of magnitude of time (see table 1).

## 2.3 Supporting Graphs

Graphs provide a succint way for representing structural properties of data models (e.g., see [GF, TTT]). Given an implementation $I_i = (D_i,R_i,f_i) \in I$, we can uniquely associate to $I_i$ an undirected multigraph $G_i=(N_i,A_i)$, called a *supporting graph*, where

1. there is a one-to-one mapping $\beta_i$ between nodes and domains;
2. for each pair of relations $[X,Y]$ and $[X,Y]$ in $R_i$ there is an arc in $A_i$ between nodes $\beta^{-1}(X)$ and $\beta^{-1}(Y)$.

Supporting graphs provide a succint "structural" representation of models and their implementations. A basic property relating hierarchical models and suppporing graphs is expressed by the following lemma [Sa]:

**Lemma 1.**    A conceptual model $I$ is hierarchical if and only if there exists an implementation $I_i \in I$ such that its supporting graph $G_i$ is a tree.

Another basic property linking hierarchical models and supporting graphs is stated below [Sa]. Given two nodes x,y in a tree $G_i$, let <x,y> denote the shortest *path* from x to y in $G_i$; and let $\beta_i$<x,y> denote the sequence of domains associated to the nodes in <x,y>.

**Lemma 2.**    Let $I$ be a hierarchical model, and let $I_i$ be an implementation whose supporting graph $G_i$ is a tree.
1.   For any two nodes x and y in $G_i$, $([\beta_i(x),\beta_i(y)], \beta_i$<x,y>$) \in C$.
2.   For any path <x,y>, $\beta_i$<x,y>=r implies $\beta_i$<y,x>=r.
3.   For any two edge-disjoint paths <x,y> and <y,z> in $G_i$,
     $( [\beta_i(x),\beta_i(z)], <[\beta_i(x),\beta_i(y)],[\beta_i(y),\beta_i(z)]> ) \in C$.

The properties expressed by the above lemma have several practical implications for the construction of time-efficient implementations. In particular, from the lemma follows that to any graph G obtained by adding arcs to $G_i$ it corresponds an implementation of $I$; furthermore, the *space* complexity of this implementation is exactly twice the number of arcs in G.

These facts can be used to construct, starting from $I_i$, a time-efficient implementation of $I$ by adding arcs (relations) to the supporting graph $G_i$ (to $R_i$). Unfortunately, the transpose of a relation does not generally coincide with its inverse (i.e., $r \cdot r$ is not necessarily the identity relation); this implies that an added arc can be used only to replace the original path. In other words, there is no a priori relationship between the *diameter* of the supporting graph of an implementation and the *time* complexity of that implementation.

## 3. AN IMPROVED IMPLEMENTATION
### 3.1 Balances and Balanced Wrapping of a Tree

Given an undirected tree T=(N,A), let d(x) denote the *degree* of node $x \in N$; i.e., the number of arcs incident on x. Any node $x \in N$ will partition T into d(x) subtrees $T_1(x), T_2(x),..., T_{d(x)}(x)$. Let $t_i(x)$ be the number of nodes in $T_i(x)$; without loss of generality, let $t_1(x) \geq t_2(x) \geq ... \geq t_{d(x)}(x)$.

Define the following measures for $x \in N$:

$$\Delta_1(x) = t_1(x) - t_{d(x)}(x)$$

that is, $\Delta_1(x)$ represents the maximum unbalance in number of nodes when partitioning T (minus x and its incident arcs) into the d(x) subtrees $T_i(x)$;

$$\Delta_2(x) = | t_1(x) - \Sigma_{j=2,d(x)} t_j(x) |$$

that is, $\Delta_2(x)$ represents the unbalance in number of nodes when partitioning T into two forests with $T_i(x)$ being one forest and all other nodes and arcs (except x and its incident arcs) forming the other; and let

$$\Delta(x) = \min \{ \Delta_1(x) , \Delta_2(x) \}.$$

A node $x \in N$ for which $\Delta(x)$ is minimum is said to be a *balance* of T; i.e., x is a balance if $\Delta(y) \geq \Delta(x)$ for all $y \in N$.

**Lemma 3.** Let $x \in N$ be a balance of T. Then $\lfloor |N| / d(x) \rfloor \leq t_1(x) \leq \lfloor |N| / 2 \rfloor$.

**Proof.** ($\lfloor |N| / d(x) \rfloor \leq t_1(x)$): it trivially follows from $t_1(x) \geq t_2(x) \geq ... \geq t_{d(x)}(x)$.

($t_1(x) \leq \lfloor |N| / 2 \rfloor$): By contraddiction, let $t_1(x) = \lfloor |N| / 2 \rfloor + v = \lceil (|N|-1) / 2 \rceil + v$ for some $v \geq 1$; then $\Sigma_{j=2,d(x)} \, t_j(x) = \lfloor (|N|-1) / 2 \rfloor - v$. Since $t_1(x) > \lfloor |N| / 2 \rfloor$, then

$$t_1(x) - t_{d(x)}(x) \geq t_1(x) - \Sigma_{j=2,d(x)} \, t_j(x);$$

thus,

$$\Delta(x) = \Delta_2(x) = |\, t_1(x) - \Sigma_{j=2,d(x)} \, t_j(x) \,| = \lceil (|N|-1) / 2 \rceil + v \; - \; (\lfloor (|N|-1) / 2 \rfloor - v).$$

Let y be the neighbour of x in $T_1(x)$. Node y partitions T into d(y) subtrees $T_1(y), T_2(y), ..., T_{d(y)}(y)$ as shown in figure 2. Let $T_a(y)$ be the subtree of T composed of $T_2(x), ..., T_{d(x)}(x)$, node x and the arcs connecting x to these subtrees. Then

$$t_a(y) = 1 + \Sigma_{j=2,d(x)} \, t_j(x) = \lfloor (|N|-1) / 2 \rfloor - v + 1$$

and

$$\Sigma_{i \neq a} \, t_i(y) = t_1(x) - 1 = \lceil (|N|-1) / 2 \rceil + v - 1.$$

Thus,

$$\Delta(y) \leq \Delta_2(y) = |\, \Sigma_{i \neq a} \, t_i(y) - t_a(y) \,| = |\, (\lceil (|N|-1) / 2 \rceil + v - 1) - (\lfloor (|N|-1) / 2 \rfloor - v + 1) \,| = \Delta(x)-2.$$

That is, $\Delta(y) < \Delta(x)$ which contraddicts the hypothesis that x is a balance of T. []

The above lemma generalizes a well known property of binary trees [Br].

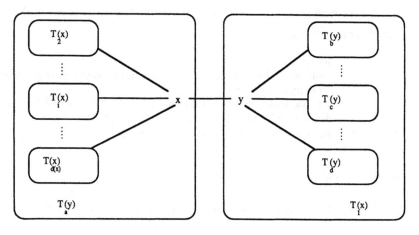

Figure 2. Graphical description of Lemma 3.

Consider now the following procedure which creates an undirected graph W from a tree T:

1. determine a balance x of T;
2. connect all nodes to x;
3. apply the same procedure recursively to each subtree $T_1(x),..., T_{d(x)}(x)$ rooted in x.

The graph W so constructed will be called a *balanced wrapping* of T.

**Lemma 4.** Let T be a tree on n≥3 nodes and let W be a balanced wrapping of T. Then the number of arcs of G which are not in T is at most
$$A(n) \leq (n-3) \log n.$$

**Proof.** By induction. The lemma can be observed to hold for n=3. Assume it to hold for 3≤n≤m-1. In the construction of G from T, all nodes (except x and its children) are connected to a chosen balance x. The same process is then applied to $T_1(x), T_2(x),..., T_{d(x)}(x)$. Therefore

$$A(m) \leq m' + \Sigma_{i=1,d(x)} A(t_i(x))$$

where $m' = m - d(x) - 1$. For simplicity, let $t_i = t_i(x)$ and $d = d(x)$. Since $t_i(x) \leq m$ for all i (1≤i≤d), by inductive hypothesis $A(t_i) \leq (t_i - 3) \log t_i$. Recall $\Sigma_i t_i = m - 1$. Therefore,

$$A(m) \leq m' + \Sigma_i (t_i - 3) \log t_i \leq m' + (\Sigma_i (t_i - 3)) \log t_1 = m' + (m - 1 - 3d) \log t_1 \leq m' + m' \log t_1.$$

Since, by Lemma 3, $t_1 \leq \lfloor m / 2 \rfloor$ it follows that

$$A(m) \leq m' + + m' \log \lfloor m / 2 \rfloor \leq m' + m' (\log m - 1) \leq m' \log m = (m-d-1) \log m.$$

Since d≥ 2, then $A(m) \leq (m-3) \log m$ and the lemma is proved. []

## 3.2 Constructing an Efficient Implementation

The method described above to construct a balanced wrapping of a tree forms the basis of the proposed technique for constructing an efficient implementation of a hierarchical conceptual model.

Given a hierarchical conceptual model $I=(D,R,C)$, let $I_i = (D,R_i,f_i)$ be an implementation such that the corresponding relational graph $G_i$ is acyclic (i.e., is a tree). Starting from $I_i$ a new implementation $I_i+ = (D,R_i+,f_i+)$ is constructed as follows:

1. set $R_i+ = \emptyset$;
2. apply the method described above to construct a balanced wrapping W of $G_i$;
3. for every arc (x,y) in W do as follows:
   3.1. add to $R_i+$ relations [X,Y] and [Y,X], where X=ß(x) and Y=ß(y);
   3.2. set $f_i+([X,Y])=<[X,Y]>$ and $f_i+([Y,X])= [Y,X]$;
4. for all relations r∈ R - $R_i+$ set $f_i+(r)= f_i(r)/ R_i+$.

**Property 1.** $s(I_i+) \leq 2 ( (|D| - 3) \log |D| + |D| - 1 )$

**Proof.** By Lemma 4 and by the fact that for each arc in W exactly two relations are added to $R_i+$. []

The time efficiency of the new implementation is derived in the following

187

**Property 2.** $t(I_i^+) \leq 2$

**Proof.** It must be shown that, for any $[X,Y] \in R$, either $[X,Y] \in R_i^+$ or there exist two relations $[X,Z]$ and $[Z,Y]$ in $R_i^+$ such that $([X,Y], < [X,Z],[Z,Y] >) \in C$. If $[X,Y] \in R_i^+$ the property trivially holds. Let this be not the case. Consider the set B of nodes in $G_i$ chosen as balances in the recursive construction of W. Since $[X,Y] \notin R_i^+$ then there exists one and only one node $\beta^{-1}(Z) \in B$ such that $\beta^{-1}(X)$ and $\beta^{-1}(Y)$ belong to two different subtrees rooted in $\beta^{-1}(Z)$. Thus, by construction of W, there are in W two arcs $(\beta^{-1}(X),\beta^{-1}(Z))$ and $(\beta^{-1}(Y),\beta^{-1}(Z))$; this implies that, by construction of $I_i^+$, the two relations $[X,Z]$ and $[Z,Y]$ are in $R_i^+$. Since $I$ is acyclic and the two paths $<\beta^{-1}(X),\beta^{-1}(Z)>$ and $< \beta^{-1}(Z),\beta^{-1}(Y)>$ are edge-disjoint, then by Lemma 2 $([X,Y], < [X,Z],[Z,Y] >) \in C$, completing the proof. []

From Properties 1 and 2, it follows:

**Property 3.** $c(I_i^+) \leq 4\ ((|D| - 3) \log |D| + |D| - 1)$

### 3.3 Special Cases

In the previous section, a general upperbound on the *space x time* complexity of the implementation $I_i^+$ have been stated. More accurate bounds can be established for special classes of $G_i$. In this section, more accurate bounds are determined for the cases when $G_i$ is a strict k-ary tree (i.e., a tree where the degree of each vertex is either 1 of k+1), a complete binary tree, or a chain.

**Property 4.** Let $G_i$ be a strict k-ary tree on $n=|D|$ nodes; then $s(I_i^+) \leq 2\ (n - k - 2) \log n + n - 1)$

The property follows from observing that, in the case of a strict k-ary tree, $d(x)=k+1$ for all x.

**Property 5.** Let $G_i$ be a complete binary tree on $n= 2^k -1$ nodes; then
$$s(I_i^+) = 2\ ((n+1) \log (n+1) - 2 n )$$

The property follows from observing that in a complete binary tree, the root is a balace of the tree; hence, $A(n)= n - 3 + 2 A((n-1)/2)$.

**Property 6.** Let $G_i$ be a chain of lenght n ; then $s(I_i^+) = 2\ ((n+1) \log(n+1) - 2 n )$

The property follows from observing that a center of a chain is also a balance; hence, $A(n) = n - 3 + A(\lceil (n-1)/2 \rceil) + A(\lfloor (n-1)/2 \rfloor)$.

## 4. EXTENSIONS

In the previous section we have described a technique for constructing an efficient implementation of hierarchical conceptual models. The technique relies on the fact that a hierarchical conceptual model has at least one implementation whose corresponding supporting graph is acyclic; furthermore, to each simple path in the supporting graph corresponds exactly one relation and its transpose. The same method can be however extended to non-hierarchical conceptual model $I$ having a supporting graph (associated to one of its implementations) with a fairly simple structure.

Consider the class of non-hierarchical conceptual model $I$ having an implementation $I_0$ whose supporting graph $G_0$ contains exactly one cycle. Choose an arbitrary direction in the cycle and call it *clockwise*. Given two nodes x and y, there exist in $G_0$ at most two simple paths $\pi$ and $\partial$ from x to y, where the direction of all cycle arcs (if any) is clockwise in $\pi$ and couter-clockwise in $\partial$. Correspondingly, for any pair of domains X and Y, there exist between them at most two distinct relations $\pi(X,Y)$ and $\partial(X,Y)$ (and their transposes $\pi(Y,X)=\pi(X,Y)$ and $\partial(X,Y)=\partial(X,Y)$), where without loss of generality $\pi(X,Y)$ and $\partial(X,Y)$ correspond to the path which use only clockwise and only conter-clockwise direction for the cycle arcs, respectively.

The proposed technique to construct an efficient implementation of $I$ is informally described below. An arbitrary node x in the cycle is chosen and "split" into two nodes x' and x" forming a tree T on n+1 nodes, where n denotes the set of original domains. A balanced wrapping W of T is then constructed with the restriction that cycle arcs can only be used in a clockwise direction. Finally, all nodes are connected to x' (if not already conneced) using cycle arcs in a couter-clockwise direction, and to x" (if not already conneced) using cycle arcs in a clockwise direction. The corresponding implementation $I_1$ (obtatained by a careful association of relations to arcs) has the following properties:

**Property 7.** $s(I_1) \leq 2 \, ( \, (n-3) \log(n+1) + 3\,n - 2)$

**Property 8.** $t(I_1) \leq 2$

**Acknowledgment**

The author whould like to thank Frank W. Tompa for introducing him to the problem and for his many helpful comments and suggestions. This research has been supported in part by the Natural Sciences and Engineering Research Council of Canada under Grant A2415.

# REFERENCES

[A]    J.R. Abrial, "Data semantics", in *Data Base Management Systems*, North Holland, 1974.

[Br]    R.P. Brent, "The parallel evaluation of general arithmetic expressions", *JACM 21* (1974).

[BFP]    G. Bracchi, A. Fedeli, P. Paolini, "A relational data base management information system", *Proc. 27th ACM Nat. Conf.*, 1972.

[GF]    C.C. Gotlieb, A.L. Furtado, "Data schemata based on directed graphs", *Int. J. Comp. Inf. Sci. 8* (1979).

[M]    R. Munz, "The WELL System: A multiuser database system based on binary relationships and graph pattern-matching", *Information Systems 3* (1978).

[Sa]    N. Santoro, "Efficient abstract implementations for relational data structures", Ph.D. Thesis, Computer Science Department, University of Waterloo, 1979.

[Se]    M.E. Senko, "DIAM II: The binary infological level", *Proc. Conf. on Data Abstraction, Definition and Management*, 1976.

[ST]    N. Santoro, F.W. Tompa, "A formalism for describing conceptual models and their abstract implementations", preliminary draft, School of Computer Science, Carleton University, 1988.

[To1]    F.W. Tompa, "Choosing an efficient internal schema", in *Systems for Large Data Bases*, North Holland, 1976.

[To2]    F.W. Tompa, "Data structure design", in *Data Structures for Computer Graphics and Pattern Recognition*, Academic Press, 1979.

[TTT]    T. Tsuji, J. Toyoda, K. Tanaka, "Relational data graphs and some properties of them", *J. Comp. Syst. Sci. 15* (1977).

# Dominance in the presence of obstacles

Mark T. de Berg          Mark H. Overmars

Department of Computer Science, University of Utrecht
P.O. Box 80.089, 3508 TB Utrecht, the Netherlands

### Abstract

Given two points $p$ and $q$ and a set of points $O$ in the plane, $p$ is said to dominate $q$ with respect to $O$ if $p$ dominates $q$ and there is no $o \in O$ such that $p$ dominates $o$ and $o$ dominates $q$. In other words, $O$ is a set of obstacles that might block the "rectangular view" from $p$ to $q$. Given a set $P$ of points and a set $O$ of obstacle points we are interested in determining all pairs $(p, q) \in P \times P$ such that $p$ dominates $q$ with respect to $O$. This generalizes notions of direct dominance and rectangular visibility that have been studied before. An algorithm is presented that solves the problem in time $O(n \log n \log \log n + k)$, where $n$ is the size of $P \cup O$ and $k$ is the number of dominance pairs found. When $P \subseteq O$ a slight change in the algorithm reduces the time bound to $O(n \log n + k)$ which is optimal.

The problem is generalized by letting $O$ exist of arbitrary objects. A reduction is presented that reduces the general case to the case in which $O$ consists of points only. For many classes of obstacles this reduction can be carried out in $O(n \log n)$ time.

## 1 Introduction

The *dominance problem* in a set of points $P$ asks for all pairs $(p, q) \in P \times P$, such that $p$ dominates $q$, denoted as $p \succ q$. (A point $p = (p_1, \ldots, p_d)$ is said to *dominate* a point $q = (q_1, \ldots, q_d)$ if $p_i \geq q_i$ for all $1 \leq i \leq d$ and $p \neq q$.) In 2-dimensional space the dominance problem can easily be solved in time $O(n \log n + k)$, where $n = |P|$ and $k$ is the number of reported pairs, using a scanline algorithm.

Since $\succ$ is a transitive relation a large number of the dominance pairs might be redundant, i.e. they might be computable from other dominance pairs. To this end the notion of *direct dominance* has been introduced. For a pair of points $(p, q)$ we say that $p$ *directly dominates* $q$, denoted as $p \stackrel{\cdot}{\succ} q$, if $p \succ q$ and there exists no point $r \in P$ such that $p \succ r$ and $r \succ q$. Güting et al.[3] have given an algorithm that computes the pairs of directly dominating points in a planar set in time $O(n \log n + k)$ where $k$ is the number of such pairs.

The direct dominance problem in the plane can also be viewed as follows: $p$ dominates $q$ if there exists an axis-parallel rectangle with $p$ as top-right vertex and $q$ as bottom-left vertex. $p$ directly dominates $q$ iff this rectangle does not contain any other point of $P$. Overmars and Wood[8] (see also [6]) consider this as a special type of visibility, called *rectangular visibility* where the points are obstacles that might block the rectangular view from $p$ to $q$. They obtain the same time bounds in a different way and consider also the query version of the problem.

In this paper we generalize this notion in the following way: we are not only given a set of points $P$, but also a set of objects, which we will call obstacles.

**Definition 1.1** *Let $P$ be a set of points, let $O$ be a set of obstacles in $d$-dimensional space, and let $p = (p_1, \ldots, p_d)$ and $q = (q_1, \ldots, q_d)$ be two points in $P$. $p$ is said to dominate $q$ with respect to $O$, denoted as $p \succ_O q$, iff $p \succ q$ and there is no point $r$ on any obstacle $o \in O$ such that $p \succ r$ and $r \succ q$.*

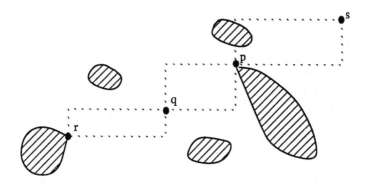

Figure 1: $p$ dominates $q$ and $q$ dominates $r$ but $s$ does not dominate $p$ with respect to the obstacles.

In other words, the rectangle with $p$ and $q$ as opposite vertices does not contain any part of any obstacle in $O$. (Note that it can contain other points from $P$.) See figure 1 for some examples. A pair $(p, q) \in P \times P$ such that $p \succ_O q$ is called a *dominance pair in $P$ with respect to $O$*, or a *dominance pair* for short.

The problem we will tackle is the following: Given a set $P$ of points and a set $O$ of obstacles, determine all dominance pairs in $P$ with respect to $O$. Note that the dominance problem is a special case by taking $O$ to be empty. The direct dominance problem can be obtained by taking $O = P$, in other words, $p \succ q$ is equivalent to $p \succ_P q$.

In the sequel of this paper we will restrict our attention to the case where $d = 2$. From now on $n$ will denote $|P| + |O|$ and $k$ the number of dominance pairs in $P$ with respect to $O$.

The paper is organized as follows.

In section 2 we present an algorithm that solves the dominance problem in the case $O$ is a planar set of points. The method is an extension of [3] and uses a combination of divide-and-conquer and plane-sweep. It runs in time $O(n \log^2 n + k)$.

Using a normalization technique (see, e.g., Karlsson and Overmars[5]) the result is improved to $O(n \log n \log \log n + k)$ in section 3.

When the number of obstacles points $n_o$ and the number of points in $P$ $n_p$ differ considerably the result can be improved. In section 4 it is shown that the problem can be solved in time $O(n \log n_p \log \log n_p + k)$ when $n_p$ is small and in time $O(n \log n + n \log n_o \log \log n_o + k)$ when $n_o$ is small.

In section 5 it is shown that in the case $P \subseteq O$ the result can be improved even further to obtain an optimal $O(n \log n + k)$ method. This immediately implies the known bounds for direct dominance and rectangular visibility in [3,8].

In section 6 we generalize the solution to the situation in which the obstacle set $O$ contains obstacles of arbitrary size and shape. It will be shown that, provided that the obstacles satisfy some simple constraints and $P$ is known, $O$ can be reduced to a set $O'$ of points only of size $O(n)$, such that for any $p, q \in P$ $p \succ_O q$ if and only if $p \succ_{O'} q$.

Finally, in section 7, we briefly review our results and indicate some directions for further research.

# 2 Points as obstacles.

Given two sets $P = \{p_1, \ldots, p_{n_p}\}$ and $O = \{o_1, \ldots, o_{n_o}\}$ of points in the plane, we are interested in computing all dominance pairs in $P$ with respect to $O$.

To this end we will develop a divide-and-conquer method. Let $V$ be the set of different $x$-coordinates of the points in $P \cup O$. If $|V| = 1$, the problem becomes 1-dimensional and is easy to solve: we sort the points in $P$ and $O$ according to $y$-coordinate and report all pairs of points in $P$ that have no point of $O$ between them in this sorted list. In this way all dominance pairs are reported in time $O(n \log n + k)$.

When $|V| > 1$, let $x_{mid} \notin V$ be such that the number of values in $V$ that are $< x_{mid}$ and the number that is $> x_{mid}$ differ by at most 1. Let $l$ be the vertical line with $x$-coordinate $x_{mid}$. $l$ divides the plane in two halves and, hence, splits $P$ and $O$ in two halves. It is obvious that whether or not two points in one half of the plane form a dominance pair cannot be influenced by an obstacle point from the other half. So we can recursively treat the two halves in the same way. After this it remains to compute the dominance pairs between the points in different halves. This merge step will be described below. We thus arrive at the following algorithm:

1. Let $V$ be the set of different $x$-coordinates in $P \cup O$ and let $n'$ be the size of $V$.

2. If $n' = 1$ solve the problem as a 1-dimensional problem as described above.

3. Otherwise, let $x_{mid} \notin V$ be chosen such that the number of elements in $V$ that are $< x_{mid}$ and the number of elements in $V > x_{mid}$ are both $\leq \lceil \frac{1}{2} n' \rceil$. Split $P$ into $P_1 = \{p \in P | p_x < x_{mid}\}$ and $P_2 = \{p \in P | p_x > x_{mid}\}$. Split $O$ into $O_1 = \{o \in O | o_x < x_{mid}\}$ and $O_2 = \{o \in O | o_x > x_{mid}\}$.

4. Report all dominance pairs in $P_1$ with respect to $O_1$ and in $P_2$ with respect to $O_2$ recursively in the same way.

5. Report all dominance pairs $(p, q) \in P_2 \times P_1$ with respect to $O$.

Step 1 of the algorithm can be performed in time $O(n)$ if we have $P$, $O$ and $V$ sorted by $x$-coordinate. After presorting on $x$-coordinate, which requires time $O(n \log n)$ these sets can be maintained sorted during the recursive calls.

When the recursion stops (in step 2) we have to sort the, say, $n_i$ points that are in the sets by $y$-coordinate. This has to be done $n'$ times, but, since we have $\sum_{i=1}^{n'} n_i = n$, the total time required for step 2 will be bounded by $O(n \log n)$ plus $O(1)$ time for every answer found.

Now let $T(n', n)$ be the time needed for the algorithm, then we have:

$$\begin{aligned} T(n', n) &= O(n \log n + k) + T'(n', n) \\ T'(n', n) &= T'(\lceil \frac{1}{2} n' \rceil, n - l) + T'(\lfloor \frac{1}{2} n' \rfloor, l) + O(n) + f(n) \end{aligned} \tag{1}$$

where $k$ is the number of dominance pairs, $0 \leq l \leq n$ and $f(n)$ is the time needed to perform the merge step (step 5). Assuming that $f(n)$ is non-decreasing and at least linear this leads to

$$T(n', n) = O(n \log n + k + \log n'(n + f(n))) = O(f(n) \log n + k) \tag{2}$$

because $n' = O(n)$.

It remains to be shown how the merge step (step 5) can be performed efficiently. First note that we only have to look at pairs $(p, q) \in P_2 \times P_1$ since all points in $P_2$ have larger $x$-coordinate than the points in $P_1$. So $p \succ q$ iff $p_y \geq q_y$.

The idea is as follows: We move a scanline downward over the plane, halting at every point in $P \cup O$. When we encounter a point $q \in P_1$, we will report all pairs $(p, q) \in P_2 \times P_1$ with

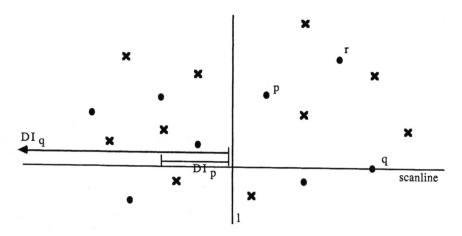

Figure 2: Some dominance intervals. $DI_r = \emptyset$. Points in $P$ are indicated with a dot, point in $O$ with a cross.

$p \succ_O q$. We know that $p$ must lie above (or on) the scanline. To find these points, for every point $p \in P_2$ above the scanline we keep track of a so-called *dominance interval* $DI_p$ at the current position $y^*$ of the scan line. This dominance interval consists of all $x$-values $< x_{mid}$ such that $x \in DI_p \Leftrightarrow p \succ_O (x, y^*)$. In other words, when the $x$-coordinate of a point $q \in P_1$ on the scanline lies in $DI_p$ then $p \succ_O q$. Note that $DI_p$ is indeed always an interval, which is of the form $]o_x, x_{mid}[$ where $o_x$ is either $-\infty$ or the $x$-coordinate of some obstacle point (or $DI_p$ is empty). See figure 2 for some examples of dominance intervals.

Suppose for the moment that no two points have the same $y$-coordinate. (If points do have the same $y$-coordinate we handle them from right to left. If a point $p \in P_2$ coincides with a point $o \in O$ then $o$ is treated first. If a point $q \in P_1$ coincides with a point $o \in O$ then $q$ is treated first. The reader can easily verify that this will be the correct order.) If we encounter a point $p \in P_2$ we must initialize $DI_p := ]-\infty, x_{mid}[$. If we encounter a point $q \in P_1$, for all $p \in P_2$ with $q_x \in DI_p$ we report the pair $(p, q)$ as a dominance pair. Obstacle points must be treated in the following way: An obstacle point $o = (o_x, o_y)$ dominates all the points to the left and below of it, so when we encounter $o$ we have to change the dominance intervals for all points $p$ that dominate $o$ in the following way: if $o \in O_1$ then for all $p$ with $o_x \in DI_p$ we must set $DI_p := ]o_x, x_{mid}[$ and if $o \in O_2$ then for all $p$ with $o_x \le p_x$ we must set $DI_p := \emptyset$.

Observe that after we have handled an obstacle point some of the dominance intervals become identical. To avoid changing all these intervals again at some later obstacle point we from now on treat them simultaneously. To this end we store identical $DI$'s only once and associate a bag with it that contains all the points $p$ for which $DI_p = DI$. This bag must allow for the following operations in constant time: inserting an element, deleting an element when we have a pointer to it and joining two bags. Moreover, all the elements should be enumerated in time the number of elements. This can be implemented e.g. as a doubly linked list.

To be able to handle obstacle points $o \in O_2$ efficiently, we must be able to determine all points in $P_2$ above the scan line that lie to the right of $o$ and remove them. A priority queue on (the $x$–coordinates of) the points in $P_2$ above the scanline will suffice for this purpose. Since we have to set $DI_p := \emptyset$ for a point $p$ to the right of $o$ in this case (i.e. remove $p$ from the bag it is in) we also store a cross pointer from the place of $p$ in the priority queue to the the place of $p$ in the bag.

We now present step 5 of the algorithm in more detail:

5. Move a scanline downward over the set of points, halting at every point $(x,y) \in P \cup O$. (To do this we need a list of points $\in P \cup O$, sorted according to $y$-coordinate. This sorted list can be obtained from the sorted lists of $P_1 \cup O_1$ and $P_2 \cup O_2$ by a simple merge. This means that we only have to sort explicitly when the recursion halts. This sorting was already performed to compute the answers in step 2.) While we move the scanline we maintain the following two data structures:

- A sorted list $L$ of the different left endpoints of the dominance intervals of points in $P_2$ above the scanline. Every left endpoint has a bag associated with it, that contains all the points in $P_2$ that have that left endpoint as the left endpoint of their dominance interval.

- A priority queue $Q$, containing the $x$-coordinates of the points in $P_2$ above the scanline.

Furthermore we maintain cross pointers from the points in $Q$ to the corresponding points in (the bags in) $L$. When we halt at a point $(x,y)$ we have the following cases:

$(x,y) \in P_1$: Walk with $x$ along $L$ as long as the left endpoint of the current bag is $< x$ and report $(p,(x,y))$ as a dominance pair for each point $p$ in these bags.

$(x,y) \in O_1$: Walk with $x$ along $L$, joining all the bags with left endpoint $\leq x$ into a new bag with $x$ as left endpoint.

$(x,y) \in P_2$: Insert $(x,y)$ in $Q$ and add it to the bag in $L$ with $-\infty$ as left endpoint (or, if necessary, create a new bag).

$(x,y) \in O_2$: Remove all points with $x$-coordinate $\geq x$ from $Q$ and, using the cross pointers, from the bags in $L$. (If a bag becomes empty, then remove the bag and its corresponding $DI$ from $L$.)

The correctness of the algorithm follows from the above discussion.

**Lemma 2.1** *The merge step can be performed in time $O(n \log n + k)$.*

**Proof.** To analyse the time complexity we have to look at the four different cases:

$(x,y) \in P_1$: We spend $O(1 + \#answers)$ time.

$(x,y) \in O_1$: We spend $O(1 + (\#(\text{bags with left endpoint} \leq x)\text{-}1))$ time. We charge the costs for joining the bags as follows to points in $P_2$: Let $q_b$ be the first point that is put in bag $b$. Then we charge the costs of joining bags $b_1, \ldots, b_s$ into bag $b_s$ to the points $q_{b_1}, \ldots, q_{b_{s-1}}$. It is clear that a point can be charged costs only once this way. Thus every point $\in P_2$ gets an extra $O(1)$ at most. The remaining $O(1)$ time of step (ii) is simply charged to $(x,y)$.

$(x,y) \in P_2$: We spend $O(\log n)$ time to insert the point in the priority queue.

$(x,y) \in O_2$: We spend $O(\#\text{deletions}(\log n + 1))$ time for deleting the points from the priority queue and from the bags. We charge these costs to the deleted points. Since a point can be deleted only once, every point $\in P_2$ is charged with this extra $O(\log n + 1)$ at most once.

The total time bound follows immediately. (Note that no sorting is required because the points were already presorted. We only have to merge the lists which requires time $O(n)$.) $\square$

So we have $f(n) = O(n \log n)$ in equation 2. This leads to the following result:

**Theorem 2.2** *All dominance pairs in a set of points $P$ with respect to a set of points $O$ can be computed in time $O(n \log^2 n + k)$, where $n = |P| + |O|$ and $k$ is the number of answers.*

# 3  Normalizing the problem.

When we take a closer look at the time analysis of the above algorithm, we see that the operations on the priority queue $Q$ form the bottleneck in the algorithm. The question thus arises wether we can use another data structure that performs these operations more efficiently. Such a data structure indeed exists for our application.

The crucial observation here is that we can normalize the problem (see e.g. [5]), i.e., convert it to the corresponding problem on a grid, without changing the dominance pairs. So we add as a preliminary step to the algorithm:

0. Replace every point $(x, y) \in P \cup O$ by $(r(x), r(y))$, where $r(x)$ and $r(y)$ are the ranks of $x$ and $y$ in the sorted order of the different $x$ and $y$-coordinates, respectively.

Since

$$r(x) \leq r(x') \Leftrightarrow x \leq x',$$

$$r(y) \leq r(y') \Leftrightarrow y \leq y' \text{ and}$$

$$(r(x), r(y)) = (r(x'), r(y')) \Leftrightarrow (x, y) = (x', y')$$

we have

$$(r(x), r(y)) \succ_O (r(x'), r(y')) \Leftrightarrow (x, y) \succ_O (x', y').$$

Hence, normalization maintains the dominance relation.

The normalization step can be performed in time $O(n \log n)$ by sorting the points by $x$- and $y$-coordinate. As this sorting was already performed by the algorithm it does not increase the time bound.

Now that we have normalized the problem we can use data structures that work on a grid. This means that we can use a VanEmdeBoas tree (see, e.g., [9],[10]) as priority queue. In such a tree INSERT and EXTRACTMAX can be performed in time $O(\log \log U)$, where $U$ is the size of the universe, in our case $U = O(n)$. Thus we can perform step 5 in $O(n \log \log n + k)$ time and we obtain the following improved result:

**Theorem 3.1** *All dominance pairs in a set of points $P$ with respect to a set of points $O$ can be computed in time $O(n \log n \log \log n + k)$, where $n = |P| + |O|$ and $k$ is the number of answers.*

# 4  Handling sets of different sizes.

In a number of applications the number of points in $P$ and the number of points in $O$ might differ considerably. In this section we will show that in such cases the method can be adapted to work more efficient. Let $n_o = |O|$ and $n_p = |P|$. We first treat the case when $n_o \gg n_p$.

**Theorem 4.1** *All dominance pairs in a set of points $P$ of size $n_p$ with respect to a set of points $O$ of size $n_o$ can be computed in time $O(n \log n_p \log \log n_p + k)$, where $k$ is the number of answers.*

**Proof.** We first perform a normalization step in the following way: Let $V_x$ be the set of different $x$-coordinates of points in $P$. We define $r'_x(x)$ as follows: If $x \in V_x$ then $r'_x(x)$ is twice the rank of $x$ in $V_x$. If $x \notin V_x$ then determine the largest $x' \in V_x$ with $x' < x$. If $x'$ does not exist $r'_x(x) = 1$, otherwise $r'_x(x) = r'_x(x') + 1$. Similar, let $V_y$ be the set of all $y$-coordinates and define $r'_y(y)$ in an analogue way. Now replace each point $p \in P \cup O$ by $p' = (r'_x(p_x), r'_y(p_y))$. It is easy to verify that in this way the dominance relation between points in $P$ with respect to $O$ is maintained. We have mapped the problem onto a grid of size at most $2n_p + 1 \times 2n_p + 1$. See

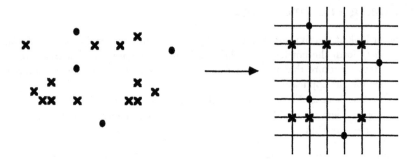

Figure 3: Normalizing the set of point when $n_p$ is small.

figure 3 for an example. As a result some obstacle points might become the same. We remove all except one of them. So the size of $O$ has become $O(n_p^2)$. This normalization can be carried out in time $O(n \log n_p)$.

Now we just run the algorithm as described above. Because the number of different $x$-coordinates is $O(n_p)$ equation 2 becomes $T(n) = O(f(n) \log n_p + k)$. Each operation on the VanEmdeBoas tree will require $O(\log \log n_p)$ and the sorting in step 2 of the method will be bounded by $O(n \log n_p)$ in total. The bound follows. □

In particular, when the number of points in $P$ is constant, the method runs in time $O(n)$ (something which can also be achieved in a more simple way by testing every pair of points in $P$ with respect to all obstacle points).

Let us now consider the case when $n_p >> n_o$.

**Theorem 4.2** *All dominance pairs in a set of points $P$ of size $n_p$ with respect to a set of points $O$ of size $n_o$ can be computed in time $O(n \log n + n \log n_o \log \log n_o + k)$, where $k$ is the number of answers.*

**Proof.** To obtain the result we use a normalization similar to the previous theorem. Let $V_x$ be the set of different $x$-coordinates of points in $O$. Define $r'_x(x)$ in a similar way as above. Now replace each point $p \in P \cup O$ by $p' = (r'_x(p_x), p_y)$. (Note that we use the normalization only on $x$-coordinate.) Unfortunately this does not maintain the dominance relation. Let $P'$ and $O'$ be the new sets. One easily verifies that $p \succ_O q \not\Leftrightarrow p' \succ_{O'} q'$ if and only if $p_x \neq q_x \wedge p'_x = q'_x$. This situation can only occur when the recursion stops, i.e., when we perform step 2 of the method. When we perform step 2 there are two possible cases: i) there is an obstacle with the $x$-coordinate. In this case also the original points were on a vertical line and we can proceed in the normal way. ii) there is no obstacle with this $x$-coordinate. In this case the original points might not lie on a vertical line. But we know that there is no obstacle point among them. So we can just compute all dominance pairs among the original points in time $O(n_i \log n_i)$ when we are left with $n_i$ points. In this way, the time spend on step 2 will be no more than a total of $O(n \log n)$. So equation 2 turns into $T(n) = O(n \log n + f(n) \log n_o + k)$. It is easily verified that $f(n) = O(n \log \log n_o)$ because the number of different $x$-coordinates at a merge step is bounded by $2n_o + 1$. □

Note that both results hold for arbitrary $n_p$ and $n_o$. Also note that it is easy to determine whether $n_p > n_o$ or not and choose the correct method in each case. Hence we can in fact solve the problem in the minimum of the two bounds without preknowledge on the sizes of $P$ and $O$.

# 5   The case $P \subseteq O$.

In some cases, like the direct dominance problem, we have the situation that $P \subseteq O$. We will now show that in this case we can improve the result to $O(n \log n + k)$. To this end we will avoid the use of a priority queue completely.

Note that, when a point $p \in P_2$ did coincide with an obstacle point $o \in O$ we treated the obstacle point first. As a result there will be no points left in the tree $Q$ to the right of $p$. But when $P \subseteq O$ this will always be the case. This means that whenever we handle a point $p \in P_2$, we have just removed all points from $Q$ with $x$-coordinate $\geq p_x$. This implies that we could as well use a stack for $Q$. When we encounter an obstacle point $o \in O_2$ we pop a number of points from the stack and when we encounter a point $p \in P_2$ we just push it on the stack. Thus we can perform both INSERT and EXTRACTMAX in $Q$ in time $O(1)$. As a result the merge step can be performed in time $O(n + k)$. (Note that the normalization is not neccessary in this case.)

**Theorem 5.1** *All dominance pairs in a set of points $P$ with respect to a set of points $O \supseteq P$ can be computed in time $O(n \log n + k)$, where $n = |P| + |O|$ and $k$ is the number of answers.*

# 6   Treating general obstacles.

We will now concentrate on the case $O$ consists of arbitrary objects rather than points. We will reduce the problem to reporting all the dominance pairs in $P$ with respect to a set $O'$ that contains only points. Then we can apply the results of the previous sections. So we are interested in finding a set $O'$ such that for all $p, q \in P : p \succ_O q \Leftrightarrow p \succ_{O'} q$.

Now let the objects in the obstacle set $O$ satisfy the following constraints:

1. The objects are connected.

2. A point inside the object can be determined in constant time.

3. The first intersection between an object $o \in O$ and an axis-parallel ray can be determined in constant time (if any). If the starting point $p$ of the ray lies in $o$ $p$ itself should be reported.

4. Whether or not a square and an object intersect can be computed in constant time.

We can compute $O'$ as follows. Consider two points $p, q \in P$ with $q$ to the left of and below $p$. As noted in the introduction, to say that $p \succ_O q$ is the same as to say that the intersection of the rectangle with $p$ and $q$ as opposite vertices with every obstacle in $O$ is empty. Therefore we define:

**Definition 6.1** *Let $p = (p_x, p_y)$ and $q = (q_x, q_y) \in \mathbf{R}^2$ with $p \succ q$. We define*

$$Rect(p, q) = \{(x, y) \in \mathbf{R}^2 | x \in [q_x, p_x], y \in [q_y, p_y]\} - \{p, q\}.$$

**Lemma 6.1** *Let $p$ and $q$ be two points in $P$, $p \succ q$, then $p \succ_O q \Leftrightarrow Rect(p, q) \cap o = \emptyset$ for all $o \in O$.*

Now note that if for two points $p$ and $q$ and an obstacle $o$ we have $Rect(p, q) \cap o \neq \emptyset$ then $o$ must either lie totally inside $Rect(p, q)$, or $o$ must intersect the boundary of $Rect(p, q)$ somewhere, or $o$ must totally cover $Rect(p, q)$. See figure 4 for some examples of the three cases. Of course we cannot afford to look at every pair of points to generate the correct obstacle points. Hence, we have to create points in a more careful way.

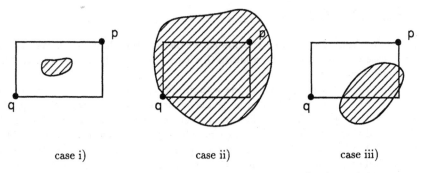

case i)  case ii)  case iii)

Figure 4: The different cases where $Rect(p,q) \cap o \neq \emptyset$.

First of all, for each obstacle $o \in O$ we take some point inside the obstacle and add it to $O'$. This will solve the first case. Secondly, for each point $p$ in $P$ we shoot a ray upwards, downwards, left and right, and determine for each obstacle the first intersection. In each direction we add to $O'$ the nearest such intersection point unequal to $p$ (if such an intersection exists). Clearly, when an object intersects the boundary of $Rect(p,q)$ one of the obstacle points created lies on this boundary and, hence, is contained in $Rect(p,q)$.

It remains to avoid the third case. When the third case occurs both $p$ and $q$ lie on (the boundary of) an obstacle. It would seem correct to remove such points but this can only be done when they do not lie on the boundary. So we choose another way of solving this. Let

$$\varepsilon = \frac{1}{2} \min(\min_{p,q \in P, p_x \neq q_x} |p_x - q_x|, \min_{p,q \in P, p_y \neq q_y} |p_y - q_y|).$$

In words, $\varepsilon$ is half of the smallest vertical or horizontal distance between points that is not equal to 0. Hence, if we draw a horizontal or vertical bar around a point $p \in P$ with width $2\varepsilon$ it does not contain any other point in $P$ with different $x$- or $y$-coordinate. For every $p \in P$ do the following: Let $S_p$ be the square with center $p$ and sides of length $2\varepsilon$. Let $NE_p$ be the north-east quadrant of $S_q$ and let $SW_p$ be the south-west quadrant, not including the boundaries (i.e., they are open squares). If $NE_p$ or $SW_p$ has a non-empty intersection with an obstacle $o \in O$ add an arbitrary point of this quadrant to $O'$. Also consider the four boundaries between quadrants, not including $p$. For each boundary, if it has a non-empty intersection with an obstacle $o \in O$ then add an arbitrary point of this boundary to $O'$. See figure 5 for some examples.

Now it is clear that when some obstacle completely covers $Rect(p,q)$ some of the points created will lie inside $Rect(p,q)$. When $Rect(p,q)$ is empty none of the points created will lie inside $Rect(p,q)$ even when $p$ or $q$ lies on the boundary of an obstacle.

Note that $|O'| \leq |O| + 10|P|$. (By choosing the points a bit more careful we could reduce this to $|O| + 5|P|$.)

**Lemma 6.2** For all $p,q \in P : p \succ_O q \Leftrightarrow p \succ_{O'} q$.

**Proof.** Follows from the above discussion. □

(Note that, using the same method it is easy to show that for any set of obstacles, not necessarily satisfying the constraints, a set $O'$ of points does exist that represents the set of obstacles. The number of points in such a set would be proportional to the total number of connected parts of all obstacles. The only problem is that it might be impossible to determine the set. Because of that we introduced the constraints.)

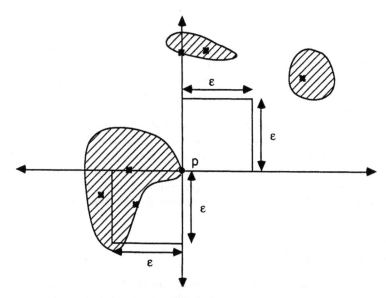

Figure 5: Points added to $O'$.

Now that we have reduced the obstacle set $O$ to a set $O'$ of size $O(n)$ that contains only points we can apply the results of the previous section. We thus find:

**Theorem 6.3** *All dominance pairs in a set of points $P$ with respect to an obstacle set $O$ can be computed in time $O(R(n) + n \log n \log \log n + k)$, where $n = |P| + |O|$, $k$ is the number of answers and $R(n)$ is the time needed to perform the reduction.*

If all points in $P$ lie on some obstacle in $O$ we can add the points in $P$ to $O'$. We then obtain $P \subseteq O'$ and obtain:

**Theorem 6.4** *All dominance pairs in a set of points $P$ with respect to an obstacle set $O$ with $P \subseteq O$ can be computed in time $O(R(n) + n \log n + k)$, where $n = |P| + |O|$, $k$ is the number of answers and $R(n)$ is the time needed to perform the reduction.*

It is obvious that $O'$ can be computed in time $O(n^2)$. But this would make the method not very efficient. In a number of cases the reduction can be performed much more efficiently.

**Theorem 6.5** *When $O$ consists of non-intersecting convex objects, satisfying the above constraints plus the constraint that the leftmost, rightmost, topmost and bottommost point can be determined in $O(1)$ time, the reduction can be performed in time $O(n \log n)$.*

**Proof.** First of all we have to determine $\varepsilon$. This can be done by sorting $P$ both horizontally and vertically in $O(n \log n)$ time. Next we determine a point in each obstacle which takes time $O(n)$.

To determine the first intersection with the rays from each point we move a scanline from left to right over the set of obstacles. With the scanline we keep a tree that holds the objects intersecting the scanline in sorted order. For each point $p \in P$ the scanline passes we can determine in $O(\log n)$ time the first intersection above and below $p$. Similar, using a vertical scan we determine the first intersection to the left and to the right of each point $p$. Obstacles can be inserted and deleted in $O(\log n)$ time.

Note that we only have to look at the square around $p$ when $p$ lies on an obstacle. Moreover this obstacle is the only one we have to consider for intersection with the square because the obstacles do not intersect. This obstacle can also be determined during the scan and points can be added if necessary. Details are left to the reader. $\square$

In fact, we only use the property that every horizontal and vertical line intersects each object in one interval and these intervals do not intersect. Hence, the result also applies to other types of obstacles.

As an example consider the case in which $O$ consists of a set of non-intersecting line segments and $P$ is the set of endpoints of these line segments. The above results show that all dominance pairs can be determined in time $O(n \log n + k)$. The dominance pairs we get this way form a kind of *rectangular visibility graph* (compare the results on normal visibility graphs, see [4,7]).

Also for other sets of obstacles the reduction can be performed efficiently. For example when we have a set of (possibly intersecting) axis-parallel line segments the reduction can be performed in time $O(n \log n)$.

# 7   Concluding Remarks.

In this paper we studied a generalization of the direct dominance problem by considering dominance pairs with respect to a set of obstacles. In the case the obstacles were points we demonstrated that all dominance pairs in a set $P$ with respect to an obstacle set $O$ can be computed in time $O(n \log n \log \log n + k)$ where $n = |P| + |O|$ and $k$ is the number of reported pairs. We improved this result in the case the two sets have unequal sizes and in the case $P \subseteq O$. In the later case we gave an $O(n \log n + k)$ solution which is optimal in the worst-case.

Finally, we generalized the result to the case the obstacle set $O$ consists of other objects than points. It was shown that given a set $P$ of points and a set $O$ of arbitrary obstacles there always exists a set $O'$ of points such that points $p$ and $q$ in $P$ form a dominance pair with respect to $O$ if and only if they form a dominance pair with respect to $O'$. When each obstacle is connected, the size of $O'$ is $O(|O| + |P|)$. If the obstacles satisfy some constraints the set $O'$ can also be found efficiently.

A number of open problems do remain. First of all it might be possible to improve out main result to $O(n \log n + k)$. The problem here is the fact that we have to maintain a priority queue for the points in $P_2$ above the scanline in the merge step of our divide-and-conquer algorithm. (Recall that $P_2$ is the subset of points in $P$ that lie to the right of the dividing line.) A way to improve the efficiency here might be to exploit the correspondence between the subsequent levels of the recursion, for example by some sort of preprocessing or by transforming the recursive algorithm into an iterative one. Our attemps have failed so far.

A second question is whether the results can be extended to higher-dimensional space. The results in [3] and [8] are not very promising here. They are able to solve the 3-dimensional direct dominance problem in time $O((n+k) \log^2 n)$ but are unable to generalize their results any further.

Other extensions might be to look at maximal elements in the presence of obstacles (i.e., points that are not dominated by any other points with repect to $O$) or counting the number of points that dominate a particular point (a kind of ECDF-counting problem, see e.g. [1,2]). Also query versions of the problems are worth studying.

## Acknowledgement

We would like to thank Derick Wood (University of Waterloo, Canada) for the helpful discussions on this topic during his stay in Utrecht in November '87.

# References

[1] Bentley, J.L., Multidimensional divide-and-conquer, *Comm. ACM* 23 (1980), pp. 214-229.

[2] Bentley, J.L., and M.I. Shamos, A problem in multivariate statistics: algorithm, data structure and applications, *Proc 15th Allerton Conference on Communication, Control and Computing*, 1977, pp. 193-201.

[3] Güting, R.H., O. Nurmi and T. Ottmann, The direct dominance problem, *Proc. 1st ACM Symp. Computational Geometry*, 1985, pp. 81-88.

[4] Ghosh, S.K., and D.M. Mount, An output sensitive algorithm for computing visibility graphs, *Proc. 28th Symp. on Foundations of Computer Science*, 1987, pp. 11-19.

[5] Karlsson, R.G., and M.H. Overmars, Normalized divide-and-conquer: A scaling technique for solving multi-dimensional problems, *Inform. Proc. Lett.* 26 (1987/88) pp. 307-312.

[6] Munro, J.I., M.H. Overmars and D. Wood, Variations on visibility, *Proc. 3rd ACM Symp. Computational Geometry*, 1987, pp. 291-299.

[7] Overmars, M.H., and E. Welzl, New methods for computing visibility graphs, *Proc. 4th ACM Symp. Computational Geometry*, 1988, pp. 164-171.

[8] Overmars, M.H., and D. Wood, On rectangular visibility, *J. Algorithms* (1988), to appear.

[9] van Emde Boas, P., Preserving order in a forest in less than logarithmic time and lineair space, *Inform. Proc. Lett.* 6 (1977) pp. 80-82.

[10] van Emde Boas, P., R. Kaas and E. Zijlstra, Design and implementation of an efficient priority queue, *Math. Systems Theory* 10 (1977) pp. 99-127.

# Separating a Polyhedron by One Translation from a Set of Obstacles*
## (Extended Abstract)

Otto Nurmi[†]

*Institut für Informatik, Universität Freiburg*
*Rheinstraße 10–12, D-7800 Freiburg i.Br.*
*Fed. Rep. of Germany*


Jörg-R. Sack[‡]

*School of Computer Science, Carleton University*
*Ottawa, Ontario, Canada K1S 5B6*

We present efficient algorithms for several problems of movable separability in 3-dimensional space. The first algorithm determines *all* directions in which a convex polyhedron can be translated through a planar convex polygonal window. The algorithm runs in linear time. This is a considerable improvement over the previous $O(n^2 \log n)$ time algorithm of [17], where $n$ is the total number of vertices in the objects. The second algorithm computes, in $O(n)$ time, *all* directions in which a convex polyhedron can be translated to infinity without collisions with a convex obstacle ($n$ is the number of vertices of the polyhedra). A generalization of the plane-sweep technique, called "sphere-sweep", is given and provides an efficient algorithm for the last problem which is: determine all directions in which a convex polyhedron can be separated from a set of convex obstacles. Our results are obtained by avoiding the standard technique of motion planning problems, the $\Omega(n^2)$ time computation of the Minkowski differences of the polyhedra.

## 1. Introduction

Motion plannig problems arise in areas like computer-aided manufacturing, computer graphics, and robotics (see the survey in [18]). One class of such problems are the "separability" problems (see, e.g., [2,17]). Given a complex object consisting of several parts, one is asked to loosen one of them by motions such as translations and/or rotations.

In this paper we consider separability problems in which the objects are convex polyhedra and one of them must be moved by one translation motion arbitrarily far away. The object that must be separated is not allowed to collide with other objects during the translation. Such problems are often called collision avoiding problems.

The *Minkowski difference* of two sets, $A$ and $B$, is $A - B = \{\, a - b \mid a \in A, b \in B \,\}$. Lozano-Perez and Wesley [6] introduced an algorithmic technique based on Minkowski differences of the obstacles and the object to be moved. They called the space containing the differences "configuration space". Since then, the configuration space technique has become a standard tool for motion planning

---

*The work was supported by Deutsche Forschungsgemeinschaft, by the Academy of Finland, and by NSERC.

[†]Present address: Department of Computer Science, University of Helsinki, Teollisuuskatu 23, SF-00510 Helsinki, Finland.

[‡]The work was done in part during a visit at the Universität Karlsruhe and the Universität Freiburg.

algorithms (see, e.g., [4,16,18]). Given an object $B$ and a set of obstacles $A_i$, $i = 1, \ldots, k$, Lozano-Perez and Wesley first construct the "grown" obstacles $A_i - B$. An object can be translated by a vector without collisions if and only if the origin can be translated by the same vector without collisions occuring between it and the grown obstacles. The Minkowski difference of two simple polygons in 3-dimensional space as well as the difference of two polyhedra may have $\Omega(n^2)$ vertices, where $n$ is the number of vertices of the objects. An example of such polyhedra is given in Section 4.

To obtain efficient solutions for the 3-dimensional translation problems discussed in this paper, we have to avoid the construction of the configuration space.

Given a convex polyhedron $A$ and a planar polygon $W$, called the *window*, our first algorithm computes all directions in which $A$ can be translated through $W$. The problem was posed by Toussaint [17] who gives an $O(n^2 \log n)$ time algorithm for it ($n$ is the number of vertices of $A$ and $W$). Our algorithm, based on the merge-step of the well-known 3-dimensional convex hull algorithm of Preparata and Hong [11], improves the time bound to $O(n)$ which is optimal. The algorithm is presented in Section 3. Related problems for 2-dimensional space have been discussed by Maddila and Yap for moving a chair through a door [19] and a polygon around the corner in a corridor [7].

Our second algorithm, presented in Section 4, determines all directions of collision-free translations of a polyhedron in the presence of one convex polyhedral obstacle. The algorithm needs $O(n)$ time, where $n$ is the total number of vertices of the object and the obstacle. It is based on a fast construction of the *extreme separating planes* of the object and the obstacle. Davis [1] defines a transformation of the convex hull of polyhedra which gives these planes. We show that the planes can be obtained in $O(n)$ time, without using a transformation ([1] does not contain a time analysis of the transformation).

Given a convex polyhedron $A$ and a set of convex polyhedral obstacles $B_1, B_2, \ldots, B_k$, our third algorithm computes *all* directions of a translation that moves $A$ arbitrarily far away such that no collisions occur between $A$ and the obstacles. This problem and related ones were solved in 2-dimensional space in [2]. In 3-dimensional space the problem has been posed by H. Seeland, Daimler Benz [14], in the following context. In the design of a car, the CAD-system should allow queries of the form: Can a given faulty part be removed easily, i.e., by means of one single translation, or not. If it is possible, direction queries posed by the designer should be answered fast while preprocessing time may be longer (an overnight job).

The simple algorithm determines all directions in which $A$ passes one obstacle. When this is done for all obstacles, the results are combined by "sphere sweep". The valid directions are obtained in a form that enables them to be stored in a data structure for planar point location. The data structure supports queries whether $A$ can be separated into a given direction by means of one single translation. The algorithm is presented in Section 5.

Finally, Section 6 contains conclusions and some further remarks.

## 2. Basic Terms

Throughout this paper, we assume that polyhedra are given by doubly-connected edge lists described, e.g., in [12]. Polygons are given by doubly-connected circular lists of their vertices. Each link of the list contains an information on which side of the corresponding edge the interior of the polygon lies.

Our algorithms output sets of directions. Because we want to translate the objects to infinity, an obvious way to represent a direction is to give a halfline originating at the origin. Such a halfline intersects the unit sphere at one point, and, conversely, each point of the unit sphere corresponds to one halfline. Thus, a set of directions can be represented as a collection of regions on the unit sphere. In our case, the regions are spherical polygons. Their sides are parts of "great circles" of

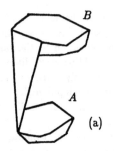 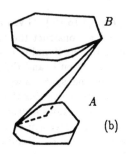

FIG. 1. (a) An extreme supporting plane of $A$ and $B$. (b) An extreme separating plane of $A$ and $B$.

the sphere. We represent such a polygon, analogously to a planar polygon, by a doubly-connected list of its vertices.

The set of halflines that corresponds to a spherical polygon forms a half-open polyhedron. It has only one vertex, the origin. We call such a polyhedron a (polygonal) *cone*. Our cones may be non-convex. The intersection of a cone and the unit sphere may be larger than a semisphere. The representation of a cone as a polygon on the unit sphere is readily constructed from any other reasonable representation of it.

Let $A$ and $B$ be disjoint polyhedra, and $P$ a plane in 3-dimensional space. We say that $P$ is *extreme* if it contains an edge of $A$ ($B$) and a vertex of $B$ ($A$), and $A$ and $B$ are contained in *one* of the closed halfspaces determined by $P$. (Note that, in contrast to [1], the faces of $A$ and $B$ need not be extreme.) A plane $P$ supports $A$ and $B$ if $A \cap P \neq \emptyset$, $B \cap P \neq \emptyset$, and $A$ and $B$ are contained in the same closed halfspace determined by $P$. If $P$ is, in addition, an extreme plane then it is an *extreme supporting plane* of $A$ and $B$ (see Fig. 1 (a)). A plane $P$ separates $A$ and $B$ if they are contained in the opposite closed halfspaces determined by $P$. It is an *extreme separating plane* of $A$ and $B$ if it is, in addition, an extreme plane (see Fig. 1 (b)).

# 3.  Passing a Polyhedron through a Convex Window

Given a convex polyhedron $A$ and a planar convex polygon $W$ (the *window*), such that $A \cap W = \emptyset$, we want to compute all directions in which $A$ can be translated through $W$. The set of such directions may, of course, be empty.

Let us assume that $A$ can be translated through $W$ in directions $d_1, d_2, \ldots, d_k$. We call them *valid* directions. Because $W$ is convex all directions that belong to the minimal convex cone containing $d_1, d_2, \ldots,$ and $d_k$, are valid directions. Our problem is to compute the maximal such cone in the sense that all directions inside the cone are valid but no direction outside the cone is valid. Following the terminology of [2], we call such a cone, $C_W(A)$, the *movability cone* of $A$ with respect to $W$.

We propose the following algorithm for computing $C_W(A)$:

1. Compute the extreme supporting planes $P_i$ ($i = 1, 2, \ldots, l$) of $A$ and $B$. Let $H_i$ denote the halfspace which is determined by $P_i$ and which contains $A$ and $B$.

2. For $i = 1, 2, \ldots, l$, translate $P_i$ to the origin; denote the corresponding translated halfspace $H_i^0$.

3. Compute the intersection $C(A, W) = \bigcap_{i=1}^{l} H_i^0$.

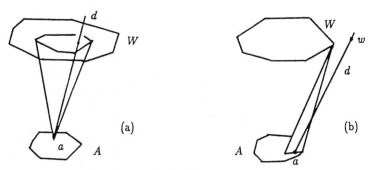

FIG. 2. Proof of Lemma 1.

The following lemma states the validity of the algorithm.

**Lemma 1.** $C(A,W) = C_W(A)$.

**Proof.** First, let $d \in C(A,W)$ be a direction. We claim that $d$ is a direction of a valid translation. Let $a \in A$ be an arbitrary point. Translate $C(A,W)$ in such a way that its apex is $a$. By the construction, the faces of the translated cone intersect $W$. Since $d$ belongs to $C(A,W)$ and $W$ is convex, $d$ intersects $W$, too. But, if $a$ is translated in the direction of $d$ then it encounters the intersection point of $d$ and $W$ (see Fig. 2 (a)). Since $a$ was chosen arbitrarily, we can conclude that all points of $A$ encounter a point of $W$ when translated in the direction $d$. Thus, $d$ is a valid translation direction and $d \in C_W(A)$.

Conversely, let $d \in C_W(A)$ be a valid direction. We show that $d \in C(A,W)$. Assume the contrary. Then there exists, for some $1 \leq i \leq l$, a halfspace $H_i^0$ that does not contain $d$. Choose a point $a \in A \cap H_i$ and translate the halfline $d$ in such a way that its endpoint is $a$. Now $d$ is not contained in $H_i$. Since $d \in C_W(A)$, $a$ will encounter a point $w \in W$ if it is translated in the direction $d$ (see Fig. 2 (b)). But since $d$ is not contained in $H_i$, $w$ cannot be contained in $H_i$. This is a contradiction since $H_i$ is a supporting plane. $\square$

The merge step of the 3-dimensional convex hull algorithm of Preparata and Hong [11] computes the "cylinder faces" of the convex hull of two convex polyhedra. The cylinder faces determine the extreme supporting planes of the polyhedra. A precondition of the step is that the orthogonal projections of the polyhedra on a plane are available, and that these projections do not intersect.

We can use the merge-step of the algorithm of [11] once its preconditions are fulfilled. To do this, we first search for a plane $S$ that separates $A$ and $W$. It is found in $O(n)$ time by solving a 3-variable linear program with $n$ constraints, where $n$ is the number of vertices of $A$ and $W$ (see [9,12]). Then we choose a projection plane $P$ which is perpendicular to $S$. The orthogonal projections of $A$ and $W$ on $P$ are separated by the line $P \cap S$. The linked list representing the contour of $A$'s projection is easy to construct in $O(n)$ time. The pre-conditions of the merge-step of Preparata and Hong's algorithm are now fulfilled. The merge-step takes $O(n)$ time. Thus the first step of our algorithm also takes $O(n)$ time.

The second step of our algorithm is linear since we have $O(n)$ planes to translate.

The computation of the intersection of $n$ halfspaces takes, in general, $\Omega(n \log n)$ time (see, e.g., [12]). The bounding planes of our halfspaces have one common point, the origin, and the first step of our algorithm renders them in a special "cylindrical" order. We shall show that the intersection of our halfspaces can be computed in linear time.

Let $a$ be a point of $A$ and $R$ the plane which contains the window $W$, translated by the vector $-a$. $R$ intersects all the halfspaces $H_i^0$, $i = 1, \ldots, l$. Each intersection is a halfplane. The cone can now be constructed by computing the common intersection of these halfplanes. The "cylindrical"

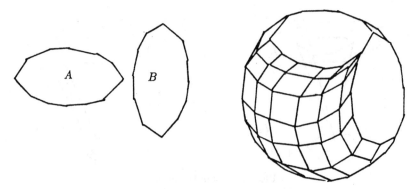

FIG. 3. The Minkowski difference of $B$ and $A$ has $O(n^2)$ vertices.

order in which the halfspaces are obtained yields the halfplanes in the order of the slopes of their bounding lines, or, more specifically, in the order of the directions of their normal vectors that point outwards from the interior of the halfspaces. The intersection is constructed by considering the halfplanes in that order. This can easily be accomplished in $O(n)$ time.

Since each step of the algorithm takes $O(n)$ time, the cone $C(A, W)$ can be computed in $O(n)$ time. From this and Lemma 1 then follows:

**Theorem 1.** *All directions in which a convex polyhedron can be translated in such a way that it passes through a planar convex polygonal window can be computed in $O(n)$ time, where $n$ is the number of vertices of the polyhedron and the polygon.*

A (non-convex) polyhedron can be passed through a convex window by one translation if and only if its convex hull can pass the window. The convex hull can be computed in $O(n \log n)$ time by the algorithm of Preparata and Hong [11]. Thus we have the corollary:

**Corollary 1.** *All directions in which a (non-convex) polyhedron can be passed through a planar convex polygon can be computed in $O(n \log n)$ time, where $n$ is the number of vertices of the polyhedron and the polygon.*

As pointed out by Toussaint [17], it is sufficient for the polyhedron to fit through the window that all its vertices fit through it. This means that some of our supporting planes are superfluous. It is easy to modify the algorithm in such a way that these planes are not computed. This does, however, not improve its asymptotic time bound.

A related problem, discussed in [17], is to find the "smallest" window $W$ on a plane $H$ through which a convex polyhedron $P$ can be passed with a single translation, orthogonal to $H$, after pre-positioning $P$ with arbitrary translations and a rotation. The problem can be solved using the result of McKenna and Seidel [8] who give an $O(n^2)$ algorithm for computing the minimum area shadow of a convex polyhedron with $n$ vertices.

# 4. Translations in the Presence of One Convex Obstacle

Given two disjoint convex polyhedra, an *object* $A$, and an *obstacle* $B$, we want to translate $A$ arbitrarily far away from $B$ without collisions occuring between $A$ and $B$. We say that such a direction is a *valid* direction of translating $A$. That is, a direction $d$ is valid if and only if, for each vector $\mathbf{v}$ whose direction is $d$, the intersection of $B$ and $A + \mathbf{v} = \{a + \mathbf{v} \mid a \in A\}$ is empty. An *invalid* direction is a direction that is not valid.

Our problem is to compute all valid directions of translating $A$ in the presence of $B$. We solve the problem by computing first all *invalid* directions of translating $A$. As in Section 3, they form a

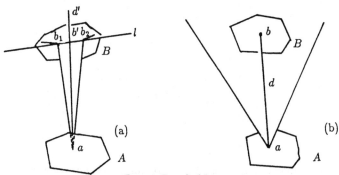

FIG. 4. Proof of Lemma 2.

polygonal cone. The cone is denoted by $C_B(A)$. By these definitions, a direction $d$ is invalid if and only if there exists points $a \in A$ and $b \in B$ such that the direction of the vector $b - a$ is $d$. Thus a direction $d$ is invalid if and only if the Minkowski difference $B - A = \{ b - a \mid b \in B, a \in A \}$ contains a vector whose direction is $d$. A standard technique for motion planning problems constructs the Minkowski difference of the obstacle and the object.

If $A$ and $B$ are convex polyhedra with in total $n$ vertices then $B - A$ may have $\Omega(n^2)$ vertices. An example of such polyhedra can be described as follows (see Fig. 3). Let $C$ be a disc in the $(x_1, x_2)$-plane and $D$ a disc in the $(x_2, x_3)$-plane. Approximate the discs by planar regular $n/2$-gons $B$ and $A$ whose vertices touch the bounding circles of the discs. It is easy to see that each vertex of $B$ introduces about $n/4$ vertices to $B - A$. Thus $B - A$ has about $n^2/8$ vertices. Thus, any algorithm that computes the differences of 3-dimensional polyhedra needs $\Omega(n^2)$ time in the worst case. (In 2-dimensional space, the Minkowski difference of two convex polygons with in total $n$ vertices can be computed in $O(n)$ time (see [3]).)

The cone $C_B(A)$ is the smallest one that contains the difference $B - A$. Notice that only a few of $(B - A)$'s edges introduce a face to $C_B(A)$. The edges form a polygonal line with $O(n)$ corners. We shall show that the following algorithm computes the cone in $O(n)$ time.

1. Compute the extreme *separating* planes $P_i$ $(i = 1, 2, \ldots, l)$ of $A$ and $B$. Let $H_i$ denote the halfspace which is determined by $P_i$ and which contains $B$.

2. For $i = 1, 2, \ldots, l$ translate $P_i$ to the origin; denote the corresponding translated halfspace $H_i^0$.

3. Compute the intersection $C(A, B) = \bigcap_{i=1}^{l} H_i^0$.

**Lemma 2.** $C(A, B) = C_B(A)$.

**Proof.** We first show that $C(A, B) \subset C_B(A)$. Since $B - A$ is convex it suffices to show that a direction $d$ is contained in $C_B(A)$ if it is contained in a face of $C(A, B)$. Thus, let $d$ be contained in the face $F$ of $C(A, B)$. For some $i$, $1 \le i \le l$, the extreme separating plane $P_i$ is parallel to $F$. Let us first assume that $P_i$ contains $A$'s vertex $a$ and $B$'s edge $(b_1, b_2)$. Consider the halfline $d'$ which is parallel to $d$ and whose endpoint is $a$. Let $b'$ be the intersection point of $d'$ and the line $l$ which contains $(b_1, b_2)$ (see Fig. 4 (a)).

Since $d$ is contained in the intersection of the separating halfspaces, $b'$ must lie in the interval $[b_1, b_2]$ of $l$. (Otherwise $d$ was cut off from $C(A, B)$ by one of the other (translated) separating planes that contain $b_1$ or $b_2$.) Thus $a$ collides with $b'$ if it is translated in direction $d'$. Hence $d \in C_B(A)$. The case where $P_i$ contains $B$'s vertex and $A$'s edge is similar.

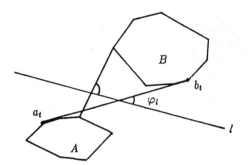

FIG. 5. The "separating tangent" of $p(A)$ and $p(B)$ has the minimal acute angle $\varphi_l$ with $l$.

Conversely, let $d \in C_B(A)$ be a direction. Then there exist points $a \in A$ and $b \in B$ such that $a$ collides with $b$ when it is translated in the direction $d$. Translate $C(A, B)$ in such a way that its vertex is $a$. By the construction, $B$ is now contained in the translated cone (see Fig. 4 (b)). Thus $b$ belongs to the translated cone and the direction $d$ also belongs to it. Since the translation does not rotate the cone the direction $d \in C(A, B)$. □

Davis [1] defines, without giving a time analysis, a transformation of two polyhedra in such a way that their extreme separating planes are obtained from the extreme supporting planes of the transformed polyhedra. We shall show how the extreme separating planes can be constructed without a transformation and in $O(n)$ time, where $n$ is the number of the vertices of the polyhedra.

Our algorithm is very similar to the one for computing the extreme supporting planes. The only difference is that the *initial plane* is chosen in a different way. The plane must be a separating plane that contains a vertex from both polyhedra.

We assume that the polyhedra $A$ and $B$ are projected orthogonally onto a plane $P$, and that their projections $p(A)$ and $p(B)$ are separated in $P$ by a line $l$. These projections are computed as explained in Section 3.

We now turn $l$ in such a way that it becomes a "separating tangent" of $p(A)$ and $p(B)$. Let $S$ denote all lines that contain a vertex of $p(A)$ and a vertex of $p(B)$. It is clear that the "tangents" are contained in $S$. If $a$ is a vertex of $p(A)$ and $b$ is a vertex of $p(B)$ then by $\varphi_l(a, b)$ we denote the size of the acute angle of $l$ and the line containing $a$ and $b$. Now the line which contains $a_t$ of $p(A)$ and $b_t$ of $p(B)$ such that $\varphi_l(a_t, b_t) = \min\{\varphi_l(a, b) \mid a$ is a vertex of $p(A)$ and $b$ is a vertex of $p(B)\}$ is a "tangent" (see Fig. 5).

Thus, we must determine the line $t \in S$ for which the acute angle is minimal. Let $a$ be a fixed vertex of $p(A)$. If the vertices of $p(B)$ are traversed circularly then the sequence of angles $\{\varphi_l(a, b) \mid b$ is a vertex of $p(B)\}$ has at most two local minima. Similarly, if a vertex of $p(B)$ is fixed and the vertices of $p(A)$ are visited circularly then the corresponding sequence of angles is bimodal. For determining the "tangent", this observation yields an algorithm such as:

1. Choose a vertex $a_0$ of $p(A)$ and a vertex $b_0$ of $p(B)$.
   Set $a \leftarrow a_0$ and $b \leftarrow b_0$.

2. Starting at $b$, traverse the vertices $b_i$ of $p(B)$ circularly in the direction of decreasing $\varphi_l(a, b_i)$ until a local minimum is found. Let $b_{\min}$ denote that node.

3. Starting at $a$, traverse the vertices $a_i$ of $p(A)$ circularly in the direction of decreasing $\varphi_l(a_i, b_{\min})$ until a local minimum is found. Let $a_{\min}$ denote that node.

4. If $a \neq a_{\min}$ or $b \neq b_{\min}$ then set $a \leftarrow a_{\min}$, $b \leftarrow b_{\min}$, and go to Step 2.

5. The "separating tangent" is the line which contains $a_{\min}$ and $b_{\min}$.

The algorithm halts at Step 5 when no new (local) minima were found. The vertices of $p(A)$ are scanned in only one direction. The vertices of $p(B)$ may be scanned in two directions. The direction can, however, change only after the first traversal. From the observation above, it follows that the vertices of $p(B)$ and $p(A)$ are visited at most twice. Thus the algorithm takes $O(n)$ time.

The *initial plane* is now the plane which contains the "tangent" and which is perpendicular to $P$.

All extreme separating planes of $A$ and $B$ can be found by "wrapping" planes around the polyhedra as in the merge-step of [11]. The "wrapping" begins from the *initial plane* which was computed above and takes $O(n)$ time. The planes are obtained in a "cylindrical" order (see Section 3). As in Section 3, steps 2 and 3 of the algorithm for computing $C_B(A)$ take $O(n)$ time. Thus the cone $C_B(A)$ is computed in $O(n)$ time.

In order to compute the valid directions of translating $A$, we traverse the edges of the spherical polygon corresponding to $C_B(A)$ and, during the traversal, we change at each edge the information that tells us on which side of the edge the interior of the polygon lies. The traversal takes $O(n)$ time. The resulting cone contains all directions of valid translations. Thus we have:

**Theorem 2.** *Given convex polyhedra $A$ and $B$ with in total $n$ vertices, all directions in which $A$ can be translated without collisions occuring between $A$ and $B$, can be computed in $O(n)$ time.*

## 5. A Polyhedron amidst a Set of Obstacles

Given a convex polyhedron $A$, the *object*, and a set of convex polyhedral *obstacles* $B_i$, $i = 1, 2, \ldots$, $m-1$, we determine all directions in which $A$ can be translated to infinity without collisions occuring between $A$ and any of the obstacles. For simplicity, we assume that each of the $m$ polyhedra has $p$ vertices.

$A$ can be translated in a direction $d$ if and only if $d$ does not belong to any cone $C_{B_i}(A)$, $(i = 1, \ldots, m-1)$ (see Section 4). We can thus compute all invalid directions by constructing the cones $C_{B_i}(A)$ for each $i = 1, \ldots, m-1$, and subsequently computing their union $\bigcup_{i=1}^{m-1} C_{B_i}(A)$. The desired valid directions are obtained, as in Section 4, by traversing the edges of the spherical polygon that correspond to the union.

The union of a set of $m$ planar $p$-gons is easy to compute by a plane-sweep algorithm in $O((mp + k)\log(mp))$ time where $k$ is the intersection of the sides of the polygons (see [10]). Our aim is to modify the plane-sweep algorithm [10] in such a way that it will compute the union of a set of spherical polygons. We call this modified technique *"sphere-sweep"*.

The edges of the polygons are parts of great circles on the unit sphere. The intersection of two edges can be computed in constant time. In constant time, one can also determine on which side of a great circle a given point of the sphere lies.

Our "sweep line", the *sweep arc* is a half circle whose endpoints are fixed in the "poles" of the sphere, i.e., the intersection points of the $x_2$-axis and the sphere. The sweep begins at the 0-meridian and advances 360° around the sphere.

Initially, the 0-meridian is tested for intersection with all edges of the polygons. The intersecting edges are stored in a balanced search tree according to the latitude values of the intersection points. The vertices of the polygons are sorted by their longitudes. They are stored in that order in a heap. The heap is the initial list of the halting points of the sweep. It is later completed by inserting intersection points of the edges of the polygons.

After these initializations, two data structures are used and the contour edges of the union are output as in [10]. The $x$-structure of [10] corresponds to our "latitude structure". The $y$-structure of [10] corresponds to our "longitude structure".

The angle between an edge and the sweep arc is, in general, not constant. But it does not have considerable effect on our algorithm because only the order of the intersections of the edges and the sweep arc is important. The order can change only at the intersections of the edges. However, we need the following observation:

**Lemma 3.** *The sweep arc encounters an edge always at one of its endpoints.*

**Proof.** The lemma follows from the fact that the sweep arc as well as the edges are parts of great circles. $\square$

The sweep takes $O((mp + k) \log(mp))$ time where $m$ is the number of the spherical $p$-gons and $k$ is the number of intersections of their edges.

The cones $C_{B_i}(A)$, $i = 1, \ldots, m - 1$ and their corresponding polygons on the unit sphere can be computed in $O(mp)$ time (see Section 4). Their union can be computed in $O((mp + k) \log(mp))$ time where $k$ is the number of the intersections the edges of polygons. The last step, the traversal along the contour of the union, takes $O(mp + k)$ time. Thus we have:

**Theorem 3.** *Given a polyhedron $A$ and a set of polyhedral obstacles, all directions in which $A$ can be translated to infinity without collisions occuring between $A$ and the obstacles, can be computed in $O((mp + k) \log(mp))$ time ($m$ is the number of given p-vertex polyhedra, and $k$ is the number of lines along which the faces of the translation cones of $A$ with respect to the obstacles intersected).*

The time bound of the algorithm cannot be improved by the method of [4] although the spherical polygons can be considered as Minkowski differences of convex polygons. The result of [4] requires that $A$ is the same polygon in all differences $B_i - A$. This does not hold in our case because the intersections of $A$ and its extreme planes depend on the obstacles.

The output of the algorithm can be stored in a data structure developed for point location in a straight line planar subdivision (see, e.g., [11]). The "lines" of the subdivision are parts of great circles. Thus, for a given direction $d$, we can answer efficiently queries whether $A$ can be translated in direction $d$ without collisions occuring between it and the obstacles. The query time depends on the chosen data structure. Without difficulty we can use, for example, the data structure of Kirkpatrick [5] which is based on a triangulation technique. Then the query time is $O(\log(mp))$. The structure needs $O(mp + k)$ space.

# 6. Conclusions

We have shown that several 3-dimensional motion planning problems can be solved more efficiently if the standard technique of computing the "grown" obstacles, i.e., the Minkowski differences of the obstacles and the object, is not used. Our results apply only to separation motions by means of a single translation. They can be considered as the first step in solving 3-dimensional problems in which several translations or rotations are allowed. Some related results already exist, for example, for the problem of computing shortest paths amidst polyhedral obstacles (see, e.g., [15]). There exist efficient solutions to the planar variants of several such problems (see, e.g., [2,17,18]). Most of them cannot be directly generalized to 3-dimensional space without an unacceptable loss of efficiency.

The representation of the output of the algorithms as polygonal regions on the unit sphere gave us the possibility to apply conventional techiques of planar algorithms.

We have used several of the few algorithmic tools for 3-dimensional problems, including the convex hull algorithm of [11] and the algorithm of [9] for 3-variable linear programming. An important and difficult problem is to develop more such tools for 3-dimensional space.

**Acknowledgments.** Both authors have discussed the topics of the paper with Rolf Klein and the second author with Emo Welzl. The authors are grateful for these discussions as well as for the support received from Th. Ottmann.

# References

1. G. Davis, Computing separating planes for a pair of disjoint polytopes. *Proc. 1st ACM Symp. Computational Geometry*, 1985, 8–14.

2. F. Dehne, and J.-R. Sack, Translation separability of sets of polygons. *Visual Computer* **3** (1987), 227–235.

3. L. Guibas, L. Ramshaw, and J. Stolfi, A kinetic framework for computational geometry. *Proc. 24th IEEE Symp. Foundations of Computer Science* 1983, 100–111.

4. K. Kedem, and M. Sharir, An efficient algorithm for planning collision-free translational motion of a convex polygonal object in 2-dimensional space amidst polygonal obstacles. *Proc. 1st ACM Symp. Computational Geometry*, 1985, 75–80.

5. D. G. Kirkpatrick, Optimal search in planar subdivisions. *SIAM J. Comput.* **12** (1983), 28–35.

6. T. Lozano-Perez, and M. Wesley, An algorithm for planning collision-free paths among polyhedral obstacles. *Communications of the ACM* **22** (1979), 560–570.

7. S. Maddila, and C.K. Yap, Moving a polygon around the corner in a corridor. *Proc. 2nd ACM Symp. Computational Geometry*, 1986, 187–192.

8. M. McKenna, and R. Seidel, Finding the optimal shadows of a convex polytope. *Proc. 1st ACM Symp. Computational Geometry*, 1985, 24–28.

9. N. Megiddo, Linear time algorithm for linear programming in $R^3$ and related problems. *SIAM J. Comput.* **12** (1983), 759–776.

10. Th. Ottmann, P. Widmayer, and D. Wood, A fast algorithm for boolean masking problem. *Comput. Vision Graphics Image Process.* **30** (1985), 249–268.

11. F. P. Preparata, and S. J. Hong, Convex hulls of finite sets of points in two and three dimensions. *Communications of the ACM* **20** (1977), 87–93.

12. F. P. Preparata, and M. I. Shamos, *Computational Geometry. An Introduction.* Springer-Verlag, New York etc., 1985.

13. J. T. Schwartz, and M. Sharir, On the piano movers' problem: V. The case of a rod moving in three-dimensional space amidst polyhedral obstacles. *Comm. Pure Appl. Math.* Vol. XXXVII (1984), 815–848.

14. H. Seeland, CAD-Abteilung, Daimler Benz AG, Stuttgart, Personal communication, 1987.

15. M. Sharir, and A. Baltsan, On shortest path amidst convex polyhedra. *Proc. 2nd ACM Symp. Computational Geometry*, 1986, 193–206.

16. M. Sharir, R. Cole, K. Kadem, D. Leven, and R. Pollack, Geometric appications of Davenport-Schinzel sequences. *Proc. 27th IEEE Symp. Foundations of Computer Science*, 1986, 77–86.

17. G. T. Toussaint, Movable separability of sets. *In:* G. T. Toussaint (ed.), *Computational Geometry*, North-Holland, Amsterdam etc., 1985, 335–376.

18. S. H. Whitesides, Computational geometry and spatial planning. *In:* G. T. Toussaint (ed.), *Computational Geometry*, North-Holland, Amsterdam etc., 1985, 377–428.

19. C.K. Yap, How to move a chair through a door. Techical Report, Courant Institute, 1984.

# Linear Time Algorithms for Testing Approximate Congruence in the Plane

Sebastian Iwanowski

Freie Universität Berlin
Institut für Mathematik III
Arnimallee 2–6
D-1000 Berlin 33
Germany

### Abstract

Let $A, B$ be two sets of $n$ points in the Euclidean plane. We want to test if they are congruent. Unfortunately, in many practical applications the input data are not given precisely, but only within a small tolerance factor $\varepsilon$. With this notion in mind, we ask if $A$ and $B$ are *approximately* congruent, i.e. if there are sets $A'$ and $B'$ consisting of points in the $\varepsilon$-neighborhoods of the points of $A$ and $B$ respectively that are exactly congruent. In this paper we give optimal algorithms for some problems of the *labelled* case, i.e. we assume that we already know which point of $A$ should be transformed to which point of $B$. First, we give a linear time algorithm for the test if two planar point sets are approximately congruent by a reflection in a line. The algorithm presented in this paper uses a generalization of the linear programming algorithm by Megiddo which is interesting in its own right. It solves the problem of finding a feasible solution for a general system of algebraic inequalities of bounded degree. Next, we derive a linear time algorithm for the test on congruence by a rotation around a fixed center. Finally, if we allow an arbitrarily small but fixed range of uncertainty, we obtain a linear time algorithm for the test on arbitrary congruence.

## 1   Introduction

The detection of similarity or congruence has many practical applications, such as in robotics, pattern recognition, and biology. The problem remains interesting even if we consider point sets as objects only, for in many cases an object is well represented by given reference points.

The problem of determining *exact* congruence has been settled by Atallah [Ata] in two dimensions. Atallah found an optimal algorithm that runs in $\Theta(n \log n)$ time to solve congruence as well as general similarity. He also generalized his algorithm for sets of geometric objects other than points. Atkinson [Atk] found a $\Theta(n \log n)$ – algorithm for congruence of point sets in three dimensions. Then Alt, Mehlhorn, Wagener, and Welzl [AMWW] showed that similarity can be easily reduced to congruence in $O(n)$ time. They gave an $O(n^{d-2} \log n)$ – algorithm for testing similarity of $d$-dimensional point sets.

Unfortunately, in most of the practical applications the data are not available exactly, but are subject to small perturbations due to inaccurate measurements or inaccurate representation of numbers by a computer. All algorithms mentioned so far use combinatorial and graph-theoretic concepts that do not tolerate any inaccuracy of the input data and would lead to wrong results if there were some inaccuracy.

This is why we use the notion of *approximate* congruence:

**Definition 1.1** Let $A = \{a_1, \ldots, a_n\}$ be a set of $n$ points in the plane. A set $A' = \{a'_1, \ldots, a'_n\}$ is called $\varepsilon$ – *approximate* to $A$ iff there exists an $n$–permutation $\pi$ such that $\| a_i - a'_{\pi(i)} \| \leq \varepsilon$ for all $i = 1, \ldots, n$.

**Remark:** Throughout this paper, $\| \ldots \|$ denotes always the Euclidean norm.

**Definition 1.2** Two sets $A, B$ of $n$ points respectively are called $\varepsilon$ – *approximately congruent* iff there exist point sets $A', B'$, $\varepsilon$ - approximate to $A, B$, that are exactly congruent.

The definitions of approximate congruence by translation, rotation around a fixed center, reflection, etc. are analogous to the above.

The concept of $\varepsilon$ – approximate congruence has been first introduced in [AMWW]. There, the authors used a definition slightly different from Definition 1.2. It can be seen easily that both definitions are equivalent if we work with $2\varepsilon$ in [AMWW] and $\varepsilon$ in Definition 1.2.

Our model of computation is a Random Access Machine (RAM) which is able to compute roots and powers of order at most four and to intersect functions that involve such operations in addition to the basic operations of a RAM. This is a model stronger than in [AMWW] where only a regular RAM and integer arithmetic is used. We will see in the following sections that some algebraic functions naturally arise from the problems considered.

In Section 2 of this paper we present the outline of a linear time algorithm for determining if two planar point sets are $\varepsilon$ – approximately congruent by a reflection in a line (labelled case). In Section 3 we give the details of the algorithm. They include a linear time algorithm to find a feasible solution for a general system of algebraic inequalities of bounded degree. This algorithm is derived from the linear programming algorithm by Megiddo [Me].

Approximate congruence by reflection has not been investigated so far. The algorithm of Section 2 will be used to derive optimal solutions for other congruence problems which have already been considered in [AMWW] with a weaker model of computation, but also with worse time bounds.

In Section 4 we derive a linear time algorithm for determining if two planar point sets are $\varepsilon$ – approximately congruent by a rotation around a fixed center (labelled case). This improves the result shown in [AMWW] where a $O(n \log n)$ – algorithm is given on that problem.

In Section 5 we give an algorithm that solves the general $\varepsilon$ – approximate congruence problem (labelled case) for all $\varepsilon$ outside an arbitrarily small range. If this range is fixed, then our algorithm runs in linear time and, therefore, is optimal. The best deterministic result derived in [AMWW] is $O(n^3 \log n)$.

## 2 Testing Approximate Reflection — Outline

We consider the following problem:

**Problem I:** Given two point sets $A = \{a_1, \ldots, a_n\}$, $B = \{b_1, \ldots, b_n\}$ in the plane. Test if there are sets $A' = \{a'_1, \ldots, a'_n\}$, $B' = \{b'_1, \ldots, b'_n\}$, $\varepsilon$ – approximate to $A, B$ respectively, and a reflection $\sigma$ in a line, such that $\sigma(a'_i) = b'_i$ for all $i = 1, \ldots, n$. If so, give a possible choice of sets $A', B'$ and a reflection line $r$.

**Algorithm I** (worst case run time $O(n)$ ):

### Step 1

Call a line $r$ $i$-admissible iff there exist points $a'_i, b'_i$ in the $\varepsilon$ – neighborhods of $a_i, b_i$ respectively with $\sigma(a'_i) = b'_i$ where $\sigma$ is the reflection in $r$. Our aim is to find a line that is $i$-admissible for all $i = 1, \ldots, n$ simultaneously. We will call such line *admissible*. In Step 1 we derive an efficient representation for the sets of those $i$-admissible lines that are candidates for an admissible line. This representation can be computed in constant time for each $i = 1, \ldots, n$ respectively. We have to distinguish two cases:

Case i): $\| a_i - b_i \| > 2\varepsilon$

Case ii): $\| a_i - b_i \| \leq 2\varepsilon$

In Case i), the $i$-admissible lines are the lines lying completely between the two branches of a certain hyperbola. In Case ii), the $i$-admissible lines are the lines that intersect a certain ellipse. In the following we will call this hyperbola or ellipse the *$i$-th conic section*. Its parameters depend on the values of $a_i, b_i$ only and can be computed in our computational model in constant time for each $i = 1, \ldots, n$.

Next, we are seeking a line $l$ that intersects all admissible lines:

If for at least one $i_0$ Case i) holds, then we just take a line $l$ that intersects both branches of the $i_0$-th hyperbola. We find such line by computing a range $S$ in which the slope value of any admissible line must lie. Then we take a line $l$ whose slope is not in $S$. Obviously, if $S = \emptyset$ there is no solution for Problem I. In Case i), an $i$-admissible line has to intersect $l$ in a possibly unbounded interval $I_i$ that is determined by the intersection points of $l$ with the $i$-th hyperbola. Let $I$ be the intersection of all $I_i$'s. Then any admissible line has to intersect $l$ in $I$.

If for all $i = 1, \ldots, n$ Case ii) holds, then we take any line $l$ and, first, test by a seperate method if there is a line $r$ parallel to $l$ that intersects all of the ellipses simoultaneously. If so, $r$ is an admissible line and we are done. If not, any admissible line has to intersect $l$. In this case, $I$ is the entire line.

In either case, we choose $l$ in such a way that it does not intersect any of the ellipses. We now transform the coordinate system for all $O(n)$ data computed so far such that $l$ becomes the new abscissa. Now, any $i$-admissible line $r$ that is a candidate for an admissible line can be represented by a value $x \in I$ (meaning that $r$ goes through $(x, 0)$) and a corresponding slope value $s \in S$ that is not zero by the choice of $l$ (since $l$ is not admissible).

## Step 2

In this step we compute two functions $g_i, h_i$ ($g_i \leq h_i$) for each $i$ with the property that for any given $x \in I$, the values between $g_i(x)$ and $h_i(x)$ serve as representation for those $i$-admissible lines through $(x, 0)$ that are candidates for an admissible line.

We will see in Section 3 that the $i$-admissible slope values for any $x$ are the values between the slopes of the tangent lines at $(x, 0)$ on the $i$-th conic section. Note that $l$ and $I$ were chosen in such a way that the tangent lines at $(x, 0)$ for each $x \in I$ on each conic section exist. Thus, for any $x \in I$, the $i$-admissible slope values define an interval $S_{xi}$ that is bounded by the slope values of the tangent lines at $(x, 0)$ on the $i$-th conic section. However, infinity may be within $S_{xi}$ so that $S_{xi}$ may consist of two components. In order to get rid of this case, we take care that zero is not in any slope interval anymore and consider the reciprocals:

If the $i$-th conic section is an ellipse, zero cannot be in $S_{xi}$, because $l$ does not intersect any of the ellipses. If some conic sections are hyperbolas, then we have already computed a range $S$ in Step 1 that contains the slope values of any admissible line. Note that $S$ does not contain zero. This means that $l$ is not admissible for *all* $i$, but $l$ may be still $i$-admissible for *some* $i$ (meaning that $S_{xi}$ contains zero for that $i$). Therefore, we intersect $S_{xi}$ with $S$ to get an interval $S'_{xi}$ that contains all $i$-admissible slope values we still need to consider and that does not contain zero. Let $g_i(x)$ and $h_i(x)$ be the reciprocals of the bounds of $S'_{xi}$. By construction, $g_i(x)$ and $h_i(x)$ are well defined for each $x$ and $i$. The values between $g_i(x)$ and $h_i(x)$ are the reciprocals of those $i$-admissible slope values that are the only candidates for admissible slope values.

We will see in Section 3 that $g_i$ and $h_i$ are convex upward and convex downward functions that can be computed in our computational model in constant time for each $i = 1, \ldots, n$ .

## Step 3

Given an $x \in I$, the $i$-admissible lines through $(x, 0)$ we need to consider are the lines with reciprocal slope value between $g_i(x)$ and $h_i(x)$. Our goal is to know if there is an $x$ with a line through $(x, 0)$ having reciprocal slope value between $g_i(x)$ and $h_i(x)$ for all $i = 1, \ldots, n$ *simultaneously*. The necessary and sufficient condition for this is that $\max\{g_i(x)\} \leq \min\{h_i(x)\}$ for some $x \in I$. The existence of an $x$ that satisfies this condition will be tested by a generalization of the linear programming algorithm by Megiddo [Me] that still runs in $O(n)$ time. Note that the $f_i, g_i$ are not linear, but they are convex piecewise differentiable functions and each pair of any such $f_i, g_i$ has only a bounded number of intersections which can be computed in constant time each. In Section 3 it will be shown that this is the condition that suffices for the programming algorithm. If there is a solution, our algorithm will output an $x$ that satisfies $\max\{g_i(x)\} \leq \min\{h_i(x)\}$. In linear time we can compute the values $\max\{g_i(x)\}$ and $\min\{h_i(x)\}$. Any value in between will serve as reciprocal of the slope of an admissible line $r$. Choose such a line and compute the reflection of the $\varepsilon$-neighborhood of $a_i$ in that line and intersect it with the $\varepsilon$-neighborhood of $b_i$ in constant time for each $i = 1, \ldots, n$. This will give us a possible choice of sets $A', B'$ in linear time.

Thus, we have solved Problem I in linear time.  $\square$

# 3   Testing Approximate Reflection — Details

In the following we will describe in detail how to carry out the three steps of Algorithm I:

## Step 1

**Case i):** $\| a_i - b_i \| > 2\varepsilon$

**Lemma 3.1** If $\| a_i - b_i \| > 2\varepsilon$, then the $i$–admissible lines are the lines that lie completely between the two branches of the hyperbola with centers $a_i, b_i$, major axis length $2\varepsilon$ , and minor axis length $\sqrt{\| a_i - b_i \|^2 - 4\varepsilon^2}$ (cf. Figure 1).

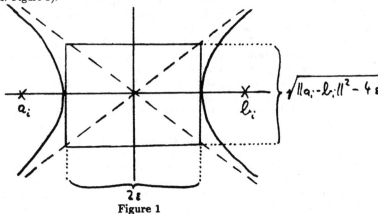

**Figure 1**

**Proof:** Confer to Figure 2:

Let $r$ be $i$–admissible and $a_i'$ be the reflection of $a_i$ in $r$ ($\| a_i' - b_i \| \leq 2\varepsilon$ by definition of "$i$–admissible"). Let $p$ be an arbitrary point on $r$. Using $\| p - a_i \| = \| p - a_i' \|$ , we get by the tringle inequality:

$$\| p - a_i \| + \| a_i' - b_i \| \geq \| p - b_i \|$$

This implies:

$$\big| \| p - a_i \| - \| p - b_i \| \big| \leq \| a_i' - b_i \| \leq 2\varepsilon \tag{1}$$

Note that (1) holds for any point $p$ on $r$.

Thus, we have shown that a necessary condition for line $r$ to be $i$–admissible is that all of its points have the property that their distances to $a_i$ and $b_i$ differ by at most the constant $2\varepsilon$. It is well known in Analytic Geometry that the set of all points $p$ satisfying (1) is the region between the two branches of a hyperbola with the parameters given above.

On the other hand, this condition is sufficient: Let $r$ be a line with all points $p$ on $r$ satisfying (1). Imply (1) to that point $c$ on $r$ that is closest to $a_i$ in order to show that the reflection in $r$ maps $a_i$ to a point $a_i'$ that lies within the $2\varepsilon$–neighborhood of $b_i$. Thus, $r$ is $i$–admissible.  □

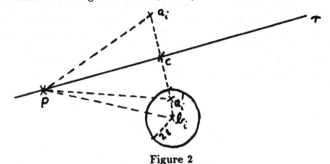

**Figure 2**

**Case ii):** $\| a_i - b_i \| \le 2\varepsilon$

**Lemma 3.2** If $\| a_i - b_i \| \le 2\varepsilon$, then the $i$-admissible lines are the lines that intersect the ellipse with centers $a_i, b_i$, major axis length $2\varepsilon$, and minor axis length $\sqrt{4\varepsilon^2 - \| a_i - b_i \|^2}$ (cf. Figure 3).

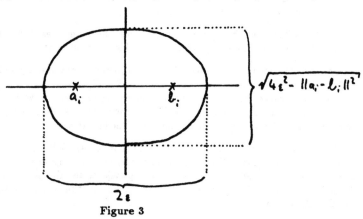

**Figure 3**

**Proof:** Confer to Figure 4:

Let $r$ be $i$-admissible and $a_i'$ be the reflection of $a_i$ in $r$ ($\| a_i' - b_i \| \le 2\varepsilon$ by definition of "$i$-admissible"). Let $p$ be the intersection point of $r$ with the line segment from $a_i'$ to $b_i$ (cf. Figure 4a). Using $\| p - a_i \| = \| p - a_i' \|$, we get:

$$\| p - a_i \| + \| p - b_i \| \le 2\varepsilon \qquad (2)$$

Note that (2) holds for at least one point $p$ on $r$.

Thus, we have shown that a necessary condition for line $r$ to be $i$-admissible is that at least one point of $r$ has the property that the sum of its distances to $a_i$ and $b_i$ is bounded by the constant $2\varepsilon$. Again, it is well known in Analytic Geometry that this means that $r$ intersects an ellipse with the parameters given above.

On the other hand, this condition is sufficient: Let $r$ be a line with at least one point $p$ satisfying (2) (cf. Figure 4b). Reflect $a_i$ in $r$ to get $a_i'$. Using $\| p - a_i' \| = \| p - a_i \|$, we get by the triangle inequality:

$$\| a_i' - b_i \| \le \| p - a_i' \| + \| p - b_i \| \le 2\varepsilon.$$

This implies that $r$ is $i$-admissible. $\qquad\qquad\qquad \square$

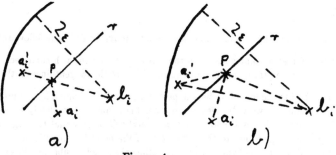

a)             b)

**Figure 4**

The general form of a conic section is given by $Ax^2 + Bxy + Cy^2 + Dx + Ey + F = 0$. Since hyperbolas and ellipses are conic sections, we can compute the parameters given in Lemmas 3.1 and 3.2, transform this into the general conic section form, and represent the hyperbola or ellipse by the coefficients $A, B, C, D, E, F$ in constant time.

Now we have to explain how to find a line $l$ with the desired properties.

Consider the case that we have to deal with at least one hyperbola: Clearly, the slope of a line lying completely between the two branches of a hyperbola is between the slopes of the asymptotes of this hyperbola. Thus, a necessary condition for a line to be $i$–admissible is that its slope is between the slopes of the asymptotes. This leads to the following

**Lemma 3.3** Let $S_i$ be the range of slopes between the asymptotes of the $i$-th hyperbola. Let $S := \bigcap S_i$ for all $i$ of Case i). Then the slope of any admissible line must be in $S$.

Once we have got the parameters of the $i$-th hyperbola, we can compute $S_i$ in constant time for each $i$. By simple interval intersection we can compute $S$ in $O(n)$ time.

Note that $S$ cannot be the full range of all slopes possible, because we deal with at least one hyperbola. Choose $l$ and compute $I$ as described in Section 2 in $O(n)$ time. Note that $I$ has to be bounded, because $l$ has to intersect both branches of at least one of the hyperbolas.

If we have to deal with ellipses only, then a line is admissible if it intersects all of the ellipses.

Let $l$ be the abscissa. Using calculus methods, we can compute the minimum and maximum value of each ellipse in constant time. Clearly, a line parallel to the abscissa is admissible iff the maximum of the minimum values is less than or equal to the minimum of the maximum values. This condition can be tested in $O(n)$ time.

In either case, implying a parallel shift to $l$, we can obtain that $l$ does not intersect any of the ellipses without losing the other desired properties. W.l.o.g. let $l$ pass through the origin.

Now we change the coordinate system such that $l$ becomes the new abscissa. This can be done with a rotation by the inclination angle of $l$. The data we need to transform to our new coordinate system are the input points and all conic sections computed so far. The new coordinates of the input points can be obtained by simple matrix multiplication. For the transformation of the parameters $A, B, C, D, E, F$ that represent a conic section with respect to the old coordinate system into parameters $A', B', C', D', E', F'$ that represent the same conic section with respect to the new coordinate system we need to perform a rotation of axes (cf. Chapter 5.1 of [KS], e.g.). For both types of transformations we need to know sin and cos of the inclination angle of $l$. These values can be obtained by the functional equation of $l$, just involving basic operations of a RAM. Once we have got sin and cos of the inclination angle of $l$, we only need basic operations to perform each transformation (cf. [KS]). Thus, we have shown that all operations needed can be performed in our computational model. Obviously, for each data involved we only need constant time. So we can perform the whole change of coordinates in linear time.

Summarizing our results, we have shown

**Theorem 3.4** Step 1 of Algorithm I can be performed in $O(n)$ time.

## Step 2

Given $x$, what are the $i$ – admissible lines going through $(x, 0)$ ? The following lemma gives the answer:

**Lemma 3.5** Given a point $p$ at which the tangent lines on the $i$-th conic section exist. Then the slope of an $i$–admissible line through $p$ is between the slopes of the tangent lines on the $i$-th conic sections at $p$.

**Proof:** Let the $i$-th conic section be a hyperbola. Instead of giving a formal proof we confer to Figure 5.

Due to Lemma 3.1 we only need to consider lines that lie completely between the two branches of the $i$-th hyperbola. It is a well known fact in Analytic Geometry that this region is exactly the region of points that have tangent lines to the hyperbola. If the point lies directly on the hyperbola then there is only one tangent line to the hyperbola which is the only $i$ – admissible line through that point (cf. Figure 5a). If the point lies properly in the interior of $R_i$ then there are always two different tangent lines to the hyperbola. To which branches of the hyperbola the tangent lines go depends on the position of the point relative to the asymptotes (cf. Figure 5b-d). If the point lies directly on an asymptote then this asymptote is one of the tangent lines (cf. Figure 5d).

The arguments are similar for the case that the $i$-th conic section is an ellipse. □

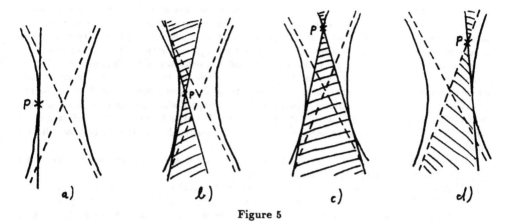

**Figure 5**

How to proceed further in Step 2 is already well described in Section 2. The only questions we have to observe in detail are:

i) How do we compute and represent $g_i$ and $h_i$ ?

ii) Why are $g_i$ and $h_i$ convex functions ?

Question i) is solved by the following:

Consider a curve given by $f(x, y) = 0$ that is differentiable. Compute a multivalued function $\tau(x_0, y_0)$ that assigns to any point $(x_0, y_0)$ the slopes of the tangent lines at $(x_0, y_0)$ on $f$. If $f$ is a conic section as in our case, then detailed computations show that $\tau$ consists of two functions $\tau_1, \tau_2$ in $(x_0, y_0)$ which involve powers and roots of order at most 4. Use the coefficients of all powers and roots that occur to represent $\tau_1$ and $\tau_2$. Obviously, these coefficients can be computed in constant time for each conic section. Analogously, we can intersect $\tau$ with the horizontal lines defined by the bounds of the slope range $S$ computed in Step 1. Thus, we can compute a representation of the following functions in constant time for each $i = 1, \dots, n$:

**Definition 3.6** Let $u_i(x), v_i(x)$ be the reciprocal slope values of the tangent lines at $(x, 0)$ on the $i$-th conic section, $u_i(x) \leq v_i(x)$. Let $u$ ($v$) be the lower (upper) bound of the reciprocals of the slope range $S$ computed in Step 1 (if we have to deal with at least one hyperbola).
Define $g_i(x) := \max\{u_i(x), u\}$ and $h_i(x) := \min\{v_i(x), v\}$ for any $x \in I$ and any $i = 1, \dots, n$.

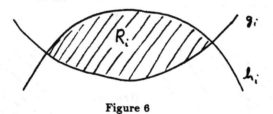

**Figure 6**

Question ii) is solved by the following:

**Lemma 3.7** $g_i$ is a *convex upward* and $h_i$ is a *convex downward* function in $I$ for any $i$.

**Proof:** Let $R_i := \{(x, s) \in I \times \Re : g_i(x) \leq s \leq h_i(x)\}$ be the region between $g_i$ and $h_i$. Note that $g_i \leq h_i$ in all of $I$. Thus, in order to prove Lemma 3.7, it suffices to prove that $R_i$ is convex (cf. Figure 6).

To see this, make the following observation: At this point, we deal with two different two-dimensional vector spaces. One is the original vector space that contains the input data. The other one is the vector space that contains the $R_i$ for each $i$. Due to our construction, a point $(r_x, r_y)$ in the vector space that contains $R_i$ corresponds to a line given by the equation $x = r_y y + r_x$ in the original vector space. This is exactly the dual transformation of points into lines and vice versa as used in the literature, only $x$ and $y$ are switched. Thus, horizontal lines are the only lines that do not have a dual representation. In the original vector space, horizontal lines need not be considered. Note that a point $r$ is in $R_i$ iff the dual line of $r$ is $i$ – admissible and its reciprocal slope value is in the range $S$ which has been computed in Step 1. This follows directly from the definition of $R_i$. The dual transformation has already been well studied, e.g. in [E], p.13. Let us restate what we need to know about it (already taking in account that we switched $x$ and $y$):

i) If three points $p_1, p_2, p_3$ are on a nonhorizontal line $l$ in the dual vector space then the lines dual to $p_1, p_2, p_3$ intersect in the point dual to $l$ in the original space.

ii) If three points $p_1, p_2, p_3$ are on a horizontal line in the dual vector space then the lines dual to $p_1, p_2, p_3$ are parallel in the original vector space (i.e. they intersect at infinity which is the "point" dual to $l$). Since the dual transformation preserves the relations "below", "above", and "between", those parallel lines are in the same horizontal order as the original points.

Having the concept of duality in mind we can now easily prove that $R_i$ is convex: Given two arbitrary points $r_1, r_2 \in R_i$ we have to prove that all points on the line segment connecting $r_1$ and $r_2$ are in $R_i$.

Case i): Let the $i$-th conic section be a hyperbola:

Consider the lines dual to $r_1$ and $r_2$. Since either one is $i$ – admissible by definition of $R_i$, we can determine a region where the $i$th hyperbola has to be located. We distinguish the cases whether the connecting line segment is horizontal (Figure 7a) or not (Figure 7b). In Figure 7, the shaded region is the region where the $i$th hyperbola has to be located. Let $\xi$ be any point on the line segment connecting $r_1$ and $r_2$. If this line segment is horizontal then the line dual to $\xi$ is a line parallel to the lines dual to $r_1$ and $r_2$ and in between them (Figure 7a). If this line segment is nonhorizontal then the line dual to $\xi$ intersects both of the lines dual to $r_1$ and $r_2$ in the same point (Figure 7b). In either case the line dual to $\xi$ cannot intersect the shaded region. Thus, it lies completely between the branches of the $i$th hyperbola. So it is $i$ – admissible. With the same argument, the inverse slope value of the line dual to $\xi$ lies between $u$ and $v$: Just use the fact that this holds for the lines dual to $r_1$ and $r_2$ and apply Figure 7.

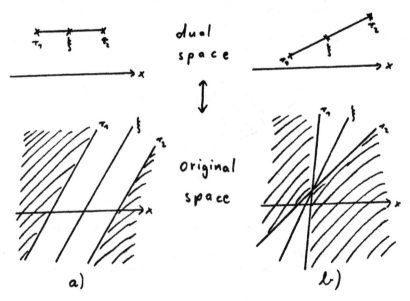

Figure 7

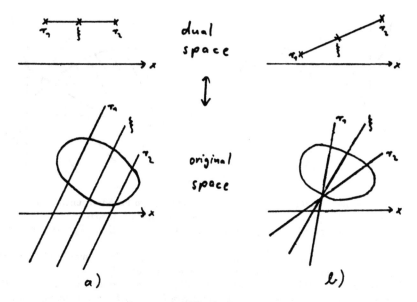

**Figure 8**

Case ii): Let the $i$-th conic section be an ellipse.

Define $r_1, \xi, r_2$ as above. Confer to Figure 8: If the lines dual to $r_1, r_2$ intersect the ellipse then also the line dual to $\xi$ intersects the ellipse.

In either case we have shown that $\xi$ is in $R_i$.                                    □

Summarizing our results, we have shown:

**Theorem 3.8** Step 2 of Algorithm I can be performed in $O(n)$ time.

## Step 3

By definition of $g_i$ and $h_i$, it follows that the right and left derivatives exist everywhere in $I$. It is either the derivative of $u_i$ $(v_i)$ or 0 depending which one of $u_i(x), u$ ( $v_i(x), v$ ) is larger. With this notion in mind, the right or left derivative can be computed in constant time for every $x \in I$.

Thus, we have reduced Problem I to the following problem:

**Problem II:** Given $n$ realvalued functions $h_j$ and $g_i$, $h_j$ being convex downward and $g_i$ being convex upward, that have a right and left derivative anywhere in their domain of definition. Any two functions intersect in a bounded number of points, and these points can be computed in constant time each. Additionally, a right or left derivative of any function of the input can be computed for any $x$ in constant time.

Test if there is an $x$ which satisfies $\max\{g_i(x)\} \leq \min\{h_j(x)\}$. If so, give a feasible value for $x$.

**Algorithm II** (worst case run time $O(n)$):

We first proceed in the same way as Megiddo [Me]:

**Lemma 3.9** Given $x$. We can check in linear time whether $\max\{g_i(x)\} \leq \min\{h_j(x)\}$ or not. If not, any feasible solution must be either greater or smaller than $x$. We can decide in linear time which case holds.

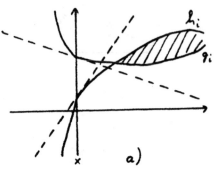 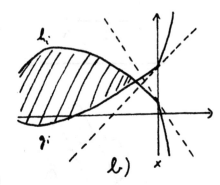

**Figure 9**

The **Proof** is the same as in [Me] where this lemma is only stated for *linear* functions $g_i, h_j$. Since the proof is not long, we will repeat it and show how it works for convex functions as well:

Compute $max\{g_i(x)\}$ and $min\{h_j(x)\}$ in linear time and compare. If $max\{g_i(x)\} \leq min\{h_j(x)\}$, we are done.

Otherwise, let $G_x = \{g_k : g_k(x) = max\{g_i(x)\}\}$ and $H_x = \{h_l : h_l(x) = min\{h_j(x)\}\}$ be the sets of functions $g_k, h_l$ that coincide with the maximum or minimum respectively.

If $\xi$ is feasible then, by definition, $g_i(\xi) \leq h_j(\xi)$ for *all* $g_i, h_j$. Let us restrict to $g_k, h_l$ of $G_x, H_x$ only.

If $\xi > x$, then the tangents belonging to the right derivatives of $g_k, h_l$ have to intersect in the domain right of $x$ for any pair $(k, l)$ (confer to Figure 9a). For this observation we only need the fact that each $g_k$ is convex upward and each $h_l$ is convex downward.

Analogously, if $\xi < x$, then the tangents belonging to the left derivative of $g_k, h_l$ have to intersect in the domain left of $x$ for any pair $(k, l)$ (cf. Figure 9b).

This observation leads to the following strategy: Compute the maximum $u_r$ of the right derivatives at $x$ for all $g_k \in G_x$ and the minimum $v_r$ of the right derivatives at $x$ for all $h_l \in H_x$.

If $u_r \leq v_r$, then only $\xi > x$ may be possible.

If $u_r > v_r$, then compute the minimum $v_l$ of the left derivatives at $x$ for all $g_k \in G_x$ and the maximum $u_l$ of the left derivatives at $x$ for all $h_l \in H_x$.

If $v_l \geq u_l$, then only $\xi < x$ may be possible.

In the remaining case ($u_r > v_r$ and $v_l < u_l$), $\xi$ cannot lie on either side of $x$, and there is no feasible solution.

Obviously, we can compute everything we need in $O(n)$ time. $\qquad\square$

Just as in [Me], order the $g_i, h_j$ to pairs $(g_{i_1}, g_{i_2})$ and $(h_{j_1}, h_{j_2})$. W.l.o.g, let there be even numbers of $g_i$ and $h_j$ respectively. Let $k$ be the upper bound for the number of intersection points between any two points. In [Me], $k$ is always one. This is why we have to modify the algorithm given in [Me] considerably by the following:

i) Compute the intersection points of each pair in $O(n)$ time. W.l.o.g. let any intersection consist of $k$ points (this is the worst case as we will see in Lemma 3.10).

ii) Use the algorithm of Lemma 3.9 to find a feasible $x$ or eliminate $(1/2^{k+1})$ $n$ of the original $n$ functions from further consideration in $O(n)$. The way how to do this will be shown in Lemma 3.10.

iii) Solve Problem II recursively for the remaining $(1 - 1/2^{k+1})$ $n$ functions.

This leads to the following recurrence relation:

$$T(n) \leq T((1 - 1/2^{k+1})n) + O(n)$$

As easily can be seen, $T(n)$ is $O(n)$ as long as $k$ is constant.

The only thing left to show for Algorithm II is

**Lemma 3.10** If $k$ is the number of intersection points between any two of the $n$ functions given in Problem II, then in $O(n)$ we can find a feasible solution for Problem II or, at least, eliminate $(1/2^{k+1})$ $n$ of those functions that are not relevant for any feasible solution.

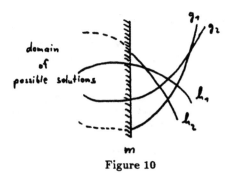

**Figure 10**

**Proof** (Induction by $k$):

$k = 1$:

This is the original case considered in [Me]. Compute the median $m$ of the intersection points of every pair of Step i) in $O(n)$ (cf. [AHU]). Use Lemma 3.9 with $m$ for $x$ to check if $m$ is feasible or, if not, to determine on which side of $m$ any feasible solution lies. According to Lemma 3.9, this costs $O(n)$ time. If the intersection of a pair is *not* in the domain where any feasible solution must lie, then one of the functions belonging to the pair is always above the other in the domain of possible feasible solutions. If the pair is of the $g$ – type, we can eliminate the lower function from further consideration. If the pair is of the $h$ – type, we can eliminate the upper function from further consideration (cf. Figure 10). Since half of the pairs intersect in the domain *not* belonging to any feasible solution, we can eliminate $n/4$ of the $n$ functions considered originally.

$k \rightarrow k+1$:

Let all pairs have $k + 1$ intersections. Compute the median $m_1$ of the *smallest* intersection points of every pair of Step i) in $O(n)$. Use Lemma 3.9 as in the case $k = 1$. If possible feasible solutions must be *smaller* than $m_1$, then we know that half of the pairs do not have any intersection in the feasible domain (those pairs with smallest intersection greater than $m_1$). Just as in the case $k = 1$, we can eliminate $n/4$ of the $n$ functions considered originally.

Otherwise, consider the pairs with smallest intersection less than $m_1$ only. Those pairs do have only $k$ intersections greater than $m_1$ where possible feasible solutions must lie. Use the inductive assumption for $k$ to eliminate $(1/2^{k+1})n'$ functions of those $n'$ still considered. Since $n' = n/2$, we have eliminated $(1/2^{k+2})n$ of the original $n$ functions what we wanted to show. □

Since induction does not give much insight how the algorithm works in sequential order, let us explain informally what the algorithm of Lemma 3.10 does:

We first compute the median $m_1$ of the smallest intersection points. If any solution is less than $m_1$, we are done. Otherwise, we compute the median $m_2$ of the second smallest intersection of those pairs whose smallest intersection is less than $m_1$. If any solution is less than $m_2$, we consider those of the pairs still considered whose second smallest intersection is greater than $m_2$. Those pairs do not intersect in the feasible domain (lying between $m_1$ and $m_2$) anymore, and we can eliminate one function of each pair. Otherwise, we have to compute $m_3$ of those of the pairs still considered whose second smallest intersection is less than $m_2$, and so on ... With any computation of a median we loose half of the pairs considered so far. Since all pairs have only constantly many intersections, after a constant number of steps we end up in a case where we can eliminate a constant fraction of the functions considered originally. Note that we can by sure eliminate one function of a pair, once we have computed all intersections of that pair. Thus, if each pair has exactly $k$ intersections, this is the worst case.

Summarizing the results of this section, we have shown

**Theorem 3.11** Problem II and, thus, Problem I can be solved in linear time.

# 4  Testing Approximate Rotation

In this section we will use the fact that in the plane any rotation in a point can be represented by two line reflections applied consecutively. This is well known in Analytic Geometry (see, e.g. [MP], p.305). Let us summarize what we need in the following:

**Theorem 4.1** In the plane, a rotation by angle $\alpha$ around center $p$ is equivalent to two consecutive reflections in arbitrary lines $l_1, l_2$ which intersect in $p$ with angle $\alpha/2$. For a given rotation, $l_1$ is an arbitrary line through $p$. Once, $l_1$ is chosen, $l_2$ is fixed.

We now consider a problem which has already been considered in [AMWW] where a $O(n \log n)$ – algorithm is given:

**Problem III:** Given two sets $A, B$, each consisting of $n$ points, and a fixed point $p$ in the plane. Test if there are sets $A', B', \varepsilon$ – approximate to $A, B$ respectively, such that $B'$ can be obtained from $A'$ by a rotation around $p$. If so, give a possible choice of sets $A', B'$ and the angle of rotation around $p$ by which $A'$ can be moved to $B'$.

**Algorithm III** (worst case run time $O(n)$):

By Theorem 4.1, we are looking for two lines $l_1, l_2$ intersecting in $p$ such that a set $A', \varepsilon$ – approximate to $A$, is moved to a set $B', \varepsilon$ – approximate to $B$, if we first reflect $A'$ in $l_1$ and then its image in $l_2$. If there is a solution, it must work with any line $l_1$ through $p$. Then line $l_2$ depends on the choice of $l_1$.

The algorithm now goes as follows:
i) Choose any line $l_1$ going through $p$ and reflect all points of $A$ in $l_1$ exactly. Let $H$ be the image of $A$ by this reflection.
ii) Test if $H, B$ are $\varepsilon$ – approximately congruent by a reflection in a line through $p$. This is Problem I with the additional restriction that the line of reflection has to pass through $p$.
For ii) we proceed as in Algorithm I. In Step 1 we do not choose *any* line $l$ that intersects all admissible lines. We rather impose the additional condition that $l$ pass through $p$. Let $p = (p_x, 0)$ in the new coordinate system (obtained by the change of coordinates in Step 1). Then in Step 3 we only test if $\max\{g_i(p_x)\} \leq \min\{h_i(p_x)\}$. By Definition 3.6, this test is equivalent to the test if there is an admissible line through $p$. The lines dual to the points between $\max\{g_i(p_x)\}$ and $\min\{h_i(p_x)\}$ are the admissible lines. Compute the angles between $l_1$ and the lines dual to $\max\{g_i(p_x)\}$ and $\min\{h_i(p_x)\}$. Multiply them by 2 to obtain the range for the angles of rotation. This is all we wanted to know. So we can omit carrying out Algorithm II. If $H, B$ turn out to be $\varepsilon$ – approximately congruent by a reflection in a line through $p$ then Algorithm I provides a possible line of reflection $l_2$ and sets $H', B', \varepsilon$ – approximate to $H, B$ respectively. Let $A'$ be the reflection of $H'$ in line $l_1$. Then we have found a possible choice of sets $A', B'$ as required. The angle of rotation by which $A'$ can be moved to $B'$ around $p$ is twice the angle between $l_1$ and $l_2$.

Our algorithm clearly runs in $O(n)$ time. In order to see that it works correctly, we just have to note that any set $A', \varepsilon$ – approximate to $A$, is mapped to a set $H', \varepsilon$ – approximate to $H$, by reflection in $l_1$ and vice versa. Thus, $A, B$ are $\varepsilon$ – approximately congruent by a rotation around $p$ iff $H, B$ are $\varepsilon$ – approximately congruent by reflection in a line through $p$. Now it is obvious that $A', B'$ can be chosen as mentioned above. $\square$

The result of this section is

**Theorem 4.2** Problem III can be solved in linear time.

# 5 Testing Arbitrary Approximate Congruence

In this section we will reduce the test on arbitrary congruence to the tests on reflection and rotation. It is easy how to do this in the exact case: Just compute the centroids $c_A, c_B$ of $A, B$ in $O(n)$ time. Clearly, if $A, B$ are congruent by any isometry, this isometry has to map $c_A$ to $c_B$. So we make a translation of $A$ such that $c_A$ is mapped to $c_B$. Let $H$ be the image of $A$ by this translation. Then $A, B$ are congruent iff $H, B$ are congruent by rotation around their common centroid $c_B$ (possibly preceded by a reflection in a line through $c_B$). Note that we can basically neglect the case that we need a reflection: Simply test if we can map $A$ to $B$ by a translation which maps $c_A$ to $c_B$ followed by some rotation around $c_B$. If this test gives a negative answer, then reflect $A$ in any line to some set $\bar{A}$ and make the first test with $\bar{A}, B$ instead of $A, B$. Clearly, $A, B$ are congruent iff either one of those tests gives a positive answer.

Considering the *approximate* case we know the following: If $A, B$ are $\varepsilon$ – approximately congruent then there must be an isometry that maps a set $A'$, $\varepsilon$ – approximate to $A$, to a set $B'$, $\varepsilon$ – approximate to $B$. By the remark above, we neglect the case that we need a reflection. Then the isometry can be described in the following way: First make a translation that maps a certain point in the $\varepsilon$ – neighborhood $N_\varepsilon(c_A)$ of $c_A$ (we know that this point has to be $c_{A'}$) to a certain point in $N_\varepsilon(c_B)$ (we know that that point has to be $c_{B'}$). Let $H$ be the image of $A$ by this translation. Then $A, B$ are $\varepsilon$ – approximately congruent iff $H, B$ are $\varepsilon$ – approximately congruent by rotation around $c_{B'}$. Our problem is that we do not know which are the sets $A'$ and $B'$, and, consequently, we do not know which translation we have to apply to $A$ and around which center we have to test on $\varepsilon$ – approximate congruence by rotation.

Our basic idea for the test whether $A, B$ are $\varepsilon$ – approximately congruent or not is the following: Just make a translation that maps $c_A$ to $c_B$. Let $H$ be the image of $A$ by this translation. Then use Algorithm III to test if $H, B$ are $\varepsilon$ – approximately congruent by a rotation around their common centroid $c_B$.

Following this strategy, we may commit two errors:

The first error is that we may not apply the correct translation and, thus, obtain an incorrect set for $H$. But we know that a certain reference point (namely $c_{A'}$) is mapped into the $2\varepsilon$ – neighborhood of the point where it should be mapped to (namely $c_{B'}$). This is why we consider the following trivial but important lemma:

**Lemma 5.1** Let $\tau_1, \tau_2$ be two different translations in the plane. If for some point $c$ in the plane the distance between $\tau_1(c)$ and $\tau_2(c)$ is a certain real number $\delta$ then for *all* points $p$ in the plane the distance between $\tau_1(p)$ and $\tau_2(p)$ is $\delta$.

The second error is that we may rotate around the wrong center. The only thing we know is that the correct center is distant from our center by at most $\varepsilon$. This is why we consider the following question: If we rotate a point $p$ by a fixed angle $\alpha$ in two different ways, first, around some center $c_1$ and, second, around some center $c_2$ which is not very far from $c_1$, how much do the images of $p$ differ depending from the distance between $c_1$ and $c_2$ ? The following lemma gives the answer:

**Lemma 5.2** Given a point $p$ in the plane. Let $p_1$ be the image of $p$ by a rotation with angle $\alpha$ and center $c_1$. Let $p_2$ be the image of $p$ by a rotation with angle $\alpha$ and center $c_2$. Then the distance between $p_1$ and $p_2$ is at most $2 \parallel c_1 - c_2 \parallel$.

**Proof:** W.l.o.g. let $c_1$ be the origin $(0,0)$ and $c_2$ be on the abscissa having coordinates $(c, 0)$. Let $p, p_1, p_2$ have coordinates $(p_x, p_y), (p'_x, p'_y), (p''_x, p''_y)$ respectively (cf. Figure 11). Using the matrix representation of a rotation by angle $\alpha$ we get:

$$\begin{pmatrix} p'_x \\ p'_y \end{pmatrix} = \begin{pmatrix} \cos\alpha & \sin\alpha \\ -\sin\alpha & \cos\alpha \end{pmatrix} \begin{pmatrix} p_x \\ p_y \end{pmatrix}$$

$$\begin{pmatrix} p''_x \\ p''_y \end{pmatrix} = \begin{pmatrix} c \\ 0 \end{pmatrix} + \begin{pmatrix} \cos\alpha & \sin\alpha \\ -\sin\alpha & \cos\alpha \end{pmatrix} \begin{pmatrix} p_x - c \\ p_y - 0 \end{pmatrix}$$

Simple computation gives that the difference vector between $p_1$ and $p_2$ is $(c - c\cos\alpha , c\sin\alpha)$. Its absolute value is $\sqrt{2 - 2\cos\alpha} \cdot | c |$. $\sqrt{2 - 2\cos\alpha}$ can be at most 2 (for $\alpha = 180°$), $| c |$ is $\parallel c_1 - c_2 \parallel$. This is all we wanted to prove. $\qquad \square$

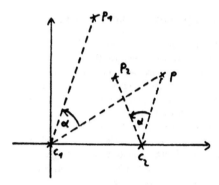

**Figure 11**

Less formally, Lemma 5.2 says: If the error in the center of rotation is small, then the error in rotating a point remains small.

Now we are ready to state the problem that we can solve efficiently:

**Problem IV:** Given two sets $A, B$, each consisting of $n$ points. Test if they are $\varepsilon$ - approximately congruent. If so, give a possible choice of sets $A', B'$, $\varepsilon$ - approximate to $A, B$ respectively, and the isometry that maps $A'$ to $B'$.

The algorithm should satisfy the following requirements: Let $\varepsilon_{opt}$ be the smallest $\varepsilon$ such that $A, B$ are $\varepsilon$ - approximately congruent. Let $\gamma$ be an arbitrarily small but fixed constant. If $\varepsilon/\varepsilon_{opt} \notin [1, 1+\gamma)$, then the algorithm should always give the correct answer. If $\varepsilon/\varepsilon_{opt} \in [1, 1+\gamma)$, then the algorithm may give a wrong answer.

**Algorithm IV** (worst case run time $O(n)$ if $\gamma$ is constant):

Compute $c_B$. Cover $N_\varepsilon(c_B)$ with $\lambda$ circles of radius $\delta = \frac{\varepsilon}{4} \cdot \frac{\gamma}{1+\gamma}$. We will see in Theorem 5.3 that $\lambda = 2\varepsilon^2/\delta^2$ suffices. Let $c_1, \ldots, c_\lambda$ be the centers of the circles of the covering.
Compute $c_A$. For each $i = 1, \ldots, \lambda$ do the following:

Apply a translation to $A$ such that $c_A$ is mapped to $c_i$. Let $H_i$ be the image of $A$ by this translation. Test if $H_i$ is $\varepsilon$ - approximately congruent to $B$ by rotation around $c_i$ using Algorithm III.

If the answer is negative for each such test, then give a negative answer to Problem IV. If the answer is positive for one of those tests, then give a positive answer to Problem IV. Let $c_{i_0}$ be the center of rotation used by the test that gave a positive answer in Algorithm III. Let $H'_{i_0}, B'$ be the possible choice of sets, $\varepsilon$ - approximate to $H_{i_0}, B$ respectively, and $\alpha$ be the angle of rotation given by Algorithm III. Then the inverse image of $H'_{i_0}$ by the translation that maps $c_A$ to $c_{i_0}$ and $B'$ are possible choices for sets $A', B'$ as wanted in Problem IV. The translation that maps $c_A$ to $c_{i_0}$ followed by a rotation around $c_{i_0}$ by angle $\alpha$ is the isometry that maps $A'$ to $B'$. $\qquad\square$

**Theorem 5.3** Algorithm IV solves Problem IV in worst case run time $O(n/\gamma^2)$.

**Proof:**
**Correctness:** If it is possible to show that some $H_{i_0}$ (which is congruent to $A$) and $B$ are $\varepsilon$ - approximately congruent by rotation around *some* center $c_{i_0}$, then $A, B$ clearly have to be $\varepsilon$ - approximately congruent. Consequently, if the answer is positive, it must be correct, since Algorithm IV does check this condition for several centers $c_i$. Thus, for any $\varepsilon < \varepsilon_{opt}$, Algorithm IV must give a negative answer, because there is no $\varepsilon$ - approximate congruence. Thus, we have proven that Algorithm IV works correctly for $\varepsilon/\varepsilon_{opt} < 1$.

Only when Algorithm IV gives a negative answer and $\varepsilon \geq \varepsilon_{opt}$, there may still be a solution, because Algorithm IV does not test on rotation around every center possible.

Assume that $A, B$ are $\varepsilon$ - approximately congruent. This implies that there is a point $p \in N_\varepsilon(c_B)$ and a set $H'$, congruent to $A$, such that $H', B$ are $\varepsilon$ - approximately congruent by rotation around $p$. Since in Algorithm IV we cover $N_\varepsilon(c_B)$ with circles of radius $\delta$, we know that there is a point $c_{i_0}$ within the $\delta$ - neighborhood of $p$ such that we test in Algorithm IV if a set $H_{i_0}$, congruent to $A$, and $B$ are $\varepsilon$ - approximately congruent by rotation around $c_{i_0}$. By Lemma 5.1, the error we do by choosing $H_{i_0}$ instead of $H'$ is at most $2\delta$ for each point. By Lemma 5.2, the error we do by choosing $c_{i_0}$ as the center of rotation instead of $p$ is at most $2\delta$. In the worst case, these two errors add up. Thus, if $\varepsilon \geq \varepsilon_{opt} + 4\delta$, Algorithm IV must give a positive answer. In Algorithm IV, $\delta$ is chosen in such a way that $\varepsilon \geq \varepsilon_{opt} + 4\delta \iff \varepsilon \geq (1+\gamma)\,\varepsilon_{opt}$. Since, by the discussion above, Algorithm IV works always correctly if the answer has to be negative, we have shown that Algorithm IV works correctly for $\varepsilon \geq (1+\gamma)\,\varepsilon_{opt}$.

**Run time:** Computing the centroids and the translation at the beginning of Algorithm IV clearly can be done in $O(n)$. In Section III we have shown that we can test on $\varepsilon$ - approximate congruence by rotation around any $c_i$ in $O(n)$. Thus, we have to show that we can cover a circle of radius $\varepsilon$ by $\lambda = 2\varepsilon^2/\delta^2$ circles of radius $\delta$ in $O(\lambda)$:

Confer to Figure 12: Consider the square with side length $2\varepsilon$ that contains the circle of radius $\varepsilon$. Cover this square by small squares with side lengths $\sqrt{2}\,\delta$. It is easy to see that we need $4\varepsilon^2/2\delta^2$ such small squares. The circle of radius $\delta$ around the center of a small square contains this square properly. Thus, it suffices to take the centers of each small square for the covering. Clearly, we can compute each such center in constant time.

So far, we have shown that we can perform Algorithm IV in time $O(\lambda n)$ with the $\lambda$ given above. If we plug in $\delta$ as chosen in Algorithm IV, we get: $\lambda = 32(\frac{1+\gamma}{\gamma})^2$ which is $O(1/\gamma^2)$. This completes the proof.

$\square$

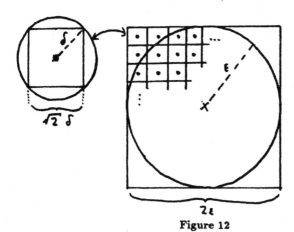

**Figure 12**

# Conclusion:

We have presented optimal algorithms for some congruence detection problems. The problems are given in a way that takes into account how the data are given in real life (using the concept of *approximate* congruence). An important restriction is that we only considered the *labelled* case. In [AMWW], some algorithms are given for the unlabelled case, but they are probably not optimal. Especially, the upper bound for an algorithm on arbitrary congruence is an interesting problem. The best bound known so far is $O(n^8)$, obtained in [AMWW].

It may be a good way in order to tackle the unlabelled case that given $A, B$ we first ask which matchings of labels are possible for any $\varepsilon$ - approximate congruence between $A$ and $B$ and then check each labelling computed by the first step with the algorithms presented in this paper. But so far it is not clear how many labellings we still have to consider.

An additional interesting question is whether it is possible to reach the optimal time bounds obtained in this paper with the model of computation used in [AMWW].

**Acknowledgement:**
I thank Helmut Alt and Emo Welzl for several helpful discussions that lead me to crucial ideas.

# 6 References

[AHU] A.V. Aho, J.E. Hopcroft, J.D. Ullman: *The Design and Analysis of Computer Algorithms*, Addison-Wesley, 1974

[AMWW] H. Alt, K. Mehlhorn, H. Wagener, E. Welzl: *Congruence, Similarity and Symmetries of Geometric Objects*, Discr. Comp. Geom. 3 (1988), pp.237-256

[Ata] M.J. Atallah: *Checking Similarity of Planar Figures*, International J. Comp. Inf. Science 13 (1984), pp. 279–290

[Atk] M.D. Atkinson: *An Optimal Algorithm for Geometrical Congruence*, J. of Algorithms 8 (1987), pp. 159–172

[E] H. Edelsbrunner: *Algorithms in Combinatorial Geometry*, EATCS Monographs on Theoretical Computer Science, Springer-Verlag, 1987

[KS] P.J. Kelly, E.G. Straus: *Elements of Analytic Geometry*, Scott, Foresman and Co., 1968

[MP] R.S. Millman, G.D. Parker: *Geometry, A Metric Approach with Models*, Springer-Verlag, 1981

[Me] N. Megiddo: *Linear Time Algorithm for Linear Programming in $R^3$ and Related Problems*, SIAM J. of Computing 12 (1983), pp. 759–776

# MOVING REGULAR $k$-GONS IN CONTACT[1]

Stephan Abramowski, Bruno Lang, Heinrich Müller
Department of Computer Science, University of Karlsruhe, Germany

## Abstract

Given $m$ circles in the plane, contacts between them can be specified by a system of quadratic distance equalities. An approximative solution for the trajectories of the circles for a system of one degree of freedom is given, by replacing the circles by translationally moving regular $k$-gons. The approximation yields trajectories that are piecewise linear. The next linear generation of the $m$ trajectories are found by an incremental algorithm in $O(m^2)$ time. Further, an algorithm is presented which finds the next collision between $m$ $k$-gons moving on lines at constant speed in time $O(k \cdot m^{2-x})$ for a constant $x > 0$ using linear space. Finally, more practical collision detection algorithms are sketched based on neighborhood information which, however, do not guarantee a nontrivial worst-case time bound.

## 1. The problem

The following problem is discussed.

## Moving Circles in Contact.

> **Input.** A set of $m$ nonintersecting circles in contact with one degree of freedom, i.e. they are are forced to move uniquely as long as no contacts are resolved. The contacts are specified by equations
>
> $$d(p_i, p_j) = d_i + d_j, (i, j) \in E \subset \{1, ..., m\}^2, \ i \neq j.$$
>
> The pair $(p_i, d_i)$ denotes a circle with center $p_i$ and radius $d_i$, $d(.,.)$ the euclidian metric.
>
> **Output.** The first collision of the configuration with itself in a chosen direction.

Figure 1 depicts a set of circles in contact which have to move deterministically. The thick drawn circles move along the thin drawn circles fixed in the plane, into the direction of the arrow. The second thick circle causes the first collision.

Applications of this problem are in motion planning of nonrigid objects. For example, the rotatory arm of a (two dimensional) manipulator can be modeled by a set of circles (figure 2). The arm is represented by a chain of circles. The circles model the material of the arm, i.e. they are not allowed to penetrate. Within this restriction, the circles can be placed arbitrarily maintaining the distance between consecutive ones. This second condition can be modeled by

---

[1] partially supported by Deutsche Forschungsgemeinschaft (DFG, Mu744/1-1)

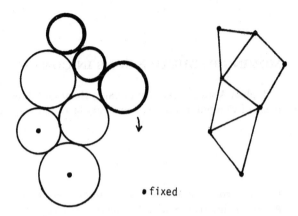

Figure 1: A set of circles in contact and its configuration graph.

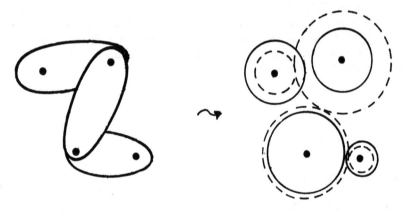

Figure 2: Modelling a 2-d manipulator's arm by circles.

another sequence of circles, drawn hatched in the figure. A typical task is now to carry over a given configuration of the arm into a second given configuration. Supposing that they are already in a stable state, this is performed by generating a sequence of forced motions carrying the initial configuration over into the final one. The sequence is obtained by successively breaking contacts.

In 3-space, spheres are used instead of circles. Sphere models were used by e.g. Badler, Smoliar (1979) for modeling and animating human bodies.

For a set of circles, the contact graph is obtained as follows. The vertices are the centers of the circles. An edge is introduced between two circles if they are are in contact (figure 1). For a stable configuration of circles in contact, counting variables and equations leads to a of number of edges at least three less than twice the number of vertices. For a uniquely deformable configuration, the number of edges is four less than twice the number of vertices (figure 1). The graph theoretic description is used in the analysis of kinematic chains in mechanics (Auerbach, Hort, 1929).

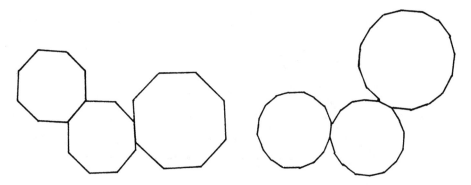

Figure 3: Symmetric regular $k$-gons replacing the circles.

## 2. Discussion

The solution of the problem consists of two steps. The first step is to calculate the paths of the moving circles. A parameter representation of the paths can be obtained by eliminating all but one variable in the defining system of equations, i.e. by transforming the given system of equations into a diagonal form. Variable elimination can be carried out using resultants (Collins 1975). The problem with this approach is that the degree of the resulting polynomials as well as the size of their coefficients may grow tremendously, i.e. exponentially. This limits the straightforward algebraic approach to small numbers of circles. A possible solution is described in sect. 3. An approximate version of the problem is introduced by replacing the circles by regular $k$-gons. This results in piecewise linear curves which can be calculated easily by incrementally applying the Gauss elimination algorithm for systems of linear equations. Due to the iteration, numerical problems are not excluded with this algorithms. Based on the observation that the linearized equations of contact can be straightforwardly translated into equations over integer combinations of roots of unity it is shown that the algorithm can be carried out exactly in integer arithmetic. This is the content of sect. 4.

Once the paths obtained, for each of them the first collision with other circles must be found. An efficient solution requires to restrict the pairs of circles tested one against another. In linear approximation, the first collision in a set of $k$-gons moving on linear trajectories at constant speed is to be found. Based on point location in a high dimensional space, a solution is presented in sect. 5 requiring $O(k \cdot m^{2-x})$ time for a constant $x > 0$ in $O(m)$ space. The intention of the algorithm is to demonstrate that collision detection can be performed in less than quadratic time. In this form, it is of minor practical interest. Sect. 6 contains further ideas how to solve the collision detection problem which, however, do not guarantee a worst case better than $O(m^2)$. The strategy is to test neighbored circles only. Testing neighbored circles requires storing and updating information about the neighborhood. In the worst case, its overhead may dominate the number of saved collision tests.

## 3. Linearization by $k$-gons

The circles are now replaced by symmetric regular $k$-gons, $k \geq 4$, $k$ even, cf. figure 3. The euclidian distance $d$ in the contact equations is replaced by $d_k$ (figure 4),

$$d_k(p, q) := (cos\phi, sin\phi) \cdot (q - p),$$

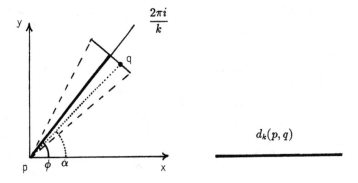

Figure 4: The distance $d_k(p, q)$.

$$\phi = \frac{2\pi j}{k} \text{ with } \alpha \in [\frac{2\pi j}{k} - \frac{\pi}{k}, \frac{2\pi j}{k} + \frac{\pi}{k}),$$

$$q_x - p_x = d(p, q) \cdot \cos\alpha,$$

$$q_y - p_y = d(p, q) \cdot \sin\alpha.$$

The indices of $q$ and $p$ denote the respective coordinates. The new equations

$$d_k(p_i, p_j) = d_i + d_j, \ (i, j) \in E \subset \{1, ..., m\}^2, \ i \neq j$$

are piecewise linear. Changing the contact at a corner $p_i, p_j$ replaces

$$(\cos\phi, \sin\phi) \cdot (p_i - p_j) = d_i + d_j$$

by

$$(\cos(\phi \pm \frac{2\pi}{k}), \sin(\phi \pm \frac{2\pi}{k})) \cdot (p_i - p_j) = d_i + d_j.$$

The problem is now reduced to solving a sequence $A_r x = b_r$, $r = 1, ...$ of systems of linear equations with a number of variables exceeding the number of equations by one (one degree of freedom for forced motion). Consecutive systems correspond to the traversal of a corner, or to a change of contact. They can be arranged to differ in one equation only. A further observation that should be considered is that the system is sparse. There are at most four nonzero coefficients in every equation. Our solution is an incremental version of the Gauss elimination algorithm.

**Proposition 1.** The continuation of the paths at a corner or in case of collision can be calculated in $O(m^2)$ time and $O(m^2)$ space.

**Proof.** Let be $M_r$ a transformation matrix diagonalizing $A_r$, i.e. $M_r \cdot A_r = D_r$, $D_r = (I, d_r)$ up to a permutation of columns, $d_r$ a not further specified column. The result $x$ is immediately obtained from $D_r \cdot x = M_r \cdot b_r$, i.e. the components of $x$ are linear expressions in the component of $x$ corresponding to $d_r$.

Let be $T_0 := M_0$, and $T_r$ recursively defined by $T_r \cdot M_{r-1} = M_r$. $T_0$ can be found in $O(m^3)$ time by Gauss elimination or possibly in less time by adapting a fast LUP decomposition algorithm (Aho, Hopcroft, Ullman, 1974). For $r > 0$, $T_r$ can be calculated from

$$D_r = T_r(D_{r-1} + E_r), \ E_r := M_{r-1} \cdot (A_r - A_{r-1}).$$

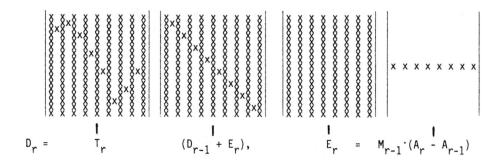

$$D_r = \qquad T_r \qquad (D_{r-1} + E_r), \qquad E_r = M_{r-1} \cdot (A_r - A_{r-1})$$

Figure 5: The nonzero entries of the matrices $T_i$.

Since $A_r - A_{r-1}$ contains at most eight nonzero entries, the matrix $D_{r-1} + E_r$ to be transformed by $T_r$ has at most nine columns with more than one nonzero entry (figure 5). Hence $T_r$ has at most nine columns with more than one nonzero entry and can be calculated in $O(m)$ time from $D_{r-1} + E_r$.

The calculation of a path consists in updating $M_{r-1}$ to $M_r$, $D_{r-1}$ to $D_r$, and $M_{r-1} \cdot b_{r-1}$ to $M_r \cdot b_r$. The most expensive step is the first update, $M_r = T_r \cdot M_{r-1}$, which costs $O(m^2)$ time. The second update costs $O(m)$ time from the discussion above. The third update goes by

$$M_r \cdot b_r = T_r \cdot (M_{r-1} \cdot b_{r-1} + M_{r-1} \cdot (b_r - b_{r-1}))$$

in $O(m)$ time from $M_{r-1} \cdot b_{r-1}$ ($b_r - b_{r-1}$ has at most one nonzero entry). The space bound is immediate.

## 4. Arithmetical considerations

The incremental algorithm may cause numerical problems when implemented in an existing floating point arithmetic. Instead of analyzing its numerical stability, we will focus on exact calculation. For this purpose, the radii $d_i$ of the circles are positive integers. The arithmetic of real numbers is replaced by the arithmetic over the field $Q[\omega_k]$, $\omega_k = e^{\frac{2\pi i}{k}}$ the k-th root of unity, $Q$ the rational numbers. Every element in $Q[\omega_k]$ has a unique representation $\sum_{i=0}^{n-1} a_i \omega^i$, $a_i \in Q$, $n = \phi(k)$, $\phi$ Euler's function whose value is the cardinality of the set $\{i : 1 \le i \le k, \gcd(i,k) = 1\}$ For simplicity we suppose $k$ a power of two for the following, and $k \ge 4$. This implies $\phi(k) = \frac{k}{2}$.

Using

$$\cos\phi = \frac{e^{i\phi} + e^{-i\phi}}{2}, \quad \sin\phi = \frac{e^{i\phi} - e^{-i\phi}}{2i}, \quad \text{and } i = e^{\frac{k}{4}},$$

the defining equalities $(\cos\frac{2\pi j}{k}, \sin\frac{2\pi j}{k})(p_i - p_j) = d_i + d_j$ can be rewritten in $Q[\omega_k]$ as

$$\left(\frac{\omega_k^j + \omega_k^{-j}}{2}, \frac{\omega_k^{j-\frac{k}{4}} - \omega_k^{-j-\frac{k}{4}}}{2}\right)(p_i - p_j) = d_i + d_j,$$

allowing to solve them exactly over $Q[\omega_k]$, if the distances $d_i$ are integers (or rationals).

In the remainder of this section the emulation of arithmetic over $Q[\omega_k]$ in integer arithmetic is sketched. Operations over $Q[\omega_k]$ are reduced to operations over $Z[\omega_k]$, $Z$ the integers. Algorithms for the arithmetic over $Z[\omega_k]$ can then be taken from e.g. Aho, Hopcroft, Ullman (1974). Multiplication and inversion are reduced to integer Fourier transforms.

### Addition and subtraction.

$$\sum_{i=0}^{n-1} \frac{a'_i}{a''_i}\omega_k^i \pm \sum_{i=0}^{n-1} \frac{b'_i}{b''_i}\omega_k^i, \ a'_i, \ a''_i, \ b'_i, \ b''_i \text{ integers, reduces to}$$

$$\sum_{i=0}^{n-1}(a_i \pm b_i)\omega_k^i, \ a_i = b''_i a'_i, \ b_i = a''_i b'_i,$$

which must be divided by $a''_i b''_i$. Integer addition is straightforward. Reduction into the unique representation requires an integer common greatest divisor algorithm (Aho, Hopcroft, Ullman, 1974, sect. 8.10).

### Multiplication.

$$\left(\sum_{i=0}^{n-1} \frac{a'_i}{a''_i}\omega_k^i\right) \cdot \left(\sum_{i=0}^{n-1} \frac{b'_i}{b''_i}\omega_k^i\right) \text{ reduces to}$$

$$\left(\sum_{i=0}^{n-1} a_i\omega_k^i\right) \cdot \left(\sum_{i=0}^{n-1} b_i\omega_k^i\right), \ a_i = a'_i \prod_{j \neq i} a''_j, \ b_i = b'_i \prod_{j \neq i} b''_j,$$

which must be divided by $\prod_{j=0}^{n-1} a''_j \cdot \prod_{j=0}^{n-1} b''_j$. The $i$-th coefficient of the product,

$$d_i = \sum_{j=0}^{i} a_j b_{i-j} - \sum_{j=i+1}^{n-1} a_j b_{n+i-j},$$

is called negative wrapped convolution. The negative wrapped convolution over integers can be calculated by e.g. a fast Fourier transform (FFT), cf. Aho, Hopcroft, Ullman, sect. 7.3.

### Inversion.

$$\frac{1}{\sum_{i=0}^{n-1} \frac{a'_i}{a''_i}\omega_k^i} \text{ is replaced by } \frac{1}{\sum_{i=0}^{n-1} a_i\omega_k^i} =: \sum_{i=0}^{n-1} B_i\omega_k^i, \ a_i = a'_i \prod_{j \neq i} a''_j.$$

The original expression is the $\prod_{j=0}^{n-1} a''_j$-multiple of the old one. The unknown rational coefficients $B_i$ of the new expression satisfy the equalities

$$\sum_{j=0}^{i} a_j B_{i-j} - \sum_{j=i+1}^{n-1} a_j B_{n+i-j} = D_i, \ i = 0, ..., n-1, \ D_0 = 1, \ D_i = 0 \text{ for } i > 0.$$

The matrix of coefficients is

$$\mathbf{A} = \begin{pmatrix} a_0 & -a_{n-1} & -a_{n-2} & -a_{n-3} & \cdots & -a_2 & -a_1 \\ a_1 & a_0 & -a_{n-1} & -a_{n-2} & \cdots & -a_3 & -a_2 \\ a_2 & a_1 & a_0 & -a_{n-1} & \cdots & -a_4 & -a_3 \\ \vdots & \vdots & \vdots & \vdots & \ddots & \vdots & \vdots \\ a_{n-1} & a_{n-2} & a_{n-3} & a_{n-4} & \cdots & a_1 & a_0 \end{pmatrix}.$$

By Cramer's rule, the unknown $B_i$ are rationals with the determinant $|\mathbf{A}|$ as denominators. Hence the result of

$$\frac{|\mathbf{A}|}{\sum_{i=0}^{n-1} a_i\omega_k^i} =: \sum_{i=0}^{n-1} b_i\omega_k^i$$

is in $Z[\omega_k]$. This expression may be calculated by inverting negative wrapped convolution using bitwise FFT and its inverse (Theorem 7.2 and sect. 7.3. of Aho, Hopcroft, Ullman, 1974). Bitwise FFT is carried out w.r.t. $Z_p$, i.e. the integers mod $p$. $p$ is chosen greater than twice the absolute amount of any integer in the result. A suitable $p$ may be found in the above reference or in the book by Lipson (1981). Further, a root of unity in $Z_p$, $\psi$, satisfying $\psi^{2n} = 1$ must exist. Then the Fourier transform is based on $\omega := \psi^2$. Using the abbreviation $f(x) := \sum_{i=0}^{n-1} a_i x^i$,

$$|\psi^{(2i+1)j}|_{i,j=0,\ldots,n-1} \cdot |\mathbf{A}| = |\psi^{(2i+1)j} \cdot f(\psi^{2i+1})|_{i,j=0,1\ldots,n-1} = |\psi^{(2i+1)j}|_{i,j=0,\ldots,n-1} \cdot \prod_{i=0}^{n-1} f(\psi^{2i+1})$$

yields

$$|\mathbf{A}| = \prod_{i=0}^{n-1} f(\psi^{2i+1}).$$

The factors of the product are just the Fourier transforms of $(a_0, a_1\psi^1, \ldots, a_{n-1}\psi^{n-1})$ used in the negative wrapped convolution formula by Aho, Hopcroft, Ullman, Theorem 7.2. Hence, using this theorem, the backward transform of

$$\Big( \prod_{j=0,j\neq i}^{n-1} f(\psi^{2j+1}) \Big)_{i=0,\ldots,n-1}$$

leads to the result $\sum_{i=0}^{n-1} b_i \omega_k^i$, with coefficients $b_i$ mod $p$. It remains the choice between the positive and the negative possibility for the $b_i$. If the least nonnegative representative of $b_i$ mod $p$ is greater than $\frac{p}{2}$, $b_i$ is negative, and nonnegative else.

If the rational arithmetic is carried out exactly, the time and space complexity depends on the maximum size of the numerators and denominators.

**Proposition 2.** Let $L_0$ be the maximum number of bits of any integer in the input. Then the number of bits of the numerators and denominators of any rational number occurring during calculation is $O(k \cdot (m + \log k) + L_0)$.

**Proof.** The solution of $A_r x = b_r$ may alternatively be obtained by Cramer's rule. Let $a_r$ be the column with $d_r = T_r \cdot a_r$, cf. the notation in the previous section. Hence $A_r$ without column $a_r$ is regular. Let $A'_r$ be the matrix $A_r$ without column $a_r$. Then the $i$-th component of $x$ is $\frac{det(A'_{r,i})}{det(A'_r)}$, with $A'_{r,i}$ obtained from $A'_r$ by replacing the $i - th$ column by $b_r - d_r$. The determinant of a $q \times q$-matrix with at most three nonzero entries in a row is a sum of $3^q$ products of $q$ factors. The entries of the determinant in the denominator are of the form $\sum_{i=0}^{n-1} a_i \cdot \omega^i$, $a_i \in \{-1, 0, 1\}$. At most two of the $a_i$ are nonzero. Hence the denominator has the same form, with $a_i$ integers of $O(m)$ bits. The entries of the determinant of the numerator are the form $\sum_{i=0}^{n-1} b_i \cdot \omega^i$, with either $b_i \in \{-1, 0, 1\}$ and all but at most two of the $b_i$ are equal to zero, or the $b_i$ are linear expressions with integer coefficients of $O(L_0)$ bits at most. The latter happens for at most one column. Hence the numerator has the same form, with $b_i$ integers of $O(m + L_0)$ bits. Division of both expressions and normalization of the factors results in a unique expression of the form $\sum_{i=0}^{n-1} C_i \cdot \omega^i$. Using Cramer's rule for the $C_i$ (cf. the discussion of division above), the numerators and denominators of the $C_i$ are integers with $O(n \log n + m \cdot n + L_0)$ bits.

Up to now we have bounded the components of $x$ for arbitrary $r$. Similarly, the entries of the matrices $T_r$ and all intermediate results occuring during their calculation can be bounded by formulating them as determinants of submatrices of $A_r$ or expressions applying only a constant number of operations on such determinants. $n = O(k)$ yields the result.

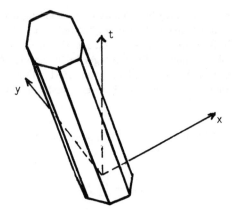

Figure 6: Moving $k$-gons in the time-plane space become $k$-sided bars

## 5. Collision detection

Now we turn to the following problem

**Collision Detection**

> **Input.** $m$ $k$-gons, moving on piecewise linear trajectories in the plane, dependent on a common time parameter.
>
> **Output.** The collisions among the $k$-gons, sorted according to their events in time.

The straightforward solution is by pairwise comparison resulting in $O(k \cdot m^2)$ time and $O(m)$ space for finding the next collision. The purpose of this section is to demonstrate that it is possible to find the next collision in less than quadratic time, i.e. in $O(k \cdot m^{2-x})$ time for some $x > 0$, and linear space. The construction is brute force, not trying to optimize $x$.

**Proposition 3.** Given $m$ nonintersecting $k$-gons in the plane, moving on linear trajectories at constant speed, the first collision can be found in $O(m)$ space and $O(k \cdot m^{2 - \frac{1}{2^{4}}} \log^2 m)$ time.

**Proof.** In the plane-time space, the $k$-gons in motion become $k$-sided bars (figure 6). The question is for two intersecting bars with the smallest nonnegative intersection coordinate on the time axis. The test between two bars is reduced to a test between a line and bar by transferring the material of one of the involved bars to the other. This goes by simply replacing the radius of the second generating $k$-gon by the sum of the two radii (figure 7). The problem is now to find the smallest intersection (with respect to one of the coordinates, the time coordinate) between a set of lines and a set of special faces in 3-space. The faces are distributed in $k$ subsets which are treated separately the same way. A subset contains all faces induced by $k$-gon edges with the same orientation. W.l.o.g. the orientation is parallel to the $x$-axis. Then the points $(x, y, t)$ of a face satisfy

$$y = a \cdot t + y_0, \quad b \cdot t + x_0 \le x \le b \cdot t + x_1,$$

for constants $a$, $b$, $x_0$, $x_1$, $y_0$. A line

$$x = c \cdot t + d, \quad y = e \cdot t + f$$

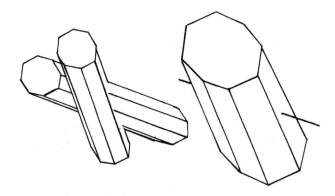

Figure 7: Replacing one bar by a line.

intersects the face if

$$b \cdot t + x_0 \leq c \cdot t + d \leq b \cdot t + x_1$$

for $t = \frac{y_0 - f}{e - a}$, $e - a \neq 0$, i.e.

$$c \cdot f - d \cdot e - y_0 \cdot c + a \cdot d + x_0 \cdot e - b \cdot f + b \cdot y_0 - a \cdot x_0 \leq 0,$$

$$c \cdot f - d \cdot e - y_0 \cdot c + a \cdot d + x_1 \cdot e - b \cdot f + b \cdot y_0 - a \cdot x_1 \geq 0,$$

for $e - a > 0$ ($e - a < 0$ analogously). Substituting

$$g := c \cdot f - d \cdot e, \ \alpha := y_0, \ \beta := -a, \gamma := -x_0,$$

$$\gamma' := -x_1, \ \delta := b, \ \epsilon := a \cdot x_0 - b \cdot y_0, \ \epsilon' := a \cdot x_1 - b \cdot y_0$$

yields two linear inequalities

$$g - \alpha \cdot c - \beta \cdot d - \gamma \cdot e - \delta \cdot f - \epsilon \leq 0$$

$$g - \alpha \cdot c - \beta \cdot d - \gamma' \cdot e - \delta \cdot f - \epsilon' \geq 0.$$

Now $c, d, e, f, g$ can be seen as coefficients of a hyperplane, and $p = (\alpha, \beta, \gamma, \delta, \epsilon)$ resp. $p' = (\alpha, \beta, \gamma', \delta, \epsilon')$ as points in a 5-space. The inequalities are satisfied for those hyperplanes intersecting the line segment $p, p'$. The hyperplanes are found by the following strategy. First, the set of lines in 5-space is partioned in subsets of lines intersecting the hyperplanes in the same order. For every subset, the hyperplanes are preprocessed into a 1-d range tree according to this order. Based on the preprocessed structure, the hyperplanes intersected by a line segment $p, p'$ are found quickly by finding the range tree belonging to the line $p, p'$, followed by a range search in the range tree with the interval $p, p'$.

The details are as follows. We are interested in lines in 5-space defined by two points $p_0 = (\alpha_0, \beta_0, \gamma_0, \delta_0, \epsilon_0)$ and $p'_0 = (\alpha_0, \beta_0, \gamma'_0, \delta_0, \epsilon'_0)$. In parameter representation, the line has the form $(1 - \lambda) \cdot p_0 + \lambda \cdot p'_0$, $\lambda$ arbitrary real numbers. The $\lambda$-value of the intersection with a hyperplane $g_i - \alpha \cdot c_i - \beta \cdot d_i - \gamma \cdot e_i - \delta \cdot f_i - \epsilon = 0$, $i = 1, ..., m$, is

$$\lambda_i = \frac{g_i - \alpha_0 \cdot c_i - \beta_0 \cdot d_i - \gamma_0 \cdot e_i - \delta_0 \cdot f_i - \epsilon_0}{e_i \cdot (\gamma'_0 - \gamma_0) + (\epsilon'_0 - \epsilon_0)}$$

using

$$g_i - \alpha_0 \cdot c_i - \beta_0 \cdot d_i - ((1 - \lambda) \cdot \gamma_0 + \lambda \cdot \gamma'_0) \cdot e_i - \delta_0 \cdot f_i - ((1 - \lambda) \cdot \epsilon_0 + \lambda \cdot \epsilon'_0) = 0.$$

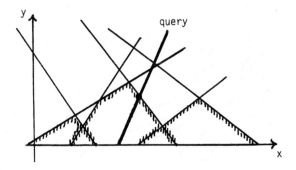

Figure 8: Finding the first intersection point.

The lines correspond one-to-one to points in 7-space with coordinates $\alpha_0, \beta_0, \gamma_0, \gamma_0', \delta_0, \epsilon_0, \epsilon_0'$. Pairwise comparison of the $\lambda_i$, i.e. $\lambda_i = \lambda_j$, $i \neq j$, $i, j = 1, ..., m$, yields $O(m^2)$ surfaces in 7-space,

$$(g_i - \alpha_0 \cdot c_i - \beta_0 \cdot d_i - \gamma_0 \cdot e_i - \delta_0 \cdot f_i - \epsilon_0) \cdot (e_j \cdot (\gamma_0' - \gamma_0) + (\epsilon_0' - \epsilon_0)) =$$

$$(g_j - \alpha_0 \cdot c_j - \beta_0 \cdot d_j - \gamma_0 \cdot e_j - \delta_0 \cdot f_j - \epsilon_0) \cdot (e_i \cdot (\gamma_0' - \gamma_0) + (\epsilon_0' - \epsilon_0)),$$

$$i \neq j, \ i, j = 1, ..., m.$$

They define a cell partition of the 7-space. Every cell corresponds to a set of lines with the same order of intersection points with the hyperplanes in 5-space. Substituting

$$u_1 := \gamma_0' - \gamma_0, \ u_2 := \alpha_0 \cdot (\gamma_0' - \gamma_0), \ u_3 := \beta_0 \cdot (\gamma_0' - \gamma_0), \ u_4 := \gamma_0 \cdot (\gamma_0' - \gamma_0),$$

$$u_5 := \delta_0 \cdot (\gamma_0' - \gamma_0), \ u_6 := \epsilon_0 \cdot (\gamma_0' - \gamma_0), \ u_7 := \epsilon_0' - \epsilon_0, \ u_8 := \alpha_0 \cdot (\epsilon_0' - \epsilon_0),$$

$$u_9 := \beta_0 \cdot (\epsilon_0' - \epsilon_0), \ u_{10} := \gamma_0 \cdot (\epsilon_0' - \epsilon_0), \ u_{11} := \delta_0 \cdot (\epsilon_0' - \epsilon_0), \ u_{12} := \epsilon_0 \cdot (\epsilon_0' - \epsilon_0),$$

replaces the surfaces by $O(m^2)$ planes

$$g_i e_j u_1 - c_i e_j u_2 - d_i e_j u_3 - e_i e_j u_4 - f_i e_j u_5 - u_6 + g_i u_7 - c_i u_8 - d_i u_9 - e_i u_{10} - f_i u_{11} - u_{12} =$$

$$g_j e_i u_1 - c_j e_i u_2 - d_j e_i u_3 - e_j e_i u_4 - f_j e_i u_5 - u_6 + g_j u_7 - c_j u_8 - d_j u_9 - e_j u_{10} - f_j u_{11} - u_{12},$$

$$i \neq j, \ i, j = 1, ..., m,$$

in 12-space $u_1, ..., u_{12}$. They induce a partition into $O((m^2)^{12}) = O(m^{24})$ cells. To every cell a 1-d range tree is assigned with the $O(m)$ hyperplanes in the $\alpha, \beta, \gamma, \delta, \epsilon$-space as keys, according to the arrangement of intersections for the cell. Using the straightforward point location algorithm of Dobkin, Lipton (1976) and the incremental algorithm described by Edelsbrunner (1986) for building up a cell decomposition, the partition is preprocessed in $O(m^{2^{24}})$ time into a data structure of size $O(m^{2^{24}})$ which allows point location in $O(\log m)$ time.

In order to find the first intersection point, a further data structure is assigned to every node of the range trees (figure 8). The hyperplanes organized in the subtree induced by a node correspond to lines in the original $x, y, t$-space. Their projections onto the $y, t$-plane induce a cell decomposition. The intersection points of the partition with the $y$-axis are stored for binary search. To every interval on the $y$-axis, the upper boundary of the intersected cell of the partition is assigned. The upper boundary is preprocessed for binary search. The space requirements for one of of these augmented range trees is $O(m \log m)$ since the node data structures are of linear size (Chazelle, Lee, Guibas, 1983). The overall space and time requirements are $O(m^{25} \log^2 m)$. They are dominated by the requirements of the point location structure.

Now, coming up with a query face, the corresponding query segment $p_0, p_0'$ is calculated in $O(1)$ time. The cell of line $p_0, p'$ is found by point location in 12-space in time $O(\log m)$ using the data structure from above. Then the nodes of the corresponding range tree in the interval $p_0, p_0'$ are found by tree traversal in time $O(\log m)$. Next, the point of intersection of the projection of line $p_0, p_0'$ onto the $y, t$-plane with the $y$-axis is calculated in $O(1)$ time. For each of the $O(\log m)$ nodes in the tree covering the interval, the interval of intersection point on the $y$-axis is found by binary search in $O(\log m)$ time. The first intersected boundary segment intersected by the infinite segment of line $p_0, p_0'$ starting at the intersection point in direction of increasing $t$ is obtained by binary search on the boundary in time $O(\log m)$. Using the $e$ and $f$ coefficients of the boundary line, and the coefficients $y_0$ and $a$ of the query face, the minimum of the $O(\log m)$ values $t = \frac{y_0 - f}{e - a}$ gives the first intersection point over time, and the line belonging to that boundary segment is that intersecting the query face first. Hence the total query time for one face is $O(\log^2 m)$.

The time complexity stated in the proposition is obtained by subdividing the set of $m$ lines into $m^{1 - \frac{1}{2^{24}}}$ subsets of $m^{\frac{1}{2^{24}}}$ lines. One subset can preprocessed in $O((m^{\frac{1}{2^{24}}})^{2^{24}}) = O(m)$ time and space. The total preprocessing time is $O(m^{2 - \frac{1}{2^{24}}})$, the preprocessing space is $O(m)$ since only one set must be held preprocessed at any time. The total time for answering the $m$ queries on one subset is $O(m \log^2 m)$, and for all subsets we get $O(m^{2 - \frac{1}{2^{24}}} \log^2 m)$. This result applies to all $k$ subsets of faces from the the beginning adding a further factor $k$ to the total result.

There is certainly a lot of redundancy in this algorithm. For example, the range trees over different partition elements might be compressed using persistent trees. For point location, the probabilistic algorithm of Clarkson (1986) reduces the exponent $2^{24}$ to $24 + \kappa$, $\kappa > 0$ any fixed constant. Linearization is superfluous if point location is carried out directly in the partition induced by the rational surfaces using the algorithm of e.g. Chazelle (1985). However, these improvements concern the local behavior only. More of interest is a good amortized behavior over all collisions, a question which is addressed in the next section.

## 6. Collision detection by updating immediate neighborhoods

In the approach of the previous section, the data structures are newly built every new path segment. No use is made of coherence between consecutive segments, and their seems no straightforward method to do so with this approach in time space. The following discussion makes use of coherence, i.e. data structures from the current path segment can be used immediately for the next segment. The idea is to maintain neighborhood information in order to test only neighbored $k$-gons. The classical structure for neighborhoods are Voronoi diagrams. The type of Voronoi diagram used here is the $k$-gon diagram. The $k$-gon Voronoi diagram can be easily defined using the interpretation of Fortune (1986) of the classical Voronoi diagrams. Voronoi diagrams can be interpreted as the projection of the intersection lines of congruent cones in 3-space. The tops of the cones are located in the $x - y$-plane at the given points. The axes of the cones are parallel to the $z$-axis. For $k$-gon Voronoi diagrams, the cones are replaced by congruent $k$-gon pyramids. They are placed so that their intersections with the $x - y$-plane are the given $k$-gons, cf. figure 9. Calculation of $k$-gon Voronoi diagrams is possible in time $O(n \log n)$ by divide-and-conquer, or possibly by space sweep analogous to the algorithm of Fortune (1986).

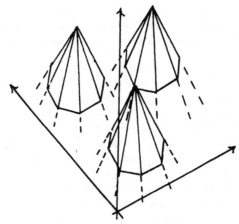

Figure 9: Defining $k$-gon Voronoi diagrams by $k$-gon pyramids.

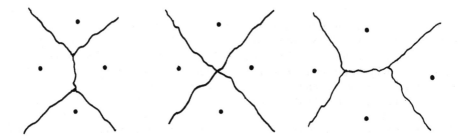

Figure 10: Four candidates where the neighborhood may change.

The first type of an event of interest is when the $k$-gons start to move are those where the neighborhoods change. Neighborhoods change when new regions in the $k$-gon Voronoi diagram become neighbored. For finite Voronoi edges, the $k$-gons belonging to the four regions incident with its end points are involved (figure 10). The neighborhood changes if four such $k$-gons become "co-cyclic". Let be $p, q, r, s$ the centers of the $k$-gons. $p, q, r, s$ are called co-cyclic if there exists a point $w$ with

$$ d_k(w, p) - d_p = d_k(w, q) - d_q = d_k(w, r) - d_r = d_k(w, s) - d_s. $$

Elimination of $w$ yields a cubic polynomial in the time parameter $t$ whose closest root can be found in constant time. Infinite Voronoi edges are treated analogously (figure 11).

The second type of an event of interest is the expected next collision of neighbored $k$-gons under the assumption that they remain neighbored and they do not change their direction of motion. It is determined by calculating the closest distance between the two considered $k$-gons. At most $O(k)$ sectors must be expected for this purpose, each in constant time.

The third type of an event of interest is when a moving $k$-gon changes its direction of motion. The update in this case may lead to new events of the first and second type. The number of new events is proportional to the number of Voronoi vertices of the $k$-gon changing its direction.

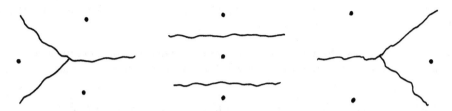

Figure 11: New neighborhoods for an infinite Voronoi edge.

As usual for sweeps, the events are inserted in a priority queue. The sweep is simulated by processing the queue sequentially. During the sweep it may further happen that events in the queue must be eliminated. This happens for example if a $k$-gon changes its direction causing the predicted collision stored in the priority queue to become obsolete. Insertion and deletion in the priority queue is logarithmic in its size, the first element is found in constant time.

## 7. Remarks

The tracking problem in sect. 3 was used by an incremental Gauss algorithm. It remains to analyse its numerical stability for larger values of $k$. Does an alternative algorithm with (at least amortized) less than quadratic time using linear space exist? A further question is the rate of convergency against the original circular case for increasing $k$. Is it possible to obtain a Fourier expansion of the trajectories of the circular case from the linear trajectories over the $k$-th root of unity?

The collision detection algorithm of sect. 6 trades bookkeeping the current neighborhoods against the alternative to test all pairs of $k$-gons. In the worst case, the number of neighborhood changes can become quadratic in the number of $k$-gons, even for straight line motions without altering the direction of movement. In practice, neighborhoods can be expected to change rarely. For a sequence of deterministically moving circles, an analysis based on Davenport-Schinzel sequences like that of Atallah (1983) and more recent research may yield a better estimation. Do better alternatives of neighborhood than Voronoi-diagrams exist?

## References

Aho, A.V., Hopcroft, J.E., Ullman, J.D. (1974) The Design and Analysis of Computer Algorithms, Addison-Wesley, Reading, Mass., 1974

Atallah, M.J. (1983) Dynamic Computational Geometry, IEEE FOCS, 92-99

Auerbach, F., Hort, W. (1929) Handbuch der physikalischen und technischen Mechanik, Verlag von John A. Barth, Leipzig

Badler, N.J., Smoliar, S.W. (1979) Digital Representation of Human Movement, ACM Computing Surveys 11, 19

Chazelle, B., Guibas, L.J., Lee, D.T. (1983) The Power of Geometric Duality, IEEE FOCS, 217-225

Chazelle, B. (1985) Fast Searching in a Real Algebraic Manifold with applications to geometric complexity, Proceedings CAAP'85, Lecture Notes in Computer Science, Springer-Verlag, 145-263

Clarkson, K.L. (1986) Further Applications of Random Sampling, ACM STOC, 414-423

Collins, G. E. (1975) Quantifier Elimination for Real Closed Fields by Cylindrical Algebraic decomposition, Second GI Conference on Automata Theory and Formal Languages, Lecture Notes in Computer Science, 33, 134-183

Dobkin, D.P., Lipton, R.J. (1976) Multidimensional Searching Problems, SIAM J. on Computing 5, 181-186

Edelsbrunner, H. (1987) Algorithms in Combinatorial Geometry, Springer-Verlag, Berlin

Fortune, S. (1986) A Sweepline Algorithm for Voronoi Diagrams, Second ACM Symposium on Computational Geometry, 313-322

Lipson, J.D. (1981) Elements of Algebra and Algebraic Computing, Addison-Wesley Publ. Comp., Reading, Mass.

# EPSILON-NETS FOR HALFPLANES

Gerhart Woeginger

Department of Mathematics, Free University Berlin
Arnimallee 2-6, D-1000 BERLIN 33

**Abstract.** Given some finite point set P in the plane and some real $\varepsilon$, $0 < \varepsilon < 1$, we want to colour a minimal subset of P red, such that the following holds: every open halfplane that contains more than $\varepsilon \cdot |P|$ of the points in P also contains at least one red point. It is shown that it always suffices to colour $\lceil 2/\varepsilon \rceil - 1$ points red (independent of the size of P). If $\varepsilon < 2/3$, we can choose these $\lceil 2/\varepsilon \rceil - 1$ points among the extreme points of P. If all red points must be extreme, our solution is optimal and it can be found in $O(n\log n)$ time. If the red points are allowed to be any elements of P, our result is almost optimal: There are point sets requiring at least $2\lceil 1/\varepsilon \rceil - 2$ red points. The both bounds differ at most by one.

## 0. Introduction

To give a precise formulation of the problems this paper deals with, we settle some terminology taken from [HW]: A *range space* is a pair $(X,R)$, where X is a (maybe infinite) set and R is a set of subsets of X. The members of X are called *points* or *elements*, members of R are called *ranges*. If P is a finite subset of X and $0 < \varepsilon < 1$, then $N \subseteq P$ is called an *$\varepsilon$-net of* P for R, if it contains one point in each range $r \in R$ for that $|P \cap r| > \varepsilon \cdot |P|$ holds.

In our case, X is the set of all points in the plane, R is the set of all open halfplanes and P is a finite set of points. Obviously, the set of red points corresponds to an $\varepsilon$-net N of P for R.

To give the main theorem on $\varepsilon$-nets, we need the notion of the *Vapnik - Chervonenkis dimension* of a range space: A subset of X is said to be *shattered* by R if all of its subsets can be obtained by intersections with ranges of R. The Vapnik-Chervonenkis dimension is the cardinality of the largest shattered subset of X.

It is easily seen that our halfplane range space is of VC-dimension 3. If we take three points which do not lie on a common line, then we can get each of the eight subsets by intersections with halfplanes. Now consider any set of four points in the plane. If one of the points is contained in the convex hull of the other three points, then we cannot cut off this single point. If none of the points is contained in the convex hull of the other points, then we cannot cut off two diametrical points. Hence, there is a shattered set of cardinality three and no set of cardinality four is shattered; thus the VC - dimension is three.

Haussler and Welzl [HW] introduced $\varepsilon$-nets and proved that, if $(X,R)$ is a range space of VC-dimension d and P is a finite subset of X, then there exists an $\varepsilon$-net of P of size at most $\lceil 8d/\varepsilon \cdot \log(8d/\varepsilon) \rceil$ (independent of the size of P). [WW] proved that there are range spaces not permitting $\varepsilon$-nets of size less than $\max(\lceil d/\varepsilon \rceil - d, d+1)$.

Thus we get that there are $\varepsilon$-nets of size at most $\lceil 24/\varepsilon \cdot \log(24/\varepsilon) \rceil$ for our halfplane range space; the lower bound cannot be applied to this case. In this paper we improve the upper bound to $\lceil 2/\varepsilon \rceil - 1$ and give a lower bound of $2 \cdot \lceil 1/\varepsilon \rceil - 2$. For some $\varepsilon$, both bounds coincide, for some $\varepsilon$ there is a gap of size one.

In the rest of the paper we will make use of the following notations and conventions: By conv(P) we denote the convex hull of a finite point set P in $\mathbb{R}^d$. All vertices of conv(P) are elements of P. If $p_1, p_2, \ldots, p_m$ is a sequence of points and we treat some $p_{i+1}$, then the index i+1 always means $(i) \bmod(m)+1$. Analogously, index i+2 means $(i+1) \bmod(m)+1$, etc.

## 1. An upper bound in all dimensions for large epsilons

We start this section with a generalization of our problem to higher dimensions: Given a finite point set P in $\mathbb{R}^d$ and some real $\varepsilon$ between 0 and 1, we want to find $\varepsilon$-nets for *halfspaces*. That means every halfspace containing more than $\varepsilon \cdot |P|$ of the points in P must contain a point of the net, too. We show that in $\mathbb{R}^d$ for $\varepsilon \geq d/(d+1)$ there exists a net of size d.

In order to prove Theorem 1.2 we will need the following proposition. (cf. Edelsbrunner [E]):

Proposition 1.1: Let P be a finite set of n points in $\mathbb{R}^d$. Then there exists a so-called centerpoint z of P: no open halfspace that does not contain z, contains more than $n \cdot d/(d+1)$ points of P. (z need not be an element of P!). □

Theorem 1.2: Let P be some finite point set in $\mathbb{R}^d$ and let $d/(d+1) \leq \varepsilon < 1$. Then there exists an $\varepsilon$-net of P of size at most d.

Proof: By Proposition 1.1, there exists a centerpoint z of P. Now let $\mathfrak{S}$ be the family of all sets $\{p_1, p_2, \ldots, p_k\}$, $k \le d+1$, for which $conv(\{p_1, p_2, \ldots, p_k\})$ contains this centerpoint z. By the Theorem of Caratheodory, the family $\mathfrak{S}$ cannot be empty. If $|S_1| \le d$ holds, for some set $S_1$ in $\mathfrak{S}$, we simply choose the set $S_1$ to form our net: every halfplane that cuts off more than $2n/3$ points of P, must contain the centerpoint and, hence, one point out of $S_1$.

If $|S_1| = d+1$ holds for all $S_1$ in $\mathfrak{S}$, then let $P_1 = \{p_1, p_2, \ldots, p_{d+1}\}$ be that element in $\mathfrak{S}$, for which $conv(\{p_1, p_2, \ldots, p_{d+1}\})$ has the smallest volume. We claim that $conv(P_1)$ does not contain any other point of P: If there is another point of P, say q, lying in $conv(P_1)$, then consider the sets $S_i = P_1 - \{p_i\} \cup \{q\}$. The centerpoint z must lie in one of these $S_i$, contradicting the minimal volume of $conv(P_1)$.

After we have found the set $P_1$, we consider the following partition of $\mathbb{R}^d$ into cones: The line through z and some point $p \in \mathbb{R}^d$ cuts the closure of one of the faces of $conv(P_1)$. All points whose lines cut the closure of the same face, belong to the same cone (if the line cuts the closure of more than one face, the point belongs to more than one cone). Obviously, one of these cones, say the cone belonging to the face $\{p_1, p_2, \ldots, p_d\}$, must contain at least $1/(d+1)$ points of P.

Then $\{p_1, p_2, \ldots, p_d\}$ is an $\varepsilon$-net for P: Take some open halfspace h. If h does not contain z, then, by the definition of a centerpoint, it contains at most $d/(d+1) \cdot |P|$ points of P. If h does contain z but none of the points $p_1, p_2, \ldots, p_d$, it avoids the $|P|/(d+1)$ points lying in the cone behind $p_1, p_2, \ldots, p_d$ (by the choice of $p_1, p_2, \ldots, p_d$, there are no points of P between z and $p_1, p_2, \ldots, p_d$, so all $|P|/(d+1)$ points lie behind $p_1, p_2, \ldots, p_d$). Therefore, every halfspace containing more than $\varepsilon \cdot |P|$ points must contain one of the points $p_1, p_2, \ldots, p_d$ and so we have found an $\varepsilon$-net of size at most d for P. $\square$

Setting d=2 in Theorem 1.2, we easily get the following corollary for point sets in the plane.

Corollary 1.3: Let P be some finite set of n points in the plane and let $2/3 \le \varepsilon < 1$. Then there exists an $\varepsilon$-net of P for halfplanes of size 2 and we can construct it in $O(n \log^5 n)$ time. $\square$

Proof: All we have to do is to show that an $\varepsilon$-net as constructed in Theorem 1.2 can be found in $O(n \log^5 n)$ time. First, we need the centerpoint of P. Cole, Sharir and Yap [CSY] show how to get it in $O(n \log^5 n)$ time. Next, we observe that we do not need the smallest size triangle containing the centerpoint: We only need an empty triangle (containing no other point of P). Hence, we triangulate the point set P in $O(n \log n)$ time (any triangulation will do it; the Delaunay triangulation, for example, can be obtained in $O(n \log n)$ time; see [E]). Then we search for the triangle that contains the centerpoint and find it in $O(n)$ time, as there are only $O(n)$ triangles.

Finally, we run through the point set and count, to which side of the triangle each single point "belongs". The triangle side "possessing" the largest number of points forms the $\varepsilon$-net. Obviously, these steps can be done all together in $O(n \log^6 n)$ time. □


## 2. An upper bound in the plane for all epsilons

In this section we show that for each $\varepsilon$ between 0 and 2/3 and for each finite point set P in the plane, there exists an $\varepsilon$-net of size at most $\lceil 2/\varepsilon \rceil - 1$. We can construct this net in $O(n \log n)$ time.

As every halfplane that contains any point of a point set P must contain at least one extreme point of P, it is very natural to try to get $\varepsilon$-nets *consisting of extreme points only*. However, this does not work for $2/3 \le \varepsilon < 1$: In this case $\lceil 2/\varepsilon \rceil - 1 = 2$ holds. Consider some point set P with three extreme points and many points in the interior. No two extreme points can form an $\varepsilon$-net, as all the other points can be cut off by a line. Fortunately, we have already treated this case in the last section and so we can restrict ourselves to $0 < \varepsilon < 2/3$.

If we speak of an $\varepsilon$-net of P in this section, or short a net of P, we always mean an $\varepsilon$-net of P for halfplanes.

**Theorem 2.1:** Let P be some finite point set in the plane and let $0 < \varepsilon < 2/3$. Then there exists an $\varepsilon$-net for P of size at most $\lceil 2/\varepsilon \rceil - 1$.

**Proof:** Let N be an $\varepsilon$-net for P such that (i) all points of N lie on the convex hull of P and (ii) N is minimal, i.e. for no $p \in N$, $N - \{p\}$ is an $\varepsilon$-net for P. Such an $\varepsilon$-net exists, as every halfplane containing any point of P must contain a point on the hull, too, and so the hull itself forms an $\varepsilon$-net for P.

Let $p_1, p_2, \ldots, p_k$ be the elements of N sorted in such a way that they lie in clockwise direction on conv(P). For $1 \le i \le k$, let $\ell(i)$ denote the line through $p_{i-1}$ and $p_{i+1}$, let $\hbar(i)$ be the open halfplane bounded by $\ell(i)$ that contains $p_i$ and let $\|\hbar(i)\|$ be the number of points in $\hbar(i) \cap P$. If k=3, we have already found an $\varepsilon$-net of size $3 \le \lceil 2/\varepsilon \rceil - 1$ and so we will assume $k \ge 4$ in the rest of our proof.

Now assume that for some $1 \le i \le k$, $\|\hbar(i)\| \le \varepsilon \cdot |P|$ holds. We prove that in this case $N - \{p_i\}$ is a smaller $\varepsilon$-net for P: Every halfplane not containing any point in conv($p_{i-1}, p_i, p_{i+1}$) cuts off the same points of the net as before. Every halfplane containing some point in conv($p_{i-1}, p_i, p_{i+1}$), but neither $p_{i-1}$ nor $p_{i+1}$, cuts the convex hull once between $p_{i-1}$ and $p_i$, a second time between $p_i$ and $p_{i+1}$ and nowhere else (or only in $p_i$ and nowhere else). Thus it contains at most the $\le \varepsilon \cdot |P|$ points in $\hbar(i)$. This makes $N-\{p_i\}$ an $\varepsilon$-net for P and so (as this contradicts the minimality of N) $\|\hbar(i)\| > \varepsilon \cdot |P|$ must hold for all i, $1 \le i \le k$.

The next we claim is that every point of P lies at most in two of the $\hbar(i)$'s: Assume, some fixed $p \in P$ lies in three different $\hbar(i)$'s. Then, as $k \ge 4$, p must lie in some $\hbar(i) \cap \hbar(j)$, where $i+1 \ne j$ and $i-1 \ne j$ holds (see Figure 1).

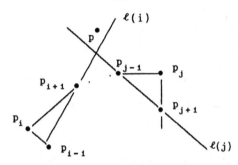

Figure 1:

The intersection of $\hbar(i)$ and $\hbar(j)$ must not contain any point of P

But in this case the points $p_{i+1}$ and $p_{j-1}$ lie in the interior of $conv(p, p_i, p_j)$ and obviously not on the convex hull of P, a contradiction. So we get that the $\hbar(i)$'s all together contain each point at most twice and therefore

$$2 \cdot |P| \ge \sum_{i=1}^{k} \|\hbar(i)\| > \sum_{i=1}^{k} \varepsilon \cdot |P| = k \cdot \varepsilon \cdot |P|$$

holds. This directly implies $2 > \varepsilon \cdot k \iff k \le \lceil 2/\varepsilon \rceil - 1$. So we have found an $\varepsilon$-net of size less or equal $\lceil 2/\varepsilon \rceil - 1$. $\square$

Now let us turn to the algorithmic aspects of Theorem 2.1. Though it only proves the pure existence of an $\varepsilon$-net of size at most $\lceil 2/\varepsilon \rceil - 1$, it also gives rise to a good algorithm running in $O(n \log n)$ time, if $0 < \varepsilon < 2/3$. Informally speaking, we start in some point on the convex hull of P and construct step by step the next point in the net performing a kind of binary search.

Algorithm 2.2: constructs an $\varepsilon$-net of size at most $\lceil 2/\varepsilon \rceil - 1$ for a finite point set P in the plane, if $0 < \varepsilon < 2/3$.

To simplify the formulation of the algorithm, we introduce the following notation: If $q_1, q_2, \ldots, q_m$ is a sequence of points in the plane, then $\hbar(q_i, q_j)$ denotes the open halfplane bounded by the line through $q_i$ and $q_j$ that contains $q_{i+1}$ and $\|\hbar(q_i, q_j)\|$ denotes the number of points in $P \cap \hbar(q_i, q_j)$.

Step 1: Construct the convex hull of P. Let $q_1, q_2, \ldots, q_m$ be the points on the hull sorted in clockwise direction.

Step 2: (Initialize)

$q_{act} := q_1$ and E-net$:= (q_1)$

Step 3: Using binary search, calculate the index hilf such that

(i) $\| \hbar(q_{act}, q_{hilf}) \| \le \varepsilon \cdot |P|$

(ii) $\| \hbar(q_{act}, q_{hilf+1}) \| > \varepsilon \cdot |P|$

(iii) act+1 $\le$ hilf $\le$ n

Step 4: E-net$:=$ E-net $+ (q_{hilf})$ and $q_{act} := q_{hilf}$

Step 5: If $\| \hbar(q_{act}, q_1) \| > \varepsilon \cdot |P|$ then goto Step 3, else goto Step 6.

Step 6: Let p be the second point in E-net (after $q_1$).

If $\| \hbar(q_{act}, p) \| \le \varepsilon \cdot |P|$ then E-net$:=$ E-net $- (q_1)$. □

We start the discussion of Algorithm 2.2 with analyzing the time it needs to construct an $\varepsilon$-net for a point set P, where $|P| = n$:

Step 1 can be done in $O(n\log n)$ time using one of the standard algorithms for the construction of convex hulls (see e.g [E]) and, obviously, Step 2 takes constant time.

Steps 3 to 5 are performed at most $\lceil 2/\varepsilon \rceil$-times, once for each point of the $\varepsilon$-net. Each binary search consists of at most $\log n$ steps. Using a good range search algorithm (see [EW]), the number of points in $\hbar(q_i, q_j)$ can be calculated in $O(n^{0.695})$ time, if the point set is preprocessed once in $O(n\log n)$ time. Hence, the Steps 3 to 5 take all together at most $O(\lceil 2/\varepsilon \rceil \cdot \log n \cdot n^{0.695} + n\log n)$ time.

At last, Step 6 is done in at most $O(n^{0.695})$ time and so the whole algorithm runs in $O(\lceil 2/\varepsilon \rceil \cdot \log n \cdot n^{0.695} + n\log n)$ time.

Obviously, Algorithm 2.2 produces a minimal $\varepsilon$-net as in Theorem 2.1, and so $|$E-net$| \le \lceil 2/\varepsilon \rceil - 1$ holds. Now one could think that the algorithm gives up too soon: Found the last point $p_k$ (let $p_1, p_2, \ldots, p_k$ denote the points of the $\varepsilon$-net sorted in clockwise direction), it could try to improve $p_1$ to a point $q_j$ (on the hull between $p_1$ and $p_2$) such that $\| \hbar(p_k, q_j) \| \le \varepsilon \cdot |P|$ and $\| \hbar(p_k, q_{j+1}) \| > \varepsilon \cdot |P|$. Afterwards, starting from $q_j$, it could try to improve $p_2$, then $p_3$ and so on until some $p_i$ is not improvable any more. Leaving aside that there is no reason that this "improvement" really reduces the cardinality of the net, we give the following example demonstrating that the improved algorithm need not even terminate.

Example 2.3: Let p be some prime $\ge 5$, let P be a set of p points on a circle and let $\varepsilon$ be some real number such that $1 < \varepsilon \cdot p < (p-1)/2$ holds.

If the improved algorithm would terminate, there would be a set $N \subseteq P$ with the following property: between two adjacent points of N there are exactly

$\lfloor \varepsilon \cdot p \rfloor$ points on the circle (else the algorithm would go on improving). But then $\lfloor \varepsilon \cdot p \rfloor + 1$, that lies between 2 and $(p-1)/2$, would be a divisor of the prime p; a contradiction. □

We finish this section with the following summarizing theorem:

**Theorem 2.4:** Given some finite point set P in the plane and some real $\varepsilon$, $0 < \varepsilon < 2/3$, there exists an $\varepsilon$-net of P for halfplanes of size at most $\lceil 2/\varepsilon \rceil - 1$ and we can construct it in $O(n\log n)$ time. □


## 3. The lower bound

In this section we give some "worst-case" point sets requiring large $\varepsilon$-nets. The case $\varepsilon \geq 2/3$ is trivial: Take n points on a circle. Then a single point can always be cut off by an open halfplane and each $\varepsilon$-net must contain at least 2 points (upper bound = lower bound).

If $\varepsilon < 2/3$, then there are point sets not allowing $\varepsilon$-nets on the convex hull of size less than $\lceil 2/\varepsilon \rceil - 1$. Moreover, the algorithm of the previous section is very close to optimal as there are point-configurations not allowing $\varepsilon$-nets of size less than $2 \cdot \lceil 1/\varepsilon \rceil - 2$, if the points of the net are allowed to be any elements of P. The both bounds differ at most by one, as the following table demonstrates:

| | $1/\varepsilon = m$ | $m < 1/\varepsilon \leq m+0.5$ | $m+0.5 < 1/\varepsilon < m+1$ |
|---|---|---|---|
| $\lceil 2/\varepsilon \rceil - 1$ | $2m-1$ | $2m$ | $2m+1$ |
| $2\lceil 1/\varepsilon \rceil - 2$ | $2m-2$ | $2m$ | $2m$ |

**Lemma 3.1:** For each $\varepsilon$, $0 < \varepsilon < 2/3$, there are finite point sets P in the plane such that each $\varepsilon$-net for P consisisting only of points on conv(P) contains at least $\lceil 2/\varepsilon \rceil - 1$ elements.

**Proof:** We will construct a point set P consisting of n points that has the following properties :

(i) there are exactly $k = \lceil 2/\varepsilon \rceil - 1$ points $p_1, p_2, \ldots, p_k$ lying on the convex hull of P

(ii) for each i, the interior of the triangle between the points $p_i$, $p_{i+1}$ and the intersection point of the lines through $p_{i-1}$, $p_{i+1}$ and $p_i$, $p_{i+2}$ contains m points of P (in the notation of Theorem 2.1 this would mean that $h(i) \cap h(i+1)$ contains exactly m points of P).

(iii) the point set P only consists of those points that are listed above, that is $|P| = n = k + k \cdot m$.

(iv) $2m+1 > \varepsilon \cdot n$ holds.

Assume, there was an $\varepsilon$-net for P, consisting only of points on conv(P), that does not contain all points on the hull. Say it does not contain some fixed point $p_i$. Then the open halfplane, that is bounded by the line through $p_{i-1}$ and $p_{i+1}$ and that contains $p_i$, cuts off 2m points in the triangles plus the point $p_i$. As $2m+1 > \varepsilon \cdot n$ holds, we get a contradiction.

Hence, the $\varepsilon$-net must contain all the $\lceil 2/\varepsilon \rceil - 1$ points on the hull and it remains to show that there exist numbers m and n fulfilling (i) through (iv). That means, $n = k+k \cdot m = (\lceil 2/\varepsilon \rceil - 1)(m+1)$ and $2m+1 > \varepsilon n$ must hold. This implies $2m+1 > \varepsilon (\lceil 2/\varepsilon \rceil - 1)(m+1)$

$$\frac{2m+1}{\varepsilon \cdot (m+1)} > \lceil 2/\varepsilon \rceil - 1$$

$$\frac{1}{\varepsilon} \cdot \left( 2 - \frac{1}{m+1} \right) > \lceil 2/\varepsilon \rceil - 1$$

If m becomes very large, the left side comes arbitrarily close to $2/\varepsilon$. Thus there exists some number m fulfilling the last inequality. Setting $n = (\lceil 2/\varepsilon \rceil - 1) \cdot (m+1)$ completes our proof. $\square$

**Lemma 3.2:** For each $\varepsilon$, $0 < \varepsilon < 2/3$, there are finite point sets P in the plane such that each $\varepsilon$-net for P contains at least $2 \lceil 1/\varepsilon \rceil - 2$ elements.

**Proof:** We will construct a point set containing n points. Consider the following configuration:

$\lfloor \varepsilon \cdot n \rfloor + 2$ points are placed on a parabel $y = x^2$, $-1 < x < +1$. Then every single point p in this group on the parabel can be cut off by a line $\ell(p)$ that lies under the other $\lfloor \varepsilon \cdot n \rfloor + 1$ points in this group. Then we place $n/(\lfloor \varepsilon \cdot n \rfloor + 2)$ of these groups on the parabel $y = -x^2$, $-\infty < x < +\infty$, in such a way that each $\ell(p)$ lies under the $\lfloor \varepsilon \cdot n \rfloor + 1$ points belonging to the same group as p and over all the other points.

Assume, an $\varepsilon$-net for P would contain less than $2 \cdot \lfloor n/(\lfloor \varepsilon \cdot n \rfloor + 2) \rfloor$ points. Then there is some group having only one point p in the net. But now $\ell(p)$ cuts off $\lfloor \varepsilon \cdot n \rfloor + 1$ points not in the net, a contradiction. Hence, each $\varepsilon$-net must contain at least $2 \cdot \lfloor n/(\lfloor \varepsilon \cdot n \rfloor + 2) \rfloor$ points.

$$2 \left\lfloor \frac{n}{\lfloor \varepsilon n \rfloor + 2} \right\rfloor \geq 2 \left\lfloor \frac{n}{\varepsilon n + 2} \right\rfloor = 2 \left\lfloor \frac{1}{\varepsilon + 2/n} \right\rfloor$$

If n becomes very large and $1/\varepsilon$ is integer, then the value of the brackets comes arbitrarily close to $1/\varepsilon$ but it does never reach it. Therefore, the right side becomes at most $2(1/\varepsilon - 1) = 2 \lceil 1/\varepsilon \rceil - 2$. If n becomes very large and $1/\varepsilon$ is not integer, then the value of the brackets comes arbitrarily close to

$1/\varepsilon$. Hence, the right side becomes at most $2\lfloor 1/\varepsilon \rfloor = 2\lceil 1/\varepsilon \rceil - 2$. In each case an $\varepsilon$-net for P must contain at least $2\lceil 1/\varepsilon \rceil - 2$ elements. □

We finish this section with a lemma improving the lower bound for $1/2 \le \varepsilon < 5/9$ (then the lower bound becomes equal to the upper bound).

**Lemma 3.3:** For each $\varepsilon$, $1/2 \le \varepsilon < 5/9$, there is a point set P such that each $\varepsilon$-net for P contains at least 3 points.

Proof: Consider the following point set P:

(i)  $|P| = 9m+3$

(ii) All points of P are placed on
  three branches, each branch contains
  $3m+1$ points

Assume, there was an $\varepsilon$-net for P consisting of two points. If both points lie on one branch, we can cut off the other two branches containing $6m+2 > \varepsilon \cdot (9m+3)$ points. Hence, one of the branches, say the branch to $p_1$, does not contain any point in the net and the other two branches each contain one point of the net (the branch to $p_2$ contains $p_2'$ and the branch to $p_3$ contains $p_3'$).

If there are at least $2m+1$ points between $p_2$ and $p_2'$ (inclusively $p_2$, exclusively $p_2'$), the line cutting off these points and the branch to $p_1$ cuts off at least $5m+2 > \varepsilon \cdot (9m+3)$ points and none of the net. The same holds for the point $p_3'$. Hence $p_2'$ must not be among the first $m$ points on the branch to $p_2$ and analogously, $p_3'$ must not be among the first $m$ points on the branch to $p_3$. Then the open halfplane that contains $p_1$ in its interior and $p_2'$ and $p_3'$ on its boundary contains at least $5m+1$ points that are all not in the net. If $5m+1 > \varepsilon \cdot (9m+3)$ holds, our proof is complete. As $(5m+1)/(9m+3) = 5/9 - 2/(27m+9)$ holds, for $m$ sufficiently large the inequality is fulfilled. □

## 4. Discussion

For all $\varepsilon$ between 0 and 1, we gave a constructive solution to get $\varepsilon$-nets for halfplanes of size at most $\lceil 2/\varepsilon \rceil - 1$. If $0 < \varepsilon < 2/3$, we can construct a net in $O(\lceil 2/\varepsilon \rceil \cdot \log n \cdot n^{0.695} + n \log n)$ time. If $2/3 \le \varepsilon < 1$, we can find it in at most $O(n \log^5 n)$ time.

The most interesting open problem concerning $\varepsilon$-nets of point sets is to find $\varepsilon$-nets for halfspaces in dimensions $\ge 3$. Our proof method does *not* generalize to higher dimensions: In $\mathbb{R}^2$ we considered minimal $\varepsilon$-nets, where the omission of any point led to a non-$\varepsilon$-net. Fortunately, the size of such a net was bounded by the constant $\lceil 2/\varepsilon \rceil - 1$, depending only on $\varepsilon$. In $\mathbb{R}^3$ there are nets not permitting such a bound: Put n points on the unitsphere in the following way: $\lfloor \varepsilon \cdot n \rfloor$ of the points are placed very close to the North pole, all the others are placed around the equator. We consider the $\varepsilon$-net N containing all points on the equator. Clearly, every halfspace containing more than $\varepsilon \cdot n$ points , contains one point of N and so N is a net. On the other hand, if we omit one of the points of the net, this point and the points at the pole can be cut off by a plane and so N is a minimal $\varepsilon$-net of size $(1-\varepsilon) \cdot n$.

**References:**

[CSY] R.Cole, M.Sharir, C.K.Yap, On k-hulls and related problems, SIAM J. Comput. Vol. 16, No. 1 (1987) 61-77.

[E] H.Edelsbrunner, *Algorithms in Combinatorial Geometry"*, EATCS Monographs in Theor. Computer Science, Springer Verlag, Berlin 1987.

[EW] H.Edelsbrunner, E.Welzl, Halfplanar range search in linear space and in $O(n^{0.695})$ query time, Inf. Proc. Letters 23 (1986) 289-293.

[HW] D.Haussler, E.Welzl, Epsilon-nets and simplex range queries, Discrete Comput. Geometry 2: 127-151 (1987).

[WW] E.Welzl, G.Woeginger, On Vapnik-Chervonenkis dimension one, in preparation.

## GREEDY TRIANGULATION
## CAN BE EFFICIENTLY IMPLEMENTED IN THE AVERAGE CASE
(detailed abstract)

Andrzej Lingas

Department of Computer and Information Science
Linköping University, 581 83 Linköping, Sweden

*Abstract:* Let $S$ be a set of $n$ points uniformly distributed in a unit square. We show that the greedy triangulation of $S$ can be computed in $O(n \log^{1.5} n)$ expected time (without bucketing). The best previously known upper-bound on the expected-time performance of an algorithm for the greedy triangulation was $O(n^2)$.

### 1. Introduction

Given a set $S$ of $n$ points in the plane, a *triangulation* of $S$ is a maximal set of non-intersecting straight-line segments (edges) whose endpoints are in $S$. In numerical applications of triangulations, one of the proposed criteria of goodness involves the minimization of the total length of the edges in a triangulation [PS85]. A *minimum weight triangulation* of $S$ (M(S) for short) is a triangulation of $S$ that achieves the smallest total edge length. The complexity status of the problem of computing $M(S)$ has been open for at least a decade [Ll77,PS85]. The two most known heuristics for $M(S)$ are the so called *greedy triangulation* and *Delaunay triangulation* ($GT(S)$ and $DT(S)$ for short, respectively) [Ll77,PS85]. The former inserts a segment into the triangulation if it is a shortest among all segments between points in $S$ not intersecting those previously inserted. The latter computes the straight-line dual of the Voronoi diagram of $S$ completing it to a full triangulation if it is necessary. Neither approximates the optimum in the general case [Ki80,MZ79,Le87]. However, if $S$ is convex, i.e. lies on its convex hull, the greedy triangulation yields a solution within a constant factor from the optimum [LL87]. Also, if $S$ is uniformly distributed in a unit square then both heuristics yield solutions within a logarithmic factor from $M(S)$ *almost certainly*, i.e. with the probability $\geq 1 - cn^{-\alpha}$, $c > 0$, $\alpha > 1$ [Li86]. The above result on the Delaunay triangulation has been strengthened by showing that the expected length of $DT(S)$ is within a constant factor from the expected length of $M(S)$ [CL84]. In Section 2, we observe that the ratio between the expected length of $GT(S)$ and the expected length of $M(S)$ is $O(\sqrt{\log n})$.

While $DT(S)$ can be computed in time $O(n \log n)$, the most efficient known algorithms for $GT(S)$ run in time $O(n^2 \log n)$ [G79,Li87,Li87a]. If $S$ is uniformly distributed in a unit square then $DT(S)$ can be computed in linear expected-time by using bucketing and the floor function [BWY80] whereas the best known upper-bound on the expected-time performance of an algorithm for $GT(S)$ is $O(n^2)$ [MZ82].

A *diagonal* of a planar straight-line graph (PSLG for short) [PS85] $G$ is an open segment that neither intersects any edge of $G$ nor includes any vertex of $G$ and that has endpoints in the set of vertices of $G$. The problem of computing $GT(S)$ easily reduces to that of computing shortest diagonals of consecutive partial greedy triangulations of $S$. In [Li87], a characterization of a shortest diagonal of a PSLG $G$ in terms of the so called Voronoi diagram with barriers of $G$, and an $O(n \log n)$ algorithm of Wang and Shubert for such a diagram [WS87] have been used to show that a shortest diagonal of $G$ can be found in time $O(n \log n)$. This immediately implies that the greedy triangulation of a PSLG *, in particular $GT(S)$, can be computed in time $O(n^2 \log n)$ and space $O(n)$. In [Li87a], a more refined algorithm for the greedy triangulation of a PSLG has been presented. This algorithm updates the Voronoi diagram with barriers of the current PSLG and the data structures supporting the selection of a shortest diagonal, every time after inserting a new diagonal. A partial result suggesting a good average-case performance of this algorithm has been proved in [Li87a]. In this paper, we show that the algorithm from [Li87a] applied to $S$ runs in $O(n \log^{1.5} n)$ expected-time if $S$ is uniformly distributed in a unit square. It is worthy to point out that neither bucketing nor the floor function are used by our algorithm. It is an intriguing question whether one can compute $GT(S)$ in quadratic or sub-quadratic (worst-case) time. A quadratic-time solution in the convex case is known [LL87]. Moreover, in the semi-circular polygon case even a linear solution is possible [LLS87].

A *unimonotone polygon* is a simple polygon whose perimeter can be decomposed into an edge and a monotone chain [FM84]. Note that a *histogram* is a special case of unimonotone polygon. In this paper, we show that also the greedy triangulation of a unimonotone polygon can be computed in quadratic time since a shortest diagonal of a unimonotone polygon can be found in linear time. The latter result is of interest in its own rights as no sub-$n \log n$ upper time-bound is known for the related problems of computing a closest pair of vertices of a histogram or the Voronoi diagram with barriers of a histogram (or equivalently, the generalized Delaunay triangulation [Li87,WS87] of a histogram).

The remainder of this paper is divided as follows. In Section 2, the ratio between the expected length of $GT(S)$ and the expected length of $M(S)$ is shown to be $O(\sqrt{\log n})$. In Section 3, the $O(n \log^{1.5} n)$ bound on the expected-time performance of the algorithm for $GT(S)$ is derrived. In Section 4, the linear-time method of computing a shortest diagonal of a unimonotone polygon is given.

## 2. Preliminaries

Let $S$ be a set of $n$ points uniformly distributed in a unit square. The following fact has been proven in Section 2 in [Li86].

*Fact 2.1:* $\mid DT(S) \mid = O(\sqrt{n \log n})$ and $\mid GT(S) \mid = O(\sqrt{n \log n})$ hold almost certainly.

We divide the unit square into cells of area about $1/n$ by forming an array of size $\lfloor \sqrt{n} \rfloor$ by $\lfloor \sqrt{n} \rfloor$. We refer to the above array of cells as to the *1-grid*. The expected length of any triangulation of $S$ is $\Omega(\sqrt{N})$. Simply, the expected number of non-empty cells in the 1-grid is $\Omega(N)$, and each of the non-empty cells contains a piece of $GT(S)$ of length $\Omega(1/\sqrt{n})$ (see p. 27 in [Li86]). Hence, we obtain the following theorem strengthening the result on the greedy triangulation from [Li86].

*Theorem 2.1:* For a set of $n$ points uniformly distributed in a unit square, the ratio between the expected length of $GT(S)$ and the expected length of $M(S)$ is $O(\sqrt{\log n})$.

---

* The greedy triangulation can be easily generalized to include PSLG's by considering the edges of the input PSLG as segments inserted in the plane *apriori*.

## 3. Probabilistic analysis of the algorithm for the greedy triangulation

To present the algorithm for $GT(S)$, we need introduce the concept of Voronoi diagram with barriers of a PSLG.

**Definition 3.1:** Let $G = (V, E)$ be a PSLG. For $v \in V$, the region $P(v)$ consists of all points $p$ in the plane for which the shortest, open straight-line segment between $p$ and a vertex of $G$ that does not intersect any edge of $G$ is $(p, v)$. The minimal set of straight-line segments and half-lines that complements $G$ to the partition of the plane into the regions $P(v)$, $v \in V$, is called the Voronoi diagram with barriers of $G$ ( $Vorb(G)$ for short ).
A diagonal of a PSLG $G$ is an open segment that neither intersects any edge of $G$ nor includes any vertex of $G$ and that has endpoints in $V$.

Consider $v \in V$, and let $m$ be the number of edges of $G$ incident to $v$. It follows that $P(v)$ is a collection of $m$ polygonal sub-regions, called sectors in [Li87], separated by the $m$ incident edges. Note that a sector contains at most one concave vertex. The following precise characterization of a shortest diagonal of a PSLG $G$ in terms of $Vorb(G)$ is given in [Li87].

**Fact 3.1 [Li86]:** A shortest diagonal of a PSLG $G = (V, E)$ either lies inside the union of the regions of its endpoints in $Vorb(G)$ or cuts off an empty triangular face from $G$ and lies within at most four different (sectors of the) regions in $Vorb(G)$.

By using the above characterization and the mentioned algorithm of Wang and Shubert [WS86] for $Vorb(G)$, it is shown in [Li87] that a shortest diagonal of a PSLG can be found in time $O(n \log n)$ and space $O(n)$. This immediately implies a straight-forward algorithm for a greedy triangulation of a PSLG running in time $O(n^2 \log n)$ and space $O(n)$.
In [Li87a], a more sophisticated algorithm has been presented. Instead of building $Vorb(G)$ from scratch every time after augmenting $G$ by a new edge, $Vorb(G)$ and accompanying data structures are updated.
The algorithm specialized to planar $n$-point set $S$ maintains five data structures for a current partial greedy triangulation $G$ of $S$. The first two data structures are respectively the DCEL (see [PS85]) and TREES representations of $Vorb(G)$.
For every vertex $w$ of $G$, and every sector of the region of $w$ induced by incident edges, TREES contains an ordered 2-3 tree with vertices assigned to the edges of the sector in clockwise order from left to right. The 2-3 tree supports the following query in logarithmic time: given a straight-line $L$, report the edge or edges of the sector touched or intersected by $L$ plus the sectors (possibly of other regions) adjacent to the above edges.
The third data structure, is a min-heap, $HSD(G)$, containing the edges of the straight-line dual of $Vorb(G)$ according to their lengths. The TREES representation of $Vorb(G)$ together with the standard DCEL one are used to update $Vorb(G)$ (i.e. themselves) and consequently $HSD(G)$.
The fourth data structure, $STAR(G)$, for every vertex $w$ of $G$ contains an ordered 2-3 tree with leaves assigned to the edges of $G$ in angular order around $w$. The fifth data structure, $HT(G)$, is again a min-heap containing the edges of $Vorb(G)$ cutting off empty triangular faces from $G$, ordered according to their lengths. For each such an edge, two-way pointers are additionally maintained to the two other edges of $G$ of the triangular face, kept in the tree of their common endpoint in $STAR(G)$. The $STAR(G)$ data structure is used to update $HT(G)$.

1) Construct the DCEL and TREES representations of $Vorb(S)$;

2) Construct $HSD(S)$;

3) Initialize $STAR(S)$;

4) Construct $HT(S)$;

5) $G \leftarrow S$;

6) **while** $G$ is not a full triangulation of $S$ **do**

**begin**

    a) Pick a shortest edge $e$ from $HSD(G)$ and $HT(G)$, delete it from its heap, and augment $G$ by $e$.

    b) Update the representations of $Vorb(G)$, $STAR(G)$, $HSD(G)$ and $HT(G)$.

**end**

The correctness of the above algorithm follows from Fact 3.1 and the definition of the straight-line dual of $Vorb(G)$.

For the new edge $e$ inserted in the current partial greedy triangulation $G$, let $k(e)$ be the total number of vertices of the parts of the regions in $Vorb(G)$ cut off from their sites by $e$. Theorem 4.1 in [Li87a] yields the following upper bound on the worst-case performance of this algorithm.

**Fact 3.2**[Li87a]: Let $S$ be a set of $n$ points in the plane. The algorithm (1-6) constructs the greedy triangulation of $G$ in time $O(n \log n + \log n \sum_{e \in T} k(e))$ and space $O(n)$.

For any PSLG on $n$ vertices, $Vorb(G)$ is of size $O(n)$ [Li87,WS87]. Hence, for all edges $e$ in $T$, $k(e) = O(n)$ trivially holds and the algorithm (1-6) computes the greedy triangulation in time $O(n^2 \log n)$ and space $O(n)$.

Since $| e |= O(1/\sqrt{n})$ in the average case, one would expect $k(e) = O(1)$, for the overwhelming majority of edges $e$ in $GT(S)$.

**Lemma 3.1:** Given a set $S$ of $n$ points uniformly distributed in a unit square, let $e_1, e_2, ..., e_k$ be a sequence of $k$ edges in $GT(S)$. The expected value of $\sum_{m=1}^{k} k(e_m)$ is $O(\sqrt{n} \sum_{m=1}^{k} | e_m |)$.

*Proof:* Let $1 \leq m \leq k$, and let $e_m = (s,t)$. Next, let $G_m$ be the partial greedy triangulation of $S$ into which $e_m$ is inserted by Algorithm (1-6). Let $TC(e_m)$ be the set of sites in $S - \{s,t\}$ whose regions in $Vorb(G_m)$ touch a part of a region in $Vorb(G_m)$ cut off from its site by $e_m$, and let $C(e_m)$ be the set of sites in $S - \{s,t\}$ whose regions in $Vorb(G_m)$ are cut by $e_m$. By Euler's formula applied to the dual of $Vorb(G)$ we have $k(e_m) = O(\#TC(e_m) + \#C(e_m))$.

In order to estimate the expected cardinality of $\#TC(e_m)$, consider a point $p$ in $TC(e_m)$. It follows that the region of $p$ in $Vorb(G_m)$ touches the boundary of some part $P$ of the region of some point $q$ in $S - \{s,t,p\}$ cut off from $q$ by $e_m$ (see Fig. 3.1). Let $D$ be the distance between $p$ and the common point $r$ of the boundaries of $P$ and the region of $p$ in $Vorb(G_m)$. Note that the distance $L$ between $p$ and $e_m$ is bounded by $2D$.

We claim that $(r,p)$ passes through $\Omega(D\sqrt{n})$ empty cells (of the 1-grid). Suppose otherwise. Then, a point $w$ in $S$ lies in one of the cells through which $(r,p)$ passes and it is closer to $r$ than $p$. This does not yield directly a contradiction with the definition of $Vorb(G_m)$ since some edges of $G$ can prevent $w$ from seeing $r$. However, note that the edges of $G_m$ do not cross $(r,p)$ and their endpoints are clearly in $S$. On the other hand, the site $w$ is in the distance $\leq \sqrt{2/n}$ from $(r,p)$. Therefore, we can prove in a standard way (see Theorem 2.1 in [Li87] for instance) that there is a point $w'$ in $S$ that can see $r$ from a smaller distance than $p$. By the above contradiction, $p$ touches a point of the region of a point $q$ in $S - \{s,t,p\}$ cut off by $e_m$ from $q$ with the probability $e^{-\Omega(L\sqrt{n})}$.

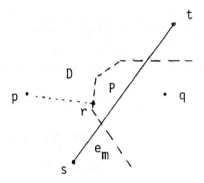

Fig. 3.1.

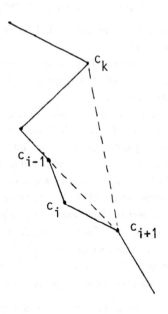

Fig. 4.1.

In order to estimate the expected cardinality of $C(e_m)$, we similarly show that the region of a site $p$ in $Vorb(G_m)$ crosses $e_m$ with the probability $e^{-\Omega(L\sqrt{n})}$.

On the other hand, the number of points in $S$ in the distance from $e_m$ between $L - 1/\sqrt{n}$ and $L$ is greater than $12(\pi L + |e_m|)\sqrt{n} + j$ with the probability not greater than

$$\binom{n}{12(\lceil(\pi L + |e_m|)\sqrt{n}\rceil + j + 1)}(\frac{(2\pi L + 2|e_m|)\sqrt{n}}{n})^{12(\pi L + |e_m|)\sqrt{n}+j+1}$$

which, by $\binom{N}{j} \le (\frac{3N}{j})^J$ where $1 \le J \le N$ (see [AV78]), is not greater than

$$(\frac{3n}{12(\lceil(\pi L + |e_m|)\sqrt{n}\rceil + j + 1)})^{12(\pi L+|e_m|)\sqrt{n}+j+1}(\frac{(2\pi L + 2|e_m|)\sqrt{n}}{n})^{12(\pi L+|e_m|)\sqrt{n}+j+1}$$

$$\le 0.5^{12(\pi L + |e_m|)\sqrt{n}+j+1}.$$

Thus, the expected value of $k(e_m) = \#TC(e_m) + \#C(e_m)$ is $O(\sum_{i=1}^{\lceil\sqrt{2n}\rceil}(i + |e_m|\sqrt{n})e^{-\Omega(i)}) + O(\sum_{i=1}^{\lceil\sqrt{2n}\rceil}\sum_{j=1}^{n}(i + j + |e_m|\sqrt{n})0.5^{(i+j+|e_m|\sqrt{n})}e^{-\Omega(i)}) = O(|e_m|\sqrt{n})$. ∎

Combining the above lemma and Fact 3.2 with the fact that the expected length of the greedy triangulation of a set of $n$ points uniformly distributed in a unit square is $O(\sqrt{n\log n})$ (Fact 2.1), we obtain the main result of this section.

**Theorem 3.1:** The expected-time performance of the algorithm (1-6) for a set of $n$ points uniformly distributed in a unit square is $O(n\log^{1.5} n)$.

## 4. The shortest diagonal problem for unimonotone polygons

To start with, we need recall the basic definitions of polygonal *diagonals, monotone chains* and *unimonotone polygons* [FM84].

A *diagonal* of a simple polygon $P$ is an open segment disjoint from the boundary of $P$ whose both endpoints are vertices of $P$. A diagonal of $P$ is *internal* (*external*, respectively) if it lies inside (outside, respectively)$P$.

A chain $C = (c_1, ..., c_n)$ is a planar straight-line graph with vertex set $\{c_1, c_2, ..., c_n\}$ and edge set $\{(c_i, c_{i+1}) \mid i = 1, .., n - 1\}$. A chain is said to be *monotone* with respect to a straight-line $l$ if any straight-line perpendicular to $l$ intersects it in at most one point. A simple polygon is unimonotone if its boundary can be decomposed into an edge and a chain monotone with respect to a straight-line. Note that a *histogram* is a special case of a unimonotone polygon. Without loss of generality we assume further that all considered monotone chains or unimonotone polygons are monotone with respect to the $Y$-axis. Given a point $p$ in the plane, $X(p)$, $Y(p)$ respectively denote the $X$ and $Y$ coordinate of $p$. For convenience, if a monotone chain is defined by a sequence of the form $(c_1, c_2, ..., c_n)$ then we assume that $Y(c_i) > Y(c_{i+1})$ holds for $i = 1, 2, ..., n - 1$.

To present our method of finding shortest internal or external diagonals of a unimonotone polygon, we need introduce different types of diagonals of monotone chains.

**Definition 4.1** Let $C = (c_1, ..., c_n)$ be a monotone chain. An open segment $(c_i, c_j)$ is a *diagonal* of $C$ if it is disjoint from the edges and vertices of $C$. A diagonal $(c_i, c_j)$ of $C$ is a *down* diagonal for $c_i$ if $i < j$, otherwise it is an *up* diagonal for $c_i$. Next, a diagonal $(c_i, c_j)$ of $C$ is a *right* diagonal if the segment $(c_i, c_j)$ lies on the right side of the sub-chain of $C$ with the endpoints $c_i$

and $c_j$, otherwise it is a *left* diagonal of $C$. Finally, a diagonal $(c_i, c_j)$ of $C$ is a *plus* diagonal for $c_i$ if $X(c_j) - X(c_i) \geq 0$, otherwise it is a *minus* diagonal for $c_i$.

Our method of finding a shortest internal (or external, respectively) diagonal of a unimonotone chain relies on the following, simple algorithm for finding shortest down-right-plus diagonals for vertices of a monotone chain. By symmetry, this algorithm can be easily modified to work for the down-left-minus, up-left-minus, up-right-plus cases of diagonals.

*Algorithm 4.1*

*input:* a monotone chain $C = (c_1, ..., c_n)$ with respect to $Y$-axis.
*output:* a subset of the set of shortest down-right-plus diagonals $drp(i)$ for $c_i$, $i = 1, ..., n$, containing the shortest down-right-plus diagonal for $C$.

for $i = 2, ..., n$ do
begin
    push $c_{i-1}$ on $STACK$;
    while $X(top(STACK)) \leq X(c_{i+1})$ do
    begin
        $c_k \leftarrow pop(STACK)$;
        $drp(k) \leftarrow$ if
        $c_{k+1}$ lies on the left side of $(c_k, c_{i+1})$ then $(c_k, c_{i+1})$ else "undefined"
    end
end
until $STACK$ empty do
begin
    $c_k \leftarrow pop(STACK)$;
    $drp(k) \leftarrow$ "undefined"
end

By induction on $i$, we can prove that $c_k$ is on the top of $STACK$ during the $i - th$ iteration of the loop in Algorithm 4.1 and $X(c_k) \leq X(c_{i+1})$ if and only if $i$ is the minimum index greater than $k$ satisfying the above inequality. By the monotonicity of $C$, $(c_k, c_{i+1})$ is a shortest down-right-plus diagonal for $c_k$ if $c_{k+1}$ lies to the left of $(c_k, c_{i+1})$ (see Fig. 4.1 ). Otherwise, no down-right-plus diagonal for $c_k$ can be a shortest down-right-plus diagonal for the whole chain $C$. This proves the correctness of Algorithm 4.1.
By symmetry between computing shortest down-right-plus diagonals and the three other cases of shortest diagonals, we have the following lemma.

*Lemma 4.1:* Given a monotone chain, we can find its shortest down-right-plus (or down-left-minus, up-left-minus, up-right-plus, respectively) diagonal in total linear time.

By the above lemma, to find, for instance, a shortest right diagonal for a monotone chain in linear time it is sufficient to have a linear-time method for computing a shortest down-right-minus diagonal for the chain. Unfortunately, the idea of Algorithm 4.1 does not work in this case. However, the following reduction of this problem to Lemma 4.1 will be sufficient for us.

*Lemma 4.2:* Any shortest right diagonal of a monotone chain $C$ is in the set of shortest up-right-plus and down-right-plus diagonals for vertices of $C$.

*Proof:* Let $(c_i, c_j)$, $i < j$, be a shortest right diagonal of $C$. We may assume without loss of generality that $X(c_j) < X(c_i)$ (otherwise, it is a down-right-plus diagonal of $C$). Now it is sufficient to note that $(c_i, c_j)$ is a up-right-plus diagonal for $c_j$ and consequently a shortest up-right-plus diagonal for $c_j$. ∎

Combining Lemma 4.1, 4.2 and using the symmetry between the cases of left and right diagonals, we obtain the following theorem:

*Theorem 4.1:* We can find all shortest right and all shortest left diagonals of a monotone chain in linear time.

Combing the above theorem with the definition of a unimonotone polygon, we immediately obtain the following corollary.

*Corollary 4.1:* We can find all shortest internal diagonals and all shortest external diagonals of a unimonotone polygon in linear time.

Corollary 4.1 yields yields the following theorem.

*Theorem 4.1:* We can find the greedy triangulation of a unimonotone $n$-vertex polygon with no three co-linear vertices in time $O(n^2)$ and space $O(n)$.

*Proof:* It is sufficient to note that any diagonal of a unimonotone polygon splits it into two unimonotone polygons and to use Corollary 4.1 recursively. ∎

*Final remarks*

1) The author is currently working on improving the upper bound on the ratio between the expected length of $GT(S)$ and the expected length of $M(S)$ for uniformly distributed point set $S$ to a constant one.

2) Our algorithm for $GT(S)$ selects a shortest diagonal of the current, partial triangulation as the next edge to insert. To obtain a linear expected-time algorithm for $GT(S)$ it would be necessary to find some local criterion for edge insertion which could be combined with the bucketing technique.

2) Aggarwal and Subashi have recently proved that given two non-overlapping monotone polygons $P_1$, $P_2$, we can find a segment realizing the shortest visible distance between a vertex of $P_1$ and a vertex of $P_2$ in linear time [O86]. Using the above result, we could easily generalize Corollary 4.1 to include monotone polygons that can be split into unimonotone polygons by drawing the segment connecting the topmost vertex with the bottom one.

## REFERENCES

[AV78] D. Angluin and L.G. Valiant, *Fast probabilistic algorithms for Hamiltonian circuits and matchings*, Proc. 9th Ann. ACM Symp. on Theory of Computing, New York.

[BWY80] J.L. Bentley, B.W. Weide, A.C. Yao, *Optimal expected-time algorithms for closest point problems*, ACM Transactions on Mathematical Software 6, pp. 563-580.

[CL84] R.C. Chang and R.C.T Lee, *On the average length of Delaunay triangulations*, BIT 24, pp. 269-273.

[FM84] A. Fournier and D.Y. Montuno, *Triangulating Simple Polygons and Equivalent Problems*, ACM Transactions on Graphics 3(2), pp. 153174.

[Gi79] P.D. Gilbert, *New Results in Planar Triangulations*, M.S. Thesis, Coordinated Science Laboratory, University of Illinois, Urbana, Illinois.

[Ki80] D.G. Kirkpatrick, *A Note on Delaunay and Optimal Triangulations*, IPL, Vol. 10, No. 3, pp. 127-131.

[Li86] A. Lingas, *The greedy and Delaunay triangulations are not bad in the average case*, IPL 22, pp. 25-31.

[Li87] A. Lingas, *The shortest diagonal problem*, submitted to IPL.

[Li87a] A. Lingas, *A space efficient algorithm for the greedy triangulation*, presented at the 13th IFIP Conference on System Modelling and Optimization, Tokyo, Japan 1987.

[LL87] C. Levcopoulos and A. Lingas, *On Approximation Behavior and Implementation of the Greedy Triangulation for Convex Polygons*, Algorithmica 2, pp. 175193.

E.L. Lloyd, *On Triangulations of a Set of Points in the Plane*, Proc. of the 18th Annual IEEE Conference on the Foundations of Computer Science, Providence.

[MZ79] G.K. Manacher, and A.L. Zorbrist, *Neither the greedy nor the Delaunay triangulation of a planar point set approximates the optimal triangulation*, Information Processing Letters, Vol.9, No. 1, pp. 31-34.

[MZ82] G.K. Manacher, and A.L. Zorbrist, *The use of probabilistic methods and of heaps for fast-average-case, space-optimal greedy algorithms*, manuscript, 1982.

[Me84] K. Mehlhorn, *Data Structures and Algorithms*, EATS Monographs on Theoretical Computer Science, Springer Verlag, New York.

[O86] J. O'Rourke, *The Computational Geometry Column*, ACM SIGACT News 18(1).

[PS85] F.P. Preparata and M.I. Shamos, *Computational Geometry, An Introduction*, Texts and Monographs in Computer Science, Springer Verlag, New York.

[WS87] C. Wang and L. Shubert, *An optimal algorithm for constructing the Delaunay triangulation of a set of line segments*, in the proceedings of the 3rd ACM Symposium on Computational Geometry, Waterloo, pp. 223-232, 1987.

# A SIMPLE SYSTOLIC METHOD TO FIND ALL BRIDGES
# OF AN UNDIRECTED GRAPH

Manfred Schimmler
*Institut für Informatik und Praktische Mathematik*
*D-2300 Kiel 1*
*West Germany*

Heiko Schröder
*Department of Engineering Physics*
*Australian National University*
*Canberra, ACT, 2601*
*Australia*

## ABSTRACT

An algorithm to find all bridges of an undirected graph on a mesh-connected processor array is presented. Asymptotically it has the same complexity in terms of time and space as the best previous known algorithm, the one of [1]. But it has a significantly simpler structure and therefore, it gains a constant factor of about two in time.

## KEYWORDS

mesh-connected processor array, systolic algorithm, bridges in graphs, transitive closure, complextiy;

## 1. INTRODUCTION

The computational model consists of N = n*n identical processors positioned on a square array (Figure 1). Each processor can communicate with its four neighbors, provided they exist. There is a single instruction stream, and a single instruction at a processor can consist of an arithmetic or boolean operation, transmitting a word to any of its neighbors and receiving a word from any of its neighbors or a constant number of combinations of these.

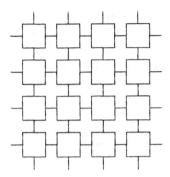

Figure 1: Mesh-connected processor array

This model has turned out to be very useful in particular in the context of VLSI algorithms or parallel algorithms in general. From the large variety of applications in the literature we will mention the graph theoretic ones only which are related to this paper.

A bridge in an undirected graph is an edge whose removal increases the number of connected components of the graph. For simplicity we assume the graph to be connected in the remainder of the paper, but the results generalize straightforeward for arbitrary graphs. In [5] a very fast and elegant algorithm to detect all bridges of a graph is presented. Unfortunately, it is based on a wrong idea, and therefore, it fails in almost all nontrivial applications. Moreover it does not seem to be repairable. In this paper we give a valid algorithm that works on a mesh-connected processor array. It is based on the method of [2] and a new simple characterization of a bridge. Section 2 contains the graph theoretic base of the algorithm which is given in section 3. In section 5 we give a short complexity analysis.

## 2. GRAPH THEORETIC BACKGROUND

Obviously the bridges of an undirected graph are exactly those edges that are not contained in any cycle of the graph. With $d(i,j)$ we denote the length of the shortest path in the graph from vertex i to vertex j. The following Lemma gives a helpful characterization of a bridge of an undirected graph $G = (V, E)$:

### Lemma 1.

There exists a cycle containing an edge $(i,j) \in E$ iff either there is a $k \in V : d(i,k) = d(k,j)$ or there is a $k \in V : d(i,k) -1 = d(k,j)$ and there is a shortest path from i to k that does not start with edge $(i,j)$.

### Proof:

'only if': Let $i,v_1,v_2,...,v_{2t-1},j,i$ be a minimal cycle containing $(i,j)$ (i.e. the minimal cycle containing $(i,j)$ is an odd cycle). Let $k = v_t$. Since the cycle is minimal there is no shorter path from i to k than $i,v_1,v_2,...,v_{t-1},k$ and there is no shorter path from k to j than $k,v_{t+1},v_{t+2},...,v_{2t-1},j$. Therefore, $d(i,k) = t$ and $d(k,j) = t$. Let $i,v_1,v_2,...,v_{2t},j,i$ be a minimal cycle containing $(i,j)$ ( i.e. the minimal cycle containing $(i,j)$ is an even cycle) and let $k = v_{t+1}$. The minimality guarantees that there is no shorter path from j to k than

$j, v_{2t}, v_{2t-1}, ..., v_{t+2}, k$. Moreover, there is no shorter path from i to k than $i, v_1, v_2, ..., v_t, k$. Therefore, $d(i,k) = t+1$ and $d(k,j) = t$ and there is a shortest path from i to k not starting with $(i,j)$.

'if ': Let $d(i,k) = d(k,j) = t$. The shortest path from i to k does not contain the vertex j because $d(i,j) = 1$ and $d(j,k) = t$. Symmetrically the shortest path from k to j does not contain i. Let v be the first common vertex of the shortest path from i to k and from j to k (possibly $k = v$). Since $d(i,k) = d(i,v) + d(v,k)$ and $d(j,k) = d(j,v) + d(v,k)$ we have $d(i,v) = d(j,v)$ and the shortest path from i to v and from j to v are disjoint except of v. Hence the shortest path from i to v, the shortest path from v to j, and the edge $(i, j)$ form a cycle of length $2 d(i,v) + 1$.

Let $d(i,k) -1 = d(k,j)$; then $d(i,k)-1 > d(k,j) = t$ and the shortest path from j to k cannot cantain the vertex i. On the other hand, the shortest path from i to k that does not start with $(i,j)$ cannot contain vertex j, because otherwise there would be a shorter path. Let v be the first common vertex of the shortest path from j to k and the shortest path from i to k that does not start with edge $(i,j)$. Since $d(i,k) = d(i,v) + d(v,k)$ and $d(j,k) = d(j,v) + d(v,k)$ we have $d(i,v) -1 = d(j,v)$ and the shortest path from i to v and from j to v are disjoint except of v. Therefore the shortest path from i to v, the shortest path from v to j, and
the edge $(i,j)$ form a cycle of length $2 d(i,v)$.

Lemma 1 characterizes a bridge as an edge not contained in a cycle of odd length nor in one of even length. The validity of the following algorithm is due to this characterization.

## 3. THE ALGORITHM

In [2] a systolic algorithm for the computation of the transitive closure has been presented. It consists of three identical systolic passes in which two swewed data matrices ( the $a_{ij}$ and the $a'_{ij}$ ) are passed through the processor mesh, as depicted in Figure 2. Both the $a_{ij}$ and the $a'_{ij}$ matrices are initially copies of the adjacency matrix of the graph. During one pass all pairs of the form $(a_{ik}, a'_{kj})$ for some k meet at processor $(i,j)$.

The computations performed at the processors are the same in each pass and at each beat. Two operations are processed at each beat:

OP1: $A_{ij} := A_{ij} \vee ( a_{ik} \wedge a'_{kj} )$;

OP2: If $k = j$ then $a_{ik} := A_{ij}$;

      If $k = i$ then $a'_{kj} := A_{ij}$;

After the first and the second pass, when the $a_{ij}$ ($a'_{ij}$) matrix has travelled through the array, it is reinput from the left (top) for the next pass. After three passes the $A_{ij}$ contain the adjacency matrix of the transitive closure of the graph.

As pointed out in [1, 3, 4] this method can be used for the computation of the length of the shortest paths between every two vertices i and j of an undirected graph with almost no additional effort (since both problems are closely related, being special cases of the algebraic path problem). Instead of the boolean operation $A_{ij} := A_{ij} \vee (a_{ik} \wedge a'_{kj})$ the integers $D_{ij}$ have to be updated by the following operation: $D_{ij} := \min ( D_{ij}, d_{ik} + d'_{kj} )$. Initially, $d_{ij}$ and $d'_{ij}$ are 1 if $(i,j)$ is an edge and $\infty$ otherwise.

To check the condition of Lemma 1 for every edge we need some more information about the shortest paths. In particular we have to keep track of the endpoints of the first edges of the shortest paths. We keep these in the variables $FIRST_{ij}$ for the shortest path from i to j.

If the first edge is not unique, i.e. if there are more than one shortest paths from vertex i to vertex j this information is kept in the boolean variables $MORETHANONE_{ij}$.

So in our algorithm instead of the $A_{ij}$ we have two log n bit registers $D_{ij}$ and $FIRST_{ij}$ and one one bit register $MORETHANONE_{ij}$. Analogously instead of the $a_{ij}$ we have $d_{ij}$, $first_{ij}$, and $morethanone_{ij}$. For the vertically moving $a'_{ij}$ we need only one log n bit register $d'_{ij}$. In addition to the $A_{ij}$ every processor $(i,j)$ has a 1 bit register $B_{ij}$. This B matrix is initialized with the adjacency matrix of the graph. After the execution of the algorithm $B_{ij}$ will be 1 iff $(i,j)$ is a bridge.

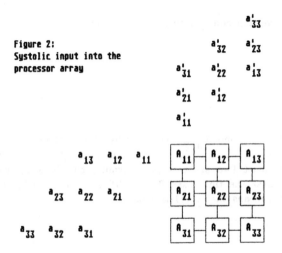

Figure 2:
Systolic input into the processor array

Now we can proceed as we do to compute the lengths $D_{ij}$ of the shortest paths. The only additional thing we have to do is to update the FIRST and MORETHANONE values whenever $d_{ik}+d'_{kj}<D_{ij}$. Initially the matrix read in from the left and the one read in from the top consist of

$$d_{ij} = d'_{ij} = D_{ij} = \begin{cases} 0 & \text{if } i=j \\ 1 & \text{if } (i,j) \in E \\ \infty & \text{else} \end{cases}$$

$$first_{ij} = FIRST_{ij} = \begin{cases} j & \text{if } (i,j) \in E \\ \infty & \text{else} \end{cases}$$

The variables $MORETHANONE_{ij}$ and $morethanone_{ij}$ are initialized with FALSE.

The computations performed at the processors are as follows:

OP1: If $d_{ik} + d'_{kj} \{ D_{ij}$ then $\langle FIRST_{ij} := first_{ik};$
$\qquad\qquad MORETHANONE_{ij} := morethanone_{ik} \rangle;$

If $d_{ik} + d'_{kj} = D_{ij}$ and $D_{ij} \neq \infty$ then $\text{MORETHANONE}_{ij} :=$

$\text{MORETHANONE}_{ij} \vee \text{morethanone}_{ik} \vee \text{FIRST}_{ij} \neq \text{FIRST}_{ik}$ ;

$D_{ij} := \min (D_{ij}, d_{ik} + d'_{kj})$ ;

OP2: If $k = j$ then $\langle d_{ik} := D_{ij};$ $\text{first}_{ik} := \text{FIRST}_{ij};$

$\quad \text{morethanone}_{ik} := \text{MORETHANONE}_{ij} \rangle$ ;

If $k = i$ then $d'_{kj} := D_{ij}$ ;

Finally we make one additional pass through the processor array, performing at each processor in each beat the operation

OP 3: If $d_{ik} = d'_{kj}$ or ( $d_{ik} - 1 = d'_{kj}$ and $\text{morethanone}_{ik}$ ) then $B_{ij} := \text{FALSE}$

## 4. VALIDITY OF THE ALGORITHM

After the first three passes the values of $d_{ij}$ and $d'_{ij}$ are the lengths of the shortest paths from i to j. A proof can be found in [2, 3, 4]. Furthermore, it is easy to see that morethanone$_{ij}$ is TRUE iff there are at least two shortest paths from i to j starting with different edges. Now due to Lemma 1 after the final pass $B_{ij}$ is 1 iff (i,j) is an edge and there is no cycle containing (i,j), which means that (i,j) is a bridge.

## 5. COMPLEXITY

Our algorithm requires four passes through the processor array. The best previous known algorithm is that of [1] which consists basically of three steps, each requiring three passes through the array, one to compute the $D_{ij}$ matrix, and two to compute the transitive closures of different graphs. There are some more operations performed but the additional effort is comparable with our handling of the FIRST and MORETHANONE variables. So our method performes better by about a factor of two.

## 6. CONCLUSIONS

We have presented a method to find all bridges of an undirected graph in O(n) steps on an n*n mesh-connected processor array. Compared to the algorithm of [1] it is significantly simpler and faster by a factor of two. Compared to the algorithm of [5] it has the advantage that it works.

# 7. REFERENCES

[1] M. J. Atallah, S. R. Kosaraju, *Graph Problems on a Mesh-Connected Processor Array*, JACM, Vol 31, No 3, July 1984, 649-667.

[2] L. J. Guibas, H. T. Kung, C. D. Thompson, *Direct VLSI Implementation of Combinatorial Algorithms*, In Proc. Caltec Conf. on VLSI, 1979, 509-525.

[3] H.-W. Lang, *Transitive Closure on an Instruction Systolic Array*, to appear in Proc. Int. Conf. on Systolic Arrays, San Diego, California, May 1988.

[4] J. D. Ullman, *Computational Aspects of VLSI*, Computer Science Press, Rockville, Maryland, 1984.

[5] M. Zubair, B. B. Madan, *Efficient Systolic Algorithm for Finding Bridges in a Connected Graph*, Parallel Computing, 6, 1988, 57-61

# COLOURING PERFECT PLANAR GRAPHS
# IN PARALLEL

**Iain A. Stewart**

**Computing Laboratory, University of Newcastle upon Tyne,**

**Newcastle upon Tyne, NE1 7RU, England.**

## Introduction

The class of planar graphs is probably the best known of all classes of graphs, partly because of its association with one of the most famous problems in mathematics, the Four Colour Problem. This problem was, of course, solved in 1977 by Appel and Haken [1]: however, the solution of this problem does nothing to diminish our interest in planar graphs in general.

The class of planar graphs is sufficiently large to warrant the study of certain sub-classes. Recently, the class of perfect planar graphs has been subjected to analysis (see, for example, [8], [9], [10], [14], [15], [16], and [17]). In this paper we consider the colouring of perfect planar graphs. In particular, we are interested in the parallel colouring of a perfect planar graph using at most 4 colours.

The problem of whether a planar graph can be sequentially coloured with 3 colours is NP-hard [6], however the problem of sequentially colouring a planar graph with 4 colours can be solved using a (complicated) polynomial time algorithm. This suggests that we might be able to find a parallel algorithm running in poly-log time and using a polynomial number of processors. The associated problems of colouring a planar graph with 8, 7, 6, and 5 colours have been considered in [2], [13], [5], and both [13] and [4], respectively, with all problems having been shown to be in NC. It is unknown whether the problem of colouring a planar graph with 4 colours is in NC.

In this paper we show that the problem of colouring a perfect planar graph using at most 4 colours is in NC. We present a recursive algorithm that runs in $O(\log^4 n)$ time using $O(n^3)$ processors when implemented on an EREW PRAM. We develop a randomized NC algorithm and modify it, using a well-known construction, to obtain a deterministic NC algorithm.

We begin by giving the definitions relevant to this paper, and then describe our randomized parallel algorithm. We then analyse its time complexity and convert the randomized parallel algorithm into a deterministic parallel algorithm. Finally, we present our conclusions.

## Definitions and Notation

The reader is referred to [7] for all standard graph-theoretic definitions. The following are non-standard. We denote by $G_{\alpha\beta}$ the subgraph of the properly-coloured graph $G=(V,E)$ induced by all those vertices of V coloured with either of the colours $\alpha$ and $\beta$. We call the connected components of $G_{\alpha\beta}$ the $\alpha$-$\beta$ *components* of G. An $\alpha$-$\beta$ *interchange* consists of swapping the colours of the vertices of some specified $\alpha$-$\beta$ component of G.

We use EO(k) to mean "the expected value is O(k)".

## The Randomized Parallel Algorithm

Initially, we present a sequential, recursive algorithm for properly colouring a perfect planar graph $G=(V,E)$ using at most 4 colours. Having done this, we transform this sequential algorithm into a parallel algorithm. We assume, for the moment, that we are given a planar drawing of G. Our algorithms consist of 3 distinct stages.

Stage 1 : Find a maximal independent set I of the subgraph $G_7$ of G induced by the vertices of degree less than 7.

Stage 2 : Delete the vertices in I from G and colour the remaining graph, $G_I$, recursively.

Stage 3 : Colour the vertices of I, bearing in mind the colouring of $G_I$, so as to obtain a proper colouring of G.

It is well-known (e.g. [7]) that every planar graph (with more than 4 vertices) has at least 4 vertices of degree not exceeding 5. However, we can say more than this.

**Lemma 1 [13]** : Let G be a planar graph with n vertices. Then there are more than n/6 vertices with degree less than 7.

Lemma 1 ensures that our algorithm always terminates with our graph G properly coloured. We shall find further use for it when we examine the complexity of our forthcoming parallel algorithm. However, we now explain how to complete Stage 3.

We begin with two simple observations.

(a) Let $v \in I$, where I is as in Stage 1 (so v has degree at most 6). Suppose that v is adjacent to two vertices which are coloured by the colours $\alpha$ and $\beta$, and are in the same $\alpha$-$\beta$ component of $G_I$. Then there is an odd-length cycle in G, consisting of a path of vertices coloured $\alpha$ or $\beta$, together with edges (x,v) and (v,y), where x and y are neighbours of v coloured $\alpha$ and $\beta$ respectively, that has no chords. By [16] planar graphs satisfy the Strong Perfect Graph Conjecture [3] and so this cycle must have length 3. Hence x and y must be joined by an edge of G. Obviously, if there is only one neighbour of v coloured $\alpha$ and only one neighbour of v coloured $\beta$, then these vertices are in different $\alpha$-$\beta$ components of $G_I$ if and only if they are not joined by an edge in $G_I$.

(b) Suppose that N consists of a collection of neighbours of an uncoloured vertex $v \in I$, where $|N| > 1$ and all vertices in N are coloured with the colour $\alpha$ or the colour $\beta$. Let $N_\alpha$ and

$N_\beta$ be the subsets of N consisting of those vertices coloured $\alpha$ and $\beta$, respectively. Then by adding vertices and altering edges of $G_I \cup \{v\}$, as in Fig. 1, we may assume that the vertices of

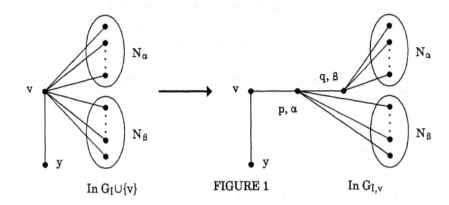

In $G_I \cup \{v\}$          FIGURE 1          In $G_{I,v}$

N are all in the same $\alpha$-$\beta$ component. Notice that the amended graph $G_{I,v}$ is still properly coloured, but may not be planar.

Suppose further that there is a neighbour y of v coloured with the colour $\alpha$ or the colour $\beta$, such that y is not in the same $\alpha$-$\beta$ component in $G_I$ as any of the vertices of N. Then clearly y is not in the same $\alpha$-$\beta$ component in $G_{I,v}$ as the vertices of N. By performing an $\alpha$-$\beta$ interchange in $G_{I,v}$ at p (and so swapping the colours of all those vertices in $N_\alpha$, $N_\beta$, and $\{p, q\}$), and identifying the vertices p, q, and v of $G_{I,v}$ (leaving the identified vertex uncoloured), we obtain our original graph $G_I$, properly coloured, with the colours of the vertices of N all swapped and the colour of y unchanged. We obtain a similar conclusion (with $\alpha$ and $\beta$ reversed) by performing an $\alpha$-$\beta$ interchange in $G_{I,v}$ at y and identifying the above vertices.

Thus, given any collection of neighbours of some uncoloured vertex $v \in I$, where these neighbours are coloured using only 2 colours $\alpha$ and $\beta$, we may assume that all of these vertices lie in the same $\alpha$-$\beta$ component. This assumption is implicit in the analysis of the possible cases that follows, as is the assumption that there must be an edge from a neighbour of v coloured $\alpha$ to a neighbour of v coloured $\beta$, for all different colours $\alpha$ and $\beta$ (or else a simple $\alpha$-$\beta$ interchange would enable us to colour v).

We now analyse the different cases that may occur for the colouring of the neighbours of v, for some vertex $v \in I$. Clearly we may assume that all 4 colours actually colour some neighbours of v, as if not, we may colour v with an unused colour. There are, essentially, 14 different cases to consider for the distribution of the colours amongst the neighbours. Notice, in the following analysis, that:

(i) we require at most 4 interchanges to ensure that v can be coloured;

(ii) we may assume that for any $\alpha$-$\beta$ interchange, for some colours $\alpha$ and $\beta$, there are exactly 2 $\alpha$-$\beta$ components adjacent to v;

(iii) it does not matter on which $\alpha$-$\beta$ component we perform the $\alpha$-$\beta$ interchange.

Suppose that v has 6 neighbours and that some colour appears 3 times amongst v's neighbours. Then, w.l.o.g., we have one of the 3 cases shown in Fig. 2 (remember that the

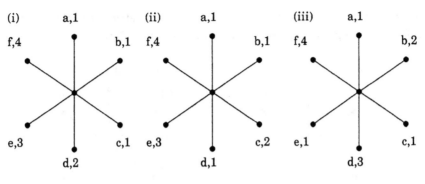

FIGURE 2

vertices appear as in the planar drawing of G).

<u>Case (i)</u> : A 1-3 interchange enables us to colour v.

<u>Case (ii)</u> : A 1-3 interchange, followed by either a 1-4 interchange or a 3-4 interchange enables us to colour v.

<u>Case (iii)</u> : W.l.o.g., by performing a 1-3 interchange the vertex a is recoloured 3. By then performing a 1-4 interchange, we may assume that the vertex c is recoloured 4. Finally, a 1-2 interchange enables us to colour v.

Consider the case when v has 6 neighbours and there is one neighbour coloured 1 and one neighbour coloured 2. Suppose further that these neighbours are separated by 2 vertices in the planar drawing. Without loss of generality, we have the 3 cases as shown in Fig. 3.

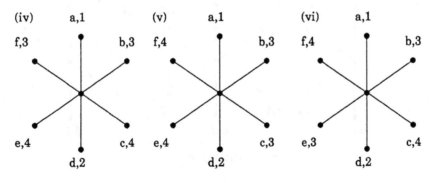

FIGURE 3

<u>Case (iv)</u> : W.l.o.g., we may perform a 2-3 interchange, so that we recolour vertex f with the colour 2. Again, we may assume that a 3-4 interchange recolours vertex e with the colour 3. A 1-4 interchange enables us to colour v.

<u>Case (v)</u> : A 3-4 interchange enables us to colour v.

<u>Case (vi)</u> : Performing a 3-4 interchange reduces us to Case (iv).

Suppose that the two vertices coloured 1 and 2 are separated by exactly 1 neighbour of v in the planar drawing. W.l.o.g., we have the 2 cases shown in Fig. 4.

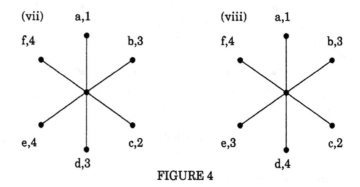

FIGURE 4

Case (vii) : A 3-4 interchange reduces us to Case (ii).

Case (viii) : A 3-4 interchange reduces us to Case (iii).

Finally, suppose that the two vertices coloured 1 and 2 are adjacent in the planar drawing (notice that if v has 4 or 5 neighbours, then we may add a dummy neighbour and reduce to a case where v has 6 neighbours). Then w.l.o.g., we have 3 cases as shown in Fig. 5.

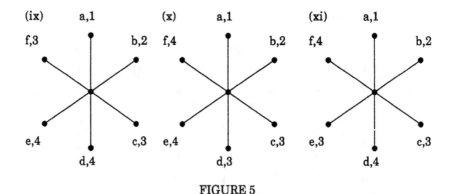

FIGURE 5

Case (ix) : By performing a 1-3 interchange, we may assume that vertex c is recoloured 1. Then, by performing a 1-4 interchange, we may assume that vertex a is recoloured 4. This reduces us to Case (ii).

Case (x) : Performing a 2-4 interchange enables us to colour v.

Case (xi) : If (b,f) is an edge, a 1-3 interchange enables us to colour v. Hence we may assume otherwise. A 1-3 interchange reduces us to Case (iv).

It is clear, assuming we can find a maximal independent set I (Stage 1) in the first place, that we have a sequential algorithm for colouring G using at most 4 colours. The time complexity of this algorithm can easily be shown to be $O(n^3 \log n)$; we use the algorithm in

[8] to find a maximal independent set in a perfect planar graph. We remark that the algorithm of [15], which properly colours a perfect planar graph, has time complexity $O(n^{3/2})$.

Having found a sequential algorithm to properly colour our graph G using at most 4 colours, we now show how to transform this sequential algorithm into a randomized parallel algorithm. The crux of the problem is how to colour a collection of vertices of I simultaneously.

Suppose that we took 2 vertices of I, say v and w, at random and performed the necessary interchanges at neighbours of v and w, hoping then to properly colour both v and w. This approach yields 2 problems. Firstly, we might be trying to change the colour of a neighbour of v or w to, say, 1 and 2 simultaneously, and secondly, an interchange at a neighbour of v might upset the colours of some of the neighbours of w. We need to be sure that these 2 situations do not occur when we colour a given collection of vertices of I.

From the sequential algorithm above, it takes at most 4 interchanges at neighbours of some vertex $v \in I$ to enable us to colour v. As mentioned earlier, we may assume that there are exactly 2 $\alpha$-$\beta$ components adjacent to v, where $\alpha$ and $\beta$ are the colours involved in the interchange, and it does not matter which one of these components is interchanged.

We may clearly assume that all 4 colours are used on the neighbours of the vertices of I. Consider, for each $v \in I$, the first interchange required, and let ($\alpha, \beta$) be the most popular pair of colours of this first interchange (and so we wish to perform an $\alpha$-$\beta$ interchange on at least $|I|/6$ vertices). Let $X_1, ..., X_t$ be unbiased, independent, random variables taking a value 0 or 1, one variable for each $\alpha$-$\beta$ component. For each $\alpha$-$\beta$ component, we perform an $\alpha$-$\beta$ interchange if and only if the corresponding random variable takes the value 1. Hence, the probability that a vertex has a successful $\alpha$-$\beta$ interchange (that is, one without external interference) is 1/2, with the expected number of vertices with a successful first interchange being at least $|I|/12$.

Let $J \subseteq I$ denote these vertices. Perform the relevant interchanges and colour vertices of J if possible. Consider those vertices $J' \subseteq J$ that remain uncoloured. By proceeding as above, with the most popular pair of colours of the next interchange, the expected number of vertices with a successful second interchange is at least $|J'|/12$.

Clearly, after 4 interchanges at any vertex of I (if necessary), we may colour this vertex. Hence, after 4 interchanges as above, the expected number of vertices of I that have been coloured is at least $|I|/12^4$. Consequently, after enough of these phases of 4 interchanges, we expect to colour all vertices of I.

By [12], there is an NC algorithm that finds a maximal independent set of an arbitrary graph, and hence we clearly have a randomized parallel algorithm that colours a perfect planar graph using at most 4 colours, given that the graph is presented in the form of a planar drawing.

**The Deterministic Algorithm and its Time Complexity**

We shall implement the algorithm using an EREW PRAM. By [12], Stage 1 can be completed in $O(\log^2 n)$ time using $O(n^3)$ processors (as, by [7], a planar graph has $O(n)$ edges), while, apart from the recursive call, Stage 2 can be completed in constant time using $O(n^2)$ processors.

Stage 3 is the most complex part of the algorithm to analyse. We can work out which case applies to a vertex $v \in I$, as well as remembering the sequence of interchanges required to colour $v$, in constant time using $O(n^2)$ processors. The phase of interchanges can be done in $O(\log n)$ time using $O(n^3)$ processors, and as, on average, there is a constant fraction of the vertices of I coloured after this phase, repeating the phase $EO(\log n)$ times will colour all the vertices of I. Hence, apart from the recursive call, the algorithm takes time $EO(\log^2 n)$ using $O(n^3)$ processors, when implemented on an EREW PRAM.

Suppose that the algorithm takes $EO(f(n))$ time using $O(n^3)$ processors, for some function $f(n)$. Then, by Lemma 1,

$$f(n) \leq f(5n/6) + k\log^2 n$$

for some constant k. It is easy to see (using induction) that $f(n) = EO(\log^4 n)$, and so we have a randomized NC algorithm for colouring a perfect planar graph using at most 4 colours, that runs in $EO(\log^4 n)$ time using $O(n^3)$ processors, when implemented on an EREW PRAM.

To convert this algorithm to a deterministic NC algorithm, with the same time and processor bounds, we use the well-known techniques of Luby [12]. The details are omitted, but our randomized NC algorithm can easily be seen to satisfy the conditions laid down by Luby for conversion into a deterministic NC algorithm.

**Conclusions**

Our algorithm assumes that the perfect planar graph is presented in the form of a planar drawing. However, this is not strictly necessary, for by [11] the problem of constructing a planar drawing of a planar graph is in NC; in particular, this problem can be solved in $O(\log^2 n)$ time using $O(n)$ processors on a CREW PRAM.

The original version of this algorithm bypassed the construction of the randomized algorithm, by using a different method of selecting a "nice" set of vertices at which to simultaneously perform $\alpha$-$\beta$ interchanges, as in the phase of interchanges in Stage 3. Here, a planar multigraph was constructed with vertices corresponding to the $\alpha$-$\beta$ components, with edges joining two vertices if and only if the corresponding $\alpha$-$\beta$ components had a common adjacent uncoloured vertex. This multigraph was then coloured using at most 5 colours and our "nice" set of vertices was chosen using this colouring. For more details, see [14]. We are indebted to an anonymous referee for pointing out Luby's technique and thus speeding up the algorithm considerably.

# References

[1]    K. APPEL AND W. HAKEN, Every planar map is four-colorable, *Illinois J.Math.* 21 (1977), 429-567.

[2]    F. BAUERNÖPPEL AND H. JUNG, Fast parallel vertex coloring, in: Fundamentals of Computation Theory, Lecture Notes in Computer Science, Vol. 199, pp. 28-35 Springer, Berlin, 1985.

[3]    C. BERGE, Sur une conjecture relative au problème des codes optimaux, *Comm. 13ieme Assemblee Gen. URSI*, Tokyo, 1962.

[4]    J. BOYAR AND H. KARLOFF, Coloring planar graphs in parallel, *J. Algorithms*, 8, 470-479(1987).

[5]    K. DIKS, A fast parallel algorithm for six-colouring of planar graphs, *12th Internat. Symp. on the Mathematical Foundations of Computer Science*, Bratislava, August, 1986.

[6]    M. R. GAREY AND D. S. JOHNSON, "Computers and Intractability: A Guide to the Theory of NP-Completeness", Freeman, San Francisco, 1979.

[7]    F. HARARY, "Graph Theory", Addison-Wesley, Reading, MA, 1969.

[8]    W. -L. HSU, An $O(n^3)$ algorithm for the maximal independent set problem on planar perfect graphs, *J. ACM*, to appear.

[9]    W. -L. HSU, Coloring planar perfect graphs by decomposition, *Combinatorica*, 6 (4), (1986), p. 381-385.

[10]   W. -L. HSU, Recognizing planar perfect graphs, *J. ACM*, Vol. 34, No.2, 1987, p. 255-288.

[11]   P. N. KLEIN, J. H. REIF, An efficient parallel algorithm for planarity, *Proc. 27th IEEE Annual Symposium on Foundations of Computer Science*, Toronto, 1986.

[12]   M. LUBY, A simple parallel algorithm for the maximal independent set, *SICOMP 15* (4) (1986).

[13]   J. NAOR, A fast parallel coloring of planar graphs with five colors, *Inf. Proc. Letters* 25 (1987) 51-53.

[14]   I. A. STEWART, A parallel algorithm to four-colour a perfect planar graph, Technical Report 245, Univ. of Newcastle upon Tyne, 1987.

[15]   I. A. STEWART, An Algorithm for Colouring Perfect Planar Graphs, *Proc. 7th Conference on Foundations of Software Technology and Theoretical Computer Science*, Lecture Notes in Computer Science 287 (Ed. K.V.Nori), Springer-Verlag, Berlin, 1987, p. 58-64.

[16]   A. TUCKER, The strong perfect graph conjecture for planar graphs, *Canad. J. Math.* 25 (1973), 103-114.

[17]   A. TUCKER AND D. WILSON, An $O(N^2)$ Algorithm for Coloring Perfect Planar Graphs, *J. Algorithms* 5, (1984) 60-68.

# AN EFFICIENT PARALLEL ALGORITHM FOR
# THE ALL PAIRS SHORTEST PATH PROBLEM

Tadao Takaoka

Department of Computer Science, University of Ibaraki

Hitachi, Ibaraki 316, JAPAN

## 1. Introduction

The all pairs shortest path problem (APSP) is to compute shortest paths between all pairs of vertices in a directed graph with non-negative edge costs. A number of sequential algorithms appeared in the past. The first approach is to apply a single source algorithm, which computes shortest paths from a vertex (called source) to all other, to n sources. If we apply the single source algorithm by Dantzig [3] or Dijkstra [5], the complexity for the all pairs problem becomes $O(n^3)$ where n is the number of vertices.

A specially tailored algorithm for the all pairs problem is Floyd's algorithm [6] whose complexity is $O(n^3)$. An algebraic approach to this problem is based on the distance semi-ring introduced in Aho, Hopcroft and Ullman [1] although the origin of this idea is a folklore. In this approach, we square the given distance matrix repeatedly so that we can raise the power to the n-th power in log n multiplications of matrices. Since matrix operations are suitable for parallelism, we take this approach in this paper and try to design an efficient parallel algorithm.

On the other hand, efficient average case sequential algorithms were reported by Spira [12], Takaoka and Moffat [13] Bloniarz [2] and Moffat and Takaoka [10], with the best average case to date being $O(n^2 \log n)$.

We will give a parallel algorithm for this problem with good expected time and a small number of processors.

In this paper we assume the parallel computational model of SIMD-SM-RW (single instruction stream multiple data stream + shared memory + concurrent read and write).

## 2. The all pairs shortest path problem

The problem is to compute the shortest path between every pair of vertices in a directed graph with edge costs. Let vertices be given by integers $1,\ldots,n$ and the edge cost of edge $(i, j)$ by $d(i,j)$, which is often regarded as the distance of edge $(i, j)$. The word 'length' is not used here for the reason given later. The distance matrix D is the matrix whose $(i,j)$-element $d_{ij}$ is $d(i,j)$. The product $C = A \cdot B$ of two distance matrices A and B is given by

$$c_{ij} = \min_{1 \le k \le n} \{a_{ik} + b_{kj}\} \qquad (1)$$

The distance semi-ring is defined by the following table (see Aho, Hopcroft and Ullman [1]).

| Semi-ring | real numbers |
|-----------|--------------|
| + | min |
| $\cdot$ | + |
| 1 | 0 |
| 0 | $\infty$ |

The square matrices of dimension n over a semi-ring form a semi-ring under the usual definition of matrix operations. Let I be the unit matrix. If the base semi-ring is a closed semi-ring, we can define the closure of D, $D^*$, by

$$D^* = I + D + D^2 + \ldots .$$

The distance semi-ring is a closed semi-ring and we have

$$D^* = I + D + D^2 + \ldots + D^{n-1} = (I+D)^{n-1}.$$

$D^*$ is the matrix whose $(i, j)$ element is the shortest distance from i to j. Since shortest paths can be computed easily from an algorithm for shortest distances, we focus on shortest distances. Now $D^*$ can be computed by the so-called repeated squaring in log n matrix multiplications. Dekel et al. [4] computes (1) in parallel in $O(\log n)$ time using a binary tree structure for comparisons, and give $O(\log^2 n)$ bound for parallel computation of $D^*$ with $O(n^3)$ processors. Quinn and Deo [11] gave an $O(\log n)$ time algorithm with $O(n^4)$ processors.

Let the length of a path be the number of edges in the path. Frieze and Rudolph [6] show that the average of the longest path length of the shortest paths in the graph is $O(\log^2 n)$, so that $D^*$ can be computed in $O(\log \log n)$ multiplications on average under some restricted probabilistic assumption. In this method we can stop squaring, if the matrix after squared is equal to the matrix before squared. They also show that the minimum of n numbers can be computed in parallel in $O(1)$ time on average using $O(n)$ processors. The time for squaring and the longest path length are mutually independent random variables. Based on this fact, their combined result is that the average parallel time is $O(\log \log n)$ using $O(n^3)$ processors.

We show that the average of the longest path length of the shortest paths is $O(\log n)$ under a wider probabilistic assumption and that (1) can be computed for $i, j = 1, \ldots, n$ in parallel in $O(\alpha)$ expected time using $O(n^{2.5+1/\alpha})$ processors. The combined result is that our expected parallel time is $O(\alpha \log \log n)$ with $O(n^{2.5+1/\alpha})$ processors where $\alpha$ is a positive integer. When $\alpha = 2$, we have Frieze and Rudolph's result. The probabilistic assumption in this paper is that elements of the input distance matrix are independent random variables drawn from an identical distribution.

## 3. Average path length

To analyze path lengths, we use a single source algorithm by Danzig or Dijkstra. By whichever algorithm, the solution set S of vertices, to which shortest distances from source have been established, expands one by one in the increasing order of distances from source. We label vertices with numbers, vertex v with j if v enters S at the j-th step. Source is labelled 1. Let $L(j)$ be the average length of the shortest path from source to the vertex whose label is j. In Gu and Takaoka [8], it is shown that

$$L(j+1) \leq \frac{1}{j} \sum_{i=1}^{j} L(i)+1, \quad L(1) = 0,$$

assuming that edge costs follow the probabilistic assumption given in Section 2. From this recurrence, it follows that $L(n) = O(\log n)$. Since it holds that for $i < j$, $L(i) \leq L(j)$, we can say that the average of maximum value of path lengths of shortest paths is $O(\log n)$.

## 4. Packing n points into space of size an

Suppose that we fill array $y[1..a*n]$ with n values from array $x[1..N]$ where $N \gg n$ and $a>1$. Specifically the values $x[i_1],...,x[i_n]$ will be packed into array y in parallel by repeated trials. See the figure below the following packing algorithm. Each iteration in the loop is called "move".

```
(1≤k≤n)     moved[iₖ]:="no"
(1≤i≤an)    occupied[i]:=0
            loop
               done:="yes"
(1≤k≤n)        if moved[iₖ]="no" then
                  generate random r_{iₖ}  (1≤r_{iₖ}≤an)
                  if occupied[r_{iₖ}]=0 then
                     occupied[r_{iₖ}]:=iₖ
                     moved[occupied[r_{iₖ}]]:="yes"
                     y[r_{iₖ}]:=x[occupied[r_{iₖ}]]
                  else
                     done:="no"
            until done="yes"
```

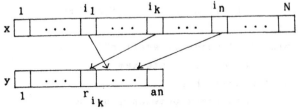

If $r_{i_k} = r_{i_l}$ for $k \neq l$ then the assignment for array occupied at the location is done at random on the first-come first-service basis. If moved$[i_k]$="no", we say $x[i_k]$ is unmoved. Occupied[i]=0 means that y(i) is not occupied. Occupied[i]=j means that y[i] is occupied by x[j]. $(1 \leq k \leq n)$ means a parallel synchronized operation by n processors, the k-th of which undertakes parameter k.

We assume that $r_{i_k}$ are mutually independent random variables with uniform distribution between 1 and an. The probability that $x[i_k]$ can not be moved is not greater than $n/(an) = 1/a$ at the first iteration.

Similarly the probability $p_j$ that $x[i_k]$ can not be moved until the j-th iteration is not greater than $(1/a)^j$. This value $p_j$ can be assumed to be equal for all $i_k$. Thus the probability $P_j$ that done="no" at the j-th iteration is not greater than $1-(1-p_j)^n$. That is,

$$P_j \leq np_j \leq n(1/a)^j$$

Let $a = n^{1/\alpha}$ and $j = 2\alpha$. Then we have $P_{2\alpha} \leq 1/n$. Similarly $P_{r\alpha} \leq (1/n)^{r-1}$. In summary, we can pack n points into space of size $n^{1+1/\alpha}$ in $O(\alpha)$ parallel steps using n processors. N and n will be replaced by n and $\sqrt{n} \cdot n^{1/\beta}$ and $\alpha$ by $\alpha\beta/(\beta-\alpha)$ for the application in Section 7. In this case, if $j = r\alpha\beta/(\beta-\alpha)$, $P_j \leq n^{-(r-1)(2+\beta)/2\beta}$. Thus, to achieve $P_j \leq 1/n^4$, we can choose $j = (9\beta+2)\alpha/(\beta-\alpha)$.

## 5. Minimum selection algorithm

We give the algorithm "find-minimum" by Frieze and Rudolph in the following. Let $m = \sqrt{n}$ .

```
            input:  n numbers  a1,....,an
            output: the minimium minval
            minval:=∞
            loop
                  *** initialize data structures ***
(1≤i≤m)     test[i]:=∞
            is-min[i]:="nobody is smaller"
                  *** choose m elements ***
(1≤i≤n)     if ai < minval then
                  ri:=random index between 1 and m
                  test[ri]:=ai
                  *** compare all pairs and record results ***
(1≤i,j≤m)   if test[i] < test[j] then
                  is-min[j]:="somebody is smaller"
                  *** pick minimum ***
(1≤i≤m)     if is-min[i]="nobody is smaller" then
                  minval:=test[i]
                  *** are we done? ***
(1≤i≤n)     if ai < minval then done:=false
            until done
```

<u>Note</u>   if $r_i = r_j$, then assignment for test[$r_i$] with $a_i$ or $a_j$ will
be done at random on the first-come first-service basis and
the first such $a_i$ will be the value of test[$r_i$].

At the first iteration about $\sqrt{n}(1-1/e^{n^{1/2}})$ positions of array
test are occupied and about $\sqrt{n}$ of $a_i$'s will remain as candidates for
minval. At the second iteration, about $\sqrt{n}(1-1/e)$ are occupied and 1-
1/e will remain in array a on average. That is, the minimum will be
found after two iterations on average using n processors. In [7] it
is shown that the probability that the fourth iteration takes place is
O(1/n). After the fourth iteration, if the number of the candidates
for minval decreases one by one, the time for that part is O(1). We
can prove that the probability of k-th iteration being followed is
$O(1/n^{ck})$ for some constant c [9]. From this we see that $n^2$ independent minimum selection algorithms executed in parallel terminate in
O(1) time if we take k=2/c. This fact will be used in Section 7.
Note that the minimum selection algorithm in this section can be
modified into a maximum selection algorithm easily.


## 6.  The algorithm

Let D be the input distance matrix.  The outermost skelton of the
algorithm is the repeated squaring  given by the following.

**Outermost algdorithm**

```
D:=D+I
loop
  D:=D'
  D':=D*D
until D'=D.
```

The next step is to design an algoritlhm for multiply two distance matrices to be used for squaring in the above.

Let us consider the problem of computing the inner product of two
distance vectors $a = (a_1,...,a_n)$ and $b = (b_1,...,b_n)$, that is,

$$\min_{1 \le i \le n} \{a_i + b_i\} \tag{2}$$

We consider hypothetical sorted vectors a' and b' of a and b and their index vectors indexa and indexb (see the following).

In the above we examine k smallest elements of **a**, that is, the first k positions of the sorted vector **a'**. Let i be the index such that i is in indexa[1..k] and $b_i$ is minimum in **b**, that is, the leftmost partner in indexb of i in indexa[1..k]. Let p(i) and q(i) be the positions of i in indexa and indexb respectively. Then we can say that the index that gives the minimum to (2) lies in indexa[1..p(i)] or indexb[1..q(i)]. If we choose k = $\sqrt{n}$, then the expected values of p(i) is $\frac{1}{2}\sqrt{n}$ and that of q(i) is less than $\sqrt{n}$.

That is, we have only to examine at most $\frac{3}{2}\sqrt{n}$ indices on average. The application of this idea to our problem is that we select $\sqrt{n}$ smallest elements from each row of A in O(1) time using n processors each. Then we compute $c_{ij}$ in (1) in $O(\alpha)$ time using $n^{2.5+1/\alpha}$ processors. This is essentially a parallel version of the $O(n^{2.5})$ expected time algorithm for distance matrix multiplication in [13]. The algorithm follows. Let $m_1 = \sqrt{n}$ and $m_2 = \sqrt{n}*n^{1/\alpha}$ for some integer $\alpha$.

**Distance matrix multiplication algorithm**

```
        *** initialization for (i) ***
    (1≤i≤n, 1≤j≤m₂) testa[i,j]:=∞
    *** (i) Find about √n smallest elements from each row of A ***
    (1≤i≤n) Pack randomly chosen m₁ elements of A[i,1],...,A[i,n]
                into array testa[i,1..m₂] using n processors
    (1≤i≤n) Find minimum aᵢ of testa[i,1..m₂] using m₂ processors
        *** initialization for (ii) ***
            clear indexa
    (1≤j≤n) uᵢⱼ:=∞, vᵢⱼ:=∞
```

```
                count:=1
            *** (ii) Set up boundary elements ***
            loop
(1≤i≤n)     Move indices k of unmoved A[i,k] such that A[i,k]≤aᵢ
                into indexa[i,1..m₂] using n processors
(1≤i,j≤n)   Find minimum B[l,j] of B[k,j] where k is in
                indexa[i,1..m₂] using m₂ processors and
                if B[l,j]<vᵢⱼ then uᵢⱼ:=A[i,l] and vᵢⱼ:=B[l,j]
            count:=count+1
            until there are no unmoved A[i,k] such that A[i,k]≤aᵢ
                or count > c (This constant will be given in
                            Section 7)
            If count > c then switch to a sequential algorithm
(1≤i≤n)     Find maximum sᵢ of uᵢ₁,...,uᵢₙ using n processors
(1≤j≤n)     Find maximum tⱼ of v₁ⱼ,...,vₙⱼ using n processors
            *** initialization for (iii)***
(1≤i,j≤n)   clear indexa and indexb
            count:=1
(1≤i,j≤n)   cᵢⱼ:=∞
            *** (iii) Find min {aᵢₖ+bₖⱼ} ***
            loop
(1≤i≤n)     Move indices k of unmoved A[i,k] such that
                A[i,k]≤sᵢ into indexa[i,1..m₂] using n processors
(1≤j≤n)     Move indices k of unmoved B[k,j] such that
                B[k,j]≤tⱼ into indexb[j,1..m₂] using n processors
(1≤i,j≤n)   Find minimum of A[i,k]+B[k,j] for k in indexa[i,1..m₂]
                using m₂ processors and update cᵢⱼ by it if possible
(1≤i,j≤n)   Find minimum of A[i,k]+B[k,j] for k in indexb[j,1..m₂]
                using m₂ processors and update cᵢⱼ by it if possible
            count:=count+1
            until there are no unmoved A[i,k] such that A[i,k]≤sᵢ
                and there are no unmoved B[k,j] such that B[k,j]≤tⱼ
                or count>c (This constant will be given in Section 7)
            If count>c then switch to a sequential algorithm
```

Note   "clear" is to make each position of array "not occupied".
       (1≤i≤n) means parallel operation for i=1,...,n.
       (1≤i,j≤n) means parallel operation for i=1,...,n and
       j=1,...,n.  Other operations are done by a single
       processor.

## 7. Analysis

In the whole framework of our algorithm, we have two major probabilistic events. One is the number of multiplications in the outermost repeated squaring process, which is O(loglog n). The other is the time for multiplying two matrices. Whatever multiplication algorithm we may use, we have the same series of matrices D, $D^2$, $D^4$, $D^8$,.... Thus the above mentioned two events are considered to be independent and the expected time for the whole algorithm is the time for multiplication times O(loglog n).

From the probabilistic assumption stated in Section 2, it holds that if each row and column of matrix D at any stage in the outermost algorithm are sorted, the corresponding index vectors are any of the permutations of (1,2,...,n) with equal probability 1/n!, where we neglect a slight deviation of the diagonal elements being the smallest. We apply this probabilistic assumption to the matrices A and B that appear in the multiplication algorithm. Before we analyze the multiplication algorithm, we state a few lemmas.

**Lemma 1.** Suppose we choose k numbers out of n numbers at random. Then the probability that the rank of the smallest of the k numbers in the n numbers is not less than $l$ is given by

$$\binom{n-l}{k} \bigg/ \binom{n}{k} .$$

**Proof.** Obvious.

**Lemma 2.** For k and $l$ such that $0 \leq k \leq n$, $0 \leq l \leq n$ and $k/n \to 0$ and $l/n \to 0$ when $n \to \infty$, we have

$$\binom{n-l}{k} \bigg/ \binom{n}{k} \sim 1/e^{kl/n}$$

**Proof.**

$$\binom{n-l}{k} \bigg/ \binom{n}{k} = \frac{(n-l)(n-l-1)...(n-l-k+1)}{n \cdot (n-1)...(n-k+1)}$$

$$= \left(1-\frac{l}{n}\right)\left(1-\frac{l}{n-1}\right) \cdots \left(1-\frac{l}{n-k+1}\right)$$

$$\sim \left(1-\frac{l}{n}\right)^k = \left(1-\frac{l}{n}\right)^{\frac{n}{l} \cdot \frac{kl}{n}} \sim 1/e^{kl/n}$$

Q.E.D.

**Lemma 3.** Suppose we have n independent events with equal probability p. If $np \to 0$ ($n \to \infty$), the probability that at least one event occurs is given by np.

**Proof.** The probability is given by $1-(1-p)^n \sim np$.          Q.E.D.

We shall not distinguish between the signs "$\sim$" and "$=$" hereafter.

We assume that find-minimum and find-maximum operations are done in $O(1)$ expected time. Let the rank of $a_i$ in the i-th row of A be $r(a_i)$. Using the analysis in Section 4, the time for phase (i) can be shown to be $O(1)$. The expected value of the rank of $a_i$ is shown to be $\sqrt{n}$. Furthermore, from lemmas 1 and 2 we have prob $\{r(a_i) \geq \sqrt{n}n^{1/\beta}\} = 1/e^{n^{1/\beta}}$ and prob $\{r(a_i) \geq \sqrt{n}n^{-1/\beta}\} = 1-n^{-1/\beta}$ for large n.

In phase (ii), the loop statement is essentially the packing algorithm in Section 4. Let $r(a_i) < \sqrt{n}n^{1/\beta}$ for $i=1,\ldots,n$. Since we have n independent move operations in the loop, for $c=(9\beta+2)\alpha/(\beta-\alpha)$, the probability that the c-th iteration is done is bounded by $n/n^4 = 1/n^3$. Hence we can have $O(1)$ expected time for the sequential algorithm even if we use an $O(n^3)$ one. On the other hand from lemma 3 we have prob $\{r(a_i) \geq \sqrt{n}n^{1/\beta}$ for some i$\} = n/e^{n^{1/\beta}} < 1/n^3$. The expected time for this case is also absorbed into the sequential algorithm with $O(1)$.

The skelton of phase (iii) is again the packing algorithm. Since $r(u_{ij}) \leq r(a_i)$, we have

$$\text{prob } \{r(s_i) \geq \sqrt{n}n^{1/\beta}\}$$

$$= \text{prob } \{r(u_{ij}) \geq \sqrt{n}n^{1/\beta} \text{ for some j}\} \leq n/e^{n^{1/\beta}},$$

and

$$\text{prob } \{r(s_i) \geq \sqrt{n}n^{1/\beta} \text{ for some i}\} \leq n^2/e^{n^{1/\beta}}.$$

Thus we have an analysis for the packing of array indices of array A into indexa similar to phase (ii).

To analyze the packing of array indices of array B into indexb, we estimate the rank of $v_{ij}$ in the j-th column of B. Let $r(a_i) > \sqrt{n}n^{-1/(2\beta)}$ for all i. Then we have

$$\text{prob } \{r(v_{ij}) \geq \sqrt{n}n^{1/\beta}\} = 1/e^{n^{1/(2\beta)}},$$

since $v_{ij}$ is the minimum of $\sqrt{n}n^{-1/(2\beta)}$ numbers, and we have

$$\text{prob } \{r(t_j) \geq \sqrt{n}n^{1/\beta}\}$$

$$= \text{prob } \{r(v_{ij}) \geq \sqrt{n}n^{1/\beta} \text{ for some i}\} = n/e^{n^{1/(2\beta)}}.$$

In this case we have an analysis similar to that for the packing of indices of A. For the case of $r(a_i) \leq nn^{-1/(2\beta)}$, we have

$$\text{prob } \{r(a_i) \leq \sqrt{n}n^{-1/(2\beta)}\} = \sqrt{n}^{-1/(2\beta)},$$

which is not small enough to be absorbed into the sequential algorithm with $O(1)$ time. To fix this problem, we modify phase (i) as follows. We repeat phase (i) $2k+1$ times and we take the median of $2k+1$ $a_i$'s for our $a_i$. Then we have

$$\text{prob } \{r(a_i) \leq \sqrt{n}n^{-1/(2\beta)}\} = n^{-k/(2\beta)}.$$

If we choose $k = 8\beta$, this probability becomes $n^{-4}$. Thus we have

$$\text{prob } \{r(a_i) \leq \sqrt{n}n^{-1/(2\beta)} \text{ for some i}\} = n^{-3},$$

which is absorbed into the sequential algorithm with $O(1)$ time. Note that the modified phase (i) takes $O(\beta)$ time.

Other operations including minimum and maximum selections terminate in $O(1)$ time, since the number of such operations in parallel is limited by $n^2$. To conclude the time analysis, let us choose $\beta = 2\alpha$. Then $c = 20\alpha$. Thus we have $O(\alpha)$ time, for the multiplication algorithm.

The maximum number of processors used is $m_2 n^2 = n^{2.5+1/\alpha}$.


## 8. Concluding remarks

Let the computing time and the number of processors for a parallel algorithm be $T(n)$ and $P(n)$ respectively. In most commercially available parallel computers, the number $p$ of processors is limited, ranging from 30 to about 65000. Under some ideal scheduling, the actual computing time will be $T(n)P(n)/p$. Hence reducing the number of processors in algorithm design is important not only for its own sake but for time efficiency when the algorithm is executed on an actual parallel computer.

Our next step would be to try to invent a parallel algorithm for APSP with O(loglog n) expected time and $O(n^{2.5})$ processors on the same computational model. Our final goal in this area is to determine whether there exists an O(1) parallel algorithm for APSP in NC, that is, with O(1) time and the number of processors being polynomial of n on any parallel computational model.


## References

[1] A. V. Aho, J.E. Hopcroft and j. D. Ullman, "The Design and Analysis of Computer Algorithms", Addison-Wesley (1974).
[2] P.A. Bloniarz, "A shortest path algorithm with expected time $O(n^2 \log n \log^* n)$, SIAM Jour. on Computing 12 (1983), pp. 588-600.
[3] B.B. Dantzig, "On the shortest route through a network", Management Science 6 (1960), pp. 187-190.
[4] E. D. Dekel, S. Nassimi and Sahni, "Parallel matrix and graph algorithms", SIAM Jour. on Computing 10 (1980).
[5] E. W. Dijkstra, "A note on two problems in connection with graphs", Numer. Math. 1 (1959), pp. 269-271
[6] R. W. Floyd, "Algorithm 97: shortest path", Comm. ACM 5 (1962), p345.
[7] A. Frieze and L. Rudolph, "A parallel algorithm for all pairs shortest paths in a random graph", Technical Report, Dept. of Com. Sci., Carnegie-Mellon Univ. (1982).
[8] Q. P. Gu and T. Takaoka, "On the average path length of O(log n) in the shortest path problem", Trans. of the IEICE of Japan E70 (1987) pp. 1155-1158.
[9] Q. P. Gu and T. Takaoka, "A parallel algorithm for the all pairs shortest path problem", submitted for publication.
[10] A. M. Moffat and T. Takaoka, "An all pairs shortest path algorithm with expected time $O(n^2 \log n)$", SIAM Jour. on Computing 6 (1987), pp. 1023-1031.
[11] J. Quinn and N. Deo, "Parallel graph algorithms", ACM Computing Surveys 16 (1984), pp. 319-348.
[12] P. M. Spira, "A new algorithm for finding all shortest paths in a graph of positive arcs in average time $O(n^2 \log^2 n)$", SIAM Jour. on Computing 2 (1973), pp. 28-32.
[13] T. Takaoka and A. M. Moffat, "An $O(n^2 \log n \log\log n)$ expected time algorithm for the all shortest distance problem", Lecture Notes in Computer Science 88, Springer (1980), pp. 642-655.

# A Parallel Algorithm for Channel Routing [†]

*John E. Savage*
*Markus G. Wloka*

Brown University
Department of Computer Science, Box 1910
Providence, Rhode Island 02912

### Abstract

We present an optimal $NC$ algorithm for 2-layer channel routing of VLSI designs. Our routing algorithm achieves channel density and runs in $O(\log n)$ time using $O(n)$ processors on an EREW P-RAM. The routing algorithm is a parallel version of the widely used Left-Edge Algorithm. It can be used to solve the maximum clique and the minimum coloring problem for interval graphs and the maximum independent set problem for co-interval graphs with optimal processor-time bounds. We give an optimizing extension to our algorithm that resolves column conflicts under certain weak conditions and runs in polylog time. The routing algorithm can easily be implemented on a multi-processor shared-memory machine so our solution has considerable practical value.

## 1   Introduction

We present an optimal $NC$ algorithm for 2-layer channel routing of VLSI designs. VLSI is concerned with the means and methods for packing large numbers of electronic components onto small chips. The computational problems that arise are complex and time consuming, and there is a great demand for algorithms to facilitate the rapid implementation of VLSI chips, for tools to implement these algorithms, and for analysis to provide an understanding of the fundamental limitations on computation with VLSI chips. Current large-scale VLSI design problems are so time-consuming that they cannot be done quickly enough with standard serial computers, even those of the supercomputer class. Consequently, there is a need to harness the power of multiple processors working in parallel. To exploit parallelism in VLSI requires however, that new algorithmic methods be developed, since traditional methods are essentially serial.

The algorithms used in VLSI chip design, as in other fields of applied computer science, suffer from sequential thinking. The designer often has not taken into account that an algorithm may have to run efficiently on a parallel machine. Most current problem-solving strategies proceed step by step towards a goal, each step depending on the outcome of the preceding step. So while a problem can be parallel, the method of solving it almost certainly is not. The naive approach of using an

---

[†]This work was supported in part by the Semiconductor Research Corporation under contract 86-07-84, the National Science Foundation under Grant DCR 83-06812, the Office of Naval Research under contract N00014-83-k-0146 and DARPA Order No. 4786.

optimizing compiler or providing a language with parallel primitives will fail because there is not enough parallelism in the statement of most algorithms.

A widespread, but flawed way of showing that an algorithm is parallel is to measure the speedup realized by adding processors. Speedup measurements fail as way to compare and develop parallel algorithms for the following reasons:

- The measurements can be and often are inaccurate, and thus the results cited in the literature are often inconclusive.

- An algorithm is often adapted to the peculiarities and the design flaws of a particular machine, and a new algorithm often must be developed for every new computer.

- Existing computers have a modest number of processors so that algorithms exhibiting good speedups on such a computer might not scale well to large increases in the number of processors.

We use an alternative approach to study parallelism. We shall develop and analyze our parallel VLSI algorithms under the P-RAM (parallel random-access machine) model of computation, a very general model based upon an idealized multiprocessor machine with shared memory. The P-RAM model has been accepted by the theoretical computer science community as the model of choice for the analysis of parallel algorithms. Classes of problems that can be solved quickly in parallel with a small number of processors have been identified. The most important such class is NC, the set of problems that can be solved in polylogarithmic time in the size of the input with a polynomial number of processors. Parallel algorithms that are in $NC$ on the P-RAM machine model can be translated into fast parallel algorithms for other parallel machine models. If a problem is Ptime-complete, it is essentially sequential in nature.

In the next Section we define our problem, optimal 2-layer routing in a rectilinear channel. We show how wires can be represented by horizontal intervals which will be the input format of our parallel algorithm. In Section 3 we briefly describe the P-RAM model of parallel computation, the class $NC$, and provide several efficient computation primitives that we shall use as subroutines. Section 4 deals with the conversion of the input connection graph into an interval representation. All preprocessing is also in $NC$. In Section 5 we give the routing algorithm, the proof of correctness upper and lower bounds on the running time, space and the number of processors required. In Section 6 we show that our routing problem is equivalent to interval graph coloring. Our algorithm can be used to find the minimum coloring, maximum clique for these graphs and to find the maximum independent set of their complements. Section 7 shows how to extend the algorithm to find an optimal routing for cell-based VLSI layout systems, and in Section 8 we conclude by stressing the practical merits of our algorithm.

## 2 Problem Description

We would now like to describe the underlying model and assumptions for our routing algorithm. We are given two sets of horizontally adjacent modules. The two sets of modules face each other across a *channel*. Each module in one set has terminals on its upper boundary. The other set has terminals on the lower boundary, and wires are used the channel area to connect terminals.

We use an undirected graph $G = (V, E)$ to abstract the connection information given by the terminals and wires. We represent a terminal by a vertex. Let $A$ contain all the vertices of the lower set of modules. Let $B$ contain all the vertices of the upper set of modules, and let $V = A \cup B$. A wire that connects two terminals is represented by an edge, and $E \subseteq V \times V$ is the set of all edges. Figure 1 shows an example of a graph $G$.

In some applications $G$ is a bipartite graph with $E \subseteq A \times B$. We do not impose this limitation and allow edges from $A \times A$ and $B \times B$ because it makes our algorithms more general without increasing their complexity. If $G$ has edges $(a, b)$ and $(b, c)$, it means that terminals $a, b, c$ are all

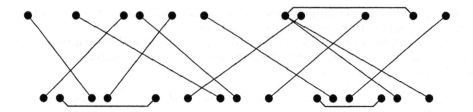

Figure 1: A Graph $G$

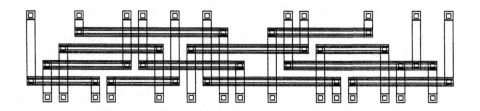

Figure 2: An Optimal Routing of $G$

electrically connected. A *net* $(a, \ldots, z)$ is the set of all vertices reachable from $a$ by traversing zero or more edges. A net is also called a connected component of $G$.

The nets of $G$ have to be mapped onto a 2-dimensional VLSI circuit. When wires cross, they have to be in different layers. Two layers can be connected by a *contact*, this contact must have a fixed minimum size. Wires must have a minimum width to comply with design rules. Wires in the same layer must keep a minimum separation between them to avoid shorts. In our examples we make wires and wire separation as wide as a contact. We assume that a wire connecting two terminals will consist of alternating horizontal and vertical segments joined by contacts. This assumption is realistic as most VLSI CAD tools, data interchange formats and mask-making facilities do not allow rectangles with arbitrary orientations.

When wires run on top of each other for a long stretch, there is the possibility of electrical interference, called *crosstalk*. To avoid crosstalk, and to further simplify the routing problem, horizontal segments are placed on one layer and vertical segments are placed on another layer.

We define *channel routing* as a mapping of $G$ onto an integer grid. The mapping of the vertex sets $A$ and $B$ onto the grid is in part determined by the fixed $x$-coordinates of the terminals. The *channel* is the portion of the grid between the two sets of vertices. A *track* is a horizontal grid line. We assume a *rectangular channel*, meaning that the vertices in each set are collinear, both sets are parallel and the channel contains no obstacles. Each vertex $v$ can be represented by its $x$-coordinate on the grid and its set membership. The $y$-coordinate of all $v \in A$ is 0. The $y$-coordinate of all $v \in B$ will be $w + 1$, where the *channel width* $w$ is defined to be the number of tracks needed to route $G$. The channel length can either be infinite, as required by the algorithm in [26] or it can be bounded by the minimal and maximal $x$-coordinates of the vertices. The goal of the routing algorithms in this paper is to minimize channel width, the number of tracks used. The layout of $G$ in Figure 2 has minimal channel width.

Once the $x$-coordinates of the vertices are fixed we can compute a lower bound on the width of the channel. Count the number of nets crossing each vertical grid line. Independently of how these

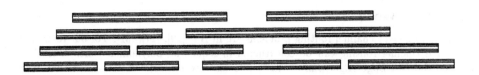

Figure 3: Minimal Solution, simplified Model

nets are mapped on the grid, they have to cross that column at least once and will take up at least one grid point on that vertical column. The maximum number of nets crossing any column is called *channel density*, with $Channel\ Density \leq Channel\ Width$

We will prove that our parallel algorithm always achieves this lower bound, when we impose the restriction that all terminals must have unique $x$-coordinates. This is done to avoid the possible overlapping of two vertical segments, also called a *column conflict*.

## 2.1 Interval Representation

A net $N_i = (a, \ldots, z)$ is a set of vertices that are all electrically connected in the circuit and are in the same connected component of $G$. Our routing algorithm operates on nets of $G$, not edges. The reason is that the operation of routing a net on a grid is as simple as routing one edge. If one net has $n$ vertices, it is connected by at least $n - 1$ edges, so an edge-based algorithm is computationally more expensive. A net $N_i = (a, \ldots, z)$ can be built by placing a vertical segment from each vertex in $N_i$ to a yet-to-be-determined track $t$, placing contacts on each end of that vertical segment, and by a horizontal wire segment between the minimum and maximum contact on that track. There will be no overlap of horizontal segments if the router works correctly. No vertical overlaps can occur if we disallow column conflicts or use a preprocessing step to remove them.

We can represent and store the net $N_i = (a, \ldots, z)$ by $I_i = (a, b, t)$, the horizontal interval $(a, b), a < b$ on track $t$. All vertices of a net must be able to determine which net they belong to so they can connect to the right track. This can be done by attaching an index field to every vertex. Computing a channel routing of $G$ under the above model is therefore the problem of assigning tracks to the set of intervals $\{I_1, \ldots, I_k\}$ so that

$$\forall_{I_i, I_j}((t_i = t_j) \land (a_i < a_j)) \rightarrow (b_i < a_j)$$

as horizontal wire segments on the same track cannot overlap. Figure 3 shows the minimal solution for the interval assignment problem that is associated with the routing problem of Figure 2.

A grid point $(x, t)$ is *in* an interval $I_k$ if $(a_k < x < b_k) \land (t = t_k)$. We will call a channel *dense below* a grid vertex $(x, t)$, if each grid vertex $(x, 1), (x, 2), \ldots, (x, t)$ is in some interval. We will call a channel *dense above* a grid vertex $(x, t)$, if each grid vertex $(x, t), (x, t + 1), \ldots, (x, w)$ is in some interval. A channel is *dense at location* $x$ if it is dense above $(x, 1)$. A channel router achieves *channel density*, if it produces a routing that is dense at one or more locations. Given an interval $I = (a, b, t)$, we will call $a$ the *start point*, $b$ the *end point* and $t$ the *track* of that interval. The *start point of a track* is the minimum of all start points in that track. The *end point of a track* is the maximum of all end points in that track.

## 2.2 Other Work

Most general 2-layer channel routing problems are in $NP$ [19], [27], [31]. For our restricted model, an optimal channel routing can be constructed in sequential polynomial time. Our algorithm is based on the *Left-Edge Algorithm* of Hashimoto and Steven [11]. Variations on that approach to channel routing are widely used and give very good results [9], [24], [28], [30]. Our parallel solution to the central problem of assigning intervals to tracks should therefore apply to other channel routers. Channel routers handle column conflicts by introducing more horizontal tracks to move the vertical segments apart, by leaving the confines of the channel or by splitting nets. Heuristics for routing in a 2-layer channel with column conflicts have been developed, they produce routings of very good quality but are not guaranteed to achieve channel density [7], [17], [24], [26]. A parallel heuristic algorithm has been developed for the Sequent, a multiprocessor machine with shared memory [32].

# 3 The P-RAM Model

Our model of parallel computation is called the P-RAM or *Parallel Random Access Machine*. The P-RAM model assumes a number of independent processors that can read, write and perform arithmetic and boolean operations. Each processor has some small amount of local memory and a unique identification number. All processors can read and write to a shared global memory. It is assumed that each processor has at initialization time its own program, which is independent of the problem size. All processors execute cycles synchronously in which they read from the common memory, compute locally and then write to common memory. P-RAMs are differentiated on whether concurrent read or write operations to the same memory location are allowed. There is a further differentiation of models depending on what action is taken in the case of a concurrent write. Note that any of the various CRCW (*Concurrent Read Concurrent Write*) P-RAMs can be simulated on an EREW (*Exclusive Read Exclusive Write*) P-RAM with an at most $O(\log n)$ increase in the time bounds. For a more detailed introduction to the P-RAM model and a survey of existing work see [15], [21] and [20].

The shared memory model of the P-RAM is close to actual implementations of parallel computers like the Encore, Sequent, Alliant and Cray machines, where the processors access memory banks by a very fast shared bus. The cost memory accesses is disregarded here, unlike other models such as the Hypercube, Mesh-of-Trees and Perfect Shuffle Network.

The algorithms that we shall present are composed of simple primitives and do not tax the P-RAM model to its limits. This section presents a collection of useful subroutines that all work in polylog time and are described in more detail in the literature: parallel prefix, sorting and transitive closure. We also provide the primitives rake(), invert(), distribute() which manage data structures in global memory.

## 3.1 Parallel Prefix

A basic and very useful tool in parallel computation is an algorithm for computing *prefix sums*. We are given $n$ elements $x_1, \ldots, x_n$ stored in a global memory array and an associative operator $*$. We would like to compute all prefix sums $S_i = x_i * \ldots * x_1, i \in 1, \ldots, n$. There is a simple parallel algorithm to compute prefix sums that runs in time $O(\log n)$ with $O(n/\log n)$ processors on a EREW [18]. The constants hidden in the asymptotic notation are very small, which is important as parallel prefix is a building block for a large number of other algorithms. Minimum, addition, multiplication and boolean operations such as AND, OR, EXOR, bitwise addition are all associative operations and are therefore optimal on an EREW.

Occasionally it is necessary to compute the prefix sums of several groups of elements. If these groups are stored in adjacent memory locations, one prefix computation suffices by replacing the operator $*$ by the following modified operator $*'$:

$$(a *' b).x \quad := \quad \text{if } (a.group == b.group) \text{ then } (a * b).x \text{ else } a.x$$
$$(a *' b).group \quad := \quad a.group$$

It is easy to show that if $*$ is associative then $*'$ is associative. This modification is sometimes called *segmented prefix* or *segmented scan* in the literature.

## 3.2 List Ranking

*List ranking* is a generalization of the prefix sum problem on linked lists. The same asymptotic bounds hold, but the algorithm is more complicated than parallel prefix. See [15] for a detailed description and further references.

## 3.3 Sorting

Recent work shows that sorting can be done in optimal $O(\log n)$ time on an $n$-processor EREW [1], [4], [22]. It should be noted that these (theoretically) optimal sorting algorithms have unrealistically large coefficients. For practical applications, Batcher's bitonic sorting algorithm [2] is still preferable. It runs in $O(\log^2 n)$ time on an $n$-processor EREW P-RAM and can be elegantly implemented to work on a butterfly or shuffle-exchange network with the same bounds. Our router depends on sorting, so the upper bounds of our algorithms change according to which sorting algorithm is used.

These algorithms rely only on the existence of a comparison function applied to two elements, or to parts of elements called keys. Sometimes we want to sort elements by a key $A$ and then sort all equivalence classes of key $A$ by key $B$. This does not require several sorting passes, since we can make the comparison function dependent on both keys. This at most doubles processing time without requiring any change to the sorting algorithm.

## 3.4 Connected Components

The connected components of a $n$-vertex, $m$-edge graph can be found in $O(\log^2 n)$ time on an $O(n+m)$ processor EREW [14]. The set of vertices forming a connected component is represented by a canonical element, the vertex with the smallest index.

## 3.5 Rake

Given an $n$-element array, let $m < n$ of the elements be marked. The `rake()` function reorders the array so that the marked elements are at the head of the array. This can be achieved by assigning a 1 to marked elements and a 0 to unmarked elements and then computing prefix sums from left to right to find new locations for the marked elements. `rake()` is used in Section 4.2 to move intervals to consecutive memory locations. The bounds of `rake()` are those of parallel prefix.

## 3.6 Invert and Distribute

For efficiency reasons we try to reduce the number of data items in a computation by having one element represent a group of components. In our case, one interval represents one net which represents a large number of terminals. Also, one track will represent all the intervals in that track. After assignments have been made by a computation to intervals and tracks, components are told of these assignments by a distribute function `distribute()`. It can be implemented to work in constant time on a CREW since all processors associated with components can simultaneously read these assignments. The running time on an EREW P-RAM is $O(\log n)$ with $O(n/\log n)$ processors. Our algorithms map elements to new locations and must allow for these elements to return to their original locations to ensure that `distribute()` works correctly. We use a function `invert()` on an array which can be implemented to run in a constant number of steps by storing with each element its old address.

# 4 Nets and Intervals

Our channel routing algorithm operates on the interval description of Section 2.1 of a channel routing problem. We provide an efficient mechanism to convert the nets of $G$ to intervals. The router is run on intervals after which the vertices in the associated nets are informed of the tracks to which they are connected. The complexity to route a channel depends on the number of terminals in a net and the format of the input data.

## 4.1 2-Terminal Nets

A graph has 2-terminal nets if every vertex is connected to at most one other vertex. If $G$ has $n$ vertices, it has at most $n$ nets. A net of $G$ contains one or two vertices. If the net contains one vertex, no routing has to be done. If the net contains two vertices, it is converted to an interval by making the smaller $x$-coordinate the start point of the interval and the larger $x$-coordinate the end point of the interval. Each interval represents its two vertices. After the router assigns tracks to intervals we use invert() and distribute() to communicate the *track* to the vertices.

**Theorem 1** *The conversion of 2-terminal nets can be converted to intervals in constant time on an n-processor EREW.*

## 4.2 N-Terminal Nets

We now remove the restriction that $G$ can only have 2-terminal nets. The complexity of converting nets to intervals is dependent on the input format of $G$. A graph $G$ is given by a *net list* or a list of $n$ vertices plus an *edge list*. A net list is the set of all connected components or nets of $G$. It is implemented by storing the vertices of a net in adjacent locations and by attaching to every vertex an index field that stores its net. We can compute the net list from an edge list by using the connected components algorithm mentioned in Section 3.3. Computing the connected components uses $O(n^2)$ processors and would be the most expensive part of the whole routing algorithm, so the net list is the preferred input format for the graph $G$, and should be used whenever possible. For example, programs that synthesize VLSI layouts, such as DeCo [5], [6] or Slap [25], [28], [29], [30], can produce net lists comprised of $(output, input, \ldots, input)$ vertex sets with an index field containing the correct net.

Once the net list is available, we have to find for every net the vertex with the minimum (maximum) $x$-coordinate which is the start (end) point of the associated interval. We assumed that all vertices of a net are stored in adjacent locations, otherwise the vertices have to be sorted by their nets. One minimum (maximum) computation is sufficient to compute all start points, if the segmented prefix operator described in Section 3.1 is used. We apply the rake() function to store the intervals in contiguous memory location. After the router assigns tracks to intervals we use invert() and distribute() to communicate the *track* to the vertices.

**Theorem 2** *n-terminal nets can be converted to intervals in $O(\log n)$ time on an $O(n/\log n)$-processor EREW. If $G$ is given as an edge list, the intervals can be computed in $O(\log^2 n)$ on a $O(n^2)$ processor EREW.*

# 5 The Parallel Routing Algorithm

There is a simple optimal sequential algorithm that assigns intervals to tracks. It takes intervals in ascending order and places an interval into the track with the smallest endpoint where it will fit or into a new track if there is no such track. The heart of this algorithm is an assignment of interval start points to smaller interval end points. We use this observation as a basis for developing a parallel version of the sequential algorithm. Our algorithm makes assignments of individual interval

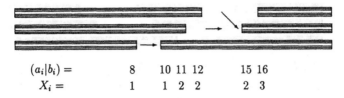

$$(a_i|b_i) = \qquad 8 \qquad 10\ 11\ 12 \qquad 15\ 16$$
$$X_i = \qquad 1 \qquad 1\ \ 2\ \ 2 \qquad 2\ \ 3$$

Figure 4: Assignments after Initialization

end points to smaller interval start points in parallel and then uses an *align* function to ensure that at most one end point is assigned to each start point.

We begin by presenting the algorithm, and then prove that certain properties hold for the assignments, from which we can deduce that the resulting channel routing has minimum width or is dense. The parallel algorithm consists of three phases, an initial assignment phase, an alignment phase, and a link phase that places chains of assignments into tracks.

## 5.1 Initialization Phase

Every interval (a,b) is converted into two halfintervals $(-\infty, b)$ and $(a, \infty)$ on which our algorithm will operate. We call these halfintervals *red* and *green* intervals respectively, and refer to them by their end points. The non-infinite coordinate imposes an order relation on the halfintervals. To get an initial assignment of red to green intervals, all halfintervals are sorted into ascending order by their non-infinite coordinates.

Note that the assumption of unique $x$-coordinates is only necessary for obtaining a correct layout. We can generalize our interval assignment algorithm by defining a red halfinterval to be larger (closed intervals) or smaller (half-open or open intervals) than a green interval if their non-infinite $x$-coordinate are equal. This ensures the correct operation of our algorithm when we use it for interval graph coloring in Section 6.

The following prefix computation is done after the sorting to initialize the assignment of red intervals to green intervals:

- If a halfinterval $I_i$ is a green interval it is given the value 1, otherwise it is given the value 0.

- The prefix sum $X_i = I_i + \ldots + I_1$ is computed for each halfinterval, with the + operator being arithmetic addition.

- If a halfinterval $I_i$ is a red interval, the value 1 is added to the prefix sum $X_i$.

The initialization has the following effect: Every red interval $b_i$ has an associated track $X_i$ which is the index of the minimum green interval $a_{X_i}$ so that $b_i < a_{X_i}$. Every green interval $a_i$ is associated with the track $X_i$ of a grid. The tracks are consecutive. Figure 4 shows the initial assignment of three red tracks to three green tracks. The notions dense, dense, below and dense above thus are defined exactly as for the channel.

## 5.2 Align

We now have to deal with the fact that several red intervals may be assigned to the same green interval $a_{X_i}$. Since we have sorted the halfintervals, it is true that $a_i < a_j, i < j$ and $b_i < b_j, i < j$. Therefore, if the assignment of $b_i$ to $a_{X_i}$ is a non-overlapping assignment, the assignment of $b_i$ to $a_{X_{i+k}}, k > 0$ is also a non-overlapping assignment. In other words, red intervals can be shifted up, that is, assigned to green intervals with a larger start point, without causing overlap.

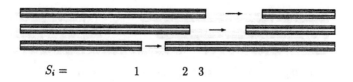

$$S_i = \qquad 1 \qquad 2 \ \ 3$$

Figure 5: Assignments after Align

Our parallel alignment algorithm performs a prefix computation on the red intervals using the alignment operator $*$ defined below. This operator is defined on the set of pairs $(top, size)$ where $top$ contains the track to which the topmost red interval has been assigned and $size$ is the number of red intervals encountered.

$$
\begin{aligned}
(a * b).size &:= a.size + b.size \\
(a * b).top &:= \mathrm{MAX}(a.top, a.size + b.top)
\end{aligned}
$$

Here $+$ is arithmetic addition. It is straightforward to show that $*$ is an associative operation.

The variables are initialized as follows: The `rake()` function is used to remove the $X_i$ associated with green intervals. $X_i'.top$ is set to $X_i$ and $X_i'.size$ is set to 1. The size field is needed to make $*$ associative. When the prefix sum $S_i = X_i' * \ldots * X_1'$ is computed, we obtain an assignment of the red interval $b_i$ to the green interval $a_{S_i.top}$. Figure 5 shows the aligned assignments of Figure 4.

## 5.3  Link

*Link* converts align's assignments of red halfintervals (interval end points) to green halfintervals (interval start points) into a placement of intervals into tracks. We add pointers that link an interval's green halfinterval to its red halfinterval and obtain lists of intervals that have to be placed in one track. We use `invert()` and `distribute()` to update these pointers after the sorting step in the initialization phase. The start point of a track is an unassigned green interval which forms the head of such a list. Tracks are assigned as follows: A value 1 is written to all green intervals, and then a value 0 is written into every green interval $b_{S_i}$. Every unassigned green interval will contain a 1, and a prefix computation on all halfintervals with addition as the operator results in unique, contiguous track numbers for the list heads. The *list ranking* function described in Section 3.2 with the associative operator defined as $a * b := a$ is then used to distribute the track number to all halfintervals. `Invert()` and `distribute()` are used to communicate the track number of each red halfinterval to its interval.

## 5.4  Proof of Correctness

We now give two definitions. Let the *LUG* be the Largest Unassigned Green halfinterval in a set of interval assignments. A *mismatch* in an assignment occurs if there exists an unassigned green halfinterval $a_j$ and a red halfinterval $b_i$ with $b_i < a_j$ and $j < S_i$. Figure 6 shows a mismatch.

We examine the assignments given by our interval assignment algorithm. We show that one-to-one assignments and the absence of mismatches hold at every step of the align algorithm. These invariants give us a concise description of the behavior of this algorithm. From these invariants we shall deduce that align's assignments are dense at the LUG. Then we show that the associated channel routing is also dense at that location.

**Lemma 1** *The align algorithm does not assign two red halfintervals to one green halfinterval.*

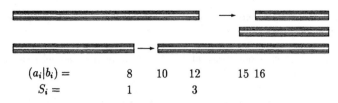

$$(a_i|b_i) = \quad\quad 8 \quad 10 \quad 12 \quad\quad 15\ 16$$
$$S_i = \quad\quad\quad 1 \quad\quad 3$$

Figure 6: A Mismatch

Proof: The values $\{S_i.top\}$ computed by align are strictly monotone. $\square$

**Lemma 2** *The align algorithm does not produce a mismatch.*

Proof: We proceed by induction on $n$, the number of red intervals assigned.

Assume there is one red interval $b_1$. The green interval $a_{X'_1.top}$ assigned to it does not create a mismatch because there is no smaller green interval that could be assigned to it, as can be seen by consulting the initialization algorithm.

Assume there are $n$ red intervals and that there is no mismatch for the first $n-1$ assigned red intervals. We have to show that the assignment of the $n$th interval to the green interval $S_n$ produces no mismatch. The assignment $S_n.top$ is equal to the maximum of $X'_n.top$ and $X'_n.size + S_{n-1}.top$. We assume that the $n$th assignment produces a mismatch and show that a contradiction results.

Case a) $S_n.top = X'_n.top$. The $n$th red interval remains assigned to its initial value $X'_n.top$. If there is a mismatch, either the red interval was initialized incorrectly or the mismatch occurs among the first $n-1$ red intervals, both of which produce a contradiction.

Case b) $S_n.top = X'_n.size + S_{n-1}.top = 1 + S_{n-1}.top$. The $n$th red interval is assigned the green interval one above the green interval assigned to the $n-1$st red interval. If a mismatch occurs, it is between the $n$th red interval and some green interval below the one assigned to the $n-1$st interval. But then there is a mismatch between the $n-1$st red interval and this green interval. $\square$

**Lemma 3** *The align algorithm leaves a LUG after all red halfintervals are assigned.*

Proof: Consider the green halfinterval $(a1, \infty)$. As $a_1 < a_i, i \in 2, \ldots, n$ and therefore $a_1 < b_i, i \in 1 \ldots n$, this green interval intersects with all red intervals. Thus there is at least one unassigned green halfinterval. $\square$

**Lemma 4** *The assignments done by the align algorithm are dense at LUG.*

Proof: Assume the assignments are not dense at LUG. The green intervals are sorted into ascending order. Therefore, the assignment is dense below the LUG. Thus, there must be a *hole* above the LUG or a track with no interval that intersects the x-coordinate of the LUG. A hole can be formed by the align algorithm in two ways:

Case a) The hole is formed by an unassigned green interval. But that green interval has a larger x-coordinate than the LUG, the largest unassigned green interval, a contradiction.

Case b) The hole is formed by a red interval with a smaller x-coordinate and (possibly) a green interval with a x-coordinate larger than the LUG. But this implies that the align algorithm has produced a mismatch, a contradiction. $\square$

**Lemma 5** *When the assignments are translated into a channel routing by the link algorithm the channel is dense at the LUG.*

Proof: Assume the channel is not dense at LUG. A hole in the channel routing can occur in two ways:

Case a) The hole occurs in a track which contains at least one interval to the right of the hole and none to the left. The first such interval must have been an unassigned green halfinterval produced by the align algorithm. But that green interval has a larger $x$-coordinate than the LUG, the largest unassigned green interval, a contradiction.

Case b) The hole occurs in a track which contains at least one interval to left of the hole. Before the link algorithm is applied there is a red halfinterval to the left of the hole. This creates a hole above the LUG produced the the align algorithm, which is a contradiction. □

## 5.5 Analysis of the Routing algorithm

Let there be $n$ intervals.

- The cost of making the initial assignment to halfintervals is the cost of sorting $2n$ integers plus a parallel prefix with addition as the associative operator. Time: $O(\log n)$, Processors: $O(n)$ on an EREW P-RAM.

- Align is a parallel prefix operation. Time: $O(\log n)$, Processors: $O(n/\log n)$ on an EREW P-RAM.

- Link is implemented by list ranking. Time: $O(\log n)$, Processors: $O(n/\log n)$ on an EREW P-RAM.

Our interval assignment algorithm is dominated by the sorting during the initialization phase.

**Theorem 3** *Channel routing of $n$ nets can be done in time $O(\log n)$ with $O(n)$ processors on a EREW P-RAM.*

To achieve the bounds of this theorem we have assumed that the sorting algorithm of [4] has been used. More practical algorithms add a factor of $\log n$ to the time.

## 5.6 Lower Bounds

With our channel routing algorithm we can solve the *INTERVAL OVERLAP* problem: *Determine whether $n$ intervals are disjoint.* One track is sufficient to route all nets if and only if there are no overlapping intervals. It has been shown that every comparison-based algorithm for *INTERVAL OVERLAP* requires $\Omega(n \log n)$ comparisons [3], [8], [23]. The processor-time product of our channel routing algorithm is optimal.

# 6 Interval Graph Coloring

We now show how our routing algorithm can find the minimum coloring of an interval graph. An undirected graph $G$ is an *interval graph* if there exists a one-to-one mapping from each vertex to an interval on the real line, and two intervals intersect if and only if there is an edge between their two vertices. Interval graphs are a subset of *perfect graphs*, which have the property that $\chi(G)$, the minimum coloring number of $G$, is equal to $\omega(G)$, the size of the maximum clique of $G$. The complement graphs $\bar{G} = (V, \bar{E})$, with $(v, w) \in E \Leftrightarrow (v, w) \notin \bar{E}$, of interval graphs form a subset of

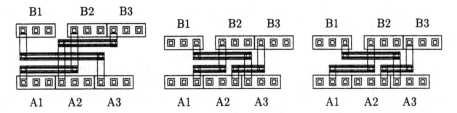

Figure 7: (a) Incorrect inefficient routing. (b) The nets are shortened. (c) The column conflicts are removed.

*comparability graphs*, which are also a subset of perfect graphs. A maximum clique in any graph is a maximum independent set in the complement graph. Perfect graphs, interval graphs, comparability graphs and their subsets have been studied extensively [10]. They are interesting because some of their properties can be computed efficiently while the corresponding properties of general graphs are $NP$-hard. So far, $NC$ algorithms have been found for maximum clique, minimum coloring in comparability graphs and maximum independent set and maximum matchings in interval and co-comparability graphs [13].

A minimum coloring of an interval graph is an assignment of vertices to a minimum number of colors so that two vertices of the same color have no edge between them. This corresponds to an assignment of intervals to a minimum number of tracks so that two intervals in the same track do not intersect. Our channel routing algorithm is therefore a minimum coloring algorithm. Note that an actual coloring is constructed here instead of just computing the coloring number $\chi(G)$.

**Corollary 1** *The minimum coloring and maximum clique of an n-node interval graph and the maximum independent set of a co-interval graph can be computed from the interval representation in time $O(\log n)$ with $O(n)$ processors on an EREW.*

The maximum clique is constructed as follows: We find the largest track start point. Property **A1** guarantees that the channel is dense at that point, therefore all intervals containing the smallest track start point form one of the maximum cliques in the interval graph. The interval representation of an interval graph can be constructed in time $O(\log^3 n)$ with $O(n^4)$ processors on a CREW [13]. Klein [16] has given efficient parallel algorithms for recognizing, finding a maximum clique, maximum independent set and optimal coloring of chordal graphs, of which interval graphs are a subset. His algorithm for finding an optimal coloring takes $O(\log^2 n)$ time using a $O(n + m)$ processor $CRCW$, where $m$ is the number of edges.

# 7    Routing Library Modules

We presented a channel routing algorithm that is in $NC$ but only works in restricted cases. We will now give an $NC$ extension that circumvents these restrictions in cell-based silicon compilers such as DeCo or Slap. This allows our channel router to produce significantly better layouts while using less computational resources that a general routing channel algorithm. When a circuit is built from standard library cells, there is usually some freedom in choosing the exact position of the terminals. Designers in the industrial world trade off slightly larger cell sizes to minimize area wasted in routing. For example, the terminals in the 3-micron CMOS cell library [12] are far apart and have some degree of freedom. A system that uses such cells can avoid column conflicts by moving terminals to unique locations. We will now describe a procedure that places terminals so that no column conflicts occur. The procedure also minimizes the horizontal length of each net, this results in shorter intervals and therefore decreases channel density.

*Example:* Consider the routing of Figure 7. We are given six modules $A1, \ldots, B3$. Each module has one terminal that can be placed in three adjacent positions. For the sake of legibility we assigned tracks so that the first two routings are legal. Our router could produce an illegal routing by interchanging track 1 and track 2. There exist cases where any track assignment would result in an illegal routing [17]. In Figure 7(a) each terminal was arbitrarily assigned to the first position. The resulting routing uses three tracks and has a conflict in column 1. In Figure 8(b) the number of tracks was reduced by shortening the horizontal segment of each net. The router can then pack more intervals into a track. There is now a conflict in column 3. The routing in Figure 7(c) is legal, the column conflict was removed by shifting one terminal.

Let us assume that a module can have any number of terminals, that the position of each module is fixed and modules can not overlap. Each terminal can be placed in two or more adjacent positions $(x, x+1 \ldots)$. We also assume that a net will be implemented by exactly one horizontal segment. Our procedure moves terminals to shorten horizontal segments. When this produces a column conflict, one of the terminals is moved to an adjacent position.

We have to avoid the following problem: The maximum terminal of a net and the minimum terminal of another have the same $x$-coordinate. We can either move the maximum terminal to the right or move the minimum terminal to the left. Either of the moves could cause another column conflict. The result might be a "ripple" of moves, at best serializing the procedure and at worst producing no solution at all. We can solve this problem simply by shortening the nets so each terminal can slide at least one position to the right. It follows that if any two terminals are at column $x$, there will be no terminal at $x + 1$. There will not be another column conflict if one terminal moves to $x + 1$.

The preprocessing to remove column conflicts starts by computing the nets of the connection graph $G$, if they are not given. Each terminal is assigned the largest value *minus 1* of its range. Then each terminal in parallel is labeled whether it is the (largest) minimum $x$-coordinate of a net. Each terminal is assigned the smallest value of its range. Then each terminal in parallel is labeled whether it is the (smallest) maximum $x$-coordinate of a net. If a terminal has the minimum $x$-coordinate of a net, it will attach to the net at its largest value $-1$. Otherwise, if the smallest value of a terminal is smaller than the net's minimum $x$-coordinate, the terminal's range must must overlap that coordinate. It can therefore attach at that minimum net coordinate. Otherwise, a terminal attaches to the net at its smallest value.

We sort the changed terminals of both sets. The sorting step is the most expensive part of the preproccessing. For every adjacent pair of two terminals we check whether they have the same $x$-coordinate. If so, we first check if they belong to the same net, and cause no column conflict. If the two terminals belong to different nets, we have to move one terminal out of the way to $x + 1$. Remember that there can be no terminals at $x + 1$. To reduce channel density, we move the upper terminal if it has the minimum $x$-coordinate of a net. Otherwise we move the lower terminal. Note that this operation shortens intervals to minimize channel density. Our algorithm then proceeds by converting nets to intervals, computes an optimal routing that achieves channel density, and produces a layout as before.

**Theorem 4** *Resolving column conflicts and minimizing channel density by allowing terminals to occupy one of several adjacent locations can be done in $O(\log n)$ time on a $O(n)$-processor EREW.*

# 8 Conclusions

We have demonstrated in this paper that massive parallelism is inherent in routing. While we had to impose some constraints on the underlying model, the algorithms are general enough to produce useful results. We believe our methods can be extended to show that other problems in VLSI design are in $NC$. The constants hidden in the $O()$ notation are very small. The algorithm and its extensions are composed of simple primitives, which can be tuned to run very fast on an actual shared-memory multiple-processor machine. This makes an implementation feasible, it will exhibit a speedup over any sequential version of the routing algorithm even for small $n$.

The authors wish to thank Dan Lopresti, Ernst Mayr and Otfried Schwarzkopf for their suggestions on this paper.

# References

[1] M. Ajtai, J. Komlos and E. Szemeredi, "An O(n log n) Sorting Network," in *15th Annual ACM Symposium on Theory of Computing*, pp. 1–9, 1983.

[2] K. E. Batcher and H. S. Stone, "Sorting Networks and Their Applications," *AFIPS Proc., Spring Joint Comput. Conf.*, vol. 32, pp. 307–314, 1968.

[3] M. Ben-Or, "Lower Bounds for Algebraic Computation Trees," in *15th Annual ACM Symposium on Theory of Computing*, pp. 80–86, May 1983.

[4] R. Cole, "Parallel Merge Sort," in *27th Annual Symposium on Foundations of Computer Science*, pp. 511–616, 1986.

[5] B. A. Dalio, "DeCo - A Hierarchical Device Compilation System," Dept. of Computer Science, Brown University, PhD Thesis CS-87-08, May 1987.

[6] B. A. Dalio and J. E. Savage, "DeCo - A Device Compilation System," in *International Workshop on Logic and Architecture Synthesis for Silicon Compilers*, May 1988.

[7] D. N. Deutsch, "A Dogleg Channel Router," in *Proc. 13th IEEE Design Automation Conf.*, pp. 425–433, 1976.

[8] D. Dobkin and R. Lipton, "On the Complexity of Computations under Varying Set of Primitives," *Journal of Computer and Systems Sciences*, vol. 18, pp. 86–91, 1979.

[9] D. Dolev, K. Karplus, A. Siegel, A. Strong and J. D. Ullman, "Optimal Wiring between Rectangles," in *13th Annual ACM Symposium on Theory of Computing*, pp. 312–317, 1981.

[10] M. Golumbic, *Algorithmic Graph Theory and Perfect Graphs*. New York, NY, Academic Press, 1980.

[11] A. Hashimoto and J. Stevens, "Wire Routing by Optimizing Channel Assignments within Large Apertures," in *Proc. 6th IEEE Design Automation Conf.*, pp. 155–163, 1971.

[12] D. V. Heinbuch, *CMOS3 Cell Library*. Reading, MA, Addison Wesley, 1988.

[13] D. Helmbold and E. Mayr, "Applications of Parallel Scheduling to Perfect Graphs," Stanford University, , 1986.

[14] D. S. Hirschberg, A. K. Chandra and D. V. Sarvate, "Computing Connected Components on a Computer," *CACM*, vol. 22, pp. 461–464, 1979.

[15] R. M. Karp and V. Ramachandran, "A Survey of Parallel Algorithms for Shared-Memory Machines," in *Handbook of Theoretical Computer Science*. North-Holland, 1988, preprint.

[16] P. N. Klein, "Efficient Parallel Algorithms for Chordal Graphs," in *29th Annual Symposium on Foundations of Computer Science*, to appear, 1988.

[17] E. S. Kuh and T. Yoshimura, "Efficient Algorithms for Channel Routing," *IEEE Trans. Computer-Aided Design*, vol. CAD-1, no. 1, pp. 25–35, Jan. 1982.

[18] R. E. Ladner and M. J. Fischer, "Parallel Prefix Computation," *JACM*, vol. 27, pp. 831–838, 1980

[19] A. S. Lapaugh, "Algorithms for Integrated Circuit Layout: an Analytic Approach," Dept. of Electrical Engineering and Computer Science, M.I.T., PhD Thesis, 1980.

[20] T. Leighton, C. E. Leiserson, B. Maggs, S. Plotkin and J. Wein, "Advanced Parallel and VLSI Computation," Dept. of Electrical Engineering and Computer Science, M.I.T., MIT/LCS/RSS 2 Mar. 1988.

[21] T. Leighton, C. E. Leiserson, B. Maggs, S. Plotkin and J. Wein, "Theory of Parallel and VLSI Computation," Dept. of Electrical Engineering and Computer Science, M.I.T., MIT/LCS/RSS 1 Mar. 1988.

[22] T. Leighton, "Tight Bounds on the Complexity of Parallel Sorting," in *16th Annual ACM Symposium on Theory of Computing*, pp. 71–80, 1984.

[23] F. P. Preparata and M. I. Shamos, *Computational Geometry*. Springer-Verlag New York Inc., 1985.

[24] J. Reed, A. Sangiovanni-Vincentelli and M. Santomauro, "A New Symbolic Channel Router: YACR2," *IEEE Trans. Computer-Aided Design*, vol. CAD-4, no. 3, pp. 208–219, July 1985.

[25] S. P. Reiss and J. E. Savage, "SLAP - A Methodology for Silicon Layout," in *Procs. Intl. Conf. on Circuits and Computers*, pp. 281–285, 1982.

[26] R. L. Rivest and C. M. Fiduccia, "A "Greedy" Channel Router," in *Proc. 19th IEEE Design Automation Conf.*, pp. 418–424, 1982.

[27] M. Sarrafzadeh, "Channel-Routing Problem in the Knock-Knee Mode is NP-Complete," *IEEE Trans. Computer-Aided Design*, vol. CAD-6, no. 4, pp. 503–506, July 1987.

[28] J. E. Savage, "Heuristics in the SLAP Layout System," in *IEEE Intl. Conf. On Computer Design*, Rye, New York, pp. 637–640, 1983.

[29] J. E. Savage, "Three VLSI Compilation Techniques: PLA's, Weinberger Arrays, and SLAP, A New Silicon Layout Program," in *Algorithmically-Specialized Computers*. Academic Press, 1983.

[30] J. E. Savage, "Heuristics for Level Graph Embeddings," in *9th Intl. Workshop on Graphtheoretic Concepts in Computer Science* , pp. 307–318, June 1983.

[31] T. G. Szymanski, "Dogleg Channel Routing is NP-Complete," *IEEE Trans. Computer-Aided Design*, vol. CAD-4, no. 1, pp. 31–40, Jan. 1985.

[32] M. R. Zargham, "Parallel Channel Routing," in *Proc. 25th IEEE Design Automation Conf.*, pp. 128–133, 1988.

# APPLICATION OF GRAPH THEORY
# TO TOPOLOGY GENERATION FOR LOGIC GATES

*Hubert Kaeslin*

Integrated Systems Laboratory
Swiss Federal Institute of Technology
8092 Zürich, Switzerland

## *ABSTRACT*

A procedure is described which given a Boolean equation of the AOI-type will generate an area-efficient layout topology for an electronic circuit implementing this function in MOS technology. n- or p-Transistors are assumed to be arranged in two horizontal rows with their gates formed by vertical polysilicon wires crossing the cell. The topology generation problem essentially consists in interconnecting these transistors using metal wires. The proposed procedure makes use of several graphs to reflect transistor network connectivity, potential conflicts between metal wires, potential embrace situations among them, and sharing of tracks between them. Classical graph algorithms, such as colouring and critical path analysis, serve to resolve the conflicts and to assign every wire a legal location.

## 1. Introduction†

One approach to VLSI design is to build a circuit from a set of previously designed and documented cells. Such a standard cell library includes a few basic latches and flip-flops and a limited number of combinational functions. In MOS technology scalar Boolean functions are obtained by generalizing the inverter circuit of Fig.1a to that of Fig.1b.

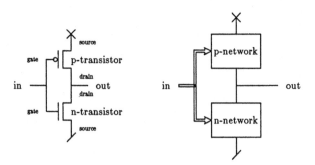

Fig.1 Schematics of CMOS inverter a. and composite logic gate b.

---

† For readers totally unfamiliar with microelectronics a simplified but sufficient summary of concepts relevant to this paper is provided as an appendix.

In the CMOS inverter we find two transistors acting as voltage controlled switches. If the input voltage is high the n-transistor is on and the p-transistor off thus connecting the output terminal to ground and yielding a low signal. For a low input one obtains the complementary situation with a high output.

In the composite circuit of Fig.1b the two transistors are substituted by a pair of dual transistor networks. In these networks there is a pair of complementary transistors having a common gate electrode for each input to the combinational function. Due to the duality of the networks when one of them is conducting the other one is not. Therefore, for any input combination there exists a conducting path from the output terminal to either one of the supply terminals but no conducting path between the supply terminals. A logic gate of this structure is called a composite gate. If the networks are restricted to the series-parallel type the logic functions become arbitrarily nested expressions of AND and OR operators with one negation at the outermost level. Equation (1) gives an example:

$$Z = \overline{A \vee (B \wedge ((C \wedge D) \vee (E \wedge F)))} \tag{1}$$

The family of logic gates implementing such functions are called functional arrays or AOI-gates‡. A schematic for a logic circuit implementing equation (1) is shown in Fig.2a.

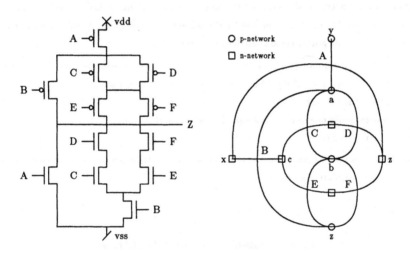

Fig.2 Schematic of AOI-gate for equation (1) a. and its pair of dual network graphs b.

Neglecting the gate electrodes n- and p-networks can obviously be drawn as graphs. In such a *network graph* transistors are represented as edges, electrical nodes as vertices. A static CMOS gate will require a pair of dual graphs to represent both of its networks, see Fig.2b.

The subsequent table is taken from [13] and lists the number of AOI-functions that can be constructed without exceeding a given number of transistors connected in series.

‡ AOI stands for And-Or-Invert and is understood to include OAI functions and functions with deeper nestings such as the one described by (1).

| maximum number of series transistors | total number of AOI-functions |
|:---:|:---:|
| 1 | 1 |
| 2 | 7 |
| 3 | 87 |
| 4 | 3521 |

Because of the effort involved in manually finding a gates topology and in designing its layout any library of standard cells will necessarily be restricted to a few dozens of combinational gates. Standard cell libraries hardly ever provide much more than the first 87 gates. An automatic tool, however, could generate gates "on the fly" as they are needed in the design process of an integrated circuit. It has been experimentally found, that the area occupied by combinational circuitry decreases with the number of available library functions [8].

There is a second motivation for automating cell generation which is even more important in an automatic design environment. While implementing the same function a logic gate may have many topological variants differing in their arrangement of inputs and the location of the output. A higher level tool can specify the one variant that best fits the needs imposed by the context. The cell generator will then construct the cell accordingly. Such module generators are described in [7, 10].

For this paper we will therefore proceed from a fixed ordering of a cells connectors, i.e. from a given ordered set of inputs, output, and gaps, if any. An acceptable connector sequence must
- cover all edges of the network
- allow for identical walks or sets of partial walks in both network graphs
- include the output connector

$A\ B\ Z\ D\ C\ E\ F$ is a valid sequence for the example of equation (1) where the gap between the two partial walks is used to accomodate the output.

## 2. Problem

| | |
|---:|:---|
| Given: | a Boolean expression of the AOI-type |
| Find: | layout topology (on a virtual grid) |
| | of the circuit implementing this logic function |
| Goal: | minimize cell area |
| Constraint: | connectors must be arranged as specified |

## 3. Assumptions made

a. MOS technology (static CMOS, dynamic CMOS, NMOS). Only static CMOS is covered in this paper since this is the most difficult case. Dual networks need to be generated that have identical connector sequences for the n- and on the p-side.

b. Transistors are assumed to be arranged in two horizontal rows, one for n-channel transistors a second for p-channel ones. Their gates are formed and interconnected by vertical polysilicon wires crossing the cell.

c. All interconnections of source and drain electrodes are done in metal. Wiring of supply connections in diffusion could easily be taken into account also (by simply omitting supply vertices in the conflict and embrace graphs).

Assumptions b) and c) imply that source and drain electrodes contacted by metal alternate with columns occupied by a poly wire for the cells output or inputs. See Fig.7 for an example of a cells topology.

No assumptions are made on where the output connector should be placed. Its location can be specified to be between any two input connectors or to the left or right of all of them. In fact our procedure can even generate cells with multiple output wires.

It should be noted that if a pair of Euler lines, closed or open, having identical edge sequences exist in the dual network graphs, all transistors of one network can be chained in one contiguous diffusion strip. The horizontal size of the cell is then guaranteed to be minimal. If no dual Euler lines exist one or more gaps will be needed. In this case minimizing the horizontal size corresponds to finding a minimum complete set of dual walks, i.e. a set of dual walks which covers all edges with a minimum number of partial walks.

Even without considering logically equivalent permutations of subgraphs this problem has been found to be np-hard for general planar graphs [14], complexity for the subclass of series-parallel graphs remains an open question, however. Both exhaustive and heuristic algorithms for synthesizing dual graphs with a minimum number of partial walks can be found in [11, 12, 9].

## 4. Terms

The signatures used to represent layout topology are shown in Fig.3 Note that a transistor forms wherever a polysilicon wire crosses diffusion.

A few terms used throughout this paper need to be introduced:

| | |
|---|---|
| Wire: | straight piece of conductor |
| Yoke: | wires forming a U-shape |
| Net: | set of electrically connected wires and yokes |
| Conflict: | yokes of different nets intersecting (on the same and only metal layer) |
| Nesting: | one yoke embracing a second one |
| Gap: | gap in the diffusion strip (for electrical separation or feedthrough of poly wire) |

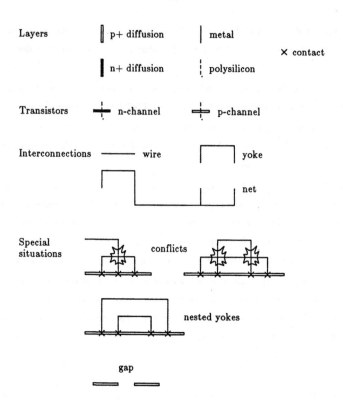

Fig.3 Explanation of signatures and terms

## 5. Solution

Our procedure for topology generation is based on graph theory and makes use of several graphs with very different interpretations. A first phase is concerned with generating the topology for one half-cell and is repeated twice, once for the n-network and once for the p-network. In a second phase the two half-cells are assembled to obtain the cells topology.

In what follows we will only describe how to determine the <u>vertical</u> location of yokes. Finding their location along the horizontal dimension and placing of transistors is straightforward and will therefore not be addressed.

### 5.1. Half-cell construction

In the network graph vertices of degree 2 reflect two transistors in series. In the layout such transistors can be made adjacent on the same diffusion strip thus requiring no metal wire to connect them. Terminal vertices and vertices of higher degree require the use of one or more metal yokes in order to interconnect all transistors incident with that electrical node. The problem consists in arranging these yokes such as to avoid the conflict situations shown Fig.3.

Lets assume for the moment we had available two graphs reflecting all potential conflict and embrace situations respectively. Later we will present an algorithm for simultaneously constructing them both. In both graphs a vertex stands for a yoke in the layout, see Fig.6. It should also be noted that these graphs need not be connected.

### 5.1.1. Resolving conflict situations

In the *conflict graph* two vertices are connected by an undirected edge if their respective yokes potentially could intersect (Fig.4 top). A short circuit can only be avoided by placing the two yokes on opposite sides of the diffusion strip. Thus in the graph we must assign different side labels to vertices connected by an edge. This is the classical graph colouring problem. Luckily using only two colours in our application makes it solvable in polynomial time [4].

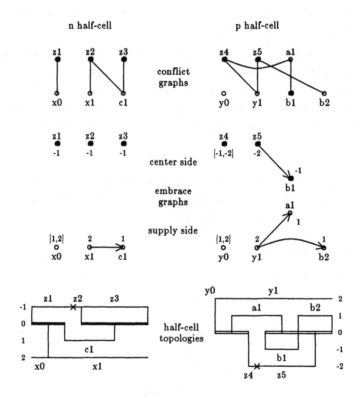

Fig.4 Topology generation from conflict and embrace graphs
(poly layer and metal-diffusion contacts omitted)

By colouring the graph a first crude decision is made. All yokes assigned one colour, say white, will be placed towards the border of the cell i.e. towards the supply rail; the ones assigned the other colour, say black, will end up towards the center. There is, by the way, an initialization step before actually starting the colouring process, in which all supply vertices are precoloured white to make sure their yokes will be placed on the periphery of the cell.

### 5.1.2. Resolving embrace situations

The *embrace graph* is a directed acyclic graph, in which an edge leads from a first vertex to a second vertex if the first yoke potentially could embrace the second yoke. However, by colouring the conflict graph we have obtained two sets of yokes and we therefore need <u>two</u> graphs to reflect potential nestings among yokes (Fig.4 center). This is taken into account by splitting the initial embrace graph (Fig.6 bottom) according to the colouring found for the conflict graph. Remember that there is a one to one correspondence between conflict and embrace graph vertices. Edges connecting vertices of different colours will get lost in the process, but since yokes placed on different sides of diffusion will never be able to embrace one another, they have lost their significance anyway.

In order to avoid intersections all yokes of the same colour must be arranged in a way to place embraced ones closer to diffusion than embracing ones. This is achieved by solving the critical path problem on the embrace graph, which involves topological sorting of the vertices and determining their earliest and latest event times [1]. All edges are assumed to have identical weights for this purpose.

After the critical path analysis is carried out each yoke is assigned a track number, according to the position of its associated vertex on the critical path. The last vertices on the critical path result in positions on the track adjacent to diffusion, the first vertices yield positions on the outermost track. Vertices off the critical path leave a certain interval to choose from. In Fig.4 the numbers associated with each vertex reflect all possible track assignements for their respective yokes. Practical considerations help to choose among alternatives.

By now vertical locations have been found for all yokes by solving one graph colouring problem and by carrying out two critical path analyses. They determine, together with the horizontal locations, the topology of one half-cell on a virtual grid (Fig.4 bottom). We will now return to

### 5.1.3. Constructing conflict and embrace graphs

For this purpose we have found an elegant solution based on a simple data structure that we will refer to as chain.

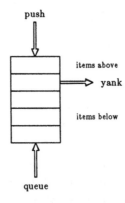

Fig.5 The chain and its operations

A chain is an ordered set of items, implemented as a linked list. Items may be entered into the chain from its top end, which operation corresponds to the push operation of a stack. The chain will also accept items to be entered from its bottom end, analogous to storing data on a queue. When withdrawing an item the item is searched for in the chain. If found the item is yanked from any position within the chain and the resulting gap is closed. Not the item removed is important, rather the two sets of items that were are above and below it are returned for use in further computation. Therefore the chain is neither a stack nor a queue nor both. The chain serves to keep track of the yokes paralleling the diffusion strip at a given transistor location. The ordering of the yokes within the chain reflects their mutual relationships in accessing the diffusion strip.

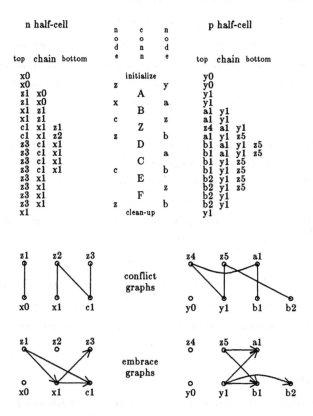

Fig.6 Evolution of the chain and construction of conflict and embrace graphs

Constructing conflict and embrace graphs is carried out while walking the network graph as specified by the connector sequence. Such a walk corresponds to sweeping over the cell topology from left to right, where vertices in the graph stand for metal columns and edges for poly columns. At every column the conflict and embrace graphs are updated by adding edges according to the situation found on the chain.

The algorithm works as follows:

0. Initialize the chain with a first vertex which will represent the leftmost supply yoke.

1. At a **metal column** register which vertex (= electrical node or net) is hit (= to be contacted). Then carry out three steps:

1a.  Search the chain for a yoke belonging to the net hit.

If such a yoke is found this means it has to be brought in to contact the diffusion strip. In doing so its vertical wire potentially could intersect with any other yoke that was pushed onto the chain later. In the conflict graph undirected edges are therefore drawn between the vertex representing the actual yoke and all vertices located above it in the chain.

As far as the nesting of yokes is concerned one finds the complementary situation. Vertices further down on the chain stand for yokes that potentially could embrace the actual yoke. In the algorithm this is expressed by adding directed edges to the embrace graph.

1b.  Yank the actual yoke from the chain if it was found there.

1c.  If the net is to be continued further towards the right push a new yoke onto the chain.

2.  Move to the adjacent **poly column** and check whether it has an edge (= transistor and input connector) associated with it in the network graph, or whether it is the specified location for a gap or the cells output connector. In the latter case carry out three steps:

2a.  Search the chain for a yoke belonging to the output net.

If such a yoke is found it has to be brought away from diffusion to the innermost track of the half-cell. In doing so its vertical wire potentially could intersect with any other yoke that was pushed onto the chain earlier. Therefore in the conflict graph undirected edges are added between the vertex representing the actual yoke and all vertices located below it in the chain.

Conversely the output yoke could embrace all other yokes further up in the chain which is reflected by adding directed edges to the embrace graph.

2b.  Yank the output [half-]yoke from the chain if it was found there.

2c.  If the output net is to be continued further towards the right queue a new [half-]yoke onto the chain.

3.  Continue with step 1 for the next metal column until the cell has been fully swept.

The evolution of the chain during execution of the algorithm and the resulting graph constructions are documented by Fig.6.

## 5.2. Assembly of half-cells

Assembling the half-cells is accomplished by simply stacking the p-topology on top of the n-topology. There are situations, however, where one track can be shared among the two half-cells thus reducing cell height. Fig.7 shows such an example. Merging the innermost tracks of both half-cells is possible if no two electrically disconnected yokes are shorted. By the way, the only node both half-cell networks have in common is the output node.

Again this problem can be solved by graph theory using the concept of interval graphs this time [5, 2]. Every yoke defines some interval along the horizontal dimension. In an interval graph each interval is represented by a vertex. Two vertices are connected by an undirected edge if their respective intervals form a non-empty intersection.

What we will refer to as *merge graph* basically is the interval graph defined by the innermost yokes of both half-cells. We have to admit one exception, however, stating that two electrically connected yokes will never be connected by an edge. In principle the chromatic number of the merge graph then determines whether track sharing is possible (=1) or not (=2). Luckily there is no need to actually colour the merge graph. Since by construction its chromatic number can only assume the values 1 or 2, the presence of any edge means track sharing is not possible.

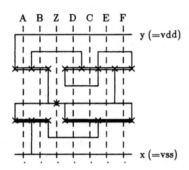

Fig.7 Final layout topology of the AOI-gate for equation (1)

## 6. Implementation

Fig.8 shows all steps in correct order.

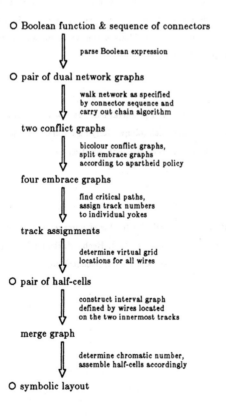

The topology generation procedure and all required graph algorithms have been programmed in Common Lisp and run on a Symbolics 3670. We make use of the Flavor system for object oriented programming.

Runtime for generating the example of Fig. 7 is approximately 7.7 s without any particular attention payed to optimizing the code. A more complex example with 32 transistors roughly took 19.5 s. Asymptotic complexity of the algorithm is not adressed in this paper since for electrical reasons the maximum number of series transistors is usually limited to fairly small numbers such as 5 or 6.

## 7. Future work

There are a few peculiar situations and problems that have not yet been studied in detail.

- Chromatic number of the conflict graph is higher than two

  According to Königs theorem [1] this case occurs whenever a conflict graph contains circuits of odd length. The set of yokes can then not be partitioned into a first subset to be routed above and a second subset to be routed below diffusion. More degrees of freedom in the placement of yokes are needed. Subdividing conflicting yokes into smaller ones, thus allowing a net to cross over diffusion without contacting it, may help in some cases, adding extra gaps may help in others. The general idea is to get rid of circuits of odd length by replacing some of their vertices by pairs of new ones before recomputing both conflict and embrace graphs.

- Conflict graph is not uniquely colourable

  The conflict graph may be coloured in different ways, if it consists of more than one component, and if there are components which do not contain any precoloured vertices. In some cases alternative colourings result in cell topologies of differing heights. There seems to be no way to find a colouring minimizing cell height that could do without backtracking.

## 8. Conclusions

Graph theory provided us with versatile and powerful concepts for expressing and solving constraints imposed by network connectivity and processing technology. Relying on classical methods from graph theory we were able to propose an algorithmic, and thus very efficient, solution to a problem, which others attack with expert systems [3, 6].

## 9. Acknowledgement

The autor wishes to thank Prof. Dr. W. Fichtner for his encouragement and for several helpful suggestions on this paper. He is also indebted to his colleagues H.R. Heeb, M. Raths, S. Seda, and H.P. Wachter for useful discussions and comments on the procedure and its implementation.

# Appendix: A quick and very simplistic introduction to MOS layout design

To design the layout of an integrated circuit means to transform circuit diagrams into multilayer structures amenable to realization in silicon.

The rules specifying the requirements for legal structures can be separated into two sets. A first category describes the topological restrictions by specifying which layers may cross without interaction, which layers may be interconnected using contacts, and which layers interact to form a transistor. A second category, usually termed "design rules", specifies all geometrical requirements imposed by the fabrication process. It is obvious, for example, that there are minimum requirements for the width of wires and the space to respect between them. Topological rules are very much the same for a large family of processes while geometrical rules vary greatly from one process to another.

*Symbolic layout* is a useful abstraction which reflects the topology of a layout but ignores its actual dimensions. Fig. 7 gives an example. Our paper aims at automatically constructing symbolic layout for logic gates, a specific but important category of digital circuits. Once topology is constructed correct sizing and spacing can be achieved automatically in a "compaction" process for which computer tools are currently available.

In most of todays MOS (Metal Oxide Silicon) technologies three conducting layers are available namely metal, poly[silicon], and diffusion. See Fig. 3. Contacts can connect a metal wire to any of the other layers. Contacts are, however, not allowed between diffusion and poly. A transistor forms where a poly wire crosses a diffusion strip. The poly wire is then called "gate", the resulting two diffusion terminals are termed "source" and "drain". The thin [silicon] oxide layer insulating polysilicon from diffusion allows the electrical field emanating from the gate to influence conductivity of the channel between drain and source, thus forming the basis for constructing an electronic switching device.

In CMOS (Complementary MOS) technology diffusion comes in two flavours called n-diffusion and p-diffusion. The resulting transistors show complementary switching behaviours. For conducting the n-channel transistor requires a positive voltage to be applied to its gate with respect to its source terminal, while the p-channel type wants a negative voltage. In static CMOS circuits n- and p-channel transistors are always used in pairs as elements of two strictly dual switching networks.

# References

1. Narsingh Deo, *Graph Theory with Applications to Engineering and Computer Science*, Prentice-Hall Inc., Englewood Cliffs NJ, 1974.

2. Peter C. Fishburn, *Interval Orders and Interval Graphs*, John Wiley & Sons, New York, 1985.

3. Daniel Gajski, "Intelligent Silicon Compilation," in *Advanced Summer Course on Logic Synthesis and Silicon Compilation for VLSI Design*, L'Aquila, 1987.

4. Michael R. Garey and David S. Johnson, *Computers and Intractability, a Guide to the Theory of NP-Completeness*, W. H. Freeman & Co., New York, 1979.

5. Michel Gondran and Michel Minoux, *Graphs and Algorithms*, John Wiley & Sons, Chichester, 1984.

6. Hansruedi Heeb, "A Rule-Based System for Polycell Generation," in *Fast-Protyping of VLSI*, ed. G. Saucier, E. Read, and J. Trilhe, pp. 109-116, Elsevier Science Publishers, Amsterdam, 1987.

7. Hansruedi Heeb and Wolfgang Fichtner, "GRAPES: A Module Generator Based on Graph-planarity," in *Proc. of the Intl. Conf. on Computer-Aided Design*, pp. 428-431, IEEE, Santa Clara CA, 1987.

8.  K. Keutzer, K. Kolwicz, and M. Lega, "Impact of Library Size on the Quality of Automated Synthesis," in *Proc. of the Intl. Conf. on Computer-Aided Design*, pp. 120-123, IEEE, Santa Clara CA, 1987.

9.  Yong-Joon Kwon and Chong-Min Kyung, "A Fast Heuristics for Optimal CMOS Layout Functional Cell Layout Generation," in *Proc. of the IEEE International Symposium on Circuits and Systems*, pp. 2423-2426, IEEE, Helsinki, 1988.

10. Y-L. Steve Lin, Daniel Gajski, and Haruyuki Tago, "A Flexible-Cell Approach for Module Generation," in *Proc. of the Custom Integrated Circuits Conf.*, pp. 9-12, IEEE, 1987.

11. Robert L. Maziasz and John P. Hayes, "Layout Optimization of CMOS Functional Cells," in *Proc. of the 24th ACM/IEEE Design Automation Conference*, pp. 544-551, IEEE, Miami Beach FL, 1987.

12. Ravi Nair and Anni Bruss, "Linear Time Algorithms for Optimal CMOS Layout," in *Algorithms and Architectures*, ed. P. Bertolazzi and F. Luccio, pp. 327-338, Elsevier Science Publishers, Amsterdam, 1985.

13. Takao Uehara and William M. vanCleemput, "Optimal Layout of CMOS Functional Arrays," *Transactions on Computers*, vol. C-30, no. 5, pp. 305-312, IEEE, May 1981.

14. Shuichi Ueno, Katsufumi Tsuji, and Yoji Kajitani, "On Dual Eulerian Paths and Circuits in Plane Graphs," in *Proc. of the IEEE International Symposium on Circuits and Systems*, pp. 1835-1838, IEEE, Helsinki, 1988.

# On the Estimate of the Size of a Directed Graph

A.Marchetti-Spaccamela

Dept. of Mathematics

University of L'Aquila

via Roma

67100 L'Aquila, Italy

**Abstract.** Given a directed graph and a source node $x_0$ we want to know the number of nodes that are connected to $x_0$, without searching the whole graph. We give biased and unbiased estimators extending previous results by Knuth and Pitt.

## 1 Introduction

Knuth ([HK65] and [K75]) showed that the number of nodes of a tree can be estimated using a random walk in the tree. Recently, Pitt [P87] has extended Knuth's estimator to the case of a rooted directed acyclic graph. In this paper we consider the more general problem of estimating the number of nodes of a directed graph that are connected to a given node.

Knuth was motivated by the problem of estimating and comparing the efficiency of backtrack-search procedures without actually running the programs [RND77]. Another application is given in [M88] where it has been shown that the problem of estimating the size of a graph can be used to estimate the size of the answers to a query in a deductive database.

In section 2 we give the basic definitions and the notation used. In section 3 we characterize the number of nodes in a directed graph that are connected to a given source node by proposing a procedure Estimate that finds an unbiased estimator for this number.

Unfortunately the termination of the procedure is not guaranteed; in section 4 we show how to modify Estimate to obtain biased estimators whose termination is guaranteed. Finally, in section 5, we extend the procedures to estimate the size of a subset of the nodes or the number of arcs that can be traversed with a path that starts from the source.

---

work partially supported by project MPI "Progetto e analisi di algoritmi".

## 2 Notations and basic definitions

A *directed graph* G=(N, A) is defined by a set N of nodes and a set A of arcs. An arc is an ordered pair (x,y) where x and y are nodes (we assume $x \neq y$). A *path* between $x_0$ and $x_m$ is a sequence of nodes < $x_0, x_1, .... x_m$> such that for each i=1,2,...m, the pair $(x_{i-1}, x_i)$ belongs to A. A *cycle* is a path such that the first and the last node coincide. A path is *acyclic* if it does not contain a subpath that is a cycle. Node x is *adjacent* to node y if (x,y) belongs to A. Node x is *connected* to node y if there is a path between x and y. A *dag* is a directed connected graph with no cycles. A dag is *rooted* if there is a node s (the root) such that there is a path between s and any other node. A *tree* is a rooted dag such that for any two nodes x and y there is at most one path between x and y. The *outdegree* out(x) of node x is the number of arcs (x,y) where y is any other node. The *indegree* in(x) of node x is the number of arcs (y,x) where y is any other node. Given a graph G=(N, A) the *reverse graph* of G is the graph rev(G)= (N, A') where A' is defined as follows: (x,y) belongs to A' if and only if (y,x) belongs to A. Clearly if x is adjacent to y in G, then y is adjacent to x in rev(G). Given a path P=<$x_0,x_1,x_2,.....x_m$> let $P_i$ be the subpath of P finishing at $x_i$ and let $\Pi_{out}(P_i)$ be the product of the outdegrees of all nodes belonging to $P_i$ with the exception of the last node $x_i$ of $P_i$; analogously let $\Pi_{in}(P_i)$ be the product of the indegrees of the nodes belonging to $P_i$ with the exception of the first node $x_0$.

An *estimator* of the value of a parameter is an algorithm that outputs a number that represents the estimated value of the parameter. A *Monte Carlo estimator* is an estimator that, during its execution, performs random choices. Clearly, by executing a Monte Carlo estimator several times with the same input, we generally obtain different outputs depending on the different choices performed. Let c be the set of all possible sequences of random choices. For a given input let m(C) be the output of the algorithm when C is the sequence of performed choices and let Pr(C) be the probability that C is chosen. The expected value of the estimate is E(m) = $\Sigma_{\forall C \varepsilon c}$ E(m(C)) Pr(C). A Monte Carlo estimator is *unbiased* if, for all possible inputs, the expected value of the estimator coincides with the value to be estimated; an estimator is *biased* if it is not unbiased.

Unbiased estimators for the number of nodes of special classes of graphs have been given: Knuth [K75] has given an unbiased estimator for the number of nodes of a tree; recently, Pitt [P87] has extended Knuth's algorithm and has obtained an unbiased estimator for the number of nodes of a rooted dag.

For clarity sake it is useful to present Knuth's unbiased estimator for the number of nodes of a tree. The algorithm starts from the root of the tree and finds a path to a leaf as follows: at step i, after having obtained a path $P_i$=< $x_0, x_1, ..., x_i$>, the algorithm chooses a randomly a node among the nodes adjacent to $x_i$. The estimates is given by the sum of the products of the outdegrees of the nodes belonging to $P_i$, for all i.

procedure Estimate_Tree

input: a tree G with root $x_0$;

output: an estimate of the number of nodes connected to $x_0$;

1. $y := x_0$;   $i := 0$;

2. while out($x_i$) <> 0

   do  begin

          let $P_i = < x_0, x_1, ..., x_i>$ be the path obtained in so far;

          choose randomly a node y such that $(x_i, y)$ is an arc;

          $i := i + 1$;

          $x_i := y$

     end;

3.  if $i > 0$

   then  Est $:= \Sigma_{\forall j < i} \Pi$ out$(P_j)$ ;

   else  Est $:= 0$.


Fig. 1 Estimate_tree.


Theorem 1 [K75] Estimate_tree is an unbiased estimator of the number of nodes that are connected to $x_0$.


## 3   Estimating the number of nodes in a graph

We now consider the problem of estimating the number of nodes of a directed graph that are connected to an initial source node $x_0$.

Procedure Estimate recieves in input a directed graph and a source node $x_0$ and gives in output an unbiased estimate of the number of nodes x connected to $x_0$. The procedure uses a procedure Random_Path that, given a graph G and a source node $x_0$ returns an acyclic path P starting from $x_0$. Initially Random_Path chooses a node $x_1$ adjacent to $x_0$, obtaining the path P= $<x_0, x_1>$. Afterwards, when the i-th iteration of the loop is completed the path P= $<x_0, x_1, ..., x_i>$ has been obtained; in order to add a new node to P the procedure randomly chooses a node $x_{i+1}$ adjacent to $x_i$. If either $x_i$ has outdegree equal to zero or $x_{i+1}$ is a node already belonging to P, then the procedure stops returning P. Otherwise $x_{i+1}$ is added to P. Procedure Estimate first calls for Random_path, and then computes for each node belonging to the returned path P two values a(i) and b(i).  b(i) is given by the product of the outdegrees over the product of the indegrees of the subpath of P finishing at $x_i$ and  c(i) is given by the number of calls to Random_Path necessary to find a path from $x_i$ to the source node $x_0$ in rev(G). The estimate is given by the sum of the products of b(i) and c(i), over all nodes $x_i$ belonging to P.

Procedure Random_Path (G, $x_0$ , P)

input: a directed graph G and a source node $x_0$;

output: an acyclic path P starting from $x_0$ ;

var flag : boolean;

1. flag := false; i := 0; P := $<x_0>$;

2. while (out($x_i$) <> 0) and not flag

   do begin

         let P= $<x_0, x_1, x_2,....x_i>$ be the path obtained in so far;

         randomly choose a node $x_{i+1}$ among the nodes connected to $x_i$;

         if ( $x_{i+1} = x_j$, for some $j \le i$)

          then flag := true

         else begin

             add $x_{i+1}$ to P;

             i := i + 1

         end

    end.

Fig. 2 Procedure Random_Path.


Procedure Estimate

input: a graph G and a source node $x_0$;

output: an estimate Est of the number of nodes connected to $x_0$;

1. Random path (G, $x_0$ , P);

2. For each node $x_i$ belonging to P , i > 0, do

   begin

      b(i) := $\Pi$ out($P_{i-1}$) / $\Pi$ in(P)

      c(i) := 0;

     2.1 repeat

          c(i) := c(i) + 1

          Random Path( rev(G), $x_i$, Q)

       until ( $x_0$ belongs to Q);

  end

3. if there is at least a node in P <> $x_0$ then Est := $\Sigma_{\forall i}$ b(i) c(i) else Est := 0.


Fig. 3 Procedure Estimate.


    Let p be the probability that the procedure Random Path applied to the reverse graph rev(G) will find an acyclic path between $x_i$ and the source node $x_0$.

<u>Lemma 1</u>  If $x_i$ is connected to $x_0$ then $E(b(i)) = p$.

<u>Proof</u>  Let $\mathcal{P}$ be the set of all acyclic paths starting from $x_0$ and finishing at $x_i$. Let $R(i)$ be an acyclic path between $x_0$ and $x_i$ and let $Pr(R(i))$ be the probability that the procedure Random Path returns a path P whose initial segment is $R(i)$. Clearly this probability is given by the reciprocal of the product of the outdegrees of the nodes of $R(i)$:

$$Pr(R(i)) = 1 / \Pi_{out}(R(i))$$

In order to evalute $E(b(i))$ we observe that, if $x_i$ belongs to the path P returned by the procedure Random_Path at step 1, then $b(i)$ is

$$\Pi_{out}(R(i)) / \Pi_{in}(R(i))$$

This implies that

$$E(b(i)) = \Sigma_{\forall \text{ acyclic paths P}} (\Pi_{out}(P_i) / \Pi_{in}(P_i)) \; Pr \; (P)$$
$$\text{passing through } x_i$$
$$= \Sigma_{\forall R(i) \, \varepsilon \, \mathcal{P}} \; \Sigma_{\forall \text{ acyclic paths P}} \; (\Pi_{out}(P_i) / \Pi_{in}(P_i)) \; Pr \; (P)$$
$$\text{with initial part } R(i)$$
$$= \Sigma_{\forall R(i) \, \varepsilon \, \mathcal{P}} \; (\Pi_{out}(R(i)) / \Pi_{in}(R(i))) \; Pr \; (R(i))$$

$$= \Sigma_{\forall R(i) \, \varepsilon \, \mathcal{P}} \; (1 / \Pi_{in}( R(i) ))$$

Let $\mathcal{Q}$ be the set of all acyclic paths from $x_i$ to $x_0$ in the reverse graph rev(G). Note that there is a one to one correspondence $v$ between paths belonging to $\mathcal{Q}$ and paths belonging to $\mathcal{P}$ and that the outdegree of a node in rev(G) is equal to the indegree of the node in G.. Let $R(i)$ be a path belonging to $\mathcal{P}$ and let $P= v(R(i))$ be the path belonging to $\mathcal{Q}$ in correspondence with $R(i)$. We have

$$1 / \Pi_{in}(R(i)) = 1/ \Pi_{out}(P)$$

(1)      $E(b(i)) = \Sigma_{\forall R(i) \, \varepsilon \, \mathcal{P}} \; (1 / \Pi_{in}( R(i) )) = \Sigma_{\forall P \, \varepsilon \, \mathcal{Q}} \; (1 / \Pi_{out}(P))$

where the rightmost product is done using the outdegrees of nodes of rev(G).

Now apply the procedure Random Path to the reverse graph rev(G) starting from $x_i$, and let Q be the returned path. If Q passes through $x_0$ then its initial part $Q(x_0)$ belongs to $\mathcal{Q}$. The probability that $Q(x_0)$ is the initial part of the returned path is given by the product of the outdegrees of the nodes of rev(G) belonging to $Q(x_0)$. By definition the sum over all paths of $\mathcal{Q}$ is given by p. Formally we have

Pr(Q($x_0$) is the initial part of the returned path) $= 1 / \Pi_{out}(Q(x_0))$

(2)    $\Sigma_{\forall P \varepsilon Q} (1 / \Pi_{out}(P)) = p$

where the product is done using the outdegrees of nodes of rev(G).

The thesis follows from (1) and (2).

QED

Theorem 2 Procedure Estimate is an unbiased estimator of the number of nodes connected to $x_0$.
Proof Suppose that there are m nodes connected to $x_0$. We want to prove that

E(Est) = m

Procedure Estimate assigns values to all nodes that belong to the path P returned by the procedure Random_Path. Without loss of generality we assume that b(i) = 0 if node $x_i$ does not belong to P. Since the expected value of a sum is equal to the sum of the expected values it is sufficient to prove that, for all nodes $x_i$ such that there is a path from $x_0$ to $x_i$ , we have

E(b(i) c(i)) = 1

Note that b(i) and c(i) are independent; in fact b(i) depends from the choice of the initial path P (step 1 of the procedure) and c(i) from the success of step 2. Hence in order to prove the theorem it is sufficient to show that if there is a path from $x_0$ to $x_i$ then

E(b(i)) E(c(i)) = 1

For a given $x_i$ let p be the probability that the procedure Random_Path applied to the reverse graph rev(G) with source $x_i$ suceeds. By lemma 1 E(b(i)) is equal to p. Now we will evaluate c(i). The probability that c(i) is equal to k is $p (1 - p)^{k-1}$; hence the expected value of c(i) is

$E(c(i)) = \Sigma_{\forall k>0} k p (1 - p)^{k-1}$

By standard techniques we obtain

$E(c(i)) = \Sigma_{\forall k>0} k p (1 - p)^{k-1} = 1/p$

QED

It is easy to see that the procedure Estimate may loop forever if it does not succeed to find a path between i and $x_0$ in rev(G) at step 2.1. We observe that if the graph is a tree or a rooted dag with root $x_0$, then all paths in rev(G) that starts from $x_i$ pass through the source node. Hence we do not need step 2.1 to compute c(i) that is 1 for all $x_i$ and the termination of the algorithm is guaranteed. We note that in these cases we obtain the procedures proposed by Knuth and Pitt, respectively, in the case of a tree and of a rooted dag.

# 4 Biased estimators

It is possible to modify the procedure Estimate to obtain biased estimators whose termination is guaranteed for all directed graphs. Namely, we give two procedures U_Estimate and L_Estimate. U_Estimate starts from an initial node $x_0$ and calls for Random Path; it gives as an estimate the product of the outdegrees of the nodes belonging to the returned path.

Procedure U_Estimate
input: a graph G and a source node $x_0$;
output: an estimate U_Est of the number of nodes connected to $x_0$;
1. Random path (G, s, P);
2. Let $P = <x_0, x_1, ..., x_m>$ be the returned path ;
   For each node $x_i$, i<m, do
   $d(i) := \Pi_{out}(P_i)$ ;
3. U_Est = $\Sigma_{i<m} d(i)$.

Fig. 4 Procedure U_Estimate.

Note that U_Estimate is essentially the procedure proposed by Knuth for computing an unbiased estimator on the number of nodes of a tree [K75]; however, if the graph is not a tree, it will return a biased estimator.

On the other side, L_Estimate is obtained from Estimate by giving in input an additional integer parameter t. Procedure L_Estimate gives in output an estimate L_Est and is obtained from procedure Estimate by modifying the stopping condition of step 2.1 as follows

until ( $x_0$ belongs to Q) or (c(i) = t)

Hence the procedure finishes after t attempts to find $x_0$ in rev(G), at most.

Theorem 3 i) The expected value of U_Est is an upper bound on the number of nodes connected to $x_0$;

ii) The expected value of L_Est is a lower bound on the number of nodes connected to $x_0$.

<u>Proof</u> i) Let $x_i$ be a node connected to $x_0$. In order to prove the theorem it is sufficient to show that $E(d(i)) \geq 1$. Let $\mathcal{P}$ be the set of all acyclic paths starting from $x_0$ and finishing at $x_i$. Let $R(i)$ be an acyclic path between $x_0$ and $x_i$ and let $Pr(R(i))$ be the probability that the procedure Random Path returns a path P whose initial part is $R(i)$. Clearly this probability is given by the reciprocal of the product of the outdegrees of the nodes of $R(i)$:

$$Pr(R(i)) = 1 / \Pi_{out}(R(i))$$

If $x_i$ is connected to $x_0$ then we have

$$E(d(i)) = \Sigma_{\forall \text{ acyclic paths } P \text{ passing through } x_i} \Pi_{out}(P_i) \; Pr(P)$$

$$= \Sigma_{\forall R(i) \, \varepsilon \, \mathcal{P}} \; \Sigma_{\forall \text{ acyclic paths } P \text{ with initial part } R(i)} \Pi_{out}(P_i) \; Pr(P)$$

$$= \Sigma_{\forall R(i) \, \varepsilon \, \mathcal{P}} \; \Pi_{out}(R(i)) \; Pr(R(i))$$

$$= \Sigma_{\forall R(i) \, \varepsilon \, \mathcal{P}} \; 1$$

$$\geq 1$$

ii) Let $x_i$ be a node connected to $x_0$; it is sufficient to show that $E(b(i) \, c(i)) \leq 1$.

We have seen in the proof of theorem 2 that b(i) and c(i) are independent and that $E(b(i))$ is equal to the probability that the procedure Random_Path will find a path between $x_i$ and $x_0$ in rev(G). Hence it is sufficient to show that $E(c(i))$ is less or equal to 1/p. We have

$$E(c(i)) = \Sigma_{\forall k<t} k \, p \, (1-p)^{k-1} + t \, (1-p)^{t-1} \leq 1/p.$$

## 5  Extensions

In some applications we are interested in estimating the number, of nodes that are connected to the source node and that verify some additional constraints. As an example of the problem we refer to [M88] where it has been shown that the problem of estimating the size of the answers of a chain of joins reduces to the problem of estimating the number of nodes that are connected to an initial source node $x_0$ and whose outdegree is zero.

It is easy to modify the procedure Estimate presented in section 3 to obtain an estimate on the number of nodes that are connected to a given source node and belong to a subset S of the nodes. We assume that there is a test that, given a node x, determines whether x belongs to S or not. Procedure Estimate_subset takes in input a graph G and the above test and estimates the number of nodes connected to $x_0$ that passes the test. The procedure is obtained form procedure Estimate by modifying step 2 as shown in figure 4.

2. For each node $x_i$ belonging to P, i>0, do
begin if $x_i$ belongs to S
      then begin
                  $b(i) := \Pi_{out}(P_{i-1}) / \Pi_{in}(P_i)$;
                  $c(i) := 0$;
                2.1 repeat
                        $c(i) := c(i) + 1$
                        Random Path( rev(G), $x_i$, Q)
                  until ($x_0$ belongs to Q);
      else begin
                  $b(i) := 0$; $c(i) := 1$
         end
  end;

Fig. 4 Modification to estimate subsets of nodes.

The proof of the following theorem is an easy modification of the proof of theorem 2 and it is left to the reader.

<u>Theorem 4</u>  Estimate_subset is an unbiased estimator of the number of nodes belonging to S connected to $x_0$.

Analougously to the procedure Estimate the termination of Estimate_subset is not guaranteed; it is possible to obtain two biased estimators U_Estimate_subset and L_estimate_subset in the same way as we already done in section 4.

Estimate_subset  can also be applied to estimate the number of arcs of a graph that can be visited from a source node. In this case  it is sufficient to assume that, for each arc, there is a dummy node; therefore  the problem of estimating the number of arcs that can be traversed from $x_0$ is reduced to the problem of estimating the number of dummy nodes that are connected to $x_0$.

# References

[HK65] Hall, M., D.E.Knuth, "Combinatorial Analysis and Computers, part II", *American Mathematical Monthly*, February 1965.

[K75] Knuth, D., "Estimating the Efficiency of Backtrack Programs", *Math. Comp.* 29, pp. 121-136, 1975.

[M88] Marchetti-Spaccamela, A., "Monte Carlo Estimates of the Size of Relations in Deductive Databases", manuscript, 1988.

[P87] Pitt, L., "A Note on Extending Knuth's Tree Estimator to Directed Acyclic Graph", *Information Processing Letters* vol. 24, pp.203-206, 1987.

[RND77] Reingold, E.M., J. Nievergelt, N. Deo, *Combinatorial Algorithms: Theory and Practice*, Prentice Hall, 1977.

# The average size of ordered binary subgraphs †

Pieter H. Hartel

Computing Science Department, University of Amsterdam
Nieuwe Achtergracht 166, 1018 WV Amsterdam

## Abstract

To analyse the demands made on the garbage collector in a graph reduction system, the change in size of an average graph is studied when an arbitrary edge is removed. In ordered binary trees the average number of deleted nodes as a result of cutting a single edge is equal to the average size of a subtree. Under the assumption that all trees with $n$ nodes are equally likely to occur, the expected size of a subtree is found to be approximately $\sqrt{\pi n}$ . The enumeration procedure can be applied to graphs by considering spanning trees in which the nodes that were shared in the graph are marked in the spanning tree. A correction to the calculation of the average is applied by ignoring subgraphs that have a marked root. Under the same assumption as above the average size of a subgraph is approximately $\sqrt{\pi n} - 2(m + 1)$, where $m$ represents the number of shared nodes and $m \ll n$ .

Key words: binary graphs  Catalan statistics  combinator graph reduction  subgraphs

## 1. Introduction

The $\lambda$-calculus[1] can be viewed as a universal programming language. Its simplicity makes $\lambda$-expressions (programmes) amenable to direct mechanical evaluation.[2] Functional programming

---

† This work is supported by the Dutch ministry of Science and Education, dienst Wetenschapsbeleid.

languages in essence are "sugared" versions of the λ-calculus. An ordered binary tree provides a natural representation for a λ-expression. A function application (a node) consists of the juxtaposition of the function (the left descendant of the node) and its argument (the right descendant of the node). A subexpression appears as a subtree, hence arbitrarily complex expressions can be represented. Graph reduction[3] is generally preferred to tree reduction because it allows subexpressions to be shared in stead of copied. This saves both space and time, since a copy of an expression can be made by creating a new pointer to the representation of the expression. This improved efficiency is not without cost.

Let us assume, that initially a λ-expression is presented for mechanical evaluation, which contains at least one reducible expression (redex). Furthermore let there be a mechanism that decides which redex is to be evaluated next. The evaluation process then consists of a number of discrete reduction steps (e.g. β-reduction, α-conversion), which take the expression to its final form, or at least a number of steps ahead. During this process the expression may take many different forms, which although semantically equivalent, require a varying amount of space for their graphical representation. Each reduction step typically causes some nodes to be deleted and some to be added to the graph. New nodes are added explicitly, for example as a result of β-reduction, but old nodes become unreachable without explicit notice. The reason is, that unless special precautions are taken, we can not know when the last reference to a possibly shared node is destroyed. This uncertainty has an important consequence for practical implementations of the λ-calculus and λ-based languages, because it necessitates a garbage collector, i.e. a device that recuperates storage occupied by parts of the graph that have become unreachable from the root. The cost of garbage collection is considerable. It may take up to an order of magnitude more time to recuperate the storage occupied by a node than it takes to allocate a node from a pool of free nodes.[4] Many methods have been proposed to control the cost of garbage collection.[5]

To arrive at a better understanding of the cost of garbage collection we will count the number of nodes that may be expected to turn into garbage during a reduction step. If we are prepared to

make some assumptions about the structure of the graphs being manipulated, interesting properties can be derived, even if the precise configuration is not known. For instance, under the
assumption that all ordered trees with $n$ nodes are equally likely to occur, it can be shown[6] that
the average height of such trees is approximately $\sqrt{\pi n}$. The pre-order spanning trees of the
graphs that occur during the evaluation of complicated expressions were found to behave in
roughly the same way: their average height is proportional to $\sqrt{\pi n}$.[7] Both ways of arriving at an
average apparently lead to comparable results.

A method is presented that allows us to extend the results about ordered binary trees (Catalan
statistics) to the graphs that occur during graph reduction. As a first approximation we regard
graphs as trees, i.e. ignore the effects of sharing completely. This allows us to apply the basic
Catalan statistics directly to our graphs (next section). In section 3 the effect of sharing is
modelled by marking the nodes in a spanning tree that correspond to shared nodes in the program graph. New counting procedures are developed to take sharing into account. The results
give a lower bound and an upper bound on the average number of nodes in a subgraph.

## 2. Enumeration of ordered binary trees

In the rest of the paper we assume, that the reader is familiar with the analysis as presented by
Knuth [8] pp. 388 - 389. To summarise the most important results, the construction method for
ordered binary trees and the formulae for $b_n$ and $B(z)$ are reproduced here. An ordered binary
tree with $n$ nodes can be built by taking a root and attaching an ordered binary tree with $i$ nodes
to the left and an ordered binary tree with $n-i-1$ nodes to the right. To construct the set of
ordered binary trees with $n$ nodes, $i$ must be varied over the range $[0..n-1]$ and the construction
must be applied recursively to the subtrees. As an example, consider the set of ordered binary
trees with 3 nodes shown in figure (1).

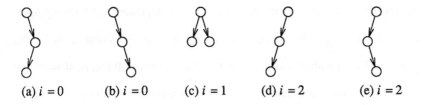

(a) $i = 0$  (b) $i = 0$  (c) $i = 1$  (d) $i = 2$  (e) $i = 2$

Figure 1 : The set of ordered binary trees with 3 nodes

The recurrence relation and the boundary conditions for $b_n \equiv$ the number of ordered binary trees

with $n$ nodes are:

$$b_n = \begin{cases} b_0 b_{n-1} + b_1 b_{n-2} + \cdots + b_{n-2} b_1 + b_{n-1} b_0, & n > 0 \\ 1, & n = 0 \end{cases} \tag{1}$$

The generating function $B(z)$ for the sequence $\langle b_n \rangle$ is:

$$B(z) = \sum_{n=0}^{\infty} b_n z^n = \frac{1 - \sqrt{1 - 4z}}{2z}$$

The closed form for the number of ordered binary trees with $n$ nodes:

$$b_n = \frac{1}{n+1} \binom{2n}{n} \tag{2}$$

With Stirling's approximation to $n!$ for large $n$ we have:

$$b_n = \frac{4^n}{n \sqrt{\pi n}} + O\left[\frac{4^n}{n^2 \sqrt{n}}\right]$$

This concludes the exposé of the basic results in Catalan statistics.

## 2.1. The size of ordered binary subtrees

The collection of nodes that become unreachable from the root when an edge is cut is called a

subtree. In addition the entire tree is considered as a subtree. As a consequence there is one sub-

tree associated with each node in a tree. Let $s_{n,k}$ be the number of subtrees with $k$ nodes of the

set of ordered binary trees with $n$ nodes. Suppose we want to find the number of subtrees with 2 nodes in the set of trees with 3 nodes. Counting subtrees in figure (1) yields the answer $s_{3,2} = 4$. The configurations (a), (b), (d) and (e) each contribute one to the total. In general it is not practical to draw all configurations in order to count subtrees, hence we must use the construction method to arrive at the desired answer. The configurations (a) and (b) were constructed by attaching all possible configurations with two nodes to the right of the root combined with all possible configurations with zero nodes to the left. Together they contribute $b_0 s_{2,2} + s_{0,2} b_2$ to the number of subtrees with two nodes. Extending this procedure to the remaining configurations yields:

$$s_{3,2} = b_0 s_{2,2} + b_1 s_{1,2} + b_2 s_{0,2} + s_{0,2} b_2 + s_{1,2} b_1 + s_{2,2} b_0$$

$$= 1 \ s_{2,2} + 1 \ s_{1,2} + 2 \ s_{0,2} + s_{0,2} 2 \ + s_{1,2} 1 \ + s_{2,2} 1 \ = 4$$

The general rule is derived from the enumeration formula for ordered binary trees: where $b_i$ is multiplied by $b_{n-i-1}$ in (1), we must now multiply $s_{i,k}$ by $b_{n-i-1}$. This corresponds to the number of subtrees with $k$ nodes in the set of trees with $i$ nodes, multiplied by the number of times this configuration occurs to the left of the root. Similarly enumeration of subtrees with $k$ nodes in the right subtree yields the term $b_i s_{n-i-1,k}$. Since a tree is regarded as a subtree we have $s_{n,n} = b_n$. The recurrence relation and the boundary conditions for $s_{n,k} \equiv$ the number of subtrees with $k$ nodes among the set of trees with $n$ nodes are:

$$s_{n,k} = \begin{cases} b_0 s_{n-1,k} + b_1 s_{n-2,k} + \dots + b_{n-2} s_{1,k} + b_{n-1} s_{0,k} \ + \\ s_{0,k} b_{n-1} + s_{1,k} b_{n-2} + \dots + s_{n-2,k} b_1 + s_{n-1,k} b_0, & 0 < k < n \\ b_n, & n \geq 0 \wedge k = n \\ 0, & \text{otherwise} \end{cases} \quad (3)$$

To derive the closed form of (3), two auxiliary results are needed. Both can be proved by induction on $n$ from (3). The first yields the number of subtrees with one node:

$$s_{n,1} = n \ b_{n-1} \qquad (4)$$

Proof: if $n=1$ we have $s_{1,1} = b_1 = 1 \ b_0$, since $b_0 = b_1 = 1$. Application of the induction hypothesis $\forall i \in 1..n-1 : s_{i,1} = i \, b_{i-1}$ to (3) yields

$$
\begin{aligned}
s_{n,1} &= b_0 \, s_{n-1,1} && + \cdots + b_{n-2} \, s_{1,1} && + b_{n-1} \, s_{0,1} + \\
&\quad s_{0,1} \, b_{n-1} + s_{1,1} \, b_{n-2} && + \cdots + s_{n-1,1} \, b_0 \\
&= b_0 \, (n-1) \, b_{n-2} + \cdots + b_{n-2} \, 1 \, b_0 && + b_{n-1} \, 0 \quad + \\
&\quad 0 \, b_{n-1} \; + 1 \, b_0 \, b_{n-2} && + \cdots + (n-1) \, b_{n-2} \, b_0 \\[4pt]
&= n \, (b_0 \, b_{n-2} + b_1 \, b_{n-3} + \cdots + b_{n-2} \, b_0) \\[4pt]
&= n \, b_{n-1}
\end{aligned}
$$

The second auxiliary result (5) can be proved in a similar fashion.

$$
s_{n,k} = s_{n-k+1,1} \; s_{k,k} , \quad 0 < k \le n \tag{5}
$$

Combining (3), (4) and (5) we find:

$$
s_{n,k} = (n-k+1) \, b_{n-k} \, b_k , \quad 0 < k \le n \tag{6}
$$

The generating function for the sequence $\langle s_{n,k} \rangle$ with a fixed value of $k$ is:

$$
S_k(z) = \sum_{n=0}^{\infty} s_{n,k} z^n = \sum_{n=0}^{\infty} (n-k+1) \, b_{n-k} \, b_k z^n = b_k z^k \frac{d}{dz}(z \, B(z)) = \frac{b_k z^k}{\sqrt{1-4z}} , \quad k > 0
$$

Since every subtree is uniquely determined by its top node, the total number of subtrees must be equal to the total number of nodes in the set of $b_n$ trees. This can be proved directly from (3).

$$
\sum_{k=1}^{n} s_{n,k} = n \, b_n
$$

The values of $s_{n,k}$ grow quickly as $n$ increases. Table (1) shows the values for small $n$ and $k$. The numbers on the diagonal are the values for $b_n$ (Catalan numbers) since $s_{n,n} = b_n$.

| n | k=1 | k=2 | k=3 | k=4 | k=5 | k=6 | k=7 | k=8 | k=9 |
|---|---|---|---|---|---|---|---|---|---|
| 1 | 1 | | | | | | | | |
| 2 | 2 | 2 | | | | | | | |
| 3 | 6 | 4 | 5 | | | | | | |
| 4 | 20 | 12 | 10 | 14 | | | | | |
| 5 | 70 | 40 | 30 | 28 | 42 | | | | |
| 6 | 252 | 140 | 100 | 84 | 84 | 132 | | | |
| 7 | 924 | 504 | 350 | 280 | 252 | 264 | 429 | | |
| 8 | 3432 | 1848 | 1260 | 980 | 840 | 792 | 858 | 1430 | |
| 9 | 12870 | 6864 | 4620 | 3528 | 2940 | 2640 | 2574 | 2860 | 4862 |

Table 1: $s_{n,k}$ for small values of $n$ and $k$.

## 2.2. Average size of subtrees

Under the assumption that all ordered binary trees with $n$ nodes occur with the same probability, we find for the size of the average subtree:

$$\overline{s_n} = \frac{\sum\limits_{k=0}^{n} k\, s_{n,k}}{n\, b_n} = \frac{\sum\limits_{k=0}^{n} k\, (n-k+1)\, b_{n-k}\, b_k}{n\, b_n} \tag{7}$$

The generating function for the sequence $\langle \sum\limits_{k=0}^{n} k\, s_{n,k} \rangle$ is:

$$\sum_{n=0}^{\infty} \sum_{k=0}^{n} (n-k+1)\, b_{n-k}\, k\, b_k z^n = z\, B'(z)\, \frac{d}{dz}(z\, B(z))$$

$$= \frac{1}{1-4z} + \frac{1}{2z} - \frac{1}{2z\, \sqrt{1-4z}}$$

$$= \sum_{n=0}^{\infty} 4^n z^n + \frac{1}{2z} - \frac{1}{2z} \sum_{n=0}^{\infty} \binom{-\tfrac{1}{2}}{n} (-4)^n z^n$$

$$= \sum_{n=0}^{\infty} \left[ 4^n - \frac{1}{2} \begin{pmatrix} -\frac{1}{2} \\ n+1 \end{pmatrix} (-4)^{n+1} \right] z^n$$

$$= \sum_{n=0}^{\infty} \left[ 4^n - (2n+1) b_n \right] z^n$$

With this result (7) becomes:

$$\overline{s_n} = \frac{4^n - (2n+1) b_n}{n \, b_n} \; , \quad n > 0$$

Let us apply this result to graph reduction. Using Stirling's approximation we find for large $n$ that $\overline{s_n} \approx \sqrt{\pi n}$. Hence a reduction step may be expected to delete a subgraph containing $\sqrt{\pi n}$ nodes from the programme graph. Unless a reduction step adds a similar number of new nodes to the graph, it is reduced in size until the average number of nodes added per reduction step and that deleted are in equilibrium. From our experiments we know that such an equilibrium may be reached when far fewer than $\sqrt{\pi n}$ nodes are added per reduction step. The effect of sharing that is completely ignored in the current approximation, should at least be taken into account.

## 3. Trees with marked nodes

Our objective is to count the number of nodes that may be deleted from a programme graph by cutting a single edge. To be able to apply the results obtained so far, we must eliminate the effect of the extra edges that turn a tree into a graph. In general there are many subsets of edges that may be considered redundant. From our experience with the determination of the average height of spanning trees we found that the results are virtually independent of the particular set of edges that is removed. We have no reason to believe that in the current case the results should be sensitive to the choice of spanning tree, since the averaging processes in determining the expected height of trees and the expected size of subtrees are so similar. The first step is therefore to consider an arbitrary spanning tree of a graph in stead of the graph itself. Such a tree has the same set of nodes as the graph and one fewer edge than the graph has nodes. In the previous section we saw that without further refinement the results with binary subtrees are still unsatisfactory. We therefore introduce a "marking" mechanism by which we can remember which nodes were originally shared. The enumeration procedures must take this marking into account.

There are two issues that deserve further attention. In the first place we have to cope with nodes that have a different number of incident edges. In the second place the origin of the edges incident upon the same node may be expected to play a role. Since we wish to study the effect of removing a single edge from a graph, we need not be too concerned about the actual number of incident edges to a node as long as it is greater than one. Usually a shared node remains reachable from the root if an incident edge is removed. A connected component in a graph may be attached to the rest of the graph via a single edge. If that edge is cut, the entire component becomes unreachable from the root. However, since our experiments indicate that cycles are rare in practical graph reduction, we ignore this effect and treat a shared node as one that can not be removed by cutting a single edge.

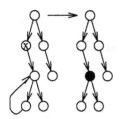

(a) removable marked node

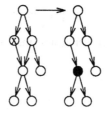

(b) permanent marked node

Figure 2 : Two binary graphs with one marked node

The origin of the edges can not be completely ignored. Figure (2-a) shows, that a marked node may originate from two different kinds of graphical structure. If the left descendant of the root (marked with the letter "x") is removed, the right subgraph of node "x" also becomes disconnected, since all edges incident upon the shared node are successively removed. The right subgraph of node "x" remains connected if it is removed from the graph shown in figure (2-b). The marked nodes in trees corresponding to the graph as in figure (2-a) are called removable, the type of marking as in figure (2-b) is called permanent. We will study both kinds of marking.

Based on the graphical configurations as illustrated in figure (2) we develop the rules by which subtrees of marked trees are enumerated. Ultimately we are interested in the expected size of subtrees if all trees are equally likely to occur. The averaging process therefore considers every node once as the root of a subtree that must be counted:

- If the root of a subtree is marked the subtree is not counted, regardless of the type of marking. This is in accordance with the origin of a marked node as a shared node in a programme graph.

- If the root of a subtree is not marked, the subtree is counted, *including* the removable marked nodes and their subtrees but *excluding* the permanent marked nodes and their subtrees.

Consider as an example the trees of figure (3). If the marking in figure (3-a) is considered permanent, there are 3 subtrees with one node, 1 subtree with two nodes and 1 subtree with four nodes. If the marking is considered removable, there are 3 subtrees with one node, 1 with four and 1 with six nodes. Similarly with permanent marking figure (3-b) shows 3 subtrees of one and 1 subtree of two nodes. With removable marking we count 3 subtrees of one, 1 of two and 1 of six nodes.

(a) one marked node                (b) two marked nodes

Figure 3 : Two marked binary trees

A more careful consideration of the graphical structure in figure (2) reveals that the permanent or removable status of a marked node is a property that varies with the choice of the first node to be deleted. For example if rather than the node "x" we remove the root in figure (2-b), even the permanent node and its subtree should disappear. However we can afford to ignore this complication, since the expected size of subtrees with only permanent or only removable marked nodes form the extremes of a range that encompasses realistic values of the expected number of nodes to be deleted in programme graphs (provided the other assumptions prove to be realistic).

It will be shown, that in a tree with only removable marked nodes, the enumeration for unmarked trees can be modified in a straight forward manner by discounting all subtrees with a marked root. Permanent marking causes considerable complication since each subtree has to be "searched" for permanent marked nodes, which "truncate" the subtree to a smaller size. First the results are generalised to reason about trees with one removable or permanent marked node. This allows us to develop the necessary intuition to solve the problem if an arbitrary number of nodes is marked.

### 3.1. The size of subtrees in trees with one removable marked node

In the collection of the $n$ $b_n$ trees with one removable marked node, each subtree appears once with a marked root and $n-1$ times with an unmarked root. Since the total number of subtrees with $k$ nodes in the $n$ $b_n$ trees with $n$ nodes is $n$ $s_{n,k}$, the number of subtrees without marked nodes is:

$$r_{n,k} = (n-1) s_{n,k} , \quad n > 0$$

The total number of subtrees must be equal to the total number of nodes in all $n$ $b_n$ trees, minus the number of marked nodes:

$$\sum_{k=1}^{n} r_{n,k} = n \ (n-1) \ b_n$$

### 3.2. Average size of subtrees in trees with one removable marked node

Making the same assumption as before about the probability distribution, we find that the average size of subtrees in trees with one removable marked node is:

$$
\overline{r_n} = \frac{\displaystyle\sum_{k=0}^{n} k\, r_{n,k}}{n^2 b_n} = \frac{(n-1) \displaystyle\sum_{k=0}^{n} k\, s_{n,k}}{n^2 b_n} = \left(1 - \frac{1}{n}\right) \overline{s_n}, \qquad n > 0 \tag{8}
$$

### 3.3. The size of subtrees in trees with one permanent marked node

The construction method for trees is extended to take the location of the marked node into account. Any one of the $n$ nodes in a tree may be marked. Therefore, the total number of trees that must be considered is $n\, b_n$. The recurrence relation of (1) is rewritten to reflect this situation:

$$
n\, b_n = \left[ b_0 (n-1)\, b_{n-1} + 0\, b_0\, b_{n-1} \right] + \left[ b_1 (n-2)\, b_{n-2} + 1\, b_1\, b_{n-2} \right] + \cdots
$$

$$
+ \left[ b_{n-1}\, 0\, b_0 + (n-1)\, b_{n-1}\, b_0 \right] + b_n \tag{9}
$$

In the pairs of terms, the first subterm corresponds to the configuration with the marked node residing in the right subtree. The second subterm has the same relation to the left subtree. Both subterms therefore represent configurations with an unmarked root. The last term of (9) enumerates the $b_n$ trees of $n$ nodes with a marked root.

The recurrence relation and boundary conditions for $p_{n,k} \equiv$ the number of subtrees with $k$ nodes in the set of trees with $n$ nodes of which one is permanent marked can now be formulated by enumerating the relevant number of subtrees in each term of (9):

$$
P_{n,k} = \begin{cases}
2(\ b_0 P_{n-1,k} + \ b_1 P_{n-2,k} + \cdots + \quad\quad b_{n-1} P_{0,k}) + \\
2\,(0\,b_0\,s_{n-1,k} + 1\,b_1\,s_{n-2,k} + \cdots + (n-1)\,b_{n-1}\,s_{0,k}) + \\
s_{n,k} + s_{n,n-k}\,, & 0 < k < n \\
1\,, & n = k = 0 \\
0\,, & \text{otherwise}
\end{cases}
\tag{10}
$$

The first two subterms of (10) correspond to configurations with an unmarked root. The marked node either resides in the left subtree or in the right subtree. The second subterm accounts for the fact that if one subtree (say with size $i$) contains the marked node, the other (with size $n-i-1$) is unmarked. This gives a contribution of $i\,b_i\,s_{n-i-1,k}$ subtrees, taken over all possible values of $i$. The factor $i$ originates from the fact that the marked node may reside at any one of the $i$ places in the subtree.

If the root of the tree is marked, the situation resembles that of the unmarked trees. This accounts for the subterm $s_{n,k}$, but with $k < n$, since there are no subtrees with $n$ nodes in any marked tree with $n$ nodes. The last subterm of (10) takes into account, that each subtree with a marked root of size $k$ "prunes" a branch of the main tree, such that an unmarked tree with $n-k$ nodes remains.

From (10) we find the open form of the generating function for the sequence $\langle p_{n,k}\rangle$ with a fixed value of $k$:

$$
P_k(z) = \sum_{n=0}^{\infty} \left[\, 2 \sum_{i=0}^{n-1} b_i\, p_{n-i-1,k} + 2 \sum_{i=0}^{n-1} i\, b_i\, s_{n-i-1,k} + s_{n,k} + s_{n,n-k} \right] z^n
$$

$$
- \left[\, s_{k,k} + s_{k,0} \right] z^k
\tag{11}
$$

The correction term in (11) is necessary to compensate for $p_{k,k}$, which if the recurrence relation in (10) is used for $k=n$ (hence outside its domain) yields $s_{k,k} + s_{k,0}$ instead of 0.

From (6) we find that:

$$s_{n,k} + s_{n,n-k} = (n-k+1)\, b_{n-k}\, b_k + (k+1)\, b_k\, b_{n-k} = (n+2)\, b_k\, b_{n-k}$$

Substitution of this result in (11) yields:

$$P_k(z) = \sum_{n=0}^{\infty} \left[ 2 \sum_{i=0}^{n-1} b_i\, P_{n-i-1,k} + 2 \sum_{i=0}^{n-1} i\, b_i\, s_{n-i-1,k} + (n+2)\, b_k\, b_{n-k} \right] z^n - (k+2)\, b_k\, b_0\, z^k$$

$$= 2 z\, B(z)\, P_k(z) + 2 z^2\, B'(z)\, S_k(z) + \frac{b_k}{z} \frac{d}{dz} \left[ z^{k+2}\, (B(z)-1) \right]$$

This equation can be solved for $P_k(z)$ yielding:

$$P_k(z) = \frac{1}{1 - 2z\, B(z)} \left[ 2 z^2\, B'(z)\, S_k(z) + \frac{b_k}{z} \frac{d}{dz} \left[ z^{k+2}\, (B(z)-1) \right] \right]$$

$$= b_k z^k \left[ \frac{k+1}{2z} \left[ \frac{1}{\sqrt{1-4z}} - 1 \right] - \frac{k+1}{\sqrt{1-4z}} + \frac{2z}{(1-4z)^{1\frac{1}{2}}} \right]$$

$$= b_k z^k \left[ \frac{k+1}{2} \sum_{n=0}^{\infty} \binom{-\frac{1}{2}}{n+1} (-4)^{n+1} z^n - (k+1) \sum_{n=0}^{\infty} \binom{-\frac{1}{2}}{n} (-4z)^n \right.$$

$$\left. + 2 \sum_{n=1}^{\infty} \binom{-1\frac{1}{2}}{n-1} (-4)^{n-1} z^n \right]$$

$$= \sum_{n=0}^{\infty} (n+2)\, (n-k)\, b_{n-k}\, b_k\, z^n$$

Hence:

$$P_{n,k} = (n+2)\, (n-k)\, b_{n-k}\, b_k\,, \quad 0 < k \le n \tag{12}$$

The total number of subtrees must be equal to the total number of nodes in all $n\ b_n$ trees minus the number of marked nodes:

$$\sum_{k=1}^{n} p_{n,k} = (n-1)\, n\, b_n$$

## 3.4. Average size of subtrees in trees with one permanent marked node

With a uniform probability distribution, the average size of subtrees in trees with one permanent marked node is:

$$\overline{p_n} = \frac{\displaystyle\sum_{k=0}^{n} k\, p_{n,k}}{n^2\, b_n} = \frac{\displaystyle\sum_{k=0}^{n} k\, (n+2)\, (n-k)\, b_{n-k}\, b_k}{n^2\, b_n} \tag{13}$$

The generating function for the sequence $\langle \displaystyle\sum_{k=0}^{n} k\, p_{n,k} \rangle$ is:

$$\sum_{n=0}^{\infty} \sum_{k=0}^{n} (n+2)\, (n-k)\, b_{n-k}\, k\, b_k\, z^n = \frac{1}{z}\, \frac{d}{dz}(z^2\, B'(z))^2$$

$$= \frac{2\,(1-2z)}{(1-4z)^2} - \frac{2}{(1-4z)^{1\!/\!2}} \tag{14}$$

$$= \sum_{n=1}^{\infty} \left[ (n+2)\, 4^n - 2 \binom{-1\!/\!2}{n} (-4)^n \right] z^n$$

After simplification of the binomial coefficient, the result can be substituted in (13) such that:

$$\overline{p_n} = \frac{(n+2)\, 4^n - (2n+1)\, (2n+2)\, b_n}{n^2\, b_n}, \quad n > 0$$

## 3.5. The size of removable marked subtrees

In the set of $\binom{n}{m}\, b_n$ trees with $m$ removable marked nodes, each subtree appears $\binom{n-1}{m-1}$ times with a marked root. The total number of subtrees with $k$ nodes in the $\binom{n}{m}\, b_n$ trees is $\binom{n}{m}\, s_{n,k}$.

Therefore $r^m_{n,k} \equiv$ the number of subtrees with $k$ nodes in the set of trees with $n$ nodes, of which $m$ are removable marked is:

$$r^m_{n,k} = \left[ \binom{n}{m} - \binom{n-1}{m-1} \right] s_{n,k} = \binom{n-1}{m} s_{n,k}, \quad n > 0$$

The total number of subtrees must be equal to the total number of nodes in the set of $\binom{n}{m} b_n$ trees, minus the number of marked nodes:

$$\sum_{k=1}^{n} r^m_{n,k} = \binom{n}{m} (n-m) b_n \qquad (15)$$

## 3.6. Average size of removable marked subtrees

With a uniform probability distribution, the average size of subtrees in removable marked trees is:

$$\overline{r^m_n} = \frac{\sum\limits_{k=0}^{n} k \, r^m_{n,k}}{\binom{n}{m} b_n} = \frac{\binom{n-1}{m} \sum\limits_{k=0}^{n} k \, s_{n,k}}{\binom{n}{m} n \, b_n} = \left[ 1 - \frac{m}{n} \right] \overline{s_n}, \quad n > 0$$

## 3.7. The size of permanent marked subtrees

The generalisation of the result to trees with an arbitrary number of permanent marked nodes follows along the lines set out in the previous paragraphs. We commence by characterising the subtrees of all $\binom{n}{m} b_n$ marked trees, by the way the $m$ permanent marked nodes are distributed over the total $n$ nodes. Using Vandermonde's convolution[8] and (1) we find that:

$$\binom{n}{m} b_n = \binom{n-1}{m} b_n + \binom{n-1}{m-1} b_n$$

$$= \sum_{i=0}^{m} \sum_{j=0}^{n-1} \binom{j}{i} b_j \binom{n-j-1}{m-i} b_{n-j-1} + \sum_{i=0}^{m-1} \sum_{j=0}^{n-1} \binom{j}{i} b_j \binom{n-j-1}{m-i-1} b_{n-j-1} \qquad (16)$$

The first sum in (16) corresponds to configurations with an unmarked root, the second to those with a marked root. In both "inner" sums, the index $j$ ranges over all possible combinations of subtrees. The "outer" sums distribute the marked nodes over the subtrees, starting with all marked nodes to the left and none to the right, one to the left and all but one to the right etc.

Looking at figure (3) again and interpreting the marked nodes as permanent, it appears that there are two kinds of subtrees eligible to be counted. If the root of the main tree is unmarked, it is counted as a subtree. This subtree is called a top-tree. It will prove useful, to extend the definition of top-trees to the case where the root of main tree *is* marked. In that case a top-tree is considered to have 0 nodes. The remaining subtrees (i.e. non top-trees) are isolated from the root by a marked node. In figure (3-a) there is one subtree of either kind and in (3-b) there are two isolated subtrees (if subtrees of top-trees are ignored). The recurrence relation and the boundary conditions for $t_{n,k}^m \equiv$ the number of top-trees with $k$ nodes in the set of trees with $n$ nodes, of which $m$ nodes are permanent marked are:

$$t_{n,k}^m = \begin{cases} \displaystyle\sum_{h=0}^{k-1} \sum_{j=0}^{n-1} \sum_{i=0}^{m} t_{j,h}^i \, t_{n-j-1,k-h-1}^{m-i}, & n > 0 \wedge m > 0 \wedge k > 0 \wedge k+m \leq n \\[2ex] \displaystyle\binom{n-1}{m-1} b_n, & n > 0 \wedge m > 0 \wedge k = 0 \\[2ex] b_n, & n \geq 0 \wedge m = 0 \wedge k = n \\[1ex] 0, & \text{otherwise} \end{cases} \qquad (17)$$

The recurrence in the above expression draws upon the "unmarked root" sum in (16), with the constraint that if the left subtree has $h$ nodes, the right subtree must host the remaining $k-h-1$ nodes. The second clause in (17) gives the number of times that the root is marked. This is the number of times a top-tree with 0 nodes must be counted in a marked tree.

We conjecture that from (17) it can be proved by induction:

$$t_{n,k}^m = \frac{(k+1)}{n-k} \binom{n+m}{m-1} \binom{2n-2k}{n-k-m} b_k , \quad n > 0 \wedge m > 0 \wedge k \geq 0 \wedge k + m \leq n \tag{18}$$

With this result we can formulate the recurrence relation and the boundary conditions for $p_{n,k}^m \equiv$ the number of subtrees with $k$ nodes in the set of trees with $n$ nodes, of which $m$ nodes are permanent marked:

$$p_{n,k}^m = \left\{ \begin{array}{ll} q_{n,k}^m + q_{n,k}^{m-1} + t_{n,k}^m , & n > 0 \wedge m \geq 0 \wedge k > 0 \wedge k+m \leq n \\ 1 , & n = 0 \wedge m = 0 \wedge k = 0 \\ 0 , & \text{otherwise} \end{array} \right\} \tag{19}$$

Where $q_{n,k}^m$ is defined as:

$$q_{n,k}^m = 2 \sum_{j=0}^{n-k-1} \sum_{i=0}^{m} \binom{j}{i} b_j \, p_{n-j-1,k}^{m-i} \tag{20}$$

The first two subterms in (19/20) exhibit a strong resemblance with (16). The first subterm corresponds to configurations with an unmarked root, the second subterm to those with a marked root. The top-trees contribute the term $t_{n,k}^m$. Since $p_{n,k}^0 \equiv s_{n,k}$, we must have that $p_{0,0}^0 = 1$. We will proof by generalised induction over $n$ and $m$ (with $k$ fixed) that:

$$p_{n,k}^m = \binom{n+m+1}{m} \binom{2n-2k}{n-k-m} b_k , \quad n > 0 \wedge m \geq 0 \wedge k > 0 \wedge k + m \leq n \tag{21}$$

In section 2.1 we have proved that (21) holds for $m = 0 \vee n = 0$. By the generalised induction principle we may assume that (21) holds for $0 < m' < m$ and $0 < n' < n$. Substitution of (21) in (20) yields:

$$q_{n,k}^m = 2 b_k \sum_{j=0}^{n-k-1} \sum_{i=0}^{m} \binom{j}{i} b_j \binom{n-j+m-i}{m-i} \binom{2n-2j-2k-2}{n-j-k-1-m+i} $$

Application of (22), see exercise 31, page 70 in Knuth's book,[8] which states that:

$$\binom{a}{c} \binom{b}{d} = \sum_x \binom{c-a+b}{x} \binom{d+a-b}{d-x} \binom{a+x}{c+d} , \quad \text{integer } c \geq 0 , \text{ integer } d \geq 0 \tag{22}$$

and changing the order of summation yields:

$$q_{n,k}^{m} = 2\, b_k \sum_{x=0}^{k+1} \sum_{j=0}^{n-k-1} \sum_{i=0}^{m} \binom{j}{i} b_j \binom{k+1}{x} \binom{n-j-2k-2}{m-x-i} \binom{2n-2j-2k-2+x}{n-j-k-1}$$

Application of Vandermonde's convolution to perform the summation over $i$ and substitution of (2) yields:

$$q_{n,k}^{m} = 2\, b_k \sum_{x=0}^{k+1} \sum_{j=0}^{n-k-1} \binom{k+1}{x} \binom{2j+1}{j} \frac{1}{2j+1} \binom{2n-2j-2k-2+x}{n-j-k-1} \binom{n-2k-2}{m-x}$$

Application of Rothe's non-symmetric addition theorem[9] to perform the summation on $j$ yields:

$$q_{n,k}^{m} = 2\, b_k \sum_{x=0}^{k+1} \binom{k+1}{x} \binom{2n-2k-1+x}{n-k-1} \binom{n-2k-2}{m-x}$$

Using (22) again but in opposite direction to perform the summation on $x$ we obtain:

$$q_{n,k}^{m} = 2\, b_k \binom{2n-2k-1}{n-k-m-1} \binom{n+m+1}{m}$$

With (18) and (19) it is readily verified that (21) holds for $n$ and $m$, which concludes the proof of (21).

The total number of subtrees in all permanent marked trees must be equal to the number of unmarked nodes:

$$\sum_{k=1}^{n} p_{n,k}^{m} = \binom{n}{m} (n-m)\, b_n$$

### 3.8. Average size of permanent marked subtrees

For a uniform probability distribution of trees we have:

$$\overline{p_n^m} = \frac{\sum\limits_{k=0}^{n} k\, p_{n,k}^{m}}{\binom{n}{m} n\, b_n} = \frac{\sum\limits_{k=0}^{n} k \binom{n+m+1}{m} \binom{2n-2k}{n-k-m} b_k}{\binom{n}{m} n\, b_n} \tag{23}$$

Unfortunately, there is no simple general solution for the sum in the enumerator of (23). To see this, let $\Delta_n^m$ be defined as:

$$\Delta_n^m = \frac{\sum\limits_{k=0}^{n} k\, p_{n,k}^{m-1}}{\binom{n+m}{m-1}} - \frac{\sum\limits_{k=0}^{n} k\, p_{n,k}^{m}}{\binom{n+m+1}{m}}, \quad 0 < m < n$$

With (21) and $m$ fixed it can be proved by induction on $n$ that:

$$\Delta_n^m = \binom{2n+1}{n-m}$$

If the sum of $\Delta_n^m$ for all possible values of $m$ is calculated, we find that all but the first and last terms cancel out:

$$\sum_{i=1}^{m} \Delta_n^i = \frac{\sum\limits_{k=0}^{n} k\, p_{n,k}^{0}}{\binom{n+1}{0}} - \frac{\sum\limits_{k=0}^{n} k\, p_{n,k}^{1}}{\binom{n+2}{1}} + \frac{\sum\limits_{k=0}^{n} k\, p_{n,k}^{1}}{\binom{n+2}{1}} - \cdots + \frac{\sum\limits_{k=0}^{n} k\, p_{n,k}^{m-1}}{\binom{n+m}{m-1}} - \frac{\sum\limits_{k=0}^{n} k\, p_{n,k}^{m}}{\binom{n+m+1}{m}}$$

Using this result and the fact that $p_{n,k}^{0} = s_{n,k}$ to calculate $\sum\limits_{k=0}^{n} k\, p_{n,k}^{m}$ we obtain:

$$\sum_{k=0}^{n} k\, p_{n,k}^{m} = \binom{n+m+1}{m}\left[\sum_{k=0}^{n} k\, s_{n,k} - \sum_{i=1}^{m} \Delta_n^i\right]$$

$$= \binom{n+m+1}{m}\left[4^n - \sum_{i=0}^{m} \binom{2n+1}{n-i}\right] \tag{24}$$

$$= \binom{n+m+1}{m}\left[4^n - \sum_{i=0}^{n} \binom{2n+1}{i} + \sum_{i=0}^{n-m-1} \binom{2n+1}{i}\right]$$

For sums of type $\sum\limits_{k=0}^{n} \binom{m}{k}$, where $n < m$ no simple solution is known (see Knuth's book, page 64). The best that can be done is to compute $\sum\limits_{k=0}^{n} k\, p_{n,k}^{m}$ for specific values of $m$. Although this may be done conveniently for small values of $m$ using (24), the method used in the previous sections is employed once more, to show why the problem has no simple solution. For instance let $m = 2$. The generating function for the sequence $\langle \sum\limits_{k=0}^{n} k\, p_{n,k}^{2} \rangle$ with a fixed value of $k$ can be found from $B(z)$ by observing that:

$$\begin{bmatrix} 2n-2k \\ n-k-2 \end{bmatrix} z^{n-k} = \frac{(n-k)(n-k-1)}{(n-k+2)} b_{n-k} z^{n-k} = z^2 \frac{d^2}{dz^2} \left[ \frac{1}{z^2} \int_0^z \zeta\, b_{n-k}\, \zeta^{n-k}\, d\zeta \right]$$

Hence:

$$\sum_{n=0}^{\infty} \left[ \sum_{k=0}^{n} \begin{bmatrix} 2n-2k \\ n-k-2 \end{bmatrix} b_k\, k \right] z^n = z\, B'(z)\, z^2 \frac{d^2}{dz^2} \left[ \frac{1}{z^2} \int_0^z \zeta\, B(\zeta)\, d\zeta \right]$$

$$= \frac{1}{1-4z} + \frac{z-1}{2z^2 \sqrt{1-4z}} - \frac{(2z+1)\sqrt{1-4z}}{4z^3} - \frac{(1-4z)^{1\!/\!2}}{4z^3} + \frac{1}{2z^3} - \frac{1}{z^2} + \frac{1}{2z} \qquad (25)$$

$$= \sum_{n=0}^{\infty} \left[ (n+2)(n+3)\, 4^n - (2n+1)(3n^2+7n+6)\, b_n \right] z^n$$

To calculate $\overline{p_n^m}$, the generating function for the sequence $\left\langle \begin{bmatrix} 2n-2k \\ n-k-m \end{bmatrix} \right\rangle$ has to be found, with $k$ and $m$ fixed. This requires $m-1$ integrations, followed by $m$ differentiation operations. Comparing (14) and (25) we may assert, that the structure of the function $B(z)$ does not permit a simple generalisation to such a procedure.

## 3.9. Approximation and numerical data

The values of $\overline{p_n^m}$ for large $n$ and comparatively small $r$ are calculated by deriving an approximation for a combination of (23) and (24):

$$\overline{p_n^m} = \frac{\begin{bmatrix} n+m+1 \\ m \end{bmatrix}}{\begin{bmatrix} n \\ m \end{bmatrix}} \left[ \frac{4^n}{n\, b_n} - \sum_{i=0}^{m} \frac{\begin{bmatrix} 2n+1 \\ n-i \end{bmatrix}}{n\, b_n} \right] \qquad (26)$$

The quotients of binomial coefficients in this equation can be written as quotients of polynomials in $n$ with integer coefficients:

$$\frac{\binom{2n+1}{n-i}}{n\,b_n} = \frac{(2n+1)}{n}\,\frac{n\,(n-1)\,\cdots\,(n-i+1)}{(n+2)\,(n+3)\,\cdots\,(n+i+1)}$$

$$= \frac{(2n+1)}{n}\,\frac{u_i(n)}{v_i(n)}\,,\quad \mathrm{lead}(u_i)=\mathrm{lead}(v_i)=1 \wedge \deg(u_i)=\deg(v_i)$$

$$\frac{\binom{n+m+1}{m}}{\binom{n}{m}} = \frac{(n+m+1)\,(n+m)\,\cdots\,(n+2)}{n\,(n-1)\,\cdots\,(n-m+1)}$$

$$= \frac{u_{m+1}(n)}{v_{m+1}(n)}\,,\quad \mathrm{lead}(u_{m+1})=\mathrm{lead}(v_{m+1})=1 \wedge \deg(u_{m+1})=\deg(v_{m+1})$$

For any two polynomials $u(n)$ and $v(n)$ over the field of rational numbers, there exists a unique pair of polynomials $q(n)$ and $r(n)$ such that:[10]

$$u(n)=v(n)\times q(n)+r(n)\,,\quad \deg(r)<\deg(v)$$

Since $\forall\, i \in 0..m+1$ the leading coefficients of $u_i(n)$ and $v_i(n)$ are 1 and the degrees of both polynomials are equal we have $q_i(n)\equiv 1$.

$$\overline{p_n^m}=\left[1+\frac{r_{m+1}(n)}{v_{m+1}(n)}\right]\left[\frac{4^n}{n\,b_n}-\frac{(2n+1)}{n}\sum_{i=0}^{m}\left[1+\frac{r_i(n)}{v_i(n)}\right]\right]$$

$$=\sqrt{\pi\,n}-2(m+1)+O\left[\frac{1}{\sqrt{n}}\right]\,,\quad n\gg m$$

Table (2) shows some values of $\overline{p_n^m}$ for small $n$ and $m$, which where calculated using (26). The last line in the table gives approximated values using the formula $\sqrt{8192\,\pi}-2(m+1)$. This shows that for small $m$ the approximation is good, but for $m$ near $\sqrt{n}$ it has no significance.

| n | m=0 | m=1 | m=2 | m=4 | m=8 | m=16 | m=32 | m=64 |
|---|------|------|------|------|------|------|------|------|
| 2 | 1.50 | .50 | | | | | | |
| 4 | 2.32 | 1.23 | .63 | | | | | |
| 8 | 3.60 | 2.38 | 1.61 | .74 | | | | |
| 16 | 5.53 | 4.16 | 3.20 | 2.00 | .84 | | | |
| 32 | 8.35 | 6.84 | 5.69 | 4.09 | 2.34 | .91 | | |
| 64 | 12.41 | 10.79 | 9.46 | 7.43 | 4.94 | 2.60 | .95 | |
| 128 | 18.22 | 16.50 | 15.01 | 12.59 | 9.25 | 5.65 | 2.77 | .97 |
| 256 | 26.48 | 24.68 | 23.07 | 20.29 | 16.09 | 10.94 | 6.19 | 2.88 |
| 512 | 38.19 | 36.34 | 34.62 | 31.54 | 26.55 | 19.69 | 12.36 | 6.54 |
| 1024 | 54.78 | 52.89 | 51.09 | 47.78 | 42.08 | 33.51 | 23.04 | 13.41 |
| 2048 | 78.26 | 76.33 | 74.48 | 70.98 | 64.72 | 54.56 | 40.65 | 25.83 |
| 4096 | 111.47 | 109.52 | 107.63 | 103.99 | 97.28 | 85.75 | 68.41 | 47.28 |
| 8192 | 158.45 | 156.48 | 154.56 | 150.82 | 143.76 | 131.12 | 110.63 | 82.60 |
| app. | 158.42 | 156.42 | 154.42 | 150.42 | 142.42 | 126.42 | 94.42 | 30.42 |

Table 2 : $\overline{p_n^m}$ for some values of $n$ and $m$.

## 4. Conclusions

The practical importance of graph reduction has provided the incentive to develop a model for the effects that reduction has on the structure of its graphs. The question that we have worked on in this paper is: how many nodes may be expected to become unreachable when an arbitrary edge is cut? In doing so a method has been developed that makes some of the standard results of Catalan statistics applicable to graphs. The method works by treating a shared node in a graph as a specially marked node in a tree. The standard enumeration method for ordered binary trees is extended to exclude subtrees with marked nodes. Disconnecting a single edge in a graph may cause shared nodes to become unreachable, because all paths to such shared nodes pass via that single edge. Assuming that all shared nodes are from this particular (removable) type, an upper bound on the expected size of a subtree is calculated ($\overline{r_n^m}$). An estimate from below is obtained by assuming that all shared nodes remain connected (permanent) if an arbitrary edge is cut. The average size of a subtree under this assumption is also derived $(\overline{p_n^m})$.

It turns out that in both cases the expected number of nodes that become unreachable when an arbitrary edge is cut is larger than we have observed in practice. One reason may be that not all edges have the same probability of being cut. Reduction probably takes place more near the leaves than near the root of the graph. However, without further investigation we can only speculate on possible probability distributions. Although the formulae that we have derived are specific with respect to the uniform distribution that we have assumed, the method allows for other distributions to be used.

## Acknowledgements

I am grateful to Henk Barendregt and Arthur Veen for comments on an earlier draft of the paper. Peter van Emde Boas and Michiel de Smid helped me sort out the literature.

## References

1. H. P. Barendregt, *The lambda calculus, its syntax and semantics,* North Holland, Amsterdam (1984).

2. P. J. Landin, "The mechanical evaluation of expressions," *Computer Journal* **6**(4) pp. 308-320 (Jan. 1964).

3. C. P. Wadsworth, *Semantics and pragmatics of the lambda calculus,* Oxford University, U.K. (1971). PhD. Thesis

4. P. H. Hartel, "A comparative study of garbage collection algorithms," PRM project internal report D-23, Computing Science Department, University of Amsterdam (Feb. 1988).

5. J. Cohen, "Garbage collection of linked structures," *Computing Surveys* **13**(3) pp. 341-367 (Sep. 1981).

6.  N. G. de Bruijn, D. E. Knuth, and S. O. Rice, "The average height of planted plane trees," pp. 15-22 in *Graph Theory and Computing*, ed. R. C. Read, Academic Press, London, U.K. (1972).

7.  P. H. Hartel and A. H. Veen, "Statistics on graph reduction of SASL programs," *Software practice and experience* **18**(3) pp. 239-253 (Mar. 1988).

8.  D. E. Knuth, *The art of computer programming, volume 1: Fundamental algorithms*, Addison Wesley, Reading, Massachusetts (1973). second edition

9.  H. W. Gould and J. Kaucky, "Evaluation of a class of binomial coefficient summations," *Journal of Combinatorial theory* **1**(2) pp. 233-247 (Sep. 1966).

10. D. E. Knuth, *The art of computer programming, volume 2: Seminumerical algorithms*, Addison Wesley, Reading, Massachusetts (1980). second edition

# $O(n^2)$ ALGORITHMS FOR GRAPH PLANARIZATION

R. Jayakumar, K. Thulasiraman, and M.N.S. Swamy

Faculty of Engineering and Computer Science
Concordia University
1455 de Maisonneuve Blvd. West
Montreal, Quebec, Canada H3G 1M8

## Abstract

In this paper we present two $O(n^2)$ planarization algorithms--
PLANARIZE and MAXIMAL-PLANARIZE. These algorithms are based on
Lempel, Even, and Cederbaum's planarity testing algorithm [9] and its
implementation using PQ-trees [8, 13]. Algorithm PLANARIZE is for the
construction of a spanning planar subgraph of an n-vertex nonplanar
graph. This algorithm proceeds by embedding one vertex at a time and,
at each step, adds the maximum number of edges possible without
creating nonplanarity of the resultant graph. Given a biconnected
spanning planar subgraph $G_p$ of a nonplanar graph G, algorithm MAXIMAL-
PLANARIZE constructs a maximal planar subgraph of G which contains $G_p$.
This latter algorithm can also be used to maximally planarize a
biconnected planar graph.

## 1. Introduction

A graph is <u>planar</u> if it can be drawn on a plane with no two edges
crossing each other except at their end vertices. A subgraph G' of a
nonplanar graph G is a <u>maximal planar subgraph</u> of G if G' is planar
and adding to G' any edge not present in G' results in a nonplanar
subgraph of G. This process of removing a set of edges from G to
obtain a maximal planar subgraph is known as <u>maximal planarization</u> of
the nonplanar graph G. On the other hand, maximal planarization of a
planar subgraph G refers to the process of adding a maximal set of
edges to G without causing nonplanarity. Determining the minimum
number of edges whose removal from a nonplanar graph will yield a
maximal planar subgraph is an NP-complete problem [1]. However, a few
algorithms which attempt to produce maximal planar subgraphs having
the largest possible number of edges have been reported [2, 3, 4].
Recently, Chiba, Nishioka, and Shirakawa [5] modified Hopcroft and
Tarjan's planarity testing algorithm [6] to construct a maximal planar

subgraph of a nonplanar graph. Their algorithm needs O(mn) time and O(mn) space for a nonplanar graph having n vertices and m edges. Ozawa and Takahashi [7] proposed another O(mn) time and O(m+n) space algorithm to planarize a nonplanar graph using the PQ-tree implementation [8] of Lempel, Even, and Cederbaum's planarity testing algorithm [9, 10]; in short, the LEC algorithm. For a general graph this algorithm may not determine a maximal planar subgraph [11, 12]. Moreover, in certain cases, this algorithm may terminate without considering all the vertices; in other words, it may not produce a spanning planar subgraph.

Whereas the planarization algorithm of [5] constructs the required planar subgraphs by considering one edge at a time, the algorithm of [7] proceeds by considering one vertex at a time. Since an O(mn) maximal planarization algorithm can be constructed in a straightforward manner by adding one edge at a time and testing for planarity at each step, these two algorithms are not significant as far as their complexities are concerned. However, the algorithm of [7] is quite interesting because at each step of this algorithm as many edges as possible are added.

It seems that no maximal planarization algorithm of complexity better than O(mn) will be possible. So, in this paper, we focus our attention on the design of $O(n^2)$ planarization algorithms. We present two planarization algorithms -- PLANARIZE and MAXIMAL-PLANARIZE of time complexity $O(n^2)$ and space complexity O(mn). Algorithm PLANARIZE is for the construction of a spanning planar subgraph of an n-vertex nonplanar graph. This algorithm proceeds by embedding one vertex at a time and, at each step, adds the maximum number of edges possible without creating nonplanarity of the resultant graph. Given a biconnected spanning planar subgraph $G_p$ of a nonplanar graph G, algorithm MAXIMAL-PLANARIZE constructs a maximal planar subgraph of G which contains $G_p$. This latter algorithm can also be used to maximally planarize a biconnected planar graph.

We refer the reader to [10, 13, 8] for discussions of the LEC algorithm and its implementation using PQ-trees. For a discussion of the concept of st-numbering, on which the LEC algorithm is based, [14] may be referred. We follow the terminology and notations used in [7] and [13].

## 2. Principle of an Approach for Planarization

In this section, we discuss the basic principle of an approach for planarization due to Ozawa and Takahashi [7]. This approach is based on the LEC algorithm for planarity testing. Let G denote a simple biconnected st-graph. Let $T_1$, $T_2$, ..., $T_{n-1}$ be the PQ-trees corresponding to the bush forms of G. For any node X in $T_i$, the frontier of X is the left-to- right order of appearance of the leaves in the subtree of $T_i$ rooted at X. Ozawa and Takahashi [7] classify the nodes of any PQ-tree according to their frontier as follows.

Type W: A node is said to be Type W if its frontier consists  of only non-pertinent leaves.

Type B: A node is said to be Type B if its frontier consists of only pertinent leaves.

Type H: A node X is said to be Type H if the subtree rooted at X can be rearranged such that all the descendant pertinent leaves of X appear consecutively at either the left or the right end of the frontier.  Note that at least one non-pertinent leaf will appear at the other end of the frontier.

Type A: A node X is said to be Type A if the subtree rooted at X can be rearranged such that all the descendant pertinent leaves of X appear consecutively in the middle of the  frontier with at least one non-pertinent leaf appearing at each end of the frontier.

The central concept of the planarization algorithm is stated in the following theorem which is essentially a reiteration  of the principle on which the LEC algorithm is based.

THEOREM 1.
    An n-vertex graph G is planar if and only if the pertinent roots in all the PQ-trees $T_2$, $T_3$, ..., $T_{n-1}$ of G are Type B, H or A.  //

We call a PQ-tree <u>reducible</u> if its pertinent root is Type B, H or A; otherwise it is <u>irreducible</u>. Theorem 1 implies that the graph G is planar if and only if all the $T_i$'s are reducible. If any $T_i$ is irreducible, we can make it reducible by appropriately deleting some of the leaves from it. Of course, we would like to delete a minimum number of leaves while trying to make $T_i$ reducible. If we make all the $T_i$'s reducible this way, then a planar subgraph can be obtained by removing from the nonplanar graph the edges corresponding to the leaves that are deleted.

It is easy to see that the PQ-tree $T_{n-1}$ is always reducible because its root is Type B. The tree $T_1$ is also reducible because it has only one pertinent leaf -- the leaf corresponding to the edge (1,2). Consider now an irreducible PQ-tree $T_i$ of  an n-vertex nonplanar graph. For a node X in $T_i$, let w, b, h, and a be the minimum number of descendant leaves of X which should be deleted from $T_i$ so that X becomes Type W, B, H, and A respectively. We denote these numbers of a node as [w,b,h,a]. Any node in $T_i$ may be made Type W, B, H, or A by appropriately deciding the types of its children. So the [w,b,h,a] number of any node can be computed from that of its children. Thus to make $T_i$ reducible, we first traverse it bottom-up from the leaves to the pertinent root and compute the [w,b,h,a] number for every node in $T_i$. Once the [w,b,h,a] number of the pertinent root is computed, we make the pertinent root Type B, H, or A depending on which one of the numbers b, h, and a of the root is the smallest. After determining the type of the pertinent root, we traverse $T_i$ top-down from the pertinent root to the leaves and decide the type of each node in the pertinent subtree of $T_i$. Note that the type of a node uniquely determines the types of its children and so the types of all the leaves in $T_i$ can be determined by this top-down traversal. This information would help us decide the nodes to be deleted from $T_i$ in order to make it reducible. After deleting these nodes from $T_i$, we can apply the reduction procedure to obtain $T_i^*$.

Repeating the above procedure for each irreducible $T_i$, we can obtain a planar subgraph of the nonplanar graph. It is easy to see that if the minimum of b, h, and a for the pertinent root in a PQ-tree $T_i$ is zero, then $T_i$ is reducible. Thus we can determine whether a $T_i$ is reducible or not from the [w,b,h,a] number of its pertinent root. In the following we summarize the above procedure.

**procedure** GRAPH-PLANARIZE(G);
**comment** procedure GRAPH-PLANARIZE determines a planar subgraph of an n-vertex nonplanar graph G by removing at each step a minimum number of edges from G.
**begin**
  construct the initial PQ-tree $T_1 = T_1^*$;
  **for** i := 2 **to** n-2 **do**
    **begin**
      construct the PQ-tree $T_i$ from $T_{i-1}^*$;
      compute the [w,b,h,a] number of each node in the pertinent subtree of $T_i$ by traversing it bottom-up;
      **if** min{b,h,a} for the pertinent root is not zero
        **then begin**
          {$T_i$ is irreducible}
          make the pertinent root Type B, H, or A depending on the minimum of b, h, and a;
          determine the type of each node in $T_i$ by traversing it

```
      top-down;
      delete the necessary nodes from T_i and make it reducible;
      remove from G the edges corresponding to the leaves that are
      deleted from T_i
    end;
    {T_i is now reducible}
    reduce T_i to obtain T_i*
  end
end GRAPH-PLANARIZE;
```

Note that the above algorithm may not determine a maximal planar subgraph. This can be explained as follows. Suppose we delete certain leaves from $T_i$ to make it reducible. In a later reduction step some of the leaves which caused the irreducibility of $T_i$ may themselves be deleted. In such a case, we may be able to return to G a subset of the edges which were removed while making $T_i$ reducible. Hence the planar subgraph obtained by procedure GRAPH-PLANARIZE may not be maximally planar.

Computing the [w,b,h,a] numbers for the nodes in a PQ-tree is a crucial step in procedure GRAPH-PLANARIZE. Ozawa and Takahashi [7] have presented formulas to compute these numbers. The main drawback of their algorithm arises from the fact that they permit deletion of both pertinent and non-pertinent leaves from a tree $T_i$ to make it reducible. Since in $T_i$, the pertinent leaves correspond to the edges entering vertex i+1 in the st-graph G and the non-pertinent leaves correspond to those entering vertices greater than i+1, it may so happen that as the algorithm proceeds, all the edges entering a vertex k > i+1 may get removed from G and thus vertex k and some of other vertices may not be present in the resulting planar subgraph. Thus the planar subgraph determined by Ozawa and Takahashi's algorithm may not be a spanning subgraph of the given nonplanar graph. However, it can be shown that in the case of a complete graph, this algorithm produces a maximal planar subgraph [11], [12].


### 3. A New Graph-Planarization Algorithm

In this section we develop an efficient algorithm to determine a spanning planar subgraph of a nonplanar graph G. The planarization approach discussed in Section 2 will form the basis of this algorithm. We modify Ozawa and Takahashi's approach so that deletion of only pertinent leaves is permitted. We first prove that with this modification, the approach of Section 2 will result in a spanning

**planar subgraph of G.**

<u>THEOREM 2</u>.

The planarization algorithm of Section 2 will determine a spanning planar subgraph of a biconnected n-vertex nonplanar graph, if only pertinent leaves are considered for deletion while making any PQ-tree $T_i$, $3 \leq i \leq n-2$, reducible.

<u>Proof</u>: Note that a PQ-tree with only one pertinent leaf is always reducible. So it follows that from no PQ-tree all the pertinent leaves will get deleted, if only pertinent leaves are to be chosen for deletion. This means that in the subgraph that results at the end of the application of the algorithm, each vertex will be connected to at least one lower numbered vertex. Thus the subgraph will be a spanning subgraph of the given nonplanar graph.  //

Let G be a nonplanar st-graph. Let $E_i$, $2 \leq i \leq n$, be the set of edges entering vertex i in G. We determine a planar subgraph of G by removing a sequence $E_4'$, $E_5'$, ..., $E_{n-1}'$ ($E_i' \subseteq E_i$) of edges such that for each i the subgraph of G obtained by removing the edges in $E_4'$, $E_5'$, ..., $E_i'$ contains a planar subgraph induced by the vertex set {1, 2, ..., i}. Thus after removing the edges in $E_4'$, $E_5'$, ..., $E_{n-1}'$, we obtain a planar subgraph of G. The edges in $E_{i+1}'$, $3 \leq i \leq n-2$, correspond to the pertinent leaves in the PQ-tree $T_i$ which should be deleted to make $T_i$ reducible. Thus $E_{i+1}'$ can be determined while making $T_i$ reducible. In order to make a PQ-tree $T_i$ reducible, we first compute the [w,b,h,a] number for each node in $T_i$. A node in $T_i$ is full if the number of leaves in the pertinent subtree rooted at the node is equal to the number of pertinent leaves. While processing $T_i$ to make it reducible, a full node and all its descendants may be made Type W, or they will remain Type B. On the other hand partial nodes may be made Type W, H, or A; but never Type B because we delete only pertinent leaves from $T_i$. Thus any pertinent node in $T_i$ may be made Type W, H, or A only. So we need to compute only the w, h, and a numbers for the pertinent nodes in $T_i$. We denote these numbers as [w,h,a].

Now we develop formulas to compute the [w,h,a] number for each pertinent node in $T_i$. We process $T_i$ bottom-up from the pertinent leaves to the pertinent root. So when a pertinent node X is processed, the [w,h,a] numbers of all its pertinent children should have already been computed. Thus we can compute the [w,h,a] number for X from the numbers for its pertinent children. In the following, P(X) denotes the set of pertinent children of X and Par(X) denotes the set of partial children of X. Along with the [w,h,a] number for each pertinent node, we also determine, for each pertinent node which is not a leaf, three children called h-child1(X), h-child2(X) and a-child(X) which will be used later to decide the type of each pertinent child of X in the reducible $T_i$.

(i) <u>X is a pertinent leaf</u>.
In this case $w = 1$, $h = 0$, and $a = 0$.

(ii) <u>X is a full node</u>.
In this case $h = 0$, $a = 0$, and

$$w = \sum_{i \in P(X)} w_i,$$

(iii) <u>X is a partial P-node</u>.
To make X Type W, all its pertinent children should be made Type W. Thus

$$w = \sum_{i \in P(X)} w_i.$$

We can make X Type H by making all its full children Type B, one partial child Type H and all other partial children Type W. Thus the h number of X is given by

$$h = \sum_{i \in Par(X)} w_i - \max_{i \in Par(X)} \{(w_i - h_i)\}.$$

In this case the partial child which is made Type H will be the h-child1(X).

We can make X Type A in two different ways. We can make one partial child of X Type A and all other pertinent children Type W. In this case

$$\alpha_1 = \sum_{i \in Par(X)} w_i - \max_{i \in P(X)} \{(w_i - a_i)\}$$

descendant pertinent leaves of X will have to be deleted. The partial child which is made Type A will be the a-child(X). On the other hand, if we make two partial children Type H, all full children Type B and all other pertinent children Type W, then

$$\alpha_2 = \sum_{i \in Par(X)} w_i - \max1_{i \in Par(X)} \{(w_i - h_i)\} - \max2_{i \in Par(X)} \{(w_i - h_i)\}$$

descendant pertinent leaves will have to be deleted from $T_i$ to make X Type A, where max1 is the first maximum and max2 is the second maximum. The partial child having $\max1\{(w_i - h_i)\}$ will be the h-child1(X) and the one having $\max2\{(w_i-h_i)\}$ will be the h-child2(X). Thus the P-node X can be made Type A by deleting

$$a = \min\{\alpha_1,\alpha_2\}$$

pertinent leaves from $T_i$. If the value of a is different from $\alpha_1$, then we make a-child(X) empty.

(iv) X is a partial Q-node.

To make X Type W, all its pertinent children should be made Type W. Thus for X

$$w = \sum_{i\in P(X)} w_i.$$

To compute the h number of X, first note that X can be made Type H only if either its leftmost child or its rightmost child is pertinent. Suppose that the leftmost child of X is pertinent. Then let us traverse the children of X from left to right and find $P_L(X)$, the maximal consecutive sequence of pertinent children such that only the rightmost node in $P_L(X)$ may be partial. If the leftmost child of X is not pertinent, then $P_L(X)$ will be empty. Suppose, on the other hand, that the rightmost child of X is pertinent. As we traverse the children of X from right to left, let $P_R(X)$ be the maximal consecutive sequence of pertinent children such that only the leftmost node in $P_R(X)$ may be partial. If the rightmost child of X is not pertinent, then $P_R(X)$ is empty. We can easily see that X can be made Type H by deleting

$$h = \sum_{i\in P(X)} w_i - \max\left\{\sum_{i\in P_L(X)} (w_i-h_i), \sum_{i\in P_R(X)} (w_i-h_i)\right\}$$

pertinent leaves from $T_i$. We let h-child1(X) be the leftmost node in $P_L(X)$ or the leftmost node in $P_R(X)$ depending on which one has the maximum $\sum(w_i-h_i)$ sum in the above formula for h.

Node X can be made Type A in two different ways. We can make one of the pertinent children of X Type A and all the other pertinent children Type W. This can be achieved by deleting

$$\beta_1 = \sum_{i\in P(X)} w_i - \max_{i\in P(X)} \{(w_i-a_i)\}$$

pertinent leaves from $T_i$. In this case the pertinent child having $\max\{(w_i-a_i)\}$ will be the a-child(X). Let $P_A(X)$ be a maximal consecutive sequence of pertinent children of X such that all the nodes in $P_A(X)$ except the leftmost and the rightmost ones are full. The endmost nodes may be full or partial. Then we can make X Type A by making all the full nodes in $P_A(X)$ Type B, the partial nodes in $P_A(X)$ Type H and all the other pertinent children of X Type W. Note that there may be more than one $P_A(X)$. Thus we can make X Type A by deleting

$$\beta_2 = \sum_{i \in P(X)} w_i - \max_{P_A(X)} \left\{ \sum_{i \in P_A(X)} (w_i - h_i) \right\}$$

pertinent leaves from $T_i$. In this case we let the leftmost node in the $P_A(X)$ selected be the h-child2(X). Thus node X can be made Type A with the deletion of

$$a = \min\{\beta_1, \beta_2\}$$

pertinent leaves from $T_i$. If the value of a is different from $\beta_1$, then we make a-child(X) empty.

Traversing $T_i$ bottom-up we can compute the [w,h,a] number for each pertinent node in $T_i$ using the above formulas. The procedure which computes these numbers for a given $T_i$ will be referred to as COMPUTE1($T_i$) [12].


LEMMA 1.
The [w,h,a] numbers for all the pertinent nodes can be correctly computed in $O(n^2)$ time.

Proof: Proof of correctness follows from our discussions so far. As regards the complexity, note that for a Q-node in $T_i$ procedure COMPUTE1($T_i$) traverses all the children of the node. Thus the amount of work done for all the Q-nodes in a $T_i$ is proportional to the number of children of all the Q-nodes in $T_i$. The children of a Q-node corresponding to a block represent vertices, except the lowest, on the outside window of the block. Moreover, any vertex in G which is represented as a child of a Q-node in $T_i$ can appear on the outside window of only one block. Thus the total number of children of all the Q-nodes in $T_i$ is less than or equal to n, the number of vertices in G.

For a P-node, the work done by procedure COMPUTE1($T_i$) is proportional to the number of its pertinent children. A pertinent child of a P-node is either a P-node or a Q-node or a leaf. Since a Q-node represents a block, there are no more than n Q-nodes in any $T_i$. Also the number of pertinent leaves in $T_i$ is in-deg(i+1), where in-deg(i+1) is the number of edges entering vertex i+1 in G. Furthermore the number of P-nodes in $T_i$ is at most i. Thus the amount of work for all the P-nodes in $T_i$ is $O(n + \text{in-deg}(i+1))$.

It follows from the above that the amount of work done by

procedure COMPUTE1($T_i$) for all the Q-nodes and P-nodes in $T_i$ is $O(n +$ in-deg($i+1$)). Summing up the work done for all $T_i$'s, we get the complexity of computing the [w,h,a] numbers as $O(m+n^2) = O(n^2)$. //

After computing the [w,h,a] number for the pertinent root of $T_i$, we can determine whether $T_i$ is reducible or not. If the minimum of h and a is zero for the pertinent root of $T_i$, then $T_i$ is reducible. If $T_i$ is not reducible, then we make the pertinent root of $T_i$ Type H or A depending on which one of h and a is minimum, and make $T_i$ reducible by deleting the necessary pertinent leaves from $T_i$. Now we need to determine the type of each pertinent node in $T_i$ to obtain a reducible $T_i$. Note that $T_i$ may have certain full nodes. If we decide to keep any such full node, then we mark it Type B.

Consider now a pertinent node X in $T_i$ whose type has been determined. To start with X is the pertinent root. We can determine the types of all the pertinent children of X uniquely from the type of X as follows.

If X is Type B, then it is a full node and we would like to keep X as well as all its descendants in $T_i$. So no action needs to be taken in this case. On the other hand, if X is not Type B, then we traverse the pertinent descendants of X to determine their type. An easy case is when X is a leaf. Then it should be Type W and so we have to delete it from $T_i$. We also have to remove the edge corresponding to X from G. Thus, in this case, the edge corresponding to X should be included in $E'_{i+1}$. If X is not a leaf, then we have the following different cases to consider.

Suppose X is Type W. Then all its pertinent children should be made Type W. Moreover, if any of these pertinent children is a full node, then the entire subtree of $T_i$ rooted at that full child should be deleted from $T_i$. If X is Type H and a P-node, then we make the partial child h-child1(X) Type H, all the full children Type B and all other partial children Type W. If X is Type H, but a Q-node, then we traverse the children of X from h-child1(X) towards the rightmost child and determine the maximal consecutive sequence of pertinent children $P_L(X)$ or $P_R(X)$. We then make all the nodes in this sequence Type B; the rightmost node in $P_L(X)$ or the leftmost node in $P_R(X)$ are made Type H and all other pertinent children of X are made Type W.

Suppose X is Type A and a P-node. Then we process the pertinent children of X as follows. If a-child(X) is not empty, then we make a-child(X) Type A and all other pertinent children Type W. On the other hand, if a-child(X) is empty, then we make the partial children h-child1(X) and h-child2(X) Type H, all full children of X Type B and all other partial children of X Type W.  If X is Type A and a Q-node, then we should process its pertinent children as follows.  If a-child(X) is not empty, then we make a-child(X) Type A and all other pertinent children Type W.  If a-child(X) is empty, then we traverse the children of X from h-child2(X) towards the rightmost child and find the maximal consecutive sequence $P_A(X)$ of pertinent children of X.  Then we make all nodes in $P_A(X)$ Type B, the endmost nodes in $P_A(X)$, if they are partial, Type H and all other pertinent children Type W.

From the above discussions it should be clear that the type of any pertinent node in $T_i$ uniquely determines the types of its pertinent children.  Hence we process the PQ-tree $T_i$ top-down from the pertinent root, and determine the set of edges $E'_{i+1}$ and delete from $T_i$ the nodes which are full and marked Type W. The procedure which achieves these will be denoted by DELETE-NODES($T_i$) [12].  Since certain pertinent leaves are deleted from $T_i$, we have to update, if necessary, for each node the number of descendant leaves.  Procedure DELETE-NODES($T_i$) performs this update also.

LEMMA 2.
   All the edges in the sets $E'_{i+1}$, $3 \leq i \leq n-2$, can be determined and removed using procedure DELETE-NODES($T_i$) in $O(n^2)$ time.

Proof: Note that for each node X in $T_i$ procedure DELETE-NODES($T_i$) traverses all the pertinent children if X is a P-node, and all the children if X is a Q-node.  Thus, as in the proof of Lemma 1 we can see that the cost of procedure DELETE-NODES($T_i$) for each i is $O(n + \text{in-deg}(i+1))$ and the lemma follows. //

Having made $T_i$ reducible, we can now reduce it to obtain $T_i^*$ using Booth and Lueker's PQ-tree reduction algorithm [8, 13].  We can then obtain the next PQ-tree $T_{i+1}$ and repeat our procedures to make $T_{i+1}$ reducible.  Note that the reduction of all the reducible PQ-trees can be performed in $O(m+n)$ time if we keep the parent pointers for all children of P-nodes and for the endmost children of Q-nodes.  Thus in Booth and Lueker's algorithm, interior children of Q-nodes in any $T_i$ are not assigned valid parent pointers and if any such interior child becomes pertinent, then its parent pointer will be determined during the bubble-up phase.  In our discussions so far, we have assumed that the correct parent pointer for every pertinent node is available.  So we have to determine the parent pointers of all the pertinent nodes in $T_i$ before processing it.  Booth and Lueker's planarity testing algorithm stops when it detects during the bubble-up phase that certain pertinent nodes cannot be assigned parent pointers, for that would imply nonplanarity of the given graph. However, since our aim is to planarize the nonplanar graph, we would like to proceed further to find parent pointers of all the pertinent nodes. As a result, our bubble-up algorithm described below is different from Booth and Lueker's.

Let X be a pertinent node in $T_i$. If X is a child of a P-node or one of the endmost children of a Q-node, then it has a valid parent pointer. On the other hand, if X is an interior child of a Q-node, then its parent pointer will be empty. To find the correct parent pointer for X, we traverse the siblings of X from X towards the rightmost child and obtain the parent pointer for X from that of the rightmost child. Let Y be the parent of X in $T_i$. If at a later time another child Z of Y is processed to find its parent pointer, then the above procedure would require traversing again all the children of Y up to the rightmost child and may result in visiting certain nodes several times. To avoid these unnecessary visits, we proceed as follows. When we traverse the children of Y from X to the rightmost child, we assign the parent pointer of the rightmost child to all the nodes traversed and store these nodes in a queue called interior-queue. So when a child Z of Y is processed, if its parent pointer is empty, then we traverse the siblings of Z until we find a node with a non-empty parent pointer. Though this path compression technique makes our bubble-up procedure efficient, many non-pertinent children of Q-nodes may be assigned parent pointer. In order to make the parent pointer of such non-pertinent nodes empty, we process the interior-queue at the end of the bubble-up. If any node in this queue is not pertinent, then its parent pointer is made empty.

The efficiencies of our procedures COMPUTE1($T_i$) and DELETE-NODES($T_i$) arise from the fact that we process only the pertinent children of any P-node. In a PQ-tree the pertinent children of a P-node may appear in any arbitrary order and so we may have to traverse all the children of a P-node to find the pertinent children. In order to avoid this, we split the children of each pertinent P-node into two groups -- one group consisting of pertinent children only and the other consisting of only non-pertinent children. The procedure which finds the parent pointer for all the pertinent nodes in a PQ-tree $T_i$ and groups the pertinent children of P-nodes together as described above will be referred to as BUBBLE-UP($T_i$) [12]. This procedure also computes the number of pertinent children as well as the number of descendant pertinent leaves of each pertinent node in the PQ-tree $T_i$.

LEMMA 3.

The total cost of procedure BUBBLE-UP($T_i$) for all $2 \leq i \leq n-2$ is $O(n^2)$.

Proof: For a PQ-tree $T_i$, the computational work done by procedure BUBBLE-UP($T_i$) for nodes which are children of Q-nodes is proportional to the number of children of all the Q-nodes in $T_i$, which is $O(n)$. The computational work done for nodes which are children of P-nodes is proportional to the number of pertinent nodes in $T_i$, which is $O(n + \text{in-deg}(i+1))$. Thus the total work required for any $T_i$ is $O(n + \text{in-deg}(i+1))$. Summing this for all the PQ-trees $T_i$ we get the time complexity of procedure BUBBLE-UP($T_i$) for all $2 \leq i \leq n-2$ as $O(m+n^2) = O(n^2)$. //

Procedures COMPUTE1($T_i$) and DELETE-NODES($T_i$) require that we should be able to determine whether a pertinent node in $T_i$ is full or partial. A pertinent node is full if the number of descendant pertinent leaves of the node is equal to the number of its descendant leaves; otherwise it is partial. Procedure BUBBLE-UP($T_i$) determines the number of descendant pertinent leaves of every pertinent node in $T_i$. Now we should find a way of determining the number of descendant leaves of every pertinent node in $T_i$. Clearly each leaf has one descendant leaf. In $T_1$, the only node which is not a leaf is the P-node corresponding to vertex 1. Thus the number of descendant leaves of this P-node is the number of edges incident out of vertex 1 in G. We determine the number of descendant leaves of any node in $T_i$, $2 \leq i \leq n-2$, from the tree $T_{i-1}$ as follows.

Assume that the number of descendant leaves of each node in $T_{i-1}$ is known. During the processing of $T_{i-1}$ we may delete some leaves from it to make it reducible. Procedure DELETE-NODES ($T_{i-1}$) also updates the number of descendant leaves of the nodes in $T_{i-1}$. Thus in $T_{i-1}^*$ the correct number of descendant leaves for each node is known. Let $E_i = \{(j_1,i), (j_2,i), \ldots, (j_k,i)\}$ be the set of edges entering vertex i in the planar subgraph obtained from G. In $T_{i-1}^*$ the leaves corresponding to the edges in $E_i$ appear as children of the same node, say X. Since these leaves are removed from $T_{i-1}^*$ to form $T_i$, the number of descendant leaves of the nodes corresponding to the vertices $j_1, j_2, \ldots, j_k$, if they are present in $T_i$, should be decreased by one and the number of descendant leaves of node X and its ancestors in $T_i$ should be decreased by in-deg(i). Moreover, we construct $T_i'$ from $T_{i-1}^*$ by adding a P-node corresponding to vertex i with leaves corresponding to the edges incident out of vertex i in G as its children. Clearly the number of descendant leaves of this P-node is equal to out-deg(i) in G. Since this node is made a child of node X, the number of descendant leaves of node X and all its ancestors in $T_i$ should be increased by out-deg(i). Thus for node X and for each one of its ancestors in $T_i$, the net increase in the number of descendant leaves is (out-deg(i) - in-deg(i)). The procedure which performs this updating will be referred to as UPDATE-DESCENDANTS($T_i$) [12].

LEMMA 4.

The total cost of procedure UPDATE-DESCENDANTS($T_i$) for all $2 \leq i \leq n-2$ is $O(n^2)$.

Proof: For a $T_i$, $2 \leq i \leq n-2$, the updates for the nodes corresponding to the vertices $j_1$, $j_2$, ..., $j_k$ require $O(in-deg(i))$ time. The updates for the other nodes may, in the worst case, result in traversing all the nodes which are not leaves in $T_i$. This would require $O(n)$ time. The total computational work required by procedure UPDATE-DESCENDANTS($T_i$) for all $T_i$'s is therefore $O(m+n^2) = O(n^2)$. //

Now we present our planarization algorithm which uses the procedures described so far. This algorithm determines a spanning planar subgraph $G_p$ of the nonplanar graph G and the sets $E_3'$, $E_4'$, ..., $E_{n-1}'$ of edges to be removed from G to obtain $G_p$.

**procedure** PLANARIZE(G);
**comment** procedure PLANARIZE determines the set of edges $E_3' = \phi, E_4'$,
         ..., $E_{n-1}'$ to be removed from a nonplanar graph G to obtain a
         spanning planar subgraph $G_p$.
**begin**
  {DESCENDANT-LEAVES(X) denotes the number of descendant leaves of node X}
  construct the initial PQ-tree $T_1 = T_1^*$;
  DESCENDANT-LEAVES(1) := out-deg(1);
  **for** each leaf X corresponding to an edge in $E_2$ **do**
    DESCENDANT-LEAVES(X) := 1;
  **for** i := 2 **to** n-2 **do**
    **begin**
      initialize $E_{i+1}'$ to be empty;
      construct the PQ-tree $T_i$ from $T_{i-1}^*$;
      UPDATE-DESCENDANTS($T_i$);
      **for** the P-node X corresponding to vertex i **do**
        DESCENDANT-LEAVES(X) := out-deg(i);
      **for** each leaf X corresponding to an edge in $E_{i+1}$ **do**
        DESCENDANT-LEAVES(X) := 1;
      BUBBLE-UP($T_i$);
      COMPUTE1($T_i$);

```
if min{h,a} for the pertinent root is not zero
  then begin
     make the pertinent root Type H or A corresponding to the
     minimum of h and a;
     DELETE-NODES(T_i)
  end;
  reduce T_i to obtain T_i*
end
end PLANARIZE;
```

THEOREM 3.

   Procedure PLANARIZE determines a spanning planar subgraph of a nonplanar graph G in $O(n^2)$ time and $O(m+n)$ space.

Proof: The fact that procedure PLANARIZE determines a spanning planar subgraph of a nonplanar graph follows from our discussions and Theorem 2. It follows from Lemmas 1 to 4 that all the procedures used in procedure PLANARIZE are of time complexity $O(m+n^2)$. The PQ-tree reduction procedure is of time complexity $O(m+n)$ [8]. Thus procedure PLANARIZE is of time complexity $O(m+n^2) = O(n^2)$.

   The space required by the procedure is bounded by the space required to store the PQ-trees, which is $O(m+n)$. Hence the theorem.   //

   As an example, we applied our graph-planarization algorithm on the nonplanar graph G shown in Fig. 1. Our algorithm determines $E_6' = \{(2,6)\}$, $E_8' = \{(2,8)\}$, and $E_9' = \{(2,9), (3,9)\}$ as the sets of edges to be removed from G to planarize it and the spanning planar subgraph $G_p$ is shown in Fig. 2. In Fig. 3 we show a planar embedding of $G_p$. From Fig. 3 we can easily see that the planar subgraph obtained is not maximally planar, since the edge (2,8) in $E_8'$ can be added to this embedding without affecting the planarity of the resultant graph. Thus the spanning subgraph determined by procedure PLANARIZE may not be maximally planar.


### 4. A Maximal Planarization Algorithm

   For a given a nonplanar graph G, let $G_p$ be a spanning planar subgraph of G obtained by the procedure PLANARIZE described in the previous section. Our interest in this section is to study the problem of constructing a maximal planar subgraph of G which contains G. For a successful application of Lempel, Even and Cederbaum's algorithm for determining the required maximal planar subgraph, it is necessary that $G_p$ have an st-numbering. This requirement necessitates

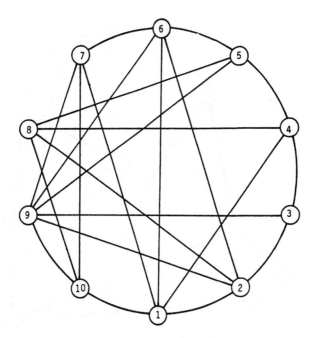

Figure 1
Nonplanar Graph G

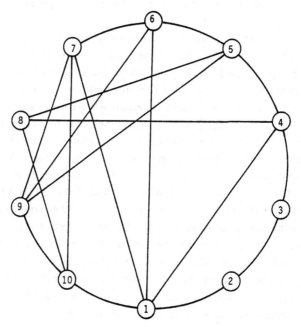

Figure 2
Spanning Planar Subgraph $G_p$

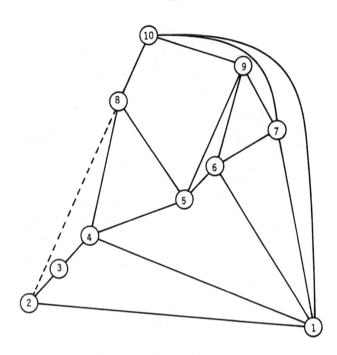

**Figure 3**

Planar Embedding of the Planar Subgraph $G_p$

that we assume that $G_p$ is biconnected, since a graph which is not biconnected may not have an st-numbering. Note that, in general, the planar subgraph produced by procedure PLANARIZE may not be connected. With the assumption that $G_p$ is biconnected, we now proceed to develop an $O(n^2)$ algorithm to construct a maximal planar subgraph of G which contains $G_p$. Let $E_3' = \phi, E_4', \ldots, E_{n-1}'$ be the sets of edges removed by procedure PLANARIZE to obtain $G_p$. Since more than one maximal planar subgraphs containing $G_p$ may exist, our aim will be to attempt to maximize the number of edges in the required graph. Thus, we attempt to add to $G_p$ as many edges as possible from the sets $E_3'$, $E_4'$, $\ldots$, $E_{n-1}'$ without affecting the planarity of the resultant graph. Our approach to maximally planarize $G_p$ is to start with G and construct its PQ-trees. After constructing a PQ-tree, say $T_i$, we make it reducible by deleting a minimum number of leaves representing the edges in $E_{i+1}'$. (Note that $T_i$ will become reducible if all the leaves from the set $E_{i+1}'$ are deleted from $T_i$.) This can be done by computing the [w,h,a] number of the pertinent nodes in $T_i$. Let $T_i(G_p)$ denote the smallest subtree of $T_i$ whose frontier contains all the pertinent leaves from $G_p$. Since we would like to include $G_p$ in the final maximal planar subgraph, we take care, during the reduction of $T_i$, not to make any node in $T_i(G_p)$ Type A except its root. This would ensure that the bottom-up reduction process proceeds at least up to the root of $T_i(G_p)$ and possibly beyond. Also, we note that, while computing the [w,h,a] numbers we ignore the presence of empty leaves from $G-G_p$.

In the following, the leaves in $T_i$ corresponding to the edges in $E'_{i+1}$ will be called the <u>new pertinent leaves</u> of $T_i$ and the other pertinent leaves of $T_i$ (corresponding to the edges entering vertex i+1 in $G_p$) will be called <u>preferred leaves</u>. A node is called full if its frontier has no empty leaf from $G_p$; it is empty if its frontier has only empty leaves from $G_p$; otherwise it is partial. We call pertinent node X a <u>preferred node</u> if it has some of the preferred leaves among its descendants. If X is not preferred, then it may either be retained in the reducible $T_i$ or it may be deleted along with all its descendants to make $T_i$ reducible. The formulas for computing the [w,h,a] numbers of pertinent nodes are the same as those given in Section 3. So, in the following we only consider the essential features of our Type assignment policy which guarantees that $G_p$ is included in the final maximal planar graph. Let X be a pertinent node in $T_i$.

## Case 1: X is a partial P-node
In this case X can have at most two partial preferred children.
(a) <u>X has no partial preferred children</u>: If X is the root of $T_i(G_p)$ or its ancestor in $T_i$, then it can be included in the reducible $T_i$ by making it Type A or Type H. Otherwise it can be included only by making it Type H. Note that in the latter situation we should set h-child2(X) and a-child(X) empty.
(b) <u>X has exactly one partial preferred child</u>: The partial preferred child has to be retained in the reducible $T_i$ and it becomes h-child1(X). Moreover, if X is the root of $T_i(G_p)$ or its ancestor in $T_i$, then it can be included by making it Type A or Type H. Otherwise it can be included only by making it Type H. Note that if X is the root of $T_i(G_p)$, then none of its children can be made Type A and so a-child(X) will be empty in this case.
(c) <u>X has two partial preferred children</u>: In this case X is the pertinent root of the reducible $T_i$. One of the partial preferred children of X becomes h-child1(X) and the other becomes h-child2(X). We set a-child(X) empty and remember that the root is processed.

## Case 2: X is a partial Q-node
In this case, all the preferred pertinent children of X should appear consecutively. We first traverse these children from the leftmost child towards the rightmost child and determine the maximal consecutive sequence P'(X) with the properties
(i) P'(X) contains all the preferred children of X;

(ii) only the leftmost node and/or the rightmost node in P'(X) may be partial; that is, its frontier may contain an empty leaf from $G_p$; and

(iii) all the other nodes in P'(X) are full; that is, the frontier of each of these nodes contains no empty leaf belonging to $G_p$.

Now X can be made Type H only when one of the following happens:

(i) P'(X) appears at the left end of X and the leftmost node in P'(X) has in its frontier no empty leaf from $G_p$. Then we set $P_L(X) = P'(X)$,

(ii) P'(X) appears at the right end of X and the rightmost node in P'(X) has in its frontier no empty leaf from $G_p$. Then we set $P_R(X) = P'(X)$.

(Note that $P_L(X)$ and $P_R(X)$ are as defined in Section 3.)

In both the above cases, we set h-childl(X) to the leftmost node in P'(X) and compute the h number for X.

Suppose neither of the above conditions is satisfied. Then if P'(X) contains only one node, this node should be made Type H or A corresponding to the minimum of h and a. If P'(X) is made Type A, then the only node in P'(X) becomes a-child(X); otherwise it becomes h-childl(X). If P'(X) has more than one node, then we set h-child2(X) to the leftmost node in P'(X) and compute the a number for X. We also remember in this case that the pertinent root is processed.

Processing the pertinent nodes of $T_i$ up to the pertinent root using the above ideas and using the formulas of Section 3, we can determine the [w,h,a] numbers of all the pertinent nodes in $T_i$. The procedure which achieves this will be referred to COMPUTE2($T_i$) [12].

LEMMA 5.

The [w,h,a] numbers of the pertinent nodes in all the PQ-trees can be computed in $O(n^2)$ time using procedure COMPUTE2($T_i$), $2 \leq i \leq n-2$.

Proof: It is easy to see that the computational work done by procedure COMPUTE2($T_i$), $2 \leq i \leq n-2$, is of the same order as that of procedure COMPUTE1($T_i$). Hence the proof follows from Lemma 1. //

Having computed the [w,h,a] numbers for the pertinent nodes in $T_i$, we can obtain a reducible $T_i$ by traversing the pertinent subtree top-down from the pertinent root using procedure DELETE-NODES($T_i$) (see Section 3). Note that while applying Procedure DELETE-NODES($T_i$), the terms "full", "empty" and "partial" should be understood as defined at the beginning of this section. During this processing some of the new pertinent leaves in $T_i$ may not be processed at all. Clearly, such pertinent leaves should be deleted from $T_i$ to make it reducible and the edges corresponding to these leaves should also be removed from the nonplanar graph G.

Recall that procedure COMPUTE2($T_i$) requires that we be able to determine whether the frontier of a node X has an empty leaf from $G_p$. Suppose the frontier of X in $G_p$ has only empty leaves. Then the procedure BUBBLE-UP($T_i$) of Section 3 will not even visit this node because this procedure traverses only the pruned pertinent subtree of $T_i$ which does not include empty leaves. In order to overcome this problem, we modify BUBBLE-UP($T_i$) so that in addition to traversing all the nodes in the pruned pertinent subtree of $T_i$, it also traverses every node whose frontier contains at least one empty leaf from $G_p$. Interestingly, as we show now, this modification does not affect the complexity of the bubble up process (given in Lemma 3), and we refer to this modified procedure as MODIFIED-BUBBLE-UP($T_i$).

LEMMA 6.

The total cost of MODIFIED-BUBBLE-UP($T_i$) for all $2 \leq i \leq n-2$ is $O(n^2)$.

Proof: Let $n_p(T_i)$ be the total number of leaves of $T_i$ belonging to $G_p$ and let UNARY($T_i$) be the number of unary nodes (nodes of degree one) traversed by MODIFIED-BUBBLE-UP($T_i$). Then the cost of MODIFIED-BUBBLE-UP($T_i$) is $O(n_p(T_i) + \text{UNARY}(T_i))$. But $n_p(T_i) = O(m_p)$ where $m_p$ is the number of edges in $G_p$ and UNARY($T_i$) = $O(n)$. Since $G_p$ is planar, $m_p = O(n)$. Hence cost of MODIFIED-BUBBLE-UP($T_i$) is $O(n)$. Summing up this cost for all $T_i$, $2 \leq i \leq n-2$, we get the result. //

Processing the PQ-trees $T_2$, $T_3$, ... $T_{n-2}$ using the different procedures described above we obtain a maximal planar subgraph of the nonplanar graph G. We now give an ALGOL-like description of the complete algorithm.

**procedure** MAXIMAL-PLANARIZE(G);
**comment** procedure MAXIMAL-PLANARIZE determines a maximal planar subgraph of the nonplanar graph G. This procedure uses the spanning planar subgraph obtained by procedure PLANARIZE.
**begin**
{Determine the spanning planar subgraph}
PLANARIZE(G);
{Maximally planarize the spanning planar subgraph}
construct the initial PQ-tree $T_1 = T_1^*$;
DESCENDANT-LEAVES(1) := out-deg(1);

```
for each leaf X corresponding to an edge in E₂ do
    DESCENDANT-LEAVES(X) := 1;
for i := 2 to n-2 do
    begin
        construct the PQ-tree Tᵢ from T*ᵢ₋₁;
        UPDATE-DESCENDANTS(Tᵢ);
        for the P-node X corresponding to vertex i do
            DESCENDANT-LEAVES(X) := out-deg(i);
        for each leaf X corresponding to an edge in Eᵢ₊₁ do
            DESCENDANT-LEAVES(X) := 1;
        MODIFIED-BUBBLE-UP(Tᵢ);
        COMPUTE2(Tᵢ);
        if min{h,a} for the pertinent root is not zero
            then begin
                make the pertinent root Type H or A corresponding to the
                minimum of h and a;
                DELETE-NODES(Tᵢ);
                delete the new pertinent leaves which are not processed from
                Tᵢ
            end;
        reduce Tᵢ and obtain T*ᵢ
    end
end MAXIMAL-PLANARIZE;
```

Clearly, when algorithm MAXIMAL-PLANARIZE is applied on a non-planar graph G, treating the edges of a planar subgraph $G_p$ as preferred edges, it produces a planar subgraph G' containing $G_p$. It can be shown that G' is indeed a maximal planar subgraph of G.

THEOREM 4.

Algorithm MAXIMAL-PLANARIZE when applied on a nonplanar graph G, treating a biconnected planar subgraph $G_p$ as the preferred graph, produces a maximal planar graph G' which contains $G_p$.

Proof: As we noted before, G' is planar and contains $G_p$. So, we need only prove that G' is a maximal planar subgraph of G. Assume the contrary. Then there exists an edge $e = (j,k) \in G$, $j < k$, such that $e \notin G'$ and $G' \cup \{e\}$ is planar. Among all such edges select the one for which k is minimum and let this edge be $e = (j, i+1)$. This means that the leaf in $T'_i$ representing e is a new pertinent one with respect to G'. Note that in $T_i$ this leaf was also new pertinent with respect to $G_p$, and it was not added while procedure MAXIMAL-PLANARIZE constructed G' starting from $G_p$. Furthermore, $T'_i$ is isomorphic to the corresponding PQ-tree $T_i$ generated when $G_p$ is treated as the preferred graph.

Also, since $G_p \subseteq G'$ all the preferred pertinent leaves of $T_i$ will also be preferred pertinent leaves in $T_i'$. Furthermore, some of the new pertinent leaves in $T_i$ may become preferred ones in $T_i'$.

Since $G' \cup \{e\}$ is planar, through a sequence of Q-node flippings and permutations of children of P-nodes, $T_i'$ can be converted into a tree $T_i''$ such that its frontier contains a maximal sequence L' with the following properties.

(i) Except the first and/or the last leaf in L', none of the others are empty leaves from G'. (Note: this means that except the first and/or the last leaf in L', none of the others are empty leaves from $G_p$.)

(ii) L' contains all the preferred pertinent leaves of G'. (Note: some of these leaves are, possibly, present in $T_i$ as new pertinent ones; the remaining leaves are preferred ones in $T_i$.)

(iii) L' contains the new pertinent leaf e and possibly a few more new pertinent ones. (Note: these leaves are all new pertinent in $T_i$ too.)

(iv) Empty leaves from G-G' may be added to L' to make it maximal. (Note: these leaves belong to $G-G_p$ too, since $G_p \subseteq G'$.)

Since $T_i'$ and $T_i$ are isomorphic, it follows that $T_i$ can also be converted into $T_i''$. Furthermore, the above observations imply that L' satisfies all the properties required to be satisfied by the sequence P'(X) (where X is the pertinent root of $T_i$) identified by COMPUTE2($T_i$). But, P'(X) is a proper subset of L' since $e \notin P'(X)$. This is a contradiction because P'(X) is not maximal as required. //

THEOREM 5.
Procedure MAXIMAL-PLANARIZE is of complexity $O(n^2)$ in time and $O(m+n)$ in space.

Proof: All the procedures used in procedure MAXIMAL-PLANARIZE are of time complexity $O(n^2)$. The PQ-tree reductions can be performed in $O(m+n)$ time [8]. Hence procedure MAXIMAL-PLANARIZE has $O(n^2)$ time complexity. Regarding the space complexity, note that the space required by the algorithm is bounded by the space required to store the different PQ-trees, which is $O(m+n)$. //

As an example, applying procedure MAXIMAL-PLANARIZE on the planar subgraph $G_p$ shown in Fig. 2, we obtain the maximal planar subgraph shown in Fig. 4.

We have implemented procedure MAXIMAL-PLANARIZE in PASCAL and tested it on several nonplanar graphs using a CDC Cyber 170. In Table 1 we show the number of edges removed by procedure PLANARIZE and the number of edges added by procedure MAXIMAL-PLANARIZE for some of the test graphs. It can be seen from Table 1 that procedure MAXIMAL-PLANARIZE adds only a very small number of edges to the spanning

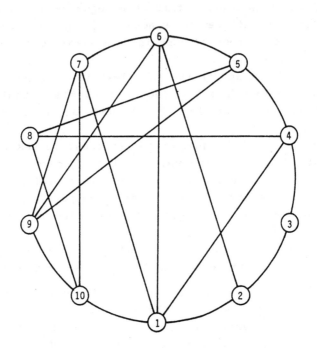

**Figure 4**

Maximal Planar Subgraph Containing $G_p$ of Fig. 2

planar subgraph. We have also shown in Table 1 the execution time required to find the maximal planar subgraph for these graphs.

Given an n-vertex biconnected graph $G_p$, it is easy to see that by applying algorithm MAXIMAL-PLANARIZE on the n-vertex complete graph $K_n$ and treating $G_p$ as the preferred graph, we can construct a maximal planar subgraph which contains $G_p$. Thus, we can maximally planarize any biconnected planar graph using algorithm MAXIMAL-PLANARIZE and we have the following theorem.

THEOREM 6.

Maximal planarization of an n-vertex biconnected planar graph can be achieved in $O(n^2)$ time. //

We now return to the question of the maximal planarization problem when the subgraph $G_p$ of G produced by algorithm PLANARIZE is not biconnected. In this case, the st-numbering of G which we used before applying algorithm PLANARIZE on the graph G may not be an st-numbering for $G_p$ because a graph which is not biconnected may not have

<u>Table 1</u>
Computational Results

| Graph | Number of vertices | Number of edges | Number of edges removed by procedure PLANARIZE | Number of edges added by procedure MAXIMAL-PLANARIZE | Execution Time (sec) |
|-------|------|------|------|------|------|
| $G_1$ | 10 | 35 | 21 | 3 | 0.263 |
| $G_2$ | 20 | 60 | 24 | 0 | 0.672 |
| $G_3$ | 30 | 95 | 42 | 5 | 0.976 |
| $G_4$ | 40 | 125 | 39 | 2 | 1.321 |
| $G_5$ | 50 | 150 | 47 | 4 | 1.985 |
| $G_6$ | 60 | 180 | 53 | 3 | 3.126 |
| $G_7$ | 70 | 225 | 57 | 0 | 4.795 |
| $G_8$ | 80 | 250 | 78 | 7 | 5.013 |
| $G_9$ | 90 | 300 | 103 | 5 | 6.792 |
| $G_{10}$ | 100 | 350 | 124 | 8 | 7.863 |

an st-numbering. Since the embedding produced by the LEC algorithm assumes that the vertices are placed at different levels dictated by the st-numbers, it follows that algorithm MAXIMAL-PLANARIZE when applied on $G_p$ produces a maximal planar subgraph (containing $G_p$) of G which is consistent with the embedding of $G_p$ as dictated by the original st-numbers. But such a subgraph may not be a maximal subgraph containing $G_p$. So, one way to proceed further for the construction of the required maximal planar graph is to first obtain the biconnected components of $G_p$. Then we should examine each edge for possible addition to one or more biconnected components of $G_p$. But, before doing so, we should obtain the new st-numbering for the graph obtained by adding the new edge to $G_p$. The complexity of such a maximal planarization algorithm will be $O(mn)$, which is the same as that of the straightforward algorithm.

Though in this case we do not obtain any improvement in the worst case complexity, by starting with the planar subgraph $G_p$ produced by algorithm PLANARIZE, we would achieve considerable savings in overall computation time because we would have already attempted to include in $G_p$ as many edges as possible and the construction of $G_p$ requires only $O(n^2)$ time.

## 5. Summary and Conclusion

In this paper we present two $O(n^2)$ planarization algorithms—PLANARIZE and MAXIMAL-PLANARIZE. These algorithms are based on Lempel, Even, and Cederbaum's planarity testing algorithm [9] and its implementation using PQ-trees [8]. Algorithm PLANARIZE is for the construction of a spanning planar subgraph of an n-vertex nonplanar graph. This algorithm proceeds by embedding one vertex at a time and, at each step, adds the maximum number of edges possible without creating nonplanarity of the resultant graph. Given a biconnected spanning planar subgraph $G_p$ of a nonplanar graph G, algorithm MAXIMAL-PLANARIZE constructs a maximal planar subgraph of G which contains $G_p$. This latter algorithm can also be used to maximally planarize a biconnected planar graph. We conclude by pointing out that no planarization algorithm of complexity $O(n^2)$ has been reported in the literature. The $O(mn)$ algorithms of [5] and [7] do not seem to lend themselves to easy modifications resulting in $O(n^2)$ planarization algorithms. Furthermore, no $O(n^2)$ algorithm for maximal planarization of a biconnected planar graph has been reported before in the literature.

We also note that the algorithms PLANARIZE and MAXIMAL-PLANARIZE can be used to construct a maximal planar subgraph even when the graph produced by PLANARIZE is not biconnected. Though the worst case complexity of such a maximal planarization algorithm will be the same as that of the algorithm in [5], we expect this algorithm to require less computation time since construction of $G_p$ requires only $O(n^2)$ time and while constructing $G_p$ we attempt to include as many edges as possible.

The question remains whether we can design an $O(n^2)$ algorithm to construct a maximal planar subgraph of a nonplanar graph. Such an algorithm will be possible provided we can find an $O(n^2)$ algorithm to construct a spanning biconnected planar subgraph of a given nonplanar graph or to construct a spanning planar connected subgraph which has an st-numbering.

# REFERENCES

[1]  M.R. Garey and D.S. Johnson, <u>Computers and Intractability: A Guide to the Theory of NP-completeness</u>, Freeman, San Francisco, 1979.

[2]  G.J. Fisher and O. Wing, "Computer Recognition and Extraction of Planar Graphs from the Incidence Matrix", IEEE Trans. on Circuit Theory, Vol. CT-13, No. 2, 154-163 (June 1966).

[3]  K. Pasedach, "Criterion and Algorithms for Determination of Bipartite Subgraphs and their Application to Planarization of Graphs", in Graphen-Sprachen und Algorithmen auf Graphen, Carl Hanser Verlag, 1976, pp. 175-183.

[4]  M. Marek-Sadowska, "Planarization Algorithm for Integrated Circuits Engineering", Proc. 1978 IEEE International Symposium on Circuits and Systems, pp. 919-923.

[5]  T. Chiba, I. Nishioka, and I. Shirakawa, "An Algorithm of Maximal Planarization of Graphs", Proc. 1979 IEEE International Symposium on Circuits and Systems, pp. 649-652.

[6]  J. Hopcroft and R. Tarjan, "Efficient Planarity Testing", J. Assoc. Comput. Mach., Vol. 21, No. 4, 549-568 (October 1974).

[7]  T. Ozawa and H. Takahashi, "A Graph-Planarization Algorithm and its Application to Random Graphs", in Graph Theory and Algorithms, Springer-Verlag Lecture Notes in Computer Science, Vol. 108, 1981, pp. 95-107.

[8]  K.S. Booth and G.S. Lueker, "Testing for the Consecutive Ones Property, Interval Graphs, and Graph Planarity Using PQ-tree Algorithms", Journal of Comp. and Syst. Sciences, Vol. 13, No. 3, 335-379 (December 1976).

[9]  A. Lempel, S. Even, and I. Cederbaum, "An Algorithm for Planarity Testing of Graphs", Theory of Graphs: International Symposium: Rome, July 1966, P. Rosenstiehl (Ed.), Gordon and Breach, New York, 1967, pp. 215-232.

[10] S. Even, <u>Graph Algorithms</u>, Computer Science Press, Potomac, Maryland, 1979.

[11] R. Jayakumar, K. Thulasiraman, and M.N.S. Swamy, "On Maximal Planarization of Non-planar Graphs", IEEE Trans. on Circuits and Systems, Vol. CAS-33, No. 8, August 1986, pp. 843-844.

[12] R. Jayakumar, "Design and Analysis of Graph Algorithms: Spanning Tree Enumeration, Planar Embedding, and Maximal Planarization", Ph.D. Thesis, Electrical Engineering Department, Concordia University, Montreal, Canada, August 1984.

[13] R. Jayakumar, K. Thulasiraman, and M.N.S. Swamy, "Planar Embedding: Linear-Time Algorithms for Vertex Placements and Edge Ordering", IEEE Trans. on Circuits and Systems, Vol. CAS-35, No. 3, March 1988, pp. 334-344.

[14] S. Even and R.E. Tarjan, "Computing an st-numbering", Theo. Comp. Sci., Vol. 2, 339-344 (1976).

# Bandwidth and Profile Minimization[*]

Manfred Wiegers and Burkhard Monien

Universität GH Paderborn, FB 17
Warburger Str. 100
D-4790 Paderborn, FRG

## Abstract

For a class of bandwidth approximation algorithms (called level algorithms), we show that the ratio of the approximation and the exact bandwidth is $\Omega(\log^{bw(G)-1}(n))$. Then we give a general approximation algorithm which tries to improve a given layout. It is based on a reordering of a previously generated layout. To generate initial layouts two further level algorithms are introduced.

To compare such algorithms we define two norms for the quality of bandwidth approximation algorithms. The first is influenced by a bound which is inherently given for all level algorithms. The second is influenced by a lower bound for the bandwidth. We have tested the improvement algorithm on many graphs and it provides substantial better bandwidth approximation than all level algorithms.

The bandwidth improvement algorithm is also used for minimizing the profile of a graph. We give a list of results obtained by our algorithm for 30 examples given by G. C. Everstine. So our algorithms becomes comparable with other algorithms which have used these examples as a test set.

# 1 Introduction

The *bandwidth minimization problem* is to find a permutation matrix $P$ of a given symmetric matrix $M$ s.t. the maximum of $|i - j|$ over all nonzero entries $p_{i,j}$ of $P * M * P^T$ is a minimum. An equivalent problem is to find for a given graph $G = (V, E)$ a one-to-one function $f : V \rightarrow \{1, 2, ...n_G\}$ s.t. the maximum of $|f(v) - f(u)|$ over all $\{u, v\} \in E$ is a minimum. This value will be denoted by $bw(G)$. The function $f$ will be called a *layout*.

We prefer the second one, which can be transformed from the first by building a graph with $sig(M)$ (the nonzero entries are replaced by 1, in the asymmetric case this must be modified appropriately) being its adjacency matrix. For a given layout $f$ of $G$ we define $bw_f(G) := \max_{\{u,v\} \in E} |f(u) - f(v)|$.

The bandwidth minimization problem can also be described as a subgraph isomorphism problem. For a graph $G = (V, E)$ we denote the set of vertices and edges by $V(G)$ and $E(G)$ respectively. The *distance* $d_G(u, v)$ of two vertices $u$ and $v$ is defined to be the length of the shortest path from $u$ to $v$ in $G$. When it is clear which graph we mean, we drop the subscript $G$. By $G^k$ we denote the *k-th power* of a graph $G = (V, E)$, where $V(G^k) = V$ and

---

[*]This research was supported by a grant from the German Research Association (DFG)

$E(G^k) = \{\{u, v\} : 1 \leq d_G(u, v) \leq k\}$. Denote by $P_n$ the path with $n$ vertices. Thus $bw(G) \leq k$ iff $G$ is a subgraph of $P_{n_G}^k$.

The bandwidth problem is one of the most explored graph problems, see the survey of P. Z. Chinn et al. and the literature list in [CCDG82]. C. H. Papadimitriou first showed that this problem is *NP-complete* in [Pap76]. The strongest result is obtained by B. Monien in [Mon86] where it is shown that this problem is *NP-complete* for caterpillars with hair length at most 3 and for caterpillars with maximal degree at most 3, respectively. On the other hand this problem can be solved for caterpillars with hair length at most 2 in time $O(n \cdot \log(n))$, see [APSZ81], and for interval graphs in time $O((n + m) \cdot \log(n))$, see [Kra86].

Since 1965, many publications on bandwidth approximation algorithms can be found. But none of these leads to an acceptable worst case approximation. One of the well-known algorithms is the Cuthill-McKee algorithm, see [CM69].

In the next section we formally introduce the class of level algorithms and show the poor worst case approximation of these for the bandwidth. Two level algorithms are presented which are needed for further considerations in section 3. In section 3 we introduce our bandwidth improvement algorithm. This algorithm starts on some given layout and tries to improve it. Further more we show that in the worst case this algorithm computes only a very poor approximation of the bandwidth for an initial layout generated by any level algorithm.

In the fourth section we present test results. These results show that our improvement algorithm performs at least as well as all the other approximation algorithms. In particular it generates substantial better layouts than all level algorithms on many graph examples. We test all algorithms mentioned here by using random graphs and a collection of 30 examples given by [Eve79b]. So the results for the 30 examples from our algorithms can be compared with other ones. Our algorithms are also used and compared for the minimization of the profile.

## 2    The Level Algorithm

The heuristic algorithm of E. Cuthill and J. M. McKee belongs to the so called class of *level algorithms*. For a graph $G = (V, E)$ a level algorithm produces a layout $f$ which satisfies:

$$\forall u, v \in V : d(f^{-1}(1), u) < d(f^{-1}(1), v) \Rightarrow f(u) < f(v). \tag{1}$$

The vertex $f^{-1}(1)$ is called *start vertex*. Formally we define for a graph $G = (V, E)$ and a given start vertex $u \in V$ a *level structure* $L(G, u)$. $L(G, u) = (L_0, L_1, ...L_k)$ is a partition of $V$ into sets, where $k := \max_{v \in V} d(u, v)$. In this structure for $0 \leq i \leq k$ the set $L_i := \{v : d(u, v) = i\}$ is called a *level*.

Let $l_i = |L_i|$ and $n_i := \sum_{j=1}^{i} l_j$. A level algorithm produces only layouts $f$, s.t. for the level structure $L(G, f^{-1}(1)) = (L_0, L_1, ...L_k)$ and for all $i \in \{0, 1, ...k\}$ is

$$f^{-1}(\{n_{i-1} + 1, n_{i-1} + 2, ...n_i\}) = L_i. \tag{2}$$

The *width* of a level structure $L(G, u)$ will be defined as $w(G, u) := \max_i |L_i|$. For a layout $f$ generated by a level algorithm we get

$$w(G, f^{-1}(1)) \leq bw_f(G) \leq 2 \cdot w(G, f^{-1}(1)) - 1. \tag{3}$$

We want to give a level algorithm called $\mathcal{BW}_0$. The algorithm lays out vertex after vertex from position 1 to $n$. We call a vertex *placed* iff it already has got a position in the layout by the algorithm. The algorithm scans the graph in BFS manner, and each time it finds a non placed vertex it will place this vertex on the next free position. Given the layout $f$, initialized with 0 for all vertices, the algorithm can decide in constant time whether a vertex is placed. So the algorithm runs in linear time in $n + m$. The algorithm depends on the graph and a second parameter, namely the start vertex $u$. The implementation is as follows:

Input: A graph $G = (V, E)$ and a vertex $u \in V$.

Output: A layout $f$ with $f^{-1}(1) = u$ satisfies (1).

BEGIN
    Let $pos := 1$;
    Lay $u$ on position $pos$ and increment $pos$;
    WHILE not all vertices are placed DO
        Let $v$ be the earliest placed vertex for which not all neighbors are placed;
        FOR all neighbors $w$ of $v$ which are not placed DO
            Lay $w$ on position $pos$ and increment $pos$
        OD
    OD
END.

In [HHH85], E. O. Hare et al. gave a level algorithm which produces a layout $f$ with bandwidth $bw_f(T) = w(T, f^{-1}(1))$ for trees $T$. Next, we want to present a similar algorithm called $\mathcal{BW}_1$. It also runs in linear time and works as follows:

Input: A graph $G = (V, E)$ and a vertex $u \in V$.

Output: A layout $f$ with $f^{-1}(1) = u$ satisfies (1).

BEGIN
    Generate a level structure $L(G, u) = (L_0, L_1, ... L_k)$;
    Lay out the vertices of level $L_k$ on positions $n_{k-1} + 1$ to $n$;
    Let $i := k$ and $pos := n_{k-1}$
    REPEAT
        WHILE not all neighbors of vertices in level $L_i$ are placed DO
            Let $v$ be the earliest placed vertex for which not all neighbors are placed;
            FOR all neighbors $w$ of $v$ which are not placed DO
                Lay out $w$ on position $pos$ and decrement $pos$
            OD
        OD;
        Decrement $i$;
        FOR all $v$ of $L_i$ which are not placed DO
            Lay out $v$ on position $pos$ and decrement $pos$
        OD
    UNTIL $i = 0$
END.

When the algorithm terminates all vertices are placed and the generated layout satisfies (1). So we can formulate the following Lemma. The proof follows directly from the results in [HHH85].

**Lemma 1** For trees $T = (V, E)$ the algorithm $\mathcal{BW}_1$ can be used to generate an optimal layout satisfying (1) in quadratic time.

The bandwidth of a layout generated by a level algorithm depends:

- on the start vertex,

- on the order of the vertices having the same distance from the start vertex.

It is an open question whether we can efficiently produce a layout which is optimal under all layouts satisfying (1). However, even if we could do so, because of (3) it would never improve the bandwidth by more than a factor of 2. On the other hand the difference between the optimal bandwidth and the best bandwidth obtainable by every level algorithm can be arbitrary large. Next we give a class of graphs with bandwidth $k$ for which every level algorithm produces a layout having bandwidth $\Omega(\log^{k-1}(n))$.

For this purpose we define trees $T_{i,j}(e)$, where $e$ is a specified vertex. For $i \geq 1$ let $T_{i,1}(e)$ be a path with $2^{i-1}$ vertices, where $e$ is one of the vertices at the end of this path. $T_{i+1,i+1}(e)$ is defined by a union of a path with $2^{i-1}$ vertices with end vertices $e$ and $e_1$ and two copies $T_{i,i}(e_2)$ and $T_{i,i}(e_3)$ respectively. Additional edges $\{e_1, e_2\}$ and $\{e_1, e_3\}$ connect this graph. $T_{i+1,j+1}(e)$ with $i > j$ is built by a union of a path with $2^{i-1}$ vertices with end vertices $e$ and $e_1$ and the two graphs $T_{i,j+1}(e_2)$ and $T_{i,j}(e_3)$. Edges $\{e_1, e_2\}$ and $\{e_1, e_3\}$ connect this graph. For an informal definition see Figure 1.

For $T_{i,j}(e)$ let the number of vertices be denoted by $n_{i,j} := |V(T_{i,j}(e))|$, the bandwidth by $bw_{i,j} := bw(T_{i,j}(e))$ and the width by $w_{i,j} := w(T_{i,j}(e), e)$. For an arbitrary $j$ we want to show that $2^{i-1} \leq n_{i,j} \leq j \cdot 2^{i-1}$, $bw_{i,j} \leq j$ and $w_{i,j} = \Omega(i^{j-1})$. First we look at $n_{i,j}$ and get the following recurrences, the first one can be proved by induction and the second one follows from the definition:

$$n_{i+1,i} = (i+1) \cdot 2^{i-2},$$
$$n_{i,j} = 2^{i-2} + n_{i-1,j} + n_{i-1,j-1}.$$

From these recurrences we get

$$n_{i,i} = (i+1) \cdot 2^{i-2}$$
$$\leq i \cdot 2^{i-1}.$$

and in general we can prove $n_{i,j} \leq j \cdot 2^{i-1}$ by the following inductive step

$$n_{i+1,j} = 2^{i-1} + n_{i,j} + n_{i,j-1}$$
$$\leq 2^{i-1} + j \cdot 2^{i-1} + (j-1) \cdot 2^{i-1}$$
$$\leq j \cdot 2^i.$$

On the other hand $T_{i,j}(e)$ contains a path with $2^{i-1}$ vertices from $e$ to one of the other leaves and so $2^{i-1} \leq n_{i,j} \leq j \cdot 2^{i-1}$. We need the lower and the upper bound to analyze the worst case approximation of this example later.

To analyze the width, notice that all the leaves except $e$ have the same distance from $e$. The number of these leaves is the width. Inductively it can be shown that:

$$w_{i,j} = \sum_{k=0}^{j-1} \binom{i-1}{k}.$$

So we have for $i \gg j$:

$$w_{i,j} = \Omega\left(\frac{i^{j-1}}{(j-1)!}\right).$$

It remains to analyze the bandwidth. We will show that $bw_{i,j} \leq j$. By induction we get $bw_{i,i} \leq i$ by a layout $f$ s.t. the marked vertex $e$ is not too far away from the last position, exactly $f(e) \geq n_{i,i} - i$. For $T_{i+1,i+1}(e)$ we must give a layout s.t. the induction invariant holds. We have layouts $f_1$ ($f_2$) for $T_{i,i}(e_2)$ ($T_{i,i}(e_3)$) s.t. the invariant holds. Using $f_1$ we lay out the tree $T_{i,i}(e_2)$ on the first $n_{i,i}$ positions. We then repeatedly lay out alternating:

- the next vertex of the path starting with vertex $e_1$, and

- $i$ vertices of $T_{i,i}(e_3)$ in reverse order of $f_2$

until all vertices of the path are placed. As the path has $2^{i-1}$ vertices and $T_{i,i}(e_3)$ has at most $i \cdot 2^{i-1}$ vertices, at most $i$ vertices of $T_{i,i}(e_3)$ have not been laid out when the last vertex $e$ of the path is placed. So the bandwidth is increased by 1 and it follows that $T_{i+1,i+1}$ has a layout $f$ s.t. $bw_f(T_{i+1,i+1}(e)) = i + 1$ and $f(e) \geq n_{i+1,i+1} - (i+1)$. By the same technique it can be verified that $bw_{i,j} \leq j$. So we get the next lemmas, where $\Delta(G)$ denotes the maximal degree.

**Lemma 2** There exist trees $T = (V, E)$ with $\Delta(T) \leq 3$ and a vertex $v \in V$, s.t. $bw(T) \leq \log(n)$ and $w(T, v) = \Omega(\frac{n}{\log(n)})$.

**Proof:** We only look at the trees $T_{i,i}$. We have $n = n_{i,i} = O(i \cdot 2^i)$ so $w_{i,i} = 2^{i-1} = \Omega(\frac{n}{\log(n)})$. Since the bandwidth is at most $i \leq \log(n)$ the result follows. ∎

**Lemma 3** For every $k$ there exist trees $T = (V, E)$ with $\Delta(T) \leq 3$ and a vertex $v \in V$, s.t. $w(T, v) = \Omega(\log^{k-1}(n))$ and $bw(T) \leq k$.

**Proof:** Looking at the trees $T_{i,k}(e)$, from $n = n_{i,k} \leq k \cdot 2^{i-1}$ it follows $i \geq \log(n) - \log(2 \cdot k)$. Previously we have shown that $w_{i,k} = \Omega(\frac{i^{k-1}}{(k-1)!})$. Because $k$ is bounded we get $w(T_{i,k}) = \Omega(\log^{k-1}(n))$. ∎

We want to show that all level algorithms have bad worst case approximation. Let $\mathcal{A}$ be a bandwidth approximation algorithm and $G$ be a graph. The algorithm $\mathcal{A}$ generates a layout $f_{\mathcal{A}}$. We define the ratio $R_{\mathcal{A}}(G)$ by:

$$R_{\mathcal{A}}(G) = \frac{bw_{f_{\mathcal{A}}}(G)}{bw(G)}. \tag{4}$$

By the lemmas above we know the ratio for selecting a bad start vertex. When a level algorithm chooses the best start vertex for the graph $T_{i,j}(e)$, then the ratio is bounded by a constant. We now show how fast the ratio can grow for a level algorithm. This is summarized in the next theorem.

**Theorem 1** For any level algorithm $\mathcal{A}$ there exist trees $T$ with $\Delta(T) = 3$ s.t.

$$
\begin{array}{lll}
(i) & R_{\mathcal{A}}(T) & = & \Omega(\dfrac{n}{\log^2(n)}), \\[2mm]
(ii) & R_{\mathcal{A}}(T) & = & \Omega(\log^{bw(T)-1}(n)).
\end{array}
$$

**Proof:** Using the trees described above we can build a tree $\overline{T_{i,j}(e)}$. For a definition of these trees see Figure 2. After the termination of the level algorithm $\mathcal{A}$ some vertex lies on position 1. W.l.o.g. this vertex is one of the vertices in $V(T_{i,j}(e_1)) \cup \{e\}$. So the algorithm has put the vertex $e_2$ of $T_{i,j}(e_2)$ into some unique level $L_l$ and partitions the other vertices of $T_{i,j}(e_2)$ according to $L(T_{i,j}(e_2), e_2)$ into the following levels. So the bandwidth of the layout generated by this algorithm must be at least $w(T_{i,j}(e_2), e_2)$. As the number of vertices is approximately 2 times the number of vertices in $T_{i,j}(e_2)$ we can use the lemmas 2 and 3 to obtain the result of the theorem 1. ∎

This theorem shows that property (1) implies a bad worst case approximation for the bandwidth. For the graphs mentioned in the above proof the algorithm of Gibbs et al. has also the same bad behavior. In the next section we present an algorithm improving given layouts.

## 3 The Bandwidth Improvement Algorithm

The level algorithm partitions the vertices in levels depending of the start vertex. But in opposite to the usual level algorithm our improvement algorithm does not place the levels one after another. The algorithm scans a given level structure with a *window* starting with the last level and laying out $k$ vertices, where $k$ is the length of the window. Whenever a window can not be filled by the vertices of one level the gap is filled up with vertices of the next lower level and so on. Going on with a new window we first lay out the vertices adjacent to those in the preceding window and fill up the rest with vertices of the actual level. We repeat this process until all vertices are placed in one of the so generated windows.

The idea of scanning the level structure with a window was the main idea for the improvement algorithm. More generally we use the same technique for scanning a given layout. For this purpose it is important to distinguish between the directions of scanning the given layout. The following algorithm does this in reverse order.

Now we give the algorithm, called $\mathcal{BW}_2$ which tries to improve a given layout. For a given layout $f$ and its inverse $f^{-1}$ our algorithm generates a class of layouts $f_k$ and $f_k^{-1}$ for $k \in \{1, 2, ...n\}$, where $k$ is the length of the window. Next we describe the algorithm to get $f_k$

and $f_k^{-1}$ for a given $k$. A vertex is called *placed* if it has got a position in the layout $f_k$ by the algorithm and it is called *last unplaced* if it is the unplaced vertex $f^{-1}(i)$ with maximal $i$.

Input: A graph $G = (V, E)$, a vertex $v \in V$, an integer $k$ and a layout $f$ with its inverse $f^{-1}$.

Output: A layout $f_k$.

```
BEGIN
    pos := n; l := k;
    oldwindow := ∅; newwindow := ∅;
    WHILE not all vertices are placed DO
        FOR i := 1 TO min(l, |{v ∈ V : v is not placed }|) DO
            Lay out the last unplaced vertex v on position pos and decrement pos;
            newwindow := newwindow ∪ {v};
        OD
        l := k; oldwindow := newwindow; newwindow := ∅;
        WHILE oldwindow ≠ ∅ DO
            Let v be the earliest placed vertex in oldwindow;
            oldwindow := oldwindow − {v};
            FOR all unplaced neighbors w of v in order of decreasing values of f DO
                Lay out the vertex w on position pos and decrement pos;
                newwindow := newwindow ∪ {w}; l := l − 1
            OD
        OD
    OD
END.
```

This algorithm also runs in linear time and space. But if we want to generate all $f_k$ we have a $O(n \cdot (n + m))$ runtime. To improve the runtime of $\mathcal{BW}_2$ we generate the layout $f_k$ for increasing values of $k$. The generating process is interrupted when $k$ has a value greater then the minimal bandwidth of all previously generated layouts. So we have reduced the runtime of the main algorithm from $O(n \cdot (n + m))$ to $O(bw_{f_{\mathcal{BW}_2}}(G) \cdot (n + m))$, where the bandwidth depends also from the initial layout $f$.

**Theorem 2** The resulting algorithm $\mathcal{BW}_2$ runs in $O(bw_{f_{\mathcal{BW}_2}}(G) \cdot (n + m))$ time and uses $O(n + m)$ space.

On the other hand we must generate layouts which the algorithm can improve. Clearly the layouts generated by $\mathcal{BW}_0$ and $\mathcal{BW}_1$ could be improved. So we get two approximation algorithms, namely $\mathcal{BW}_{20}$ and $\mathcal{BW}_{21}$, which trie to improve a layout generated by $\mathcal{BW}_0$ and $\mathcal{BW}_1$, respectively. The next problem is to find a good start vertex for $\mathcal{BW}_0$ which is also used by $\mathcal{BW}_1$.

One heuristic for the level algorithms to select a start vertex is to take some vertex s.t. the number of levels is maximal. A vertex having this property is called a *diameter vertex*. Computing such a vertex needs $O(n \cdot (n + m))$ time.

For our algorithms we used a start vertex computed in the following way. Take some vertex $v_1$, compute a level structure with start vertex $v_1$ and choose some vertex $v_2$ out of the last level. Compute a new level structure with start vertex $v_2$, and take again some vertex out of the last level of this structure. A vertex computed by this process is called *pseudo diameter vertex* and can be found in time $O(n + m)$. Notice that this vertex can differ from the pseudo diameter vertex defined in [GPS76]. For our algorithms we used such a vertex as a start vertex.

Next we want to give a lower bound for the worst case complexity of the two algorithms $\mathcal{BW}_{20}$ and $\mathcal{BW}_{21}$. It seems that the two algorithm have the same worst case approximation, they could only differ by a multiplicative constant. In the next lemma we give a bound for bandwidth restricted graphs.

**Lemma 4** There exists graphs $G$ with $\Delta(G) = 3$ s.t. for $x \in \{0, 1\}$

$$R_{\mathcal{BW}_{2x}}(G) = \Omega\left(\sqrt{\log^{bw(G)-3}(n)}\right).$$

**Proof:** Take two copies $\overline{T_{i,j}(e_1)}$ and $\overline{T_{i,j}(e_2)}$, then paste the corresponding leaves of the first and the second copy together. Denote this graph by $G_{i,j}(e_1, e_2)$. Assume now we have a layout $l$ generated by $\mathcal{BW}_0$ with start vertex $e_1$. The layout generated by $\mathcal{BW}_1$ can differ from $l$ but it has the same structure by isomorphic arguments.

In this layout $e_2$ lies on the last position. Now the improvement algorithm works on this layout with some $k$. But the generated layout becomes likewise level structure and so the bandwidth of the generated layout is in $\Omega(w(\overline{T_{i,j}(e_2)}, e_2))$.

Take two graphs $G_{i,j}(e_{11}, e_{12})$ and $G_{i,j}(e_{21}, e_{22})$ and connect them by a path starting from $e_{11}$ to $e_{21}$ having $2^i$ new vertices, so we are independent from the start vertex. So we have :

(i) the bandwidth of this graph is at most $2 \cdot j + 1$ and

(ii) the approximation bandwidth of this graph is in $\Omega(\log^{j-1}(n))$. ∎

Indeed, we have no argument which says that this is the worst case for bandwidth restricted graphs, else we have found an algorithm having a better worst case approximation than the other approximation algorithms. But when the bandwidth is not bounded, we have also the same asymptotical behavior as in (i) of theorem 1 see the following theorem.

**Theorem 3** There exists graphs $G$ with $\Delta(G) = 3$ s.t. for $x \in \{0, 1\}$

$$R_{\mathcal{BW}_{2x}}(G) = \Omega\left(\frac{n}{\log^2(n)}\right).$$

## 4   Test Results

To compare our algorithms with other bandwidth algorithms, we have to analyze all existing algorithms. Up to now no one has given a worst case approximation of a bandwidth algorithm, only the lower bound in part (i) of theorem 1 is known. Another measure to compare bandwidth algorithms is the average approximation. For the class of level algorithms it is known that they have a good average case approximation, see [Tur86].

We have tested this algorithm on randomly generated graphs of five distinct graph classes and got good improvements of the layouts generated by $\mathcal{BW}_0$ and $\mathcal{BW}_1$. Given an integer $n$ for the number of vertices, $0 \leq p \leq 1$ a *random value*, and $1 \leq bw \leq n$ a *bandwidth restriction*. The graphs in the classes were generated as follows:

$\mathcal{RG}_1$ (Random graphs)

Generate a graph in the *constant density model* (edges are generated independently, an edge occurs with probability $p$). The generating process needs $O(n^2)$ time.

$\mathcal{RG}_2$ (Fast generated "random graphs")

Order all possible edges, and start with position 0. Repeatedly guess an integer between 1 and $1 + \lfloor \frac{1}{p} \rfloor$ and add this value to the starting position. Then insert the corresponding edge specified by the position to the graph with probability $\frac{1}{2}$, until we have scanned all edges. The generating process needs $O(n + m)$ time in the average case.

$\mathcal{RG}_3$ (Bandwidth restricted random graphs)

Order the vertices and generate edges independently only when the distance of the incident vertices is at most $bw$, under this edges an edge occurs with probability $p$. The generating process needs $O(bw \cdot n)$ time.

$\mathcal{RT}_1$ (Random trees)

By verifying the famous result of A. Cayley about counting the number of trees, H. Prüfer found an one-to-one function between the words having $n$ letters of length $n - 2$ and all trees of $n$ vertices, see [Eve79a]. To generate a random tree we generate a string with uniformly distributed probability for all strings of length $n - 2$ having $n$ letters and use the function to generate the tree of $n$ vertices specified by the string. Our implementation of the generating process needs $O(n \cdot \log \log(n))$ time, because the process uses a *priority queue* to sort numbers between 1 and $n$. For the implementation of this queue see [Meh84].

$\mathcal{RT}_2$ ("Random trees" with smaller diameter)

Recursively generate a tree with $n - 1$ vertices. Select one of these vertices with uniformly distributed probability and join this vertex with a new vertex. The generating process needs $O(n)$ time.

For our tests we generated 40 examples with $n = 200$ vertices for each random class. The random value for $\mathcal{RG}_1$ and $\mathcal{RG}_2$ is $p = \frac{5}{n-1}$. For $\mathcal{RG}_3$ we set the bandwidth restriction to $bw = 40$ and the random value to $p = \frac{5 \cdot n}{bw \cdot (2 \cdot n - bw - 1)}$. So the resulting graphs of these three random graph classes have average degree 5.

Also we used the following test set:

$\mathcal{ES}$ (Available test set of practical relevant graphs)

This set contains 30 graph examples collected from applications of the finite element method with $59 \leq n \leq 2680$. For a detailed description and Figures of these examples see [Eve79b].

To get an overview of the average behavior we computed the mean value of the ratio between the bandwidth and the width of the graphs depending from the start vertex. In the columns of the following table we give these values computed by our 4 given algorithms, where the lines contain the results obtained by the graphs in one of the random classes. Since not all of the graphs generated by $\mathcal{RG}_1$, $\mathcal{RG}_2$, and $\mathcal{RG}_3$ are connected, we only considered the connected component with the most vertices in every graph. In all cases this connected component had at least 190 vertices, and so all the other connected components were of trivial size. Some of the 30 examples are not connected so we take only the component having the largest approximate bandwidth.

|  | $BW_0$ | $BW_1$ | $BW_{20}$ | $BW_{21}$ |
|---|---|---|---|---|
| $\mathcal{RG}_1$ | 1.114 | 1.122 | 0.867 | 0.960 |
| $\mathcal{RG}_2$ | 1.091 | 1.103 | 0.844 | 0.940 |
| $\mathcal{RG}_3$ | 1.164 | 1.030 | 0.765 | 0.818 |
| $\mathcal{RT}_1$ | 1.064 | 1.000 | 0.705 | 0.734 |
| $\mathcal{RT}_2$ | 1.083 | 1.000 | 0.620 | 0.648 |
| $\mathcal{ES}$ | 1.111 | 1.053 | 0.825 | 0.848 |

The table does not tell something about the distance between the exact bandwidth and the one obtained by the approximation algorithms. We will show this in the next table. Therefore, we need some more definitions. Independently, V. Chvátal and A. K. Dewdeny gave a formula to compute a lower bound for the bandwidth of graphs, namely the *lower bound bandwidth* $lbb(G) = \lceil \frac{n-1}{D(G)} \rceil$, see [CCDG82], where $D(G) = \max_{u,v \in V(G)} d(u,v)$ is the *diameter* of the graph $G$. It is simple to see that the ratio between the lower bound bandwidth and the exact bandwidth can be in $\Omega(n)$. To get a more precise value we define :

$$LBB(G) = \max_{v \in V(G)} \max_l lbb(G[v,l]), \qquad (5)$$

where $G[v,l]$ is the subgraph of $G$ induced by a vertex set containing all vertices with distance of at most $l$ from the vertex $v$. For more information about lower bound bandwidth functions see [Wie88]. Computing $LBB$ is rather complex, by ad hoc analysis it needs time $O(D(G) \cdot n_G^2 \cdot (n_G + m_G))$.

This value can be arbitrary far away from the exact bandwidth for some graphs. But it seems that the ration $bw(G)/LBB(G)$ is bounded by some slow increasing function in the number of the vertices. In the next table an overview of the average behavior of the ratio between the bandwidth and the lower bound bandwidth is given.

|  | $BW_0$ | $BW_1$ | $BW_{20}$ | $BW_{21}$ |
|---|---|---|---|---|
| $\mathcal{RG}_1$ | 3.261 | 3.291 | 2.540 | 2.808 |
| $\mathcal{RG}_2$ | 3.327 | 3.371 | 2.573 | 2.867 |
| $\mathcal{RG}_3$ | 2.271 | 2.015 | 1.482 | 1.581 |
| $\mathcal{RT}_1$ | 2.319 | 2.189 | 1.515 | 1.574 |
| $\mathcal{RT}_2$ | 2.469 | 2.280 | 1.398 | 1.467 |
| $\mathcal{ES}$ | 1.935 | 1.845 | 1.429 | 1.468 |

In Figure 3 is given detailed information of the collection of 30 graphs to compare our algorithms with well known algorithms for the bandwidth minimization, where NA stands for not available. The first two relevant columns contain partial results given by [Eve88]. The results were taken from runs of the Cuthill-McKee algorithm $\mathcal{CM}$ and Gibbs-Poole-Stockmeyer algorithm $\mathcal{GPS}$. Notice that the run of $\mathcal{CM}$, which is a level algorithm, takes many vertices as start vertex so this results can be less than our width. All the bandwidth results of $\mathcal{GPS}$ are available, but the average ratio of our first norm depends from the start vertex. But the second norm is not, so we calculate the value for the $\mathcal{GPS}$ algorithm and the 30 examples which is 1.713.

G. C. Everstine used also this test set for a related minimization problems, called *profile*. For a graph $G$ and a layout $f$ the *row bandwidth* $b_i$ for all $i \in \{1, 2, ...n\}$ is defined as :

$$b_i = 1 + i - \min(i, \min\{j : \{f^{-1}(i), f^{-1}(j)\} \in E(G)\}). \tag{6}$$

For the profile $P(G, f)$ we now have :

$$P(G, f) = \sum_{i=1}^{n} b_i. \tag{7}$$

We used G. C. Everstine's definition of the profile. The profile is a measure for the compact storage of a matrix with efficient access in $O(1)$. This formula assumes that all diagonal elements are nonzero. An exact formula must consider the diagonal elements. However, we only have the results of G. C. Everstine using formula (7) in connection with (6). In applications of the finite element method, the diagonal elements are all nonzero. At last we define the average profile :

$$\overline{P(G, f)} = \frac{P(G, f)}{n_G}. \tag{8}$$

Two of the three algorithms used by Everstine are bandwidth minimization algorithms, namely the algorithms of Cuthill and McKee and of Gibbs et al.. The other one is the algorithm of Levy in [Lev71] whose main topic is the minimization of the *maximal wavefront*, not defined here.

We have tested Everstine's collection of graph examples twice using the improvement algorithm; first with an initial layout of $\mathcal{BW}_0$ and then with an initial layout of $\mathcal{BW}_1$. Because the profile of a layout can differ from that of the reverse layout (which we can obtain easily) we minimize the profile over the layout and its reversal. A. George suggested that this can improve the profile for the bandwidth algorithm of Cuthill and McKee, while W.-A. Liu and A. H. Sherman showed that the profile of the reverse layout is at most the profile of the original layout, see [LS76]. Based on this fact the so called *reverse* Cuthill-McKee algorithm ($\mathcal{RCM}$) becomes an interesting aspect for the profile minimization.

To compare our algorithms with the results obtained by Everstine, see Figure 4, where NI stands for no improvement and - for not run. Column (a) contains the profile of one layout $f_k$ with best bandwidth under all layouts generated by the improvement algorithm, where (b) contains the best profiles under all generated layouts by the improvement algorithm.

# 5 Conclusion

In the second section we have shown that the level algorithms have poor worst case approximation. Even for bounded bandwidth graphs one cannot get acceptable results in the worst case. There exist dynamic programming algorithms for bandwidth restricted graphs in time $O(n^{bw(G)})$, see [Sax80,GS84]. Even for graphs with small bandwidth the runtime is not acceptable.

In the third section we presented a bandwidth approximation algorithm for improving a given layout. Clearly, the layout $f_n$ computed by this algorithm is the same as the initial layout $f$. To get a better runtime we gave a heuristic to break up the generating process. One can also generate layouts using binary search for $k$ in the range from 1 to $n$, but it may happen that this only leads to a local minimum. However, if we generate all layouts $f_k$ this would not improve the worst case approximation as shown.

In the next section we presented the test results and gave two norms for the quality of bandwidth approximation algorithms. As shown in the tables one can see that in the average $\mathcal{BW}_{20}$ has better improvements than $\mathcal{BW}_{21}$. Indeed, this holds for nearly all examples. In the beginning we expected that the algorithm $\mathcal{BW}_{21}$ would be much better then $\mathcal{BW}_{20}$, but this is not true. In contrast to the poor worst case approximation it is shown that there is only a small gap between the approximation and the lower bound bandwidth for the tested graphs.

At last we gave a comparison of the results given in [Eve79b] and those of our improvement algorithms for the profile. The table shows that $\mathcal{BW}_{20}$ and $\mathcal{BW}_{21}$ have a good behavior. In most cases the results are at least as good as the best result of G. C. Everstine.

# Acknowledgement

The authors wish to thank J. Rickert for his help in implementing the algorithms and reading preliminary versions. Thanks also to G. C. Everstine for his examples and comments on sparse matrices.

# References

[APSZ81]   S. F. Assmann, G. W. Peck, M. M. Syslo, and J. Zak. The bandwidth of catterpilars with hairs of length 1 and 2. *SIAM J. Algebraic and Discret Methodes*, 2:387 – 393, 1981.

[CCDG82]   P. Z. Chinn, J. Chvátalová, A. K. Dewedney, and N. E. Gibbs. The bandwidth problem for graphs and matrices - a survey. *Journal of Graph Theory*, 6:223 – 254, 1982.

[CM69]   E. Cuthill and J. Mckee. Reducing the bandwidth of matrices. In *Proc. 24th Nat. Conf. ACM*, pages 157 – 166, 1969.

[Eve79a]   S. Even. *Graph Algorithms*. Computer Science Press, Inc., 1979.

[Eve79b]   G. C. Everstine. A comparsion of three resequencing algorithm for the reduction of matrix profile and wavefront. *International journal for numerical methods in engineering*, 14:837 – 853, 1979.

[Eve88]   G. C. Everstine. Bandwidth test results. *Personal communication*, 1988.

[GPS76]   N. E. Gibbs, W. G. Pool Jr., and P. K. Stockmeyer. An algorithm for reducing the bandwidth and profile of a sparse matrix. *SIAM J. Numerical Analysis*, 13:236 – 250, 1976.

[GS84]   E. T. Gurari and I. H. Sudborough. Improved dynamic programming algorithms for bandwidth minimization and the mincut linear arrangement problem. *Journal of Algorithms*, 5:531 – 546, 1984.

[HHH85]   E. O. Hare, W. R. Hare, and S. T. Hendetniemi. Another upper bound for the bandwidths of trees. *Congressus Numerantium 50*, 77 – 83, 1985.

[Kra86]   D. Kratsch. *Finding the Minimum Bandwidth of an Interval Graph.* Manuskript N/86/18, Friedrich-Schiller-Universität Jena, 1986.

[Lev71]   R. Levy. Resequencing of the structural stiffness matrix to improve computational efficiency. *Jet Propulsion Laboratory Quart. Tech. Review*, 1:61 – 70, 1971.

[LS76]    W.-H. Liu and A. H. Sherman. Comparative analysis of the Cuthill-McKee and the reverse Cuthill-McKee ordering algorithms for sparse matrices. *SIAM J. Numerical Analysis*, 13:198 – 213, 1976.

[Meh84]   K. Mehlhorn. *Data Structures and Algorithms.* Volume 1, Sorting and Searching, EATCS Monographs on Theoretical Computer Sience, 1984.

[Mon86]   B. Monien. The bandwidth - minimization problem for caterpillars with hair length 3 is NP-Complete. *SIAM J. Algebraic and Discret Methodes*, 7:505 – 512, 1986.

[Pap76]   C. H. Papadimitriou. The NP-Completeness of the bandwidth minimization problem. *Computing*, 16:263 – 270, 1976.

[Sax80]   J. B. Saxe. Dynamic programming algorithms for recognizing small bandwidth graphs in polynominal time. *SIAM J. Algebraic and Discret Methodes*, 363 – 369, 1980.

[Tur86]   J. S. Turner. On the probable performance of heuristics for bandwidth minimization. *SIAM J. Computing*, 15:561 – 580, 1986.

[Wie88]   M. Wiegers. *Computing Lower Bounds of the Bandwidth.* Manuscript, Universität GH Paderborn, 1988.

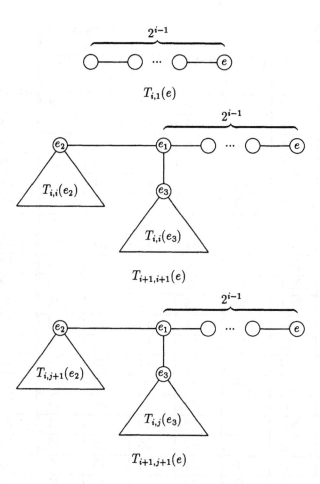

Figure 1: Definition of $T_{i,1}(e)$, $T_{i+1,i+1}(e)$ and $T_{i+1,j+1}(e)$

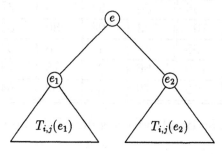

Figure 2: Definition of $\overline{T_{i,j}(e)}$

| $n$ | $\mathcal{CM}$ | $\mathcal{GPS}$ | $\mathcal{BW}_0$ | $\mathcal{BW}_1$ | $\mathcal{BW}_{20}$ | $\mathcal{BW}_{21}$ | $w$ | $LBB$ |
|------|-----|-----|-----|-----|-----|-----|-----|-----|
| 59 | 9 | 9 | 11 | 8 | 8 | 8 | 8 | 5 |
| 66 | 4 | 4 | 4 | 3 | 3 | 3 | 3 | 3 |
| 72 | 8 | 7 | 8 | 8 | 6 | 6 | 8 | 4 |
| 87 | 18 | 20 | 21 | 19 | 13 | 15 | 19 | 10 |
| 162 | 18 | 14 | 21 | 19 | 14 | 15 | 18 | 9 |
| 193 | 43 | 43 | 55 | 56 | 34 | 37 | 54 | 30 |
| 198 | NA | 12 | 15 | 13 | 8 | 9 | 11 | 8 |
| 209 | 34 | 43 | 37 | 38 | 29 | 29 | 35 | 20 |
| 221 | NA | 17 | 16 | 16 | 14 | 14 | 14 | 11 |
| 234 | NA | 17 | 22 | 22 | 15 | 15 | 22 | 10 |
| 245 | NA | 40 | 45 | 39 | 28 | 28 | 38 | 21 |
| 307 | 41 | 45 | 41 | 38 | 31 | 32 | 37 | 22 |
| 310 | 15 | 15 | 13 | 13 | 13 | 13 | 13 | 11 |
| 346 | 44 | 47 | 50 | 44 | 36 | 41 | 44 | 27 |
| 361 | 16 | 15 | 26 | 26 | 14 | 14 | 25 | 14 |
| 419 | NA | 34 | 35 | 34 | 30 | 31 | 32 | 19 |
| 492 | NA | 28 | 32 | 29 | 23 | 21 | 27 | 17 |
| 503 | 54 | 55 | 67 | 65 | 58 | 59 | 55 | 29 |
| 512 | 29 | 30 | 31 | 36 | 25 | 25 | 30 | 18 |
| 592 | 41 | 37 | 42 | 41 | 33 | 35 | 40 | 22 |
| 607 | NA | 50 | 55 | 49 | 44 | 45 | 46 | 32 |
| 758 | 27 | 26 | 26 | 26 | 20 | 22 | 25 | 15 |
| 869 | NA | 39 | 41 | 43 | 38 | 39 | 41 | 22 |
| 878 | NA | 28 | 37 | 37 | 28 | 30 | 37 | 23 |
| 918 | 47 | 50 | 73 | 63 | 45 | 44 | 61 | 27 |
| 992 | 53 | 36 | 64 | 64 | 35 | 37 | 62 | 34 |
| 1005 | NA | 107 | 103 | 108 | 80 | 85 | 101 | 33 |
| 1007 | NA | 35 | 44 | 44 | 38 | 33 | 44 | 23 |
| 1242 | 85 | 100 | 96 | 91 | 75 | 74 | 89 | 44 |
| 2680 | 69 | 69 | 70 | 68 | 59 | 64 | 60 | 44 |

Figure 3: Bandwidth results of the 30 examples

| $n$ | $\mathcal{RCM}$ | $\mathcal{GPS}$ | Levy | (a) | | (b) | |
|---|---|---|---|---|---|---|---|
| | | | | $\mathcal{BW}_{20}$ | $\mathcal{BW}_{21}$ | $\mathcal{BW}_{20}$ | $\mathcal{BW}_{21}$ |
| 59 | 5.3 | 5.8 | 5.9 | 5.49 | 5.49 | 5.49 | 5.49 |
| 66 | 3.2 | 2.9 | 2.9 | 3.18 | 2.92 | 3.18 | 2.92 |
| 72 | NI | NI | NI | 4.72 | 4.72 | 4.68 | 4.71 |
| 87 | 7.9 | 8.4 | 8.5 | 8.05 | 7.13 | 7.25 | 7.13 |
| 162 | 9.9 | 10.3 | 8.5 | 9.90 | 9.98 | 9.80 | 9.83 |
| 193 | 25.1 | 26.0 | 23.8 | 25.22 | 24.16 | 24.64 | 24.04 |
| 198 | 6.9 | 6.7 | 6.9 | 6.74 | 6.82 | 6.64 | 6.75 |
| 209 | 18.9 | 22.7 | 17.5 | 17.33 | 17.96 | 17.22 | 17.81 |
| 221 | 9.8 | 9.9 | 12.6 | 9.86 | 9.91 | 9.79 | 9.81 |
| 234 | 6.6 | 6.4 | 4.8 | 6.10 | 6.66 | 6.04 | 6.16 |
| 245 | 16.4 | NI | 12.8 | 14.51 | 15.25 | 14.45 | 14.64 |
| 307 | NI | NI | 24.9 | 25.27 | 26.07 | 25.25 | 25.55 |
| 310 | 9.7 | 9.7 | 9.5 | 9.78 | 9.76 | 9.64 | 9.76 |
| 346 | 21.7 | 23.1 | 20.7 | 23.23 | 21.27 | 21.24 | 21.23 |
| 361 | 14.1 | 14.0 | 14.2 | 14.02 | 14.02 | 14.02 | 14.02 |
| 419 | 21.6 | 21.4 | 19.0 | 21.15 | 20.67 | 20.41 | 20.45 |
| 492 | 13.6 | 12.2 | 10.0 | 10.71 | 10.63 | 10.71 | 10.51 |
| 503 | 31.7 | 32.0 | 40.0 | 33.98 | 29.55 | 29.50 | 29.55 |
| 512 | 10.4 | 10.1 | 10.6 | 10.06 | 10.00 | 9.74 | 9.74 |
| 592 | 24.6 | 19.1 | 20.4 | 19.30 | 19.23 | 18.73 | 18.95 |
| 607 | 25.9 | 25.8 | 32.8 | 25.66 | 25.98 | 25.35 | 25.29 |
| 758 | 15.0 | 10.8 | 14.1 | 10.72 | 11.05 | 10.68 | 10.84 |
| 869 | 18.6 | 18.8 | 19.0 | 18.84 | 18.77 | 18.58 | 18.69 |
| 878 | 23.4 | 22.7 | NI | 22.61 | 22.85 | 22.59 | 22.79 |
| 918 | 24.4 | 23.2 | 47.2 | 24.67 | 23.80 | 23.54 | 23.73 |
| 992 | 35.3 | 34.3 | 37.7 | 34.34 | 34.71 | 34.20 | 34.56 |
| 1005 | 40.9 | 42.9 | 43.1 | 42.11 | 42.12 | 39.38 | 40.88 |
| 1007 | 24.0 | 22.6 | NI | 24.52 | 22.62 | 22.35 | 22.59 |
| 1242 | 41.4 | 44.9 | 37.1 | 40.10 | 40.70 | 39.33 | 39.18 |
| 2680 | 39.3 | 38.9 | - | 38.86 | 38.44 | 38.38 | 37.71 |

Figure 4: Average profile of the 30 examples

# On the Spanning Trees of Weighted Graphs[*]

## Ernst W. Mayr        C. Greg Plaxton[†]

### Abstract

Given a weighted graph, let $W_1, W_2, W_3, \ldots$ denote the increasing sequence of all possible distinct spanning tree weights. Settling a conjecture due to Kano, we prove that every spanning tree of weight $W_1$ is at most $k - 1$ edge swaps away from some spanning tree of weight $W_k$. Three other conjectures posed by Kano are proven for two special classes of graphs. Finally, we consider the algorithmic complexity of generating a spanning tree of weight $W_k$.

## 1   Introduction

The minimum spanning tree problem is a classic problem in computer science for which a large number of sequential, parallel and distributed algorithms have been devised. To the graph theorist, however, this body of work has provided little beyond the underlying proof of correctness of the original greedy algorithms due to Kruskal and Prim [Kr][Pr]. Evidently, the advances have occurred in the areas of data structures, parallel processing techniques and distributed protocols. In contrast, this paper attacks questions arising from a generalization of the minimum spanning tree concept that requires additional insight at the graph-theoretic level.

If one considers partitioning the spanning trees of a weighted graph into weight classes, a number of natural questions arise with regard to relationships between the classes. In a recent paper, Kano [Kan] posed four conjectures that were motivated by the previous work of Kawamoto, Kajitani and Shinoda [KKS]. Our main result is a proof that every minimum spanning tree is at most $k - 1$ edge swaps away from some representative of the $k$th weight class, settling the first of Kano's conjectures. With regard to the three remaining conjectures, we offer a stronger unified conjecture and prove that it holds for two non-trivial families of graphs.

We also consider the algorithmic complexity of generating a representative of the $k$th weight class. For fixed $k$, we obtain a polynomial time algorithm. When $k$ is part of the input, the associated decision problem (KMST) is seen to be NP-hard using a

---

[*]This work was supported in part by a grant from the AT&T Foundation and NSF grant DCR-8351757.

[†]Primarily supported by a 1967 Science and Engineering Scholarship from the Natural Sciences and Engineering Research Council of Canada.

reduction due to Johnson and Kashdan [JK] for the related $k$th best spanning tree problem (KBST). Lawler [La] gave a simple branch-and-bound algorithm for KBST with a running time that is pseudo-polynomial in $k$. The existence of such an algorithm provides a hopeful sign that a similar result might be possible for KMST. We do not achieve this goal, but by extending a determinant method due to Okada and Onodera [OO], we obtain an algorithm for which the running time is pseudo-polynomial in the edge weights as well as $k$.

# 2   Preliminaries

Let $G = (V, E)$ be a connected undirected graph with a real-valued weight $w(e)$ assigned to each edge $e \in E$. Let $e^-$ and $e^+$ denote the endpoints of edge $e$. Parallel edges and self-loops are permissible. For any subset $E'$ of $E$ define the weight of $E'$, denoted $w(E')$, as the sum of the weights of the edges of $E'$, that is,

$$w(E') = \sum_{e \in E'} w(e).$$

A *spanning tree* $P$ of $G$ is any subset of $E$ for which the graph $(V, P)$ is acyclic and connected; in order to satisfy both of these properties simultaneously it is necessary that $|P| = |V| - 1$. In order to discuss the partition of the spanning trees of $G$ into weight classes, we will make use of the following notation.

$$
\begin{aligned}
T(G) &= \{P \mid P \text{ is a spanning tree of } G\} \\
W(G) &= \{x \mid w(P) = x \text{ for some } P \in T(G)\} \\
N(G) &= |W(G)| \\
W_i(G) &= \text{the } i\text{th smallest element of } W(G), 1 \leq i \leq N(G) \\
T_i(G) &= \{P \in T(G) \mid w(P) = W_i(G)\} \\
\rho(G, P) &= i, \text{ if and only if } P \in T_i(G) \\
\sigma(G, P) &= |T_i(G)|, \text{ if and only if } P \in T_i(G)
\end{aligned}
$$

A spanning tree of weight $W_1(G)$ is a *minimum spanning tree* of $G$. In general, a spanning tree of weight $W_k(G)$ will be referred to as a *$k$th minimal spanning tree* ($k$-MST) of $G$. Note that some previous authors have preferred to formulate their results in terms of $k$th *maximal* spanning trees [Kan][KKS], an equivalent concept.

Let $P$ and $Q$ be spanning trees of $G$. Then $|P \setminus Q|$ is the *distance* between $P$ and $Q$ and will be denoted by $d(P, Q)$. For any nonnegative integer $k$, let $L_k(G, P)$ represent the set of all spanning trees $Q$ of $G$ such that $d(P, Q) \leq k$. Notice that $d(P, Q) = d(Q, P)$ so that $Q \in L_k(G, P)$ if and only if $P \in L_k(G, Q)$. For every $e \notin P$, let $Cyc(P, e)$ denote the fundamental cycle of $G$ defined by $e$ with respect to $P$. Given distinct edges $a, b$ such that $a \in Cyc(P, b)$, the spanning tree $P \setminus \{a\} \cup \{b\}$ is defined to be a single *edge swap* away from $P$. For any two spanning trees $P$ and $Q$, we note that the length of the shortest sequence of edge swaps required to transform $P$ into $Q$ is precisely $d(P, Q)$.

We will have occasion to make use of several well-known facts about 1-MSTs. The validity of each of these statements follows easily from the proof of correctness of Kruskal's greedy algorithm for computing a 1-MST [Kr].

**Fact 1** *A spanning tree is a 1-MST of $G$ if and only if it is a 1-MST in $L_1(G, P)$.*

**Fact 2** *The unique heaviest edge in some cycle cannot belong to any 1-MST. A heaviest edge in some cycle cannot belong to every 1-MST.*

**Fact 3** *The unique lightest edge in some cutset must belong to every 1-MST. A lightest edge in some cutset must belong to some 1-MST.*

**Fact 4** *Every 1-MST contains the same distribution of edge weights.*

**Fact 5** *Every 1-MST can be generated by Kruskal's algorithm.*

Given Fact 1, it is straightforward to verify that for any $i$-MST $P$ of $G$, $L_{i-1}(G, P)$ contains a 1-MST. However, going in the other direction turns out to be much more difficult. Kawamoto, Kajitani and Shinoda [KKS] studied 2-MSTs and proved that for every 1-MST $P$ there is a 2-MST $Q$ such that $d(P, Q) = 1$. This led Kano to pose the following conjecture, for which the result of Kawamoto et al. corresponds to the special case $i = 2$.

**Conjecture 1** *If $P$ is a 1-MST of $G$ then $L_{i-1}(G, P)$ contains an $i$-MST, $1 \leq i \leq N(G)$.*

Kano was able to prove that Conjecture 1 holds for $i = 3$ and 4. We will show that Conjecture 1 is, indeed, a theorem. Kano also proposed the following three conjectures, proving them for values of $i$ less than or equal to 3, 4 and 3, respectively.

**Conjecture 2** *If $P$ is an $i$-MST in $L_i(G, P)$ then $P$ is an $i$-MST of $G$.*

**Conjecture 3** *If $P$ is an $i$-MST of $G$ then $P$ is an $i$-MST in $L_{i-1}(G, P)$.*

**Conjecture 4** *Let $\Gamma(i, j)$ denote the graph with vertex set $T_i(G)$ and an edge between each pair of $i$-MSTs $P, Q$ such that $d(P, Q) \leq j$. Then $\Gamma(i, i)$ is connected.*

## 3    Proofs

In this section, we prove Conjecture 1 along with a number of other results.

**Lemma 3.1** *Let $P$ and $Q$ be spanning trees of a given graph $G$. Then for each edge $p \in P \setminus Q$ there is an edge $q \in Q \setminus P$ such that $p \in Cyc(P, q)$ and $q \in Cyc(Q, p)$.*

**Proof:**    Let $P$, $Q$ and $p$ be as defined above and assume that there is no edge $q$ satisfying the requirements of the lemma. Then there must be a path made up of edges from $P \setminus \{p\}$ that connects the two endpoints of $p$. But then $P$ contains a cycle, contradicting the assumption that it is a spanning tree. □

**Definition 3.1** *Given a graph* $G = (V, E)$, *let* $\xi$ *represent* $G$ *or any subset of* $E$. *If* $C$ *and* $D$ *are disjoint subsets of* $E$, *then the graph or set of edges produced from* $\xi$ *by contracting the edges of* $C$ *and discarding the edges of* $D$ *will be denoted* $\xi[C, D]$. *Notice that there is a 1-1 correspondence between the edges of* $\xi[C, D]$ *and those edges of* $\xi$ *that do not belong to* $C \cup D$. *We identify the pairs of edges determined by this correspondence, inheriting edge weights in the weighted case.*

We note that if $P$ is a spanning tree of $G$, then $P[c, d]$ is a spanning tree of $G[c, d]$ if and only if $c \in P$ and $d \notin P$, or $c \notin P$ and $d \in Cyc(P, c)$.

**Lemma 3.2** *Let* $P$ *be a spanning trees of a weighted graph* $G = (V, E)$. *For any edge* $e \notin P$, *let* $G' = G[, e]$ *and* $P' = P$. *For any edge* $e \in P$, *let* $G' = G[e, ]$ *and* $P' = P[e, ]$. *In either case, the following statements hold:*

1. *$P'$ is a spanning tree of $G'$.*
2. *$\rho(G', P') \leq \rho(G, P)$.*
3. *$\sigma(G', P') \leq \sigma(G, P)$.*

**Proof:** The above statements follow easily once we exhibit an injective map $\Phi$ from $T(G')$ to $T(G)$ that takes $P'$ to $P$ and also satisfies a "constant displacement" property, namely, there exists a real value $\Delta$ such that for all spanning trees $R' \in T(G')$,

$$w(\Phi(R')) = w(R') + \Delta. \tag{1}$$

For the case $e \notin P$, take $\Phi(R')$ to be simply $R'$ so that equation 1 holds with $\Delta = 0$. For $e \in P$, let $\Phi(R')$ be the unique spanning tree $R \in T(G)$ such that $R[e, ] = R'$ and set $\Delta = w(e)$. $\square$

**Lemma 3.3** *Let* $P$ *be a 1-MST of a given weighted graph* $G = (V, E)$. *For any edge* $e \in P$, *let* $P' = P[e, ]$. *For any edge* $e \notin P$, *let* $P' = P[e, h]$ *where* $h$ *is a heaviest edge on* $Cyc(P, e) \setminus \{e\}$. *In either case,* $P'$ *is a 1-MST of* $G[e, ]$.

**Proof:** If $e \in P$ then Lemma 3.2 applies. Assume that $e \notin P$. Appealing to Fact 5, consider an execution of Kruskal's algorithm on input $G$ that generates the 1-MST $P$. As each edge $p \in P$ gets added to $P$, Kruskal's algorithm running on input $G[e, ]$ can correctly select $p$ with one exception: the edge that first puts $e^-$ and $e^+$ into the same component must be omitted. This edge must have weight $w(h)$ since it is a heaviest edge on the path from $e^-$ to $e^+$ in $P$. $\square$

**Definition 3.2** *A graph* $G = (V, E)$ *will be called a* bispanning graph *if* $E$ *is the union of two disjoint spanning trees* $P$ *and* $Q$. *Such a bispanning graph will be denoted by the triple* $(V, P, Q)$.

**Definition 3.3** *A* simple bispanning graph *is a bispanning graph that contains only cycles of length 2.*

The following result was obtained previously by Kano using Hall's Theorem [Kan]. Here we provide an alternative proof that is explicitly constructive and introduces some of the ideas that will be used to prove Conjecture 1.

**Theorem 1** *Let $P$ be a 1-MST and $Q$ be an arbitrary spanning tree of a given weighted graph $G = (V, E)$. Then there exists a bijection $\Phi$ from $P \setminus Q$ to $Q \setminus P$ such that for every edge $e \in P \setminus Q$, $\Phi(e) \in Cyc(Q, e)$ and $w(\Phi(e)) \geq w(e)$.*

**Proof:** It is sufficient to determine a bijection between the disjoint spanning trees $P' = P \setminus Q$ and $Q' = Q \setminus P$ of the bispanning graph $G' = G[P \cap Q, E \setminus P \setminus Q]$. This is due to the observation that $Cyc(Q', e) \subseteq Cyc(Q, e)$ for all $e \in P'$.

Let $p$ be a heaviest edge in $P'$ and let $q$ be any edge in $Q'$ for which $p \in Cyc(P', q)$ and $q \in Cyc(Q', p)$. Lemma 3.1 guarantees that we can find such a $q$. As $P'$ is a 1-MST, we must have $w(Q) \geq w(P)$. Let $\Phi(p) = q$.

To determine the next component of the bijection, repeat this procedure on the bispanning graph $G'[q, p]$ with disjoint spanning trees $P'[q, p]$ and $Q'[q, p]$. $P'[q, p]$ is a 1-MST of $G'[q, p]$ by Lemma 3.3. Furthermore, for all $e \in P'[q, p]$,

$$Cyc(Q'[q, p], e) = Cyc(Q', e) \setminus \{q\} \subseteq Cyc(Q', e) \subseteq Cyc(Q, e)$$

so that subsequent assignments to $\Phi$ will be guaranteed to satisfy the condition $\Phi(\cdot) \in Cyc(Q, \cdot)$. $\square$

**Lemma 3.4** *Conjecture 1 holds if and only if there is no weighted bispanning graph $B = (V, P, Q)$ such that $d(P, Q) \geq \rho(B, Q) > \rho(B, P) = 1$ and $\sigma(B, Q) = 1$.*

**Proof:** The "only if" directon is easy. To establish the "if" direction, we will prove the contrapositive. Given a counterexample $(G, P, i)$ to Conjecture 1, let $Q$ be a closest $i$-MST to $P$. Then repeated application of Lemma 3.2 proves that the bispanning graph $G' = G[P \cap Q, E \setminus P \setminus Q]$ with disjoint spanning trees $P' = P \setminus Q$ and $Q' = Q \setminus P$ also violates the conjecture, that is, $d(P', Q') \geq \rho(G', Q') > \rho(G', P') = 1$.

It remains to be proven that $\sigma(G', Q') = 1$. If not, there must be a spanning tree $Q''$ of $G'$ such that $w(Q'') = w(Q')$ and $Q'' \neq Q'$. Then $d(P', Q'') \geq d(P', Q')$ by the definition of $Q$. On the other hand, $d(P', Q'') < d(P', Q')$ since $Q''$ must have at least one edge in common with $P'$, whereas $Q'$ has none. Thus, $Q'$ must have unique weight in $T(G')$, as required. $\square$

**Theorem 2** *There is no weighted bispanning graph $B = (V, P, Q)$ such that $d(P, Q) \geq \rho(B, Q) > \rho(B, P) = 1$ and $\sigma(B, Q) = 1$. Hence, Conjecture 1 holds by Lemma 3.4.*

**Proof:** Assume the theorem is false and let $B = (V, P, Q)$ be a counterexample with smallest possible $|V|$. We will establish a contradiction by exhibiting a smaller counterexample $B'$.

As in the proof of Theorem 1, let $p$ be a heaviest edge in $P$ and let $q$ be any edge in $Q$ for which $p \in Cyc(P, q)$ and $q \in Cyc(Q, p)$. Since $P$ is a 1-MST and $\sigma(B, Q) = 1$, we must have $w(Q) > w(P)$. Therefore, $q$ is the unique heaviest edge in $Cyc(P, q)$ and does not belong to any 1-MST of $B$ by Fact 2.

Now consider the smaller bispanning graph $B' = B[q, p]$ with disjoint spanning trees $P' = P[q, p]$ and $Q' = Q[q, p]$. Clearly, $d(P', Q') = d(P, Q) - 1$. Furthermore, $\rho(B', P') = 1$ and $\sigma(B', Q') = 1$ by Lemmas 3.3 and 3.2, respectively. In order to show that $B'$ is a counterexample to the theorem, it is sufficient to prove that $\rho(B', Q') < \rho(B, Q)$. Lemma 3.2 gives us only $\rho(B', Q') \le \rho(B, Q)$, but in the present case that argument can be strengthened to yield the desired strict inequality. Namely, let $\Phi$ be the injective map taking $R' \in T(B')$ to $R \in T(B)$ such that $R' = R[q, p]$ and observe that $\Phi$ does not map any spanning tree of $B'$ into $T_1(B)$. □

The statements proven by the next three lemmas all have the structure of Lemma 3.4, underscoring the similarities between Kano's various conjectures. Note, however, that Lemma 3.5 is proven in one direction only.

**Lemma 3.5** *Conjecture 2 holds if there is no weighted bispanning graph $B = (V, P, Q)$ such that $d(P, Q) \ge \rho(B, Q) > \rho(B, P)$ and $\sigma(B, Q) = 1$.*

**Proof:** We prove the contrapositive. Accordingly, let $(G, R, i)$ be a counterexample to Conjecture 2, that is, $R$ is an $i$-MST in $L_i(G, R)$ but not an $i$-MST of $G$. There must be a least integer $j < i$ such that $L_i(G, R)$ does not contain a $j$-MST. By Fact 1, there must be a $k$-MST $P$ such that $d(P, R) \le i - k$ for some $k < j$. Let $Q$ be a closest $j$-MST to $P$. Since $Q$ is not contained in $L_i(G, R)$ we have

$$d(P, Q) \ge i - d(P, R) \ge k.$$

Now consider the bispanning graph $G' = G[P \cap Q, E \setminus P \setminus Q]$ with disjoint spanning trees $P' = P \setminus Q$ and $Q' = Q \setminus P$. By repeated application of Lemma 3.2,

$$d(P', Q') \ge \rho(G', Q') > \rho(G', P') \text{ and } \sigma(G', Q') = 1$$

so the proof is complete. □

**Lemma 3.6** *Conjecture 3 holds if and only if there is no weighted bispanning graph $B = (V, P, Q)$ such that $d(P, Q) \ge \rho(B, P) > \rho(B, Q)$ and $\sigma(B, Q) = 1$.*

**Proof:** The proof is similar to that given for Lemma 3.4. In this case, assume that $(G, P, i)$ is a counterexample to Conjecture 3, let $j$ be any integer less than $i$ such that $L_{i-1}(G, P)$ does not contain a $j$-MST, and let $Q$ be a closest $j$-MST to $P$. □

**Lemma 3.7** *Conjecture 4 holds if and only if there is no weighted bispanning graph $B = (V, P, Q)$ such that $d(P, Q) > \rho(B, P) = \rho(B, Q)$ and $\sigma(B, Q) = 2$.*

**Proof:** Once again, the proof is similar to that given for Lemma 3.4. Assume that $(G, i)$ is a counterexample to Conjecture 4, and let $P, Q$ be a closest pair of $i$-MSTs belonging to different connected components of $\Gamma(i, i)$. $\square$

Given a weighted bispanning graph $B = (V, P, Q)$ such that $Q$ has unique weight in $T(B) \setminus \{P\}$, let $i = \rho(B, P)$, $j = \rho(B, Q)$, $a = \max\{i, j\}$ and $b = \min\{i, j\}$. Unifying the "if" statements of Lemmas 3.4 through 3.7, we find that all of Kano's conjectures hold if we can guarantee that $d(P, Q) \leq \max\{a - 1, b\}$. This bound does not appear to be attainable for all values of $a$ and $b$. We expect that it can be strengthened to the inequality stated below.

**Conjecture 5** *Let $P, Q, a, b$ be as defined in the preceding paragraph. Then*

$$d(P, Q) \leq \begin{cases} a, & \text{if } a = b; \\ a - 1, & \text{if } a > b = 1; \\ \left\lfloor \frac{a+b-1}{2} \right\rfloor, & \text{otherwise.} \end{cases}$$

This conjecture is strongest possible if $a - b$ is odd, $a = b$ or $b = 1$, as may be easily seen by assigning appropriate edge weights to simple bispanning graphs. Furthermore, Conjecture 5 implies the following bound for general graphs that is *optimal* in the sense that equality can be achieved for all values of $a$ and $b$.

**Lemma 3.8** *Let $P$ be an $i$-MST of a given weighted graph $G$ and suppose that $Q$ is a closest $j$-MST to $P$ in $T(G) \setminus \{P\}$. Let $a = \max\{i, j\}$ and $b = \min\{i, j\}$. If Conjecture 5 holds, then*

$$d(P, Q) \leq \begin{cases} a, & \text{if } a = b; \\ \max\left\{a - b, \left\lfloor \frac{a+b-1}{2} \right\rfloor\right\}, & \text{otherwise.} \end{cases}$$

**Proof:** Apply Lemma 3.2 until the bispanning graph $G[P \cap Q, E \setminus P \setminus Q]$ is obtained, then use Conjecture 5. $\square$

Thus far, we have been able to prove Conjecture 5 only for certain special classes of graphs. We will now examine two such families that were obtained by restricting: the graph structure, in the first case, and the set of allowable edge weights, in the second case.

**Definition 3.4** *Let $P$ and $Q$ be spanning trees of a given graph $G$. Then $P$ and $Q$ are related by a* parallel swap *if and only if there exists a bijection $\Phi$ from $P \setminus Q = \{e_1, \ldots, e_t\}$ to $Q \setminus P = \{\Phi(e_1), \ldots, \Phi(e_t)\}$ such that each of the $2^t$ sets of $|P|$ edges containing $P \cap Q$ and exactly one edge from each pair $\{e_i, \Phi(e_i)\}, 1 \leq i \leq t$, is a spanning tree of $G$.*

As an aside, it is possible to prove that spanning trees $P$ and $Q$ are related by a parallel swap if and only if Lemma 3.1 is satisfied by a unique $q \in Q \setminus P$ for each $p \in P \setminus Q$. This means that one may readily determine whether or not two particular spanning trees are related by a parallel swap.

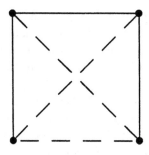

Figure 1: The smallest bispanning graph without a parallel swap.

**Theorem 3** *Let $B = (V, P, Q)$ be a bispanning graph for which $P$ and $Q$ are related by a parallel swap. Then Conjecture 5 holds.*

**Proof:** Let $B$ be a counterexample with smallest possible $|V|$. Without loss of generality, we can assume that $B$ is a simple bispanning graph. Observe that for all $R \in T(B)$, $\overline{R} = E \setminus R$ also belongs to $T(B)$ and $\rho(B, R) + \rho(B, \overline{R}) = N(B) + 1$. Hence, $P$ has unique weight in $T(B) \setminus \{Q\}$. If either $P$ or $Q$ is a 1-MST then the result follows easily by Fact 1. If $w(P) = w(Q)$ then contract any cycle to obtain a smaller counterexample. If two possible edge swaps increase (decrease) the weight of $P$ by different amounts, then contract the cycle corresponding to the larger change in order to obtain a smaller counterexample.

The preceding argument allows us to assume that $s$ of the possible edge swaps in $B$ increase the weight of $P$ by some amount $\delta_1$ while each of the remaining $|V| - 1 - s$ edge swaps causes a decrease of $\delta_2$. It is straightforward to prove that a weighted bispanning graph of this form cannot provide a counterexample to Conjecture 5. □

If Conjecture 5 holds for all bispanning graphs with up to $n$ vertices, then Conjectures 2, 3 and 4 must hold for all values of $i$ less than or equal to $n - 1$, $n$ and $n - 1$, respectively. It is easy to check that every bispanning graph $B = (V, P, Q)$ with $|V| \leq 3$ admits a parallel swap between $P$ and $Q$. For $|V| = 4$, only the bispanning graph in Figure 1 does not have this property, but we can show by case analysis that Conjecture 5 holds for this particular graph as well. These observations re-establish the results obtained by Kano with respect to the small cases of Conjectures 2 through 4.

Another interesting class of weighted graphs for which we can prove Conjecture 5 are those with edge weights drawn from an arithmetic sequence of length 3. Weighted graphs of this sort arise in the analysis of LCR networks [Kaj], and Kano was able to prove Conjecture 1 for such graphs.

**Theorem 4** *Let $B = (V, P, Q)$ be a weighted bispanning graph with edge weights drawn from the set $\{a, a + d, a + 2d\}$. Then Conjecture 5 holds.*

**Proof Sketch:** Assume without loss of generality that $a = 0$ and $d = 1$, and let $B = (V, P, Q)$ be a counterexample with smallest possible $|V|$. We will briefly outline the proof of the hardest case, which turns out to occur when $W(Q) > W(P) > W_1(B)$. Let $Q'$ be a lightest spanning tree in $L_1(Q)$. Either $W(Q') = W(Q) - 1$ or $W(Q') = W(Q) - 2$. If $W(Q') = W(Q) - 1$, it is possible to prove that $\sigma(B, Q) > 1$, a contradiction. If $W(Q') = W(Q) - 2$ then we show that $P$ must be a 2-MST with $W(P) = W_1(B) + 1$ or else $\sigma(B, Q) > 1$. It is now sufficient to prove that there is an edge $q \in Q$ of weight 2 such that $Cyc(P, q)$ contains an edge $p$ of weight 0, since this implies the existence of a smaller counterexample $B[q, p]$. $\square$

## 4 The Generation Problem

For fixed $k$, Theorem 2 yields a polynomial time algorithm for computing a $k$-MST. In particular, if $n = |V|$ and $m = |E|$ we have

**Corollary 4.1** *Given a weighted graph $G$, a $k$-MST of $G$ can be generated in $O((mn)^{k-1})$ time.*

**Proof:** First compute a 1-MST $M$. This can be done in $O(m + n \log n)$ time using Fibonacci heaps [FT], although for the present argument it would suffice to use one of the simpler 1-MST algorithms with a higher asymptotic running time. Then compute the weight of all spanning trees that are strictly fewer than $k$ edge swaps away from $M$. There are at most

$$\sum_{i=0}^{k-1} \binom{n-1}{i}\binom{m-n+1}{i} = O((mn)^{k-1}) \tag{2}$$

such spanning trees and the weight of each tree can be computed incrementally in constant time by performing the enumeration in a depth-first manner. At any point in the computation we need to remember the $k$ largest distinct weights found so far. This only increases the running time by a constant factor since $k$ is fixed. $\square$

In order to discuss complexity issues, we introduce the language problem associated with the problem of finding a $k$-MST.

**Definition 4.1** *Given a graph $G = (V, E)$, positive integer weights $w(e)$ for each $e \in E$, positive integers $k$ and $B$. The KMST problem is to determine whether or not there are $k$ spanning trees of $G$ with distinct weights less than or equal to $B$.*

By omitting the word "distinct" from Definition 4.1, we get the definition of the $k$th best spanning tree problem (KBST).

**Theorem 5** *KMST is NP-hard.*

**Proof:** The proof is identical to the one given in [JK] for KBST, which is not known to be in NP. The reduction is from HAMILTON CIRCUIT. □

Note that the $O((mn)^{k-1})$ algorithm described earlier does not solve KMST in time pseudo-polynomial in $k$. The running time of such an algorithm would have to be polynomial in $|V|$, $k$, $\log B$ and $\log w_{max}$, where $w_{max}$ is the weight of the heaviest edge. Interestingly, a pseudo-polynomial time solution *is* known for KBST; the approach is to generate all spanning trees up to the $k$th [La]. Unfortunately, this method is not powerful enough for KMST since there may be exponentially many spanning trees belonging to classes below the $k$th.

We will now present a determinant-based KMST algorithm for which the running time is pseudo-polynomial in the edge weights as well as $k$. Let $G$ be a graph with $n+1$ vertices $v_0, \ldots, v_n$ and $m$ edges $e_1, \ldots, e_m$. Construct the $m \times (n+1)$ matrix A with

$$a_{ij} = \begin{cases} +1, & \text{if } e_i = (v_j, v_k) \text{ for some } k > j; \\ -1, & \text{if } e_i = (v_j, v_k) \text{ for some } k < j; \\ 0, & \text{otherwise.} \end{cases}$$

and let $A_0$ be the $m \times n$ matrix $A$ with column 0 deleted. Letting $A_S$ denote the matrix $(a_{ij})$ for $i \in S$ and $j$ ranging from 1 to $n$, we have

$$\det A_0^T A_0 = \sum_{\substack{S \subseteq \{1,\ldots,m\} \\ |S|=n}} [\det A_S]^2$$

by the Binet-Cauchy formula (see [Kn]). Furthermore,

$$\det A_S = \begin{cases} \pm 1, & \text{if } (V, S) \text{ is a tree}; \\ 0, & \text{otherwise.} \end{cases} \qquad (3)$$

so that the determinant of $A_0^T A_0$ is the number of spanning trees of $G$. This result is due to Okada and Onodera [OO].

Similary, we can construct the $m \times (n+1)$ matrix B with

$$b_{ij} = \begin{cases} +x^{w(e_i)}, & \text{if } e_i = (v_j, v_k) \text{ for some } k > j; \\ -x^{w(e_i)}, & \text{if } e_i = (v_j, v_k) \text{ for some } k < j; \\ 0, & \text{otherwise.} \end{cases}$$

and let $B_0$ be the $m \times n$ matrix $A$ with column 0 deleted. Let $p(x)$ be the determinant of $B_0^T A_0$, and let $p_k$ be the coefficient of $x^k$ in $p(x)$.

$$\begin{aligned} \det B_0^T A_0 &= \det\left[(x^{w(e_1)} \cdots x^{w(e_m)}) A_0^T A_0\right] \\ &= \sum_{\substack{S \subseteq \{1,\ldots,m\} \\ |S|=n}} [\det A_S]^2 x^{w(S)} \\ &= \sum p_k x^k \end{aligned}$$

Using equation 3 we find that $p_k$ is the number of spanning trees of $G$ with weight $k$, that is, $p(x)$ is the generating function for the number of spanning trees in each weight class of $G$.

This relationship leads us to a pair of algorithms for solving KMST. The first idea is to compute $\det B_0^T A_0$ for all integer values of $x$ from 0 up to $W_{max}$, where $W_{max} = W_{N(G)}(G)$, and then interpolate to obtain $p(x)$. The interpolation is the most costly step, and it involves solving a system of $W_{max} + 1$ equations with $O(W_{max}^2)$ digit integer coefficients. This can be done in time polynomial in $W_{max}$ by the result of Edmonds [Ed], who proved that Gaussian elimination with pivoting is in $\mathcal{P}$ for exact rational arithmetic.

A second approach is to obtain an upper bound for $p_k$ and then compute $\det B_0^T A_0$ for a single value of $x$ that is large (or small) enough to ensure that all of the coefficients $p_k$ can be extracted from the final result. Letting $C = B_0^T A_0$ with $x = 1$, we can derive a suitable bound on $p_k$ as follows.

$$p_k \leq \det C$$
$$\leq \prod_{1 \leq i \leq n} \left( \sum_{1 \leq j \leq n} c_{ij}^2 \right)^{1/2}$$

The first inequality follows from the fact that $p_k \geq 0$ for all $k$; the second is Hadamard's inequality.

Either of these methods gives a KMST algorithm with running time pseudo-polynomial in the edge weights as well as $k$. At the expense of an extra polynomial factor, we can easily convert them into algorithms for generating a $k$-MST.

# 5   Open Problems

We have succeeded in proving one of Kano's four conjectures; the remaining three would be established by a proof of Conjecture 5. The existence of a pseudo-polynomial (in $k$ alone) algorithm for generating a $k$-MST remains open.

# References

[Ed]   J. Edmonds. Systems of distinct representatives and linear algebra. *J. of Research and the National Bureau of Standards*, **71B** (1967), 241-245.

[FT]   M. L. Fredman and R. E. Tarjan. Fibonacci heaps and their uses in improved network optimization algorithms. *JACM*, **34** (1987), 596-615.

[JK]   D. B. Johnson and S. D. Kashdan. Lower bounds for selection in $X + Y$ and other multisets. *JACM*, **25** (1978), 556-570.

[Kaj] Y. Kajitani. Graph theoretical properties of the node determinant of an LCR network. *IEEE Trans. Circuit Theory*, CT-**18** (1971), 343-350.

[Kan] M. Kano. Maximum and $k$th maximal spanning trees of a weighted graph. Combinatorica, **7** (1987), 205-214.

[KKS] T. Kawamoto, Y. Kajitani and S. Shinoda. On the second maximal spanning trees of a weighted graph (in Japanese). *Trans. IECE of Japan*, **61**-A (1978), 988-995.

[Kn] D. E. Knuth. *The Art of Computer Programming Vol. I: Fundamental Algorithms*, Addison-Wesley, Reading, Mass.

[Kr] J. B. Kruskal. On the shortest spanning subtree of a graph and the traveling salesman problem. *Proc. Amer. Math. Soc.*, **7** (1956), 48-50.

[La] E. L. Lawler. A procedure for computing the $K$ best solutions to discrete optimization problems and its application to the shortest path problem. *Management Sci.*, **18** (1972), 401-405.

[OO] Okada and Onodera. *Bull. Yamagata Univ.*, **2** (1952), 89-117 (cited in [Kn]).

[Pr] R. C. Prim. Shortest connection networks and some generalizations. *Bell System Technical J.*, **36** (1957), 1389-1401.

# ON PATHS IN SEARCH OR DECISION TREES WHICH REQUIRE ALMOST WORST-CASE TIME

Ulrich Huckenbeck

Lehrstuhl für Informatik I, Universität Würzburg,
Am Hubland, D-8700 Würzburg, West Germany

## Abstract

We prove the existence of a particular path p in a search or decision tree; this path will symbolize a computation requiring almost worst-case time.

The result about the existence of p is the following:

Given a finite tree $\mathcal{T}$ with the root r and the set B of leaves. Let every $b \in B$ be attached by a weight $w(b) \geq 0$ and let $w(\mathcal{T})$ be the sum $\sum\limits_{b \in B} w(b)$. Then there exists a path $p = (v_0, \ldots, v_\ell)$ from the root $r = v_0$ to a leaf $v_\ell \in B$ such that $g^+(v_0) \cdot \ldots \cdot g^+(v_{\ell-1}) \cdot w(v_\ell) \geq w(\mathcal{T})$ (where $g^+(v_\lambda)$ = out-degree of $v_\lambda$).

We shall use this lemma to obtain the following complexity theoretical results:

1) Searching in sorted multi-way trees requires $\Omega(\log(n))$ time.

2) Let finding a rotational minimum among j points take $\Theta(j)$ time units where T belongs to a particular class of sublogarithmic functions. Then the worst-case complexity of the Plane Convex Hull Problem is in $\Omega(nT(n))$.

3) The worst-case complexity of the Convex Hull Problem is in $\Omega(n\log T(n))$ if the following operations altogether take $\Theta(T(n))$ time units: Taking an oriented straight line $\vec{g}$ and deciding which of the n input points are on the right of it.

4) The worst case complexity of the Sorting Problem is also in $\Omega(n\log T(n))$ if the following operations can be executed within $\Theta(T(n))$ time units: Taking a real t and deciding which of the n input reals are smaller than t.

In the Applications 2) – 4) we shall realize that $n \cdot T(n)$, $n\log(T(n))$ resp. is even a tight bound.

## I) Introduction

There are a lot of complexity theoretical investigations which can be reduced to the following situation: We have a rooted decision tree $\mathcal{T}$ where each leaf b has a weight $w(b) \geq 0$. Moreover, the costs of each node v are monotonous increasing with respect to the out degree $g^+(v)$ (the weight $w(v)$ resp.).

The most transparent example of such a situation is that of a sorted multi-way tree: Let $w(v)$ be the number of keys stored in the leaf b; then the search within a node v becomes very complicated and expensive, if

- v is an interior node and $g^+(v)$ is very great, or
- v is a leaf and $w(v)$ is very great.

Another typical case is a decision tree $\mathcal{T}$ where the vertices of different out–de-grees represent different types of operations; then the costs of each vertex are the time reqired by the corresponding operation.

It is obvious that the worst case arises if we have a path $p = (v_0,...,v_\ell)$ from the root $v_0$ to a leaf $v_\ell$ such that $g^+(v_0),...,g^+(v_{\ell-1})$ and $w(v_\ell)$ are very great. Consequently, p is a good candidate for an especially expensive path, if the pro-duct of these numbers is as great as possible. Thus we can comprehend that our graph theoretical lemma described in the Abstract is an important tool to calcu-late lower bounds of worst-case complexities, and we shall indeed demonstrate it in four applications.

Our treatise is organized as follows: After the definitions in **Chapter II)**, we shall present our graph theoretical result in **Chapter III)**; this lemma is used in **Chapter IV)** to calculate lower bounds for worst-case complexities: **Application 1)** deals with the well-known Searching Problem in sorted multi-way trees. In the **Applications 2) – 4)** of our graph theoretical lemma, we consider some new and in-teresting aspects of the Convex Hull Problem and the Sorting Problem: A particular type of operations can be executed extremly fast (e.g. with the help of a parallel processor); thus we obtain lower (and even tight) complexity bounds which are dif-ferent from the usual $n \cdot \log(n)$-results. Finally, in **Chapter V)** we give a survey of our results and compare the proofs of Chapter IV with each other.

## II) Definitions and Notations

### Graph Theoretical Notations

A **(di–)graph** is a pair $G = (V,R)$ where $V \neq \emptyset$ is the set of its **vertices** and $R \subseteq V \times V$ is the set of its (directed) **edges**. For every $v \in V$ let $g^+(v)$ denote the **out–degree** of v.

### Geometric Terms

Let $\vec{g}$ be a directed (straight) line. Then we make the convention that the points on $\vec{g}$ itself are situated on the **right** of $\vec{g}$; consequently, the right half plane with respect to $\vec{g}$ is closed, while the left one is open.

For every point $P \in E^2$ and every directed lines $\vec{g}_1,...,\vec{g}_j$, we define the **intersec-tion word** $W(\vec{g}_1,...,\vec{g}_j;P) := W(P) = z_1....z_j$, where

$$z_i = \begin{cases} r, & \text{if P is on the \underline{right} of } \vec{g}_i, \\ \ell, & \text{if P is on the \underline{left} of } \vec{g}_i \end{cases} \quad (i = 1,...,j). \quad (\text{ see [1], p. 83 })$$

Moreover, for every subset $S \subseteq E^2$, we define $W(\vec{g}_1,...,\vec{g}_j;S) := W(S) := \{W(P) \mid P \in S\}$.

In Figure 1 we have $W(\vec{g}_1,...,\vec{g}_4;P) = \ell r r \ell$ and $W(S) = \{ \ell r r r , r r r r , r r \ell r , r r \ell \ell , r r r \ell , \ell r r \ell \}$.

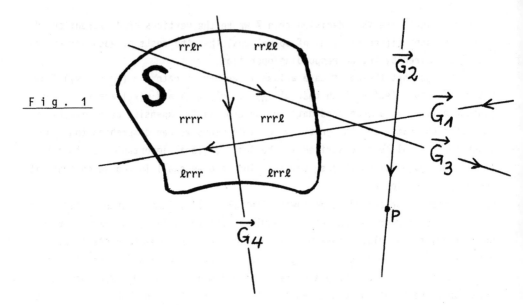

<u>F i g . 1</u>

Let $P_1,\ldots,P_n \in E^2$ with pairwise distinct x-values. Then **CONVEX**$(P_1,\ldots,P_n)$ is that sequence of extremal points which is scanned counterclockwise beginning with the point of minimal x-value; the points of this sequence have to be pairwise distinct.

## Our Underlying Model of Computation

Our underlying abstract automaton is a modified RAM working on the universe $U = \mathbb{R}$ (Applications 1, 4) or on the universe $U = E^2$ (Applications 2, 3). Our machines have the following capabilities:

- comparing reals (in the case of $U = \mathbb{R}$),
- comparing x- and y-coordinates of points (in the case of $U = E^2$),
- deciding whether triangles (A,B,C) are oriented clockwise (in the case
  of $U = E^2$),
- applying '+','-','×','/' to <u>integers</u> (in order to handle arrays).

On the other hand, our RAM's are **not** able to apply any arithmetical operations to reals.

## III)  Our Graph Theoretical Lemma

The objective of this paragraph is the proof of that graph theoretical result this treatise is based on:

## LEMMA G:

Let $\mathcal{T} = (V,R)$ be a finite rooted tree with height h; let r be the root of $\mathcal{T}$ and $B \subseteq V$ the set of its leaves.

We assume that every leaf $b \in B$ is labelled with a weight $w(b) \geq 0$, and we define the total weight of $\mathcal{T}$ as $w(\mathcal{T}) := \sum_{b \in B} w(b)$ .

Then there is a path $p = (v_0, \ldots, v_\ell)$ from the root $v_0 = r$ to a leaf $v_\ell \in B$ such that

$$\prod_{\lambda=0}^{\ell-1} g^+(v_\lambda) \cdot w(v_\ell) \geq w(\mathcal{T}).$$

( In the case of $\ell = 0$ we define $\prod_{\lambda=0}^{-1} g^+(v_\lambda) := 1$.

Note that our result generalizes the following theorem: The height of every binary tree $\mathcal{T}$ is in $\Omega(\log(|B|))$. In order to prove this statement, we define $w(b) = 1$ ($b \in B$) so that $w(\mathcal{T}) = |B|$. Then the path $p$ according to **LEMMA G** must indeed have a length in $\Omega(\log(|B|))$, because $g^+(v_\lambda) \leq 2$ ($\lambda = 0, \ldots, \ell-1$). )

Proof:

We apply an induction on the height h:

In the case of $h = 0$, our tree $\mathcal{T}$ only consists of the root $v_0 = r$, and our statement is true for the path $p = (v_0)$.

Let now $\mathcal{T}$ be a tree with height $h > 0$ and let our theorem be true for every height $h' < h$. If the root $r$ has the out-degree $m := g^+(r)$, then $r$ is the father of m subtrees $\mathcal{T}_1, \ldots, \mathcal{T}_m$. Let us now take that $\mu$ for which $w(\mathcal{T}_\mu)$ is maximal among the weights $w(\mathcal{T}_1), \ldots, w(\mathcal{T}_m)$. Then we have

$$g^+(r) \cdot w(\mathcal{T}_\mu) = m \cdot w(\mathcal{T}_\mu) \geq w(\mathcal{T}_1) + \ldots + w(\mathcal{T}_m) = w(\mathcal{T}). \quad (1)$$

Then we can find a path $\tilde{p} := (v_1, \ldots, v_\ell)$ in $\mathcal{T}_\mu$ such that

$$\prod_{\lambda=1}^{\ell-1} g^+(v_\lambda) \cdot w(v_\ell) \geq w(\mathcal{T}_\mu). \quad (2)$$

Consequently, the path $p := (v_0 = r, v_1, \ldots, v_\ell)$ has the desired property because of (1) and (2):

$$\prod_{\lambda=0}^{\ell-1} g^+(v_\lambda) \cdot w(v_\ell) \underset{(2)}{\geq} g^+(v_0) \cdot w(\mathcal{T}_\mu) \underset{(1)}{\geq} w(\mathcal{T}).$$

Remark:

Note that this proof is constructive; we obtain the following algorithm constructing p:

1) For every vertex $v \in V$ compute $w(v)$ which is the weight of the subtree rooted in v. This can be done as follows:

$$h := \text{height of } \mathcal{T};$$

for $j := h$ down to 0 do

for every interior vertex of height j do

$$w(v) := \sum_{\substack{v' \text{ is a} \\ \text{son of } v}} w(v').$$

2) Construct the path p := (v₀,...,v_ℓ) as follows:

$$v_0 := r \ (= \text{root of } T);$$
$$\lambda := 0;$$

while $v_\lambda$ is an interior vertex of $T$ do

begin

choose $v_{\lambda+1}$ such that $w(v_{\lambda+1}) = \max \{w(v') \mid v' \text{ is a son of } v_\lambda\};$
$$\lambda := \lambda + 1;$$

end;

This algorithm is illustrated in the next figure where the leaves are rectangular and the interior vertices are circles. The numbers in the vertices v mean w(v).

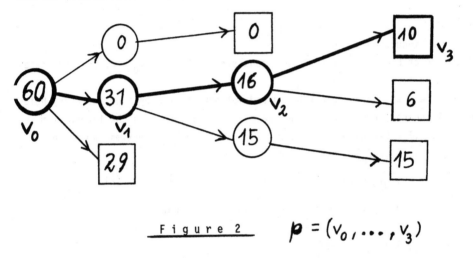

F i g u r e   2        $p = (v_0, \ldots, v_3)$

## IV)  Four Applications of LEMMA G

The first of our theorems which is proven with the help of **LEMMA G** is the following:

### Theorem 1.1
Let $T$ be a sorted multiway tree (e.g. a B-tree) where every vertex contains at least one key. Let n be the number of keys stored in $T$.
Then the complexity of searching is in $\Omega(\log(n))$, if **binary search** is applied in each node.
(Remark: We do not require that all leaves of $T$ have the same height.)

In order to prepare the proof of this theorem, we treat the following result:

### Lemma 1.2
Every sorted tree $T$ according to Theorem 1.1 has $\beta \geq (\frac{1}{2} n + \frac{1}{2})$  keys in its leaves.

## Proof:

We make an induction on the height h of $\mathcal{T}$:

The case of h = 0 is trivial, because the root must contain n ≥ 1 keys.

In order to make the conclusion 'h-1 ⟶ h', we consider the subtrees $\mathcal{T}_1,\ldots,\mathcal{T}_m$ (m ≥ 2) connected with the root r of $\mathcal{T}$. For every μ = 1,...,m, let $n_\mu$ ($\beta_\mu$ resp.) be the number of keys stored in $\mathcal{T}_\mu$ (in the leaves of $\mathcal{T}_\mu$ resp.). Then the root r contains (m-1) keys so that $n = (m-1) + n_1 + \ldots + n_m$. Thus we obtain:

$$\beta = \sum_{\mu=1}^{m} \beta_\mu \geq \frac{1}{2} \sum_{\mu=1}^{m} (n_\mu + 1) = \frac{1}{2}\left(\sum_{\mu=1}^{m} n_\mu + (m-1)\right) + \frac{1}{2} = \frac{1}{2} n + \frac{1}{2} .$$

(Note that this lemma is wrong if we permit nodes without any key: Then we can construct counterexamples by exclusively taking empty leaves.)

After that we are able to consider the

## Proof to Theorem 1.1:

For every leaf b' of $\mathcal{T}$ let w(b') be the number of the keys stored in b'. Then the application of **LEMMA G** yields a path $p = (v_0,\ldots,v_\ell = b)$ such that

$$g^+(v_0) \cdot \ldots \cdot g^+(v_{\ell-1}) \cdot w(b) \geq w(\mathcal{T}) \geq \frac{1}{2} n \quad \text{(see Lemma 1.2).} \qquad (1)$$

When applying binary search, we need at least

$0.5 \log_2(g^+(v))$ comparisons for treating interior nodes v,
$0.5 \log_2(w(v))$ " for searching in the leaves v.

Consequently the costs C of the path p are greater or equal to

$$0.5 \log_2(g^+(v_0)) + \ldots + 0.5 \log_2(g^+(v_{\ell-1})) + 0.5 \log_2(w(b)) \quad \overset{\geq}{_{(1)}}$$

$$0.5 \log_2(0.5n) \in \Omega(\log(n)).$$

## Remark 1.3

Here the question arises whether log(n) is also a general **upper** time bound to searching in sorted multiway trees; it may be surprising that this is not the case. We now construct a counterexample according to the following principles:

1) There is one path $p = (v_0,\ldots,v_h)$ where every $v_\lambda$ contains a great number g of keys; consequently p is extremely expensive.

2) Every vertex which does not occur in p stores only one key; hence the number n ( and log(n) ) are very small with respect to the costs of p.

For this end we take an arbitrary g ≥ 2. Then every vertex $v_\lambda$ (0 ≤ λ < h) has (g+1) successors, and one of them is $v_{\lambda+1}$. Hence we can connect $v_\lambda$ with g trees $T_\lambda$

which are binary trees of height h-λ-1. Thus we obtain a tree $\mathcal{T}$ where every leaf has the same height (see Fig. 3)

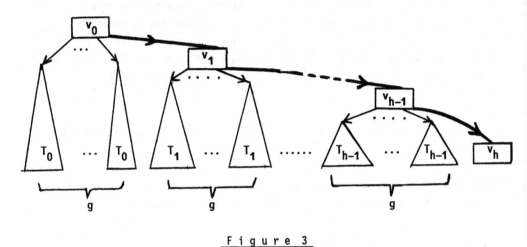

$$\underline{F\ i\ g\ u\ r\ e\ \ 3}$$

Let's now estimate the number of keys stored in $\mathcal{T}$:

For $\lambda= 0,\ldots,h-1$ we constructed g trees $T_\lambda$; each of them contains at most $2^{h+1}$ keys. Hence we have

$$n \leq g \cdot 2^{h+1} + g \cdot (h+1),$$

where the last summand is effected by the keys in $v_0,\ldots,v_h$. Consequently,

$$n \leq 4 \cdot g \cdot 2^h$$

if h is great enough.

Now the costs C(p) of the path p are at least $\frac{1}{2}(h+1) \cdot \log_2(g)$. Thus we obtain:

$$\frac{C(p)}{\log_2(n)} \leq \frac{0.5 \cdot h \cdot \log_2(g)}{\log_2(4 \cdot g \cdot 2^h)} = \frac{0.5 \cdot h \cdot \log_2(g)}{2 + h + \log_2(g)} \quad \bullet$$

But this quotient is not bounded by a constant if h and g are arbitrary. Consequently, $C(p) \notin O(\log(n))$.

These investigations yield the following result:

Searching in sorted multiway trees is <u>not</u> in O(log(n)) since we can find classes of trees which require a worst-case time which is not logarithmically bounded. On the other hand, the advantage of the time bound $\Omega(\log(n))$ is the generality, i.e., our Theorem 1.1 is true for <u>every</u> sorted multiway tree and not only for subclasses.

## Application 2  (Convex Hull)

In this section we investigate the complexity of the Plane Convex Hull Problem if
the computation of rotational mimima is supported by a parallel processor.
For this end, we point out the following observation: If $CONVEX(P_1,\ldots,P_n) =$
$(P_{i_1},\ldots,P_{i_r})$, then for any $\varsigma \geq 3$ the point $P_{i_\varsigma}$ is the first which is met when rotating the line $(P_{i_{\varsigma-2}},P_{i_{\varsigma-1}})$ around $P_{i_{\varsigma-1}}$. This fact immediately yields a gift-
wrapping-method for the Convex Hull Problem (Fig. 4) which is similar to Jarvis'
March described in [6], p. 104 - 106.

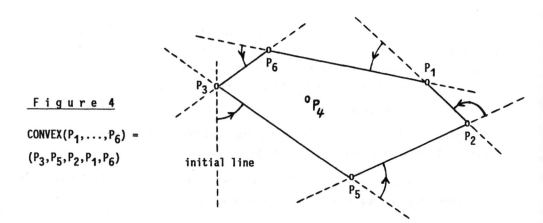

<u>**F i g u r e  4**</u>

$CONVEX(P_1,\ldots,P_6) =$

$(P_3,P_5,P_2,P_1,P_6)$

initial line

We now assume that finding rotational mimima among j points is sped up by a
parallel processor and can be done within $\Theta(T(j))$ ( instead of $\Theta(j)$ ) time units.
Then our gift-wrapping algorithm takes $O(nT(n))$ time units. In the next theorem we
discuss a class of time functions T for which $n \cdot T(n))$ is also a **lower** bound for
the Convex Hull Problem:

## Theorem 2.1  (cf. [3], p.193, Theorem 3.1)

Let $T : [1,\infty) \longrightarrow [1,\infty)$ a strictly monotonous increasing (time-)function with
$T(1) = 1$ and

$\quad$ (+) $(\forall a,b)\ T(a \cdot b) \leq T(a) + T(b)$.

Let M be a machine according to II) with the additional capability to find out
rotational minima among j points in $\Theta(T(j))$ time units.
**Then M requires** $\Omega(n \cdot T(n))$ time  to compute $CONVEX(P_1,\ldots,P_n)$, n > 0.

## Proof (Sketch):

We only line out the main ideas; more details can be found in [3], Theorem 3.1.
At first, we represent the algorithm of M in a decision tree $\mathcal{T}$, where the computations of the rotational minimum Q among the points $Q_1,\ldots,Q_j$ is symbolized as
follows:

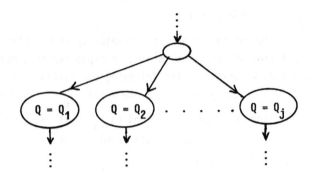

**Figure 5**

Then we have a constant $c > 0$ such that every interior vertex $v$ of $T$ costs at least $c \cdot T(g^+(v))$ time units.

Let us now label every leaf $b$ of $T$ with the following weight:

$$w(b) \; := \; \begin{cases} 1 \text{ if } b \text{ can be reached by some input } (P_1, \ldots, P_n), \\ 0 \text{ otherwise.} \end{cases}$$

( Note that we distinguish between the relevant and the irrelevant leaves; the useless leaves are labelled with zero. )

Then the well-known reduction to the sorting problem implies that $T$ has at least $n!$ leaves which can reached by some input; this means that

$$w(T) \geq n! \qquad (1)$$

We next consider the path $p = (v_0, \ldots, v_\ell)$ according to **LEMMA G**. Then we have

$$g^+(v_0) \cdot \ldots \cdot g^+(v_{\ell-1}) \; = \; g^+(v_0) \cdot \ldots \cdot g^+(v_{\ell-1}) \cdot w(v_\ell) \; \geq \; w(T) \; \geq \; n!. \quad (2)$$

Since M cannot generate new points, all polar minima are computed within subsets of the input points $\{P_1, \ldots, P_n\}$. Therefore, $g^+(v_\lambda) \leq n$ ($\lambda = 1, \ldots, \ell-1$), and we can decompose $I := \{1, \ldots, \ell-1\}$ into subsets $I_1, \ldots, I_{s-1}, I_s$ such that

$$\pi_\sigma := \prod_{\lambda \in I_\sigma} g^+(v_\lambda) \in [n, n^2] \; (\sigma = 1, \ldots, s-1) \quad \text{and} \quad \pi_s := \prod_{\lambda \in I_s} g^+(v_\lambda) \in [1, n].$$

Then (2) implies that $s \in \Omega(n)$, and we obtain the following lower bound for the costs C of the path $p$:

$$C \geq \sum_{\lambda=1}^{\ell} c \cdot T(g^+(v_\lambda)) \; \geq \; c \cdot \sum_{\sigma=1}^{s-1} \sum_{\lambda \in I_\sigma} T(g^+(v_\lambda)) \underset{(+)}{\geq} c \cdot \sum_{\sigma=1}^{s-1} T(\pi_\sigma) \; \geq c \cdot (s-1) \cdot T(n)$$

$$\in \Omega(n \cdot T(n)).$$

Note now that this expensive path p actually is used by some input $(P_1,...,P_n)$; this is a consequence of (2)( i.e. w(b) > 0 ) and the definition of the weight w. Therefore the worst-case time required by M is really in $\Omega(n \cdot T(n))$.


## Application 3   (Convex Hull)

In this part of our treatise we apply **LEMMA G** to the Convex Hull Problem again, but instead of finding polar minima, the following problem can be solved extremly fast: deciding which of the input points are on the right of a directed line and which are on the left.


## Theorem 3.1
For every n > 0 let $M_n$ be an automaton according to II which computes $CONVEX(P_1,...$ $...,P_n)$. Moreover, $M_n$ has the following properties:
- α) In the beginning the input points $P_1,...,P_n$ are stored in the array P[1],..., P[n], and the string variables w[1],...,w[n] are initialized with the empty word ε.
- β) $M_n$ can nondeterministically take a directed line $\vec{g}$ and update the words w[ν] as follows:
  
      for ν := 1 to n do
  
          w[ν] := w[ν]r if the current point P[ν] on the **right** of $\vec{g}$,
  
          w[ν] := w[ν]ℓ if the current point P[ν] on the **left** of $\vec{g}$.
  
  This operation takes $\Theta(T(n))$ (instead of $\Theta(n)$) time units where
  
      **(*)** $(\forall n \geq 60)$ $n \geq T(n) \geq 14 \cdot \ln(n)$.
- γ) $M_n$ can use strings as indices of arrays (e.g. A[w[3]]).

**Then** the machines $M_n$ require $\Omega(n\log(n))$ time units.


## Proof:
Our main ideas will be the following: We shall pick out a path p in the query tree of $M_n$ according to **LEMMA G**. Then we shall replace all the nondeterminisms along p by a single one in the beginning of p.

Let us now begin our detailed proof by defining the parabola $S := \{(x,x^2) \mid x \in \mathbb{R}\}$ and the points $D_1,...,D_n \in S$ where $D_\nu := (\nu \cdot \sqrt[3]{2}, \nu^2 \cdot \sqrt[3]{4})$, $\nu = 1,...,n$; every input $(P_1,...,P_n) = (D_{\pi(1)},...,D_{\pi(n)})$ is called a '$(D_\nu)$-permutation', if π is a permutation.
Then we represent the behaviour of $M_n$ as a decision tree $\mathcal{T}$ with the root r; the queries correspond to the vertices with out-degree 2, while all evaluations ( inclusive of those in β) ) are symbolized by vertices of out-degree 1. Every leaf b of $\mathcal{T}$ is labelled with the following number w(b):

I f  the path from r to b is an optimal path for some $(D_\nu)$-permutation
(with respect to the time of computation) then w(b) is the number of
those $(D_\nu)$-permutations which can arrive at b.
O t h e r w i s e  w(b) := 0.

(Note that we give a positive weight to the relevant leaves while the irrelevant
ones are labelled with zero. This kind of definition also occurred in the proof
to Theorem 2.1.)

We now want to estimate $w(T)$. For this end we observe that for every $(D_\nu)$-permu-
tation we can fine an optimal path $\tilde{p}$ within $T$. Let $\tilde{b}$ be the last vertex of $\tilde{p}$.
Then this $(D_\nu)$-permutation is counted once when computing $w(\tilde{b})$. Thus every $(D_\nu)$-
-permutation has an effect on $w(T)$ so that

$$w(T) \geq n! \qquad (1)$$

❨ Note that the nondeterminisms permit that
- two permutations arrive at the same leaf b ( $\implies$ w(b) $\geq$ 1 ),
- one permutation can arrive at two different leaves ( $\implies w(T) > n!$ ). ❩

Let us now take some path p from r to some leaf b' according to **LEMMA G** and
let k be the number of queries between r and b'. Then we have $2^k \cdot w(b') \geq n!$, i.e.

$$w(b') \geq \frac{n!}{2^k} \qquad (2)$$

Let j be the number of operations β) along p. If j = 0, then p can only treat one
$(D_\nu)$-permutation so that we have $1 \geq w(b') \geq n!/2^k$ and $k \geq \log_2(n!) \in \Omega(n\log T(n))$.
Therefore we shall from now on assume that

$$j > 0 . \qquad (3)$$

We next apply our trick mentioned in the beginning: For this end we realize that
the j nondeterministic operations β) can be replaced by (deterministic) updates of
the strings w[1],...,w[n], if the position of every input point $P_\nu$ with respect to
each line $\vec{G}_i$ (i = 1,...,j) is known at the beginning of p; consequently we can
transform p as follows:

Let $u_1,...,u_s$ be those n-tuples $\left(W(\vec{G}_1,...,\vec{G}_j;P_1),...,W(\vec{G}_1,...,\vec{G}_j;P_n)\right)$ which can be
generated by arbitrary directed lines $\vec{G}_1,...,\vec{G}_j$ and points $P_1,...,P_n \in S$. (Note that
all of these n-tuples $u_\sigma$ are elements of $(\{+,-\}^j)^n$ so that the set of these tuples
is indeed  f i n i t e .) Then we create a new root $\tilde{r}$ which symbolizes the nondeterministic choice of $u_\sigma$. After that we connect $\tilde{r}$ with every deterministic path
$p_\sigma$ (σ = 1,...,s) which is based on the assumption that $u_\sigma$ occurs (Fig. 6).

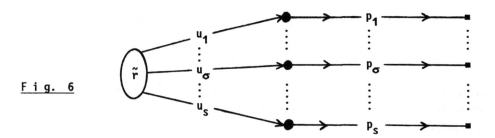

Fig. 6

It is obvious that now different $(D_\nu)$-permutations require different paths $p_\sigma$ so that

$$s = g^+(\tilde{r}) \geq w(b') \geq \frac{n!}{2^k} \geq \frac{n^n}{e^{n+k}} . \qquad (4)$$

We shall now give an upper bound for s:
First we realize that $\vec{G}_1,\ldots,\vec{G}_j$ divide the parabola S into $\leq 2j$ points of intersection and $\leq (2j+1)$ open segments. Therefore, the cardinality of any given set X = $W(\vec{G}_1,\ldots,\vec{G}_j;S)$ is $\leq 4j + 1$. Every n-tuple $u_\sigma$ is an element of some $X^n$, and because of Lemma 3.2 (see below) and (3), there are not more than $(12j)!$ possible sets X. Consequently, $s \leq (12j)! \cdot (4j+1)^n \leq (13j)^{13j+n}$. This inequality and that in (4) imply that $(13j + n)\cdot\ln(13j) \geq \ln(s) \geq n\cdot\ln(n) - n - k$ so that

$$k \geq n\cdot\ln(n) - n - (13j + n)\cdot\ln(13j). \qquad (5)$$

( Note that (5) implies a trade-off between the number k of queries and the number j of operations β). )

We next want to estimate the costs C of p. First we know that each operation β) requires at least $c_0 \cdot T(n)$ time where $c_0 > 0$. Hence $C \geq k + j \cdot c_0 \cdot T(n)$. Defining $c := \min(1,c_0)$ we obtain:

$$C \geq c\cdot(k + j\cdot T(n)); \qquad (6)$$

then (5) and (6) imply:

$$C \geq c\cdot(n\ln(n) - n - (13j + n)\cdot\ln(13j) + j\cdot T(n)). \qquad (7)$$

These results (6) and (7) make it possible to calculate the costs C of the path p:

In the case of $j \geq \frac{n}{13}$ we can conclude that $C \geq \frac{c}{13}\cdot n\cdot T(n) \in \Omega(n\log T(n))$ because of (6).

Let now $j \leq \frac{n}{13}$ . Then $13\cdot j\cdot\ln(13j) \leq 13\cdot j\cdot\ln(n) \leq \frac{13}{14}\cdot j\cdot T(n)$ because of Assumption (*). This and (7) imply:

$$C \geq c \cdot \left( n\ln(n) - n - \tfrac{13}{14} \cdot j \cdot T(n) - n\ln(13j) + j \cdot T(n) \right) =$$

$$c \cdot \left( n\ln(n) - n - n\ln(13j) + \tfrac{1}{14} j \cdot T(n) \right) =: U(j).$$

Then we consider the first derivative $U'(j) = -c \cdot \left( \tfrac{n}{j} + \tfrac{1}{14} T(n) \right)$. Since this function is stricly monotonous increasing, we have a <u>minimum</u> of U in $j_0 := \frac{14 \cdot n}{T(n)}$. This implies:

$$C \geq U(j_0) = c \cdot \left( n\ln(n) - n - n \cdot \ln\left( \tfrac{182n}{T(n)} \right) + n \right) =$$

$$c \cdot \left( n\ln(n) - n \cdot [\ln(182) + \ln(n) - \ln(T(n))] \right) =$$

$$c \cdot \left( n \cdot \ln T(n) - n \cdot \ln(182) \right) \in \Omega(n\log T(n)).$$

Consequently, the costs of p are in $\Omega(n\log T(n))$ which is independent from the number j of operations β).

We now want to see that the costs of p are indeed a lower bound to the worst-case time of the machine $M_n$. Note that this conclusion is not trivial since the following situation might occur: Every input of $M_n$ could take a less expensive path than p so that p would be useless for us. But this situation is  n o t  possible: The inequality (4) yields that w(b') > 0. Then our definition of w implies that p must indeed be an optimal path for some $(D_\nu)$-permutation. Consequently, $M_n$ really requires $\Omega(n\log(n))$ time units when treating this $(D_\nu)$-permutation. (Note that a similar idea already occurred at the end of the proof to 2.1).

In order to complete the proof of 3.1, we consider the following statement:

<u>Lemma 3.2</u>
Given $\lambda \geq 2$ and $j \geq 1$; let $p: \mathbb{R} \to \mathbb{R}$ be a polynomial with a degree $\in \{2, \ldots, \lambda\}$, and let $S := \{(x|p(x)) \mid x \in \mathbb{R}\}$.
**Then there exist at most $(6\lambda j)!$ sets of the form** $W(\vec{a}_1, \ldots, \vec{a}_j; S)$ basing on arbitrary directed lines $\vec{a}_1, \ldots, \vec{a}_j$.

**(** In the situation of Theorem 3.1 we have $\lambda = 2$; note that in this case $(12j)! \leq 2^{(2^j)} = |\text{Pot}(\{r, \ell\}^j)|$, i.e. the bound $(12j)!$ is not trivial. **)**

<u>Proof (Sketch)</u> (see [2], p. 241 - 244 for more details):
The main idea of our proof is linearizing the possible sets $W(\vec{a}_1, \ldots, \vec{a}_j; S)$. For this end we define $\Sigma := \{C_1, \ldots, C_j\}$ and make the following conventions: For every word $w \in \{r, \ell\}^j$ let $w^{C_i}$ be the result of changing the i-th symbol, and for every $\zeta', \zeta'' \in \Sigma^*$ let $w^{\zeta'\zeta''} := (w^{\zeta'})^{\zeta''}$.

We now consider some lines $\vec{a}_1, \ldots, \vec{a}_j$ and scan the curve S from the left to the right. Then we make the following observations:

We start with an initial word $w_0$ which occurs in the leftmost infinite segment of S.

The current intersection word can only change when S enters a point of $S \cap \vec{G}_i$ or leaves it; consequently, each point of $S \cap \vec{G}_i$ causes $\leq 2\lambda$ applications of the operator $C_i$ (Fig. 7).

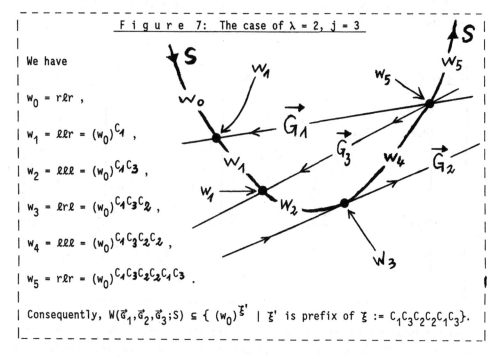

F i g u r e 7: The case of $\lambda = 2$, $j = 3$

We have

$w_0 = r\ell r$ ,

$w_1 = \ell\ell r = (w_0)^{C_1}$ ,

$w_2 = \ell\ell\ell = (w_0)^{C_1 C_3}$ ,

$w_3 = \ell r\ell = (w_0)^{C_1 C_3 C_2}$ ,

$w_4 = \ell\ell\ell = (w_0)^{C_1 C_3 C_2 C_2}$ ,

$w_5 = r\ell r = (w_0)^{C_1 C_3 C_2 C_2 C_1 C_3}$ .

Consequently, $W(\vec{G}_1, \vec{G}_2, \vec{G}_3; S) \subseteq \{ (w_0)^{\xi'} \mid \xi'$ is prefix of $\xi := C_1 C_3 C_2 C_2 C_1 C_3\}$.

Since the degree of the polynomial p is $\leq \lambda$, we know that S has at most $\lambda$ common points with each line $\vec{G}_i$. Therefore, the behaviour of the current intersection word can be described by an iterative application of the operators $C_1, \ldots, C_j$ where each of them occurs at most $2\lambda$ times. Hence there is a word $\xi \in \Sigma^*$ consisting of exactly $2\lambda$ $C_i$'s ($i = 1, \ldots, j$) such that

$$W(\vec{G}_1, \ldots, \vec{G}_j; S) \subseteq \{(w_0)^{\xi'} \mid \xi' \text{ is a prefix of } \xi\} =: X_\xi \quad \text{(see Fig. 7)}.$$

Every set $X_\xi$ of words has exactly $2^{2\lambda+1}$ subsets, and it is determined by one of the $2^j$ possible starting words $w_0$ and one of the $\dfrac{(2\lambda j)!}{[(2\lambda)!]^j}$ possible words $\xi$.

Therefore we have at most $2^{2\lambda+1} \cdot 2^j \cdot \dfrac{(2\lambda j)!}{[(2\lambda)!]^j} \leq (6\lambda j)!$ posibilities to construct a set $W(\vec{G}_1, \ldots, \vec{G}_j; S)$.

Remark 3.3:

The bound $n \cdot \log(T(n))$ is even tight and can be obtained as follows:

Let $\vec{G}(x)$ ($x \in \mathbb{R}$) be the vertical line through $(x,0)$ directed upwards; moreover, let $j := \left\lceil \dfrac{n}{T(n)} \right\rceil$. Then our machine $M_n$ executes $(j+1)$ operations $\beta$ according to the lines $\vec{G}(x_0), \ldots, \vec{G}(x_j)$ which have the following properties (see Fig. 8):

a) $x_0 < \ldots < x_j$.

b) In every vertical slab $S_i$ between $\vec{G}(x_{i-1})$ and $\vec{G}(x_i)$ ($i = 1,\ldots,j$) we have $x_i$ elements of the set $\{P_1,\ldots,P_n\}$ where $x_i \in \{\lfloor \frac{n}{j} \rfloor , \lfloor \frac{n}{j} \rfloor + 1\}$.

(Note that b) can be obtained as follows: We can find a $\tilde{j} < j$ such that $n = j \cdot \lfloor \frac{n}{j} \rfloor + \tilde{j}$. Then we define $x_i := \lfloor \frac{n}{j} \rfloor + 1$ for $i \leq \tilde{j}$ and $x_i := \lfloor \frac{n}{j} \rfloor$ for $i > \tilde{j}$. Then the lines $\vec{G}(x_i)$ are choosen according to the numbers $x_i$.

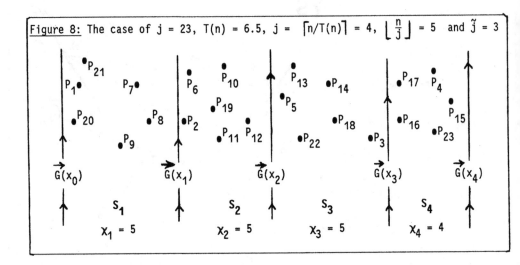

Figure 8: The case of $j = 23$, $T(n) = 6.5$, $j = \lceil n/T(n) \rceil = 4$, $\lfloor \frac{n}{j} \rfloor = 5$ and $\tilde{j} = 3$

Then the intersection words allow to scan the slabs from the left to the right; consequently, $M_n$ can sort the n input points with respect to their x-values after doing so within each slab $S_i$. After that, the Convex Hull Problem can be solved in $O(n)$ steps similar to [6], p. 103.

Our algorithm reqires $O(j \cdot T(n)) \subseteq O(n)$ time units for the operations $\beta$); moreover, the sorting within the j slabs can be done in $O( j \cdot (n/j) \cdot \log(n/j) )$ = $O( \frac{n}{T(n)} \cdot T(n) \cdot \log(T(n)) )$ = $O(n\log T(n))$ steps so that $M_n$ indeed comes out with $O(n\log T(n))$ time units.


## Application 4) (QUICKSORT)

The ideas of Application 3) can also be used to obtain an $n\log T(n)$-result if QUICK-SORT is supported by a parallel processor.


## Theorem 4.1

Let $T : \mathbb{N} \longrightarrow \mathbb{R}$ such that

(*) $(\forall n \geq 60)$ $14 \ln(n) \leq T(n) \leq n$.

For every $n \geq 60$ let $M_n$ be a machine according to paragraph II which sorts every n-tuple $(s_1,...,s_n) \in \mathbb{R}^n$ of pairwise distinct keys. Moreover, these machines have the following properties:

  $\alpha$) In the beginning, the keys are stored in an array $s[1],...,s[n]$ of real; more-
     over we have an array $w[1],...,w[n]$ of integer where all $w[\nu]$'s are ini-
     tialized with 0.
  $\beta$) $M_n$ can nondeterministically take some $t \notin \{s_1,...,s_n\}$ and can apply the fol-
     lowing operation to the **current** numbers $s[1],...,s[n]$:
        for $\nu := 1$ to n do if $s[\nu] > t$ then $w[\nu] := w[\nu] + 1$; .
     These operations altogether take $\Theta(T(n))$ instead of $\Theta(n)$ time units.
**Then** the automata $M_n$ require $\Omega(n\log T(n))$ time.


Proof:

We consider the n-tuple $(d_1,...,d_n) := (1 \cdot \sqrt{2}, 2 \cdot \sqrt{2}, ..., n \cdot \sqrt{2})$. Then every
n-tuple $(d_{\pi(1)},...,d_{\pi(n)}) = (s_1,...,s_n)$ is called by '$(d_\nu)$-permutation', if $\pi$ is
a permutation.
Thus we can get the same facts (1),(2) and (3) as in the proof to 3.1, i.e. we have
 - the query tree $\mathcal{T}$,
 - the weight $w(b)$ of every leaf $b$ = number of $(d_\nu)$-permutations which can arrive
    there ( if $b$ is the last vertex of an optimal path of computation for some
    $(d_\nu)$-permutation; otherwise $w(b) := 0$ ),
 - a path p from the root r to a leaf b' which is the result of **LEMMA G**,
 - k queries and j operations $\beta$) along p
such that
$$w(\mathcal{T}) \geq n! \quad \textbf{(1)}$$

$$w(b') \geq \frac{n!}{2^k} \quad \textbf{(2)} .$$

Moreover, we can assume that

$$j > 0 . \quad \textbf{(3)}$$

We next apply the same trick as in 3.1, i.e., we replace the (nondeterministic)
path p by deterministic modifications of it. For this end we realize that all the
operations along p can be substituted by deterministic updates of the $w[1],...,$
$w[n]$ , if we know the following data about the numbers $t_1,...,t_j$ generated nondeterministically:
 - for every $i = 1,...,j$ the cardinality $\chi(i)$ of the set $\{i' \mid t_{i'} \leq t_i\}$,
 - for every $\nu = 1,...,n$ the cardinality $\psi(\nu)$ of the set $\{i' \mid t_{i'} < s_\nu\}$.

( Then the update of $w[\nu]$ in the i-th operation $\beta$) can be done as follows:
  Let $\tilde{\nu}$ be such that $s_{\tilde{\nu}}$ is the current contents of $s[\nu]$. Then the following implications are true:

If $s_{\tilde{\nu}} > t_i$ then $\psi(\tilde{\nu}) \geq \chi(i)$, because $\{i' \mid t_{i'} < s_{\tilde{\nu}}\} \supseteq \{i' \mid t_{i'} \leq t_i\}$;
if $s_{\tilde{\nu}} < t_i$ then $\psi(\tilde{\nu}) < \chi(i)$, because in this case $\{i' \mid t_{i'} < s_{\tilde{\nu}}\} \subseteq$
$$\{i' \mid t_{i'} \leq t_i\} \setminus \{i\} .$$
This means that $\psi(\tilde{\nu}) \geq \chi(i) \Longleftrightarrow s_{\tilde{\nu}} > t_i$ so that we can update the integer $w[\nu]$
correctly by comparing $\psi(\tilde{\nu})$ and $\chi(i)$. $\big)$

Consequently, for every functions $\chi: \{1,\ldots,j\} \longrightarrow \{1,\ldots,j\}$ and $\psi: \{1,\ldots,n\} \longrightarrow$
$\{1,\ldots,j\}$, we obtain a path $p_{(\chi,\psi)}$ which can only treat one $(d_\nu)$-permutation.
Therefore we have $(j^j \cdot j^n)$ paths $p_{(\chi,\psi)}$. This and statement (2) imply that

$$j^{n+j} \geq w(b') \geq n! \cdot 2^{-k} \geq n^n \cdot e^{-n-k}$$

so that

$$k \geq n\ln(n) - n - (n+j)\ln(j) \geq n\ln(n) - n - (n+13j) \cdot \ln(13j). \quad (4)$$

But this is the result (5) in the proof of Theorem 3.1; therefore we can also in
this situation conclude that the costs C of the path p are in $\Omega(n\log T(n))$; hence
same must be true for the time required by the machines $M_n$ because $w(b') > 0$ (see (4))
and p must be an optimal path for some $(d_\nu)$-permutation.

### Remark 4.2
Also in this case we can give an $O(n\log T(n))$ algorithm: It runs similar to that
of Remark 3.3; we only have to consider the the inputs $s_1,\ldots,s_n$, the reals $t_0,\ldots,$
$t_{j+1}$ and the intervals $I_i := (t_{i-1},t_i)$ resp. instead of the points $P_1,\ldots,P_n$,
the lines $\vec{g}(x_0),\ldots,\vec{g}(x_{j+1})$ and the slabs $S_i$.

## V) Concluding Remarks

In our treatise, we gave the simple and even constructive proof to a graph theoretical result: We showed that in every rooted tree $T$ there is a path p such that
the product of the out-degrees along p and the weight of the terminal leaf is
greater or equal to the total weight of $T$.
After that, we gave four applications of this lemma dealing with Searching (Application 1), the Convex Hull (Appl. 2,3) and QICKSORT (Appl. 4). It is very interesting to compare the trees $T$ and the weight functions w ocurring in the
corresponding proofs:
The **tree** $T$ was a sorted multi-way tree in Appl. 1) and an (extended) decision tree
in the Applications 2) - 4).
The **weight** w(b) was the number of keys stored in b (Appl. 1) or the number of inputs which were able to arrive at b (Appl. 2 - 4); in Application 2, w(b) was
identical to 1.
The **total weight** w($T$) was closely related to the number of keys within $T$ (Appl. 1),
or to the cardinality of a set of possible input data (Appl. 2) - 4).

Finally, the proofs to the Theorems 3.1 and 4.1 based on the following trick:
We replaced a **sequence of nondeterministic operations** by a **single** one.

Hence the diversity of our trees $T$, the weight function w and the manipulations
in $T$ promises a lot of further applications of our graph theoretical **LEMMA G**.

## REFERENCES

[1] H.Edelsbrunner, J.O'Rourke, R.Seidel: CONSTRUCTING ARRANGEMENTS OF LINES AND
    HYPERPLANES WITH APPLICATIONS. 24th Annual Symposium on Foundations of Com-
    puter Science (1983), p. 83 - 91.

[2] U.Huckenbeck, GEOMETRISCHE MASCHINENMODELLE. Ph.D.Thesis, University of Würz-
    burg (1986).

[3] U.Huckenbeck, ON THE COMPLEXITY OF CONVEX HULL ALGORITHMS IF ROTATIONAL MINIMA
    CAN BE FOUND VERY FAST. Zeitschrift für Operations Research (ZOR), Vol. 32,
    Issue 3/4 (1988), Physica Verlag.

[4] K.Mehlhorn, DATA STRUCTURES AND ALGORITHMS 1: SORTING AND SEARCHING. Springer
    (1984).

[5] H.Noltemeier, GRAPHENTHEORIE. W. de Gruyter (1976).

[6] F.P.Preparata, M.I.Shamos, COMPUTATIONAL GEOMETRY, AN INTRODUCTION. Springer
    (1985).

# A time-optimal parallel algorithm for the computing of Voronoi-diagrams

W.Preilowski      W.Mumbeck

### Abstract

We present an $O(log(n))$-time-algorithm for computing Voronoi-diagrams using a CREW-PRAM with $n^3$ processors. We also show,that the same idea works for Power-diagrams using the same time, number of processors and model of computation.

## 1   Introduction

In the last years computational geometry has risen more and more in the center of interest. Since 1975 many sequential algorithms for geometric problems like Convex Hull, Voronoi-diagrams, Polygon-Optimization-problems, etc. have been developed. (For examples: see [1].)

But until 1985 there was not much work had been done on the parallelisation of these problems. A.Aggerwal et.al. [2] illustrated in 1985 some parallel algorithms on a CREW-PRAM [1] for the above mentioned problems.

In 1987 they expanded their work by giving more efficient algorithms [3]. Especially they gave a parallel $O(log^2(|S|))$-algorithm for constructing a Voronoi-Diagram [2] of a finite point-set $S$ using only $O(|S|)$ processors. Using the same model of parallel computation, we will present here an $O(log(|S|))$-algorithm for computing a Voronoi-diagram using $O(|S|^3)$ processors. This result is time-optimal, because sorting can be reduced to the Voronoi-diagram-problem.

Using the the CRCW-PRAM[3] , we get an $O(1)$- algorithm with $O(|S|^4)$ processors to compute all edges of the Voronoi-diagram of $S$. Only for a sorted output we need $O(log(M))$ time, where $M$ is the maximum number of edges a Voronoi-cell can have. In section 6 we show that our idea for parallel computing of Voronoi-diagrams also works for construction Power-diagrams. Power-diagrams are a generalization of Voronoi-diagrams on a set of circles, lying in the Euclidian plane.(For sequentiell Power-diagram-algorithm see [4].)

---

[1]A CREW-PRAM is a random-access-machine, where two different processors are able to read the same location, but they may not write in the same location at the same time.

[2]It is well-known that a Voronoi-diagram of a finite point-set $S \subset R^2$ ,where $R^2$ denotes the Euclidian plane, is a partition of the plane into $|S|$ disjoint polygonal konvex regions, one for each point $x \in S$. Every of these convex regions consists of all points of the plane, which are as close as or closer to $x$ than to any other point of $S$.

[3]A CRCW-PRAM is a random-access-machine, where both, simultaneous writing and reading in the same location, is allowed.

# 2  Definitions and terms

Let $R^2$ denote the two-dimensional Euclidian plane. Given a finite nonempty set $S \subset R^2$, the Voronoi-Diagram of $S$, say $VD(S)$, is defined as follows:

For each point $x \in S$, define the Voronoi-cell $VC(x)$ as the set of points $p \in R^2$ such that $p$ is as close as or closer to $x$ as to any other point in $S$.

Note that the Voronoi-cells are all nonempty closed convex sets with polygonal boundaries and two cells can meet only along their common boundary.

The Voronoi-Diagram $VD(S)$ is defined as the planar point-set formed by the union of the boundaries of all the Voronoi-cells.

$VD(S)$ is a planar graph with $O(|S|)$ edges and vertices, all the edges are straight line-segments, perhaps unbounded, and two cells can have at most one edge in common.

The point $x_j$ is called a neighbour-point of $x_i$, $i \neq j$, if and only if the $VC(x_j)$ is neighbouring $VC(x_i)$ that means they have a common boundary.

# 3  A time-optimal Voronoi-diagram-algorithm

## 3.1  Previously appeared algorithms

The previously appeared algorithms for computing a Voronoi-diagram of a finite point-set $S \subset R^2$ are working with the following idea:

-  First sort the points of $S$ according to the $x$-coordinate.

-  Distribute the points to $\frac{|S|}{2}$ processors:
   Processor 1 gets the first and the second, processor 2 gets the third
   and the fourth point, etc. .

-  All processors compute in parallel the Voronoi-diagram for the obtained
   points. That is only the computation of the perpendicular bisector
   of the two points.

-  Now $log(|S|) - 1$ iterations follow:
   In the $i$-th iteration, $1 \leq i \leq log(|S|) - 1$, there exist $\frac{|S|}{2^i}$ Voronoi-diagrams
   with pairwise disjoint points, say VD(1), VD(2),...,VD($\frac{|S|}{2^i}$).
   In the $j$-th Voronoi-diagram, $1 \leq j \leq \frac{|S|}{2^i}$ are the points:
   $j \cdot \frac{|S|}{2^i} + 1$, $j \cdot \frac{|S|}{2^i} + 2$,..., $(j+1) \cdot \frac{|S|}{2^i}$
   Merge now VD(1) and VD(2), VD(3) and VD(4),...,VD($\frac{|S|}{2^i}-1$) and VD($\frac{|S|}{2^i}$)
   in parallel by computing the contour between the two Voronoi-diagrams,
   which is a polygonal path between them, separating them.(**Figure 1**)

Every iteration can be done in parallel in $O(log(|S|))$ time. After $log(|S|)$ iterations the whole Voronoi-diagram of $S$ is computed.

## 3.2 The new idea:

The idea we present here is quite different to this method. We compute for each point of $x_i \in S$, $1 \le i \le |S|$, its own Voronoi-cell $VC(x_i)$ in parallel. We will show here, that this can be done with an easy criterion, which decides whether a point $x_j$ is a neighbour-point to $x_i$. If so, it is clear that a part of the perpendicular bisector between $x_i$ and $x_j$ is the common boundary of the Voronoi-cells $VC(x_i)$ and $VC(x_j)$.

The nice and so easy criterion to decide this, which is proofed in the following Theorem, can be computed in time $O(log(|S|))$ using $O(|S|)$ processors. It supplies the two points on the perpendicular bisector of $x_i$ and $x_j$, which build the boundary-points of the common part of $VC(x_i)$ and $VC(x_j)$. Computing now for all pairs $x_i, x_j \in S, 1 \le i < j \le |S|$, their common part of their Voronoi-cells in parallel, we get a time-optimal $O(log(|S|))$-algorithm using $O(|S|^3)$ processors for the whole Voronoi-diagram of $S$.

## 3.3 Neighbour-point-Theorem:

Before we give the whole algorithm we will proof the Theorem which makes the short time possible.

First we need some definitions:

For $x_i, x_j \in S$ we define $\mathbf{L}(x_i, x_j)$ to be the line passing through $x_i$ and $x_j$ and $\overline{x_1, x_2}$ to be the straight-line from $x_1$ to $x_2$.

Let $\mathbf{PB}(x_i, x_j)$ be the perpendicular bisector of $\overline{x_i, x_j}$.

$L \subset S$ is the set of all points lying left of and $R \subset S$ the set of all points lying right of $\mathbf{L}(x_i, x_j)$ looking on it from point $x_i$ in the direction to $x_j$.

Let $\mathbf{S_{left}} := \{s | s$ is the intersection-point of $PB(x_i, z)$ and $PB(x_i, x_j)$ for $z \in L\}$ and let $\mathbf{S_{right}} := \{s' | s'$ is the intersection-point of $PB(x_i, z')$ and $PB(x_i, x_j)$ for $z' \in R\}$.

Define the following order "$\prec$" on the points of $S_{left}$ and $S_{right}$:

Let $p \ne q \in S_{right} \cup S_{left}$:

$p \prec q$ if and only if $p$ lies left of $q$ on $PB(x_i, x_j)$ looking on it from point $x_i$ in the direction to $x_j$.(**Figure 2**)

Before we will formulate the theorem we can make two assumptions, which have no influence on the result of the theorem:

**w.l.o.g:** On the $L(x_i, x_j)$ doesn't lie another point $x \in S$.

Assume $x \notin \{x_i, x_j\}$ lies on $L(x_i, x_j)$, then one of the following conditions holds:

a) $x$ lies upstairs to $x_j$ or downstairs to $x_i$: Then the point x is unimportant for deciding whether $x_j$ is neighbour-point to $x_i$.

b) $x$ lies between $x_i$ and $x_j$: then clearly $x_j$ is no neighbour-point to $x_i$.

**w.l.o.g.:** $S_{left} \ne \emptyset$ and $S_{right} \ne \emptyset$.

If not so, then $x_i$ and $x_j$ must be cyclical consecutive points on the convex hull of $S$. Then clearly $x_i$ and $x_j$ are neighbour-points.

Now

## Theorem 1:

Let $x_i$ and $x_j \in S, i \neq j$.
Then holds: $x_j$ is a neigbour-point of $x_i$ if and only if $\max(S_{left}) \prec \min(S_{right})$.

### Proof:
" $\Longrightarrow$ "

**Let $x_j$ be neighbour-point to $x_i$:**
Then a part of $PB(x_i, x_j)$ is a part of $VC(x_i)$.
Let $\overline{x,y}$, (w.l.o.g. $x < y$), be this part of $PB(x_i, x_j)$.
It is sufficient to show:
$x = \max(S_{left})$ and $y = \min(S_{right})$.

1) $x \in S_{left}$
This is trivial, because $VC(x_i)$ is a convex polygon and the point $x_i$ is lying in $VC(x_i)$ and so the point $x'$, with $PB(x_i, x')$ is intersecting in $x$, must lie left of $L(x_i, x_j)$.

2) For all $s \in S_{left}$ holds $s \preceq x$:

Assume their exists $z \in S_{left}$ holding $z \succ x$ and the generating-point $z'$ lies left of $L(x_i, x_j)$.
Clearly, $PB(z', x_i)$ intersects $\overline{x_i, x}$.
[Otherwise the $PB(z', x_j)$ is parallel to $L(x_i, x)$ or it passes below $x_i$ or above $x$.
In each case follows $z'$ lies right of $L(x_i, x_j)$]
So $PB(z', x_i)$ must intersect the boundary of $VC(x_i)$ in two points, say $x'$ and $y'$, and it follows, that $\overline{x', y'}$ must be a part of $VC(x_i)$ and it cuts off a part of $VC(x_i)$, in contradictory to the definition of $VC(x_i)$.

From 1) and 2) follows: $x = \max(S_{left})$.

Analogously we can show: $y = \min(S_{right})$, what finishes the proof of the first direction.

" $\Longleftarrow$ "

**Assume now $x_j$ is not a neighbour-point of $x_i$:**
Then there exists no part of $VC(x_i)$ lying on $PB(x_i, x_j)$. Let $PB(x_i, z)$ and $PB(x_i, z')$ be the two parts of the $VC(x_i)$ which intersect in p between $PB(x_i, x_j)$ and $x_i$.**(Figure 3)**
Clearly $z$ lies left and $z'$ right of $\overline{x_i, x_j}$. For the intersection-points $t_1$ of $PB(z, x_i)$ and $PB(x_i, x_j)$ and $t_2$ of $PB(z', x_i)$ and $PB(x_i, x_j)$ the condition holds: $t_1 \succ t_2$.
By following $\max(S_{left}) \succ \min(S_{right})$ we finishe the proof of the Theorem.

# 4    The algorithm:

Now we give the parallel algorithm for computing the Voronoi-diagram of a finite set of points $S \subset R^2$.
**Voronoi-diagram-algorithm:**

1)    For all $x_i, x_j \in S, i < j$, decide in parallel whether $x_j$ is a neighbour-point to $x_i$ with the help of the Theorem!
       If so, then $max(S_{left}), min(S_{right})$ is the common boundary of $VC(x_i)$ and $VC(x_j)$.
2)    For all $x_i, 1 \leq i \leq |S|$ compute the $VC(x_i)$ by sorting the pieces got in step 1.

The algorithm needs time $O(log(|S|))$, what is optimal, using $O(|S|^3)$ processors. The first step needs $|S|-2$ processors for computing all intersection-points on $PB(x_i, x_j)$ for every pair $x_i, x_j$. This can be done in time $O(1)$. For computing $\max(S_{left})$ and $\min(S_{right})$ we need time $O(log(|S|))$ with $O(|S|)$ processors. Because their exist $O(|S|^2)$ pairs, we need $O(|S|^3)$ processors for step 1. The second step requires $O(log(|S|))$ time with $O(|S|)$ processors for all computed edges, because the Voronoi-diagram is a planar graph and the number of edges is bounded by $O(|S|)$.

It follows, that the whole algorithm takes time $O(log(|S|))$ with $O(|S|^3)$ processors.

# 5   Application on a CRCW-PRAM

Using a CRCW-PRAM we can compute the first step of our algorithm in constant time by using $O(|S|^4)$ processors, because the maximum and the minimum of a finite point-set with $n$ elements requieres time $O(1)$ with $O(n^2)$ processors.

Only the second step need logarithmic time for computing the Voronoi-cell of the points of $S$. This time is bounded by $O(log(M))$, where $M$ is the maximal number of edges one Voronoi-cell has, looking at all Voronoi-cells of $S$.

# 6   Power-diagrams

Power-diagrams are a generalization of Voronoi-diagrams on a set of circles. Given a finite set of $n \in IN$ circles:
$C := \{(c_1, r_1), (c_2, r_2), ..., (c_n, r_n)\}$, where $c_i \in R^2$ for $1 \leq i \leq n$ are the centerpoints and $r_j \in R_0^+$, for $1 \leq j \leq n$ are the radien of the $n$ circles.

Let $\mathbf{pow} : \mathbf{R}^2 \times \mathbf{C} \to \mathbf{R}$ with: $pow(p, (c, r)) := d^2(p, c) - r^2$, where $d(x, y)$ denotes the Euclidian distance of $x$ and $y$.

For constructing the Power-diagram we have to construct for every circle $c \in C$ its own Power-cell, which is defined as: $PC(c) := \{p \in R^2 | pow(p, c) \leq pow(p, c')$ for every $c' \in C\}$

Clearly every Power-cell is a closed konvex region, but the centerpoint of the generating circle $c$ needs not lie in its own Power-cell and there can exist circles, which have no Power-cell, because they are powered from other bigger circles! (For an example see **Figure 4** circle $S_2$.) In the same way as Voronoi-diagrams, we can define the Power-diagram as the union of all $k \leq |C|$ Power-cells of $C$. Remark, that for a set $C$ with $n$ circles and every circle has the same radius, then the Power-diagram is the Voronoi-diagram of the $n$ center-points! For an example of a Power-diagram see **Figure 4**.

In this section we want to give a similar Theorem for constructing a Power-diagram as in section 3. For $c \neq c' \in C$ define $\mathbf{PL}(c, c') := \{p \in R^2 | pow(p, c) = pow(p, c')\}$ to be the Power-line of the circles $c$ and $c'$. $c$ is a neighbour-circle to $c'$ if and only if $PC(c)$ is neighbouring $PC(c')$, what means they have a common boundary.

A part of the power-line is a part of the Power-cells of $c$ and $c'$. $PL(c, c')$ stands perpendicular on the straightline passing the center-points $cp_c$ and $cp_{c'}$ of $c$ and $c'$ and the intersection-point lies at $ip(cp_c - cp_{c'}) + c_c$ where $ip := \frac{1}{2}(\frac{r_c^2 - r_{c'}^2}{d^2(cp_c, cp_{c'})} + 1)$ , $c = (cp_c, r_1)$ and $c' = (cp_{c'}, r_2)$. The place of the intersection-point depends on the radien of $c$ and $c'$ (see **Figures 5, 6 and 7** .)

Clearly for all points $p \in R^2$ follows: If $p$ lies left of $PL(c, c')$, then $pow(p, c) < pow(p, c')$ and similarly if it lies right of $PL(c, c')$, then $pow(p, c) > pow(p, c')$.

It follows: The only difference between Voronoi-diagrams and Power-diagrams is the movement of the perpendicular bisector of the center-points of $c$ and $c'$ with the factor $cp$ and so we can use the same idea as in section 3 to decide whether a circle is a neighbour-circle to another.

Define $S_{left}$ and $S_{right}$ in the same way as in section three. For example $S_{left} := \{s | s$ is the intersection-point of $PL(c'', c)$ and $PL(c, c')$, $c'' \notin \{c, c'\}$, where the center-point of circle c" lies in the left of $\overline{cp_c, cp_{c'}}\}$.

Now clearly follows a similar Theorem as in section 3:

### Theorem 2:

Let $c \neq c' \in C$:

$c'$ is a neighbour-circle to $c$ if and only if $\max\{S_{left}\} \prec \min\{S_{right}\}$.

The proof is similar to Theorem 1.

Using a similar algorithm as in section 4 by replacing Theorem 1 by Theorem 2 in step 1 , we get an $O(log(|C|))$-algorithm for constructing a Power-diagram using $O(|C|^3)$ processors.

# Acknowledgement

We thank Burkhard Monien for motivating us to study this problems and for many helpful discussions.

# Figures

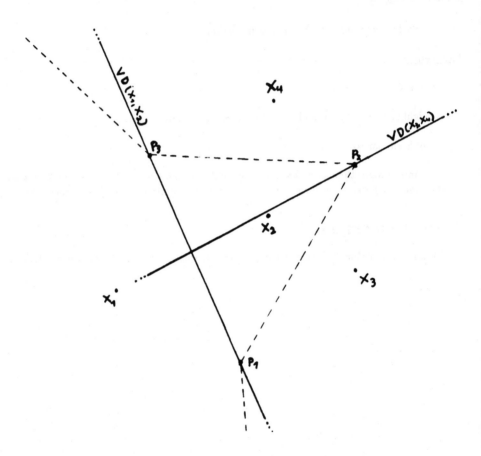

Figure 1: The doted line is the contour between VD($x_1, x_2$) and VD($x_3, x_4$). The new Voronoi-diagram is built by the doted line, the part of $VD(x_1, x_2)$, which lies to the left of the contour (the straightline $\overline{P_1, P_3}$) and the part of $VD(x_3, x_4)$, which lies to the right of the contour (the unbounded line extending from $P_2$).

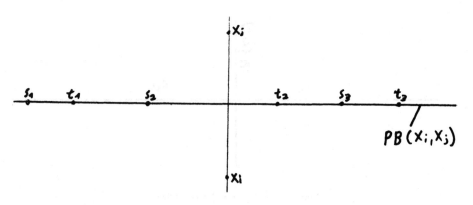

Figure 2: $s_1 \prec t_1 \prec s_2 \prec t_2 \prec s_3 \prec t_3$

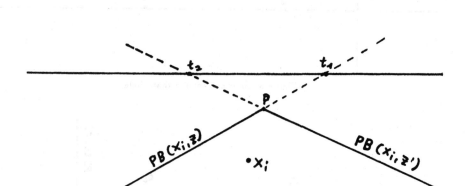

Figure 3

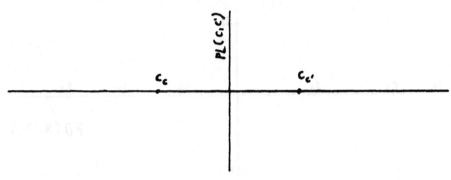

Figure 5: circle c is as big as circle c'

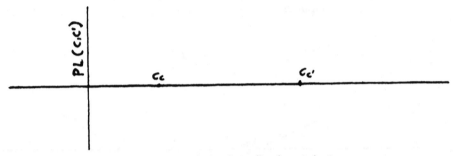

Figure 6: circle c is smaller than circle c'

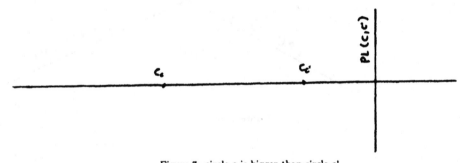

Figure 7: circle c is bigger than circle c'

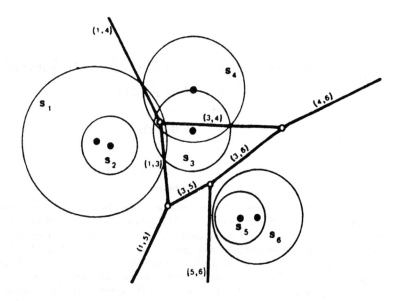

Figure 4.: A Power-diagram consisting of six circles

# References

[1]  K.Mehlhorn:
Data Structures and Algorithms 3: Multi-dimensional Searching and Computatonal Geometry,
EATCS Monographs on Theoretical Computer Science, 1984, Springer-Verlag.

[2]  A.Aggarwal, B.Chazelle, L.Guibas, C.Ó'Dúnlaing, C.Yap:
Parallel Computational Geometry, extendet abstract.

[3]  A.Aggarwal, B.Chazelle, L.Guibas, C.Ó'Dúnlaing, C.Yap:
Parallel Computational Geometry, submitted for publication in 1988.

[4]  F.Aurenhammer:
Power Diagrams: Properties, Algorithms and Applications,
SIAM J. Comput. Vol. 16 , February 1987, p. 78-96

# Voronoi Diagrams in the Moscow Metric*
## (Extended Abstract)

Rolf Klein

### Abstract

Most of the streets of Moscow are either radii emanating from the Kremlin, or pieces of circles around it. We show that Voronoi diagrams for $n$ points based on this metric can be computed in optimal $O(n \log n)$ time and linear space. To this end, we prove a general theorem stating that bisectors of suitably separated point sets do not contain loops if, beside other properties, there are no holes in the circles of the underlying metric. Then the Voronoi diagrams can be computed within $O(n \log n)$ steps, using a divide–and–conquer algorithm. Our theorem not only applies to the Moscow metric but to a large class of metrics including the symmetric convex distance functions and all composite metrics obtained by assigning the $L_1$ or the $L_2$ metric to the regions of a planar map.

**Keywords:** Bisector, computational geometry, convex distance function, metric, norm, robotics, shortest path, Voronoi diagram.

## 1  Introduction

Given a set of points $p_i$ in the plane, its *Voronoi diagram* divides the plane into *regions*, one to each point $p_i$, such that the region of $p_i$ contains all points in the plane that are closer to $p_i$ than to any other point $p_j$. Here *distance* can be measured by a *metric* in the plane, i.e., by a nonnegative real-valued function $d$ satisfying

$$
\begin{aligned}
d(a,b) &= 0 & \text{iff } a = b, \\
d(a,b) &= d(b,a), \\
\text{and} \quad d(a,b) &\leq d(a,c) + d(c,b) & \text{(Triangle inequality)},
\end{aligned}
$$

for arbitrary points $a$, $b$, and $c$.

Voronoi diagrams belong to the most useful data structures in computational geometry. One application is finding the nearest neighbor to a given query point $z$. The chief of a fire–brigade, for example, might draw on his map the Voronoi diagram for the fire–stations he is in charge of. When a fire at point $z$ is reported he looks up $z$ to determine the Voronoi region containing $z$ and thus, the station closest by.

Using airborne equipment, the underlying metric should be $L_2$, the Euclidean metric. Optimal algorithms for the computation of Euclidean Voronoi diagrams have been presented in [ShHo], [Br], and [F]. Using trucks in midtown Manhattan, distances must be measured in the $L_1$ metric. Algorithms for the $L_1$ metric were given in [H], [LeWo], and generalized in [WiWuWo] and [L].

All these metrics belong to the subclass of *norms* or symmetric *convex distance functions* that can be characterized by the following properties: Euclidean topology, invariance under translations, and additivity along straight lines. For the class of convex distance functions a divide–and–conquer

---

*This work was partially supported by the grant Ot64/4-2 from the Deutsche Forschungsgemeinschaft
†Institut für Informatik, Universität Freiburg, Rheinstr. 10-12, 7800 Freiburg, Fed. Rep. of Germany

algorithm has been suggested in [ChDr] that runs in $O(n \log n)$ steps (relative to the bisector complexity).

However, in many situations the underlying metric is *not* a convex distance function because the distance of two points may change under translations, or because straightline segments are no longer shortest paths. Examples are the Moscow metric (section 1), metrics defined by obstacles in the plane ([LePr]; [Ar]), or composite metrics ([MiPa]; section 3), to mention but a few. Therefore, an investigation of Voronoi diagrams based on *general metrics* has been begun in [KlWo]. It has been shown that in "nice" metrics *d-straight paths* exist, a substitute for straight line segments in that the distances between consecutive points on such a path add up. The existence of $d$-straight paths causes the Voronoi regions to be connected sets. Furhermore, we have shown there how to compute in $O(n \log n)$ steps Voronoi diagrams based on all those "nice" metrics $d$ whose *circles*

$$C_r^d(v) = \{z; d(v,z) \le r\}$$

admit two supporting lines of fixed orientations in fixed directions from the center, $v$.

In this paper the investigation is continued. First we introduce the Moscow metric (that fails to have the property mentioned above). Next we prove the following theorem. If $d$ is a nice metric none of whose circles has a hole, and if $l$ is a simple curve separating two point sets, $L$ and $R$, such that the intersection of $l$ with an arbitrary $d$-circle is connected, then the bisector $B_0(L, R)$ does not contain loops. This enables the computation of Voronoi diagrams within optimal $O(n \log n)$ steps. In section 3 we apply this result to the Moscow metric and to compositions of the $L_1$ and the $L_2$ metric in arbitrary maps in the plane whose borderlines are piecewise smooth.

## 2 The Moscow metric

The Moscow metric reflects a city layout where the streets are either straight lines radially emanating from a fixed center, or circles around this center. This pattern can be found in the map of numerous cities, among them Moscow and Karlsruhe, see Figure 1.

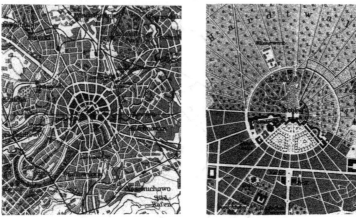

Figure 1: The maps of Moscow and Karlsruhe

**Definition 2.1** A curve connecting two points, $a$ and $b$, in the plane is called *admissible* if it consists only of segments of straight lines passing through the origin, and of segments of circles centered at the origin. The minimum Euclidean length of all admissible connecting curves is called the distance of $a$ and $b$ in the *Moscow metric*, $d_M(a,b)$.

Figure 2i) displays an admissible curve. Clearly, the $d_M$-distance of two points is *not* invariant under translations.

*Straight (=shortest) curves* in $d_M$ can be characterized as follows.

**Lemma 2.2** *A curve connecting $a$ and $b$ is $d_M$-shortest iff it is part of the unbounded curves depicted in Figure 2ii) and 2iii).*

Thus, if the angle $\alpha$ between $a$ and $b$ is greater than 2 ($\simeq 114.59...$ degrees) then the shortest connecting path follows the radii through the origin. If $\alpha < 2$ then the shortest path goes a round-about way along the circle of the point closer to 0. If $\alpha = 2$ then infinitely many shortest paths from $a$ to $b$ exist. Next we address the $d_M$-*circles*.

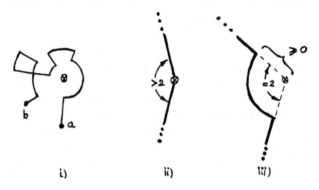

i)                                    ii)                                    iii)

Figure 2: An admissible curve and shortest path extensions

**Lemma 2.3** *Let $l$ be a straight line through the origin. Then for each circle $C_r^{d_M}(v)$ the intersection $C_r^{d_M}(v) \cap l$ is connected (or empty).*

This assertion is illustrated by Figure 3.

Figure 3: $d_M$-circles expanding from $v$

Note, however, that $d_M$-circles are in general not $d_M$-*convex*: the unique shortest path connecting the points $a$ and $b$ of the innermost circle in Figure 3 is not itself fully contained in this circle.

Similarly, Figure 6ii) shows that the Voronoi regions need not be $d_M$-convex. Figure 3 also illustrates that the $d_M$-*bisectors*

$$B(v, w) = \{z; d_M(v, z) = d_M(w, z)\}$$

are curves, not regions, except when $v$ and $w$ have the same distance to the center and the angle between them is less than $2\pi - 4$, due to the shape of the circles.

# 3 Voronoi diagrams and bisectors of sets

Rather than computing Voronoi diagrams in the Moscow metric directly, we want to develop a universal tool that works for a large *class* of metrics. Let $d$ denote a metric in the plane. If some $d$-bisectors $B(v, w)$ are regions instead of curves then parts of their boundaries, denoted by $B_0(v, w)$, can be chosen in a consistent way for defining the *normalized Voronoi diagram*, $\hat{V}^d(S)$, of a point set $S$ in the metric $d$, see [KlWo].

**Definition 3.1** A metric $d$ in the plane is called *nice* if it fulfils the following properties.

- $d$ induces the Euclidean topology in the plane

- each $d$-circle is bounded

- for all points $a \neq b$ there exists a point $c \notin \{a, b\}$ such that $d(a, b) = d(a, c) + d(c, b)$

- each $B_0(v, w)$ is a simple curve; two such bisectors intersect only finitely often

The first three properties alone imply that two points are always connectable by a $d$-*straight* curve. Here a curve $\pi$ is called $d$-straight if for each consecutive points a, b, and c on $\pi$ the equality $d(a, b) + d(b, c) = d(a, c)$ holds. Each $d$-straight curve is of minimum $d$-lenght. If $d$-straight curves exist then the Voronoi regions are *connected* sets, because they are $d$-*star-shaped* in the following sense: If $v$ lies in the region of $p$ and is connected with $p$ by a $d$-straight path $\pi$, then $\pi$ is contained in the region of $p$. For all nice metrics the Voronoi diagram of points is a planar graph with $n$ faces and $O(n)$ edges and vertices. See [KlWo] for details.

Computing Voronoi diagrams recursively, the difficulty is in merging two diagrams, $\hat{V}^d(L)$ and $\hat{V}^d(R)$, giving $\hat{V}^d(S)$, where $S = L \cup R$. To this end, the *bisector of the sets $L$ and $R$*, $B_0(L, R)$, must be computed. $B_0(L, R)$ is the union of all edges of $\hat{V}^d(S)$ that belong to the common boundary of both the region of a point in $L$ and to the region of a point in $R$. It can consist of unbounded chains and loops.

**Theorem 3.2** *Assume that $d$ is nice, that none of the bisector curves $B_0(v, w)$ is a loop, and that for all subsets $L' \subseteq L$, $R' \subseteq R$, the bisector $B_0(L', R')$ does not contain loops. Then $\hat{V}^d(L)$ and $\hat{V}^d(R)$ can be merged within $O(n)$ steps, where $n = |L \cup R|$.*

Here the complexity measure is *relative* to the bisector complexity, in that elementary operations like determining the first intersection of two bisectors etc., are accounted for by one step.

**Sketch of proof:** "Starter" segments of the bisector chains can be found using a refinement of the method suggested in [ChDr]. Processing curves instead of straight lines, it is not clear from the beginning that scanning the region boundaries in clockwise/counterclockwise order will still work, since the regions may fail to be convex: pieces of $B_0(L, R)$ can intersect a region of $\hat{V}^d(L)$ or of $\hat{V}^d(R)$ several times, and the prolongations of two segments of $B_0(L, R)$ belonging to $B_0(p, q_1)$ and $B_0(p, q_2)$ may *cross* before hitting the boundary of the region of $p$ in $\hat{V}^d(L)$. Fortunately, it turns out that it is the *connectedness* of the Voronoi regions rather than their convexity that makes the Shamos - Hoey algorithm work. □

In order to apply Theorem 3.2 we have to make sure that the bisectors of suitably separated sets don't contain loops.

**Theorem 3.3** *Assume that d is nice, and that the d-circle have no holes. Let l be an unbounded, simple curve such that the intersection of l with an arbitrary d-circle is empty or connected. If two point sets, L and R, are separated by l then $B_0(L,R)$ does not contain loops.*

**Sketch of proof:** Assume that $B = B_0(L,R')$ is a closed loop, encircling the regions of the points in $R' \subseteq R$. For each point $v$ of $B$ there exists a d-straight curve $\pi$ to a nearest neighbor $p$ in $L$ that doesn't intersect the interior of $B$, because the region of $p$ is d-star-shaped, see Figure 4i). Now the main step is in proving the following Lemma.

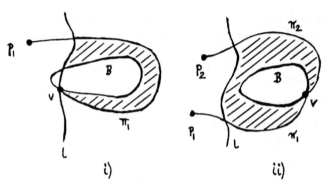

i)                    ii)

Figure 4: A loop surrounded

**Lemma 3.4** *There exists a point $v$ in $B$ admitting one or two d-straight curves, $\pi_1$ and $\pi_2$, to closest point sites $p_1$ and $p_2$ in $L$ such that $B$ is surrounded by $\pi_1$, or by $\pi_1$ and $\pi_2$, and l, see Figure 4i) and 4ii).*

This Lemma was first used in [ChDr] for convex distance functions $d$. In that case the proof is trivial because usual line segments are d-straight curves. In the present case, however, $v$ and the curves $\pi_1$ and $\pi_2$ must be constructed using a non-trivial limit theorem on the convergency of d-straight curves.

In case of Figure 4ii) we proceed as follows. Since $v$ belongs to the bisector of $L$ and $R'$ there is a point $q$ in the interior of $B$ such that $d(v,q) = d(v,p_1) = d(v,p_2)$, see Figure 5. Let $w_1$ and $w_2$

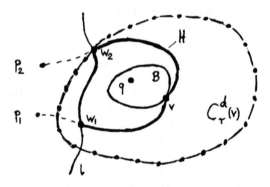

Figure 5: A hole in a circle

denote the first intersection points of $\pi_1$ and $\pi_2$ with $l$, respectively. Let $r := max(d(v,w_1),d(v,w_2))$; then $r < d(v,p_1) = d(v,p_2)$ because $p_1$ und $p_2$ are not on $l$. Now the circle $C_r^d(v)$ fully contains the closed curve $H$ consisting of segments of $\pi_1$, $\pi_2$, and $l$, as shown in Figure 5, but not the point $q$ in $H$'s interior. Thus, $C_r^d(v)$ does have a hole! $\square$

**Corollary 3.5** *Let d be a nice metric in the plane such that for each point v and for each real r ≥ 0 the set {z; d(v,z) = r} is a simple closed curve. Then none of the bisector curves $B_0(v,w)$ is a loop.*

# 4 Applications

**Theorem 4.1** *The Voronoi diagram of n points in the Moscow metric can be computed in optimal $O(n \log n)$ time, using linear space.*

**Proof:** Clearly, the Moscow metric is nice (Definition 3.1). The $d_M$-circles have no holes because each $d_M$-straight curve can be prolonged towards infinity, by Lemma 2.2. Given n points, we compute their polar coordinates and sort them in lexicographic order. Then we run a divide-and-conquer algorithm on the polar angles, using straight lines through the origin as separators. If L and R are point sets separated this way then $\hat{V}(L)$ and $\hat{V}(R)$ can be merged in time $O(|L \bigcup R|)$, by Lemma 2.3, Theorem 3.3, and Theorem 3.2. The Voronoi diagram of k sorted points of the same polar angle can be computed in time $O(k)$, see Figure 6i). Figure 6ii) depicts the Voronoi diagram of 8 points in the Moscow metric. □

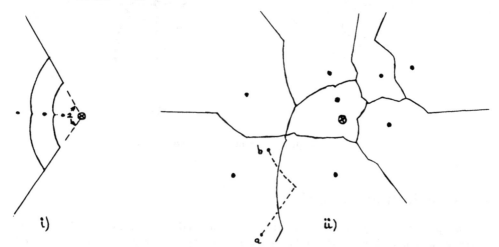

Figure 6: Voronoi diagrams in the Moscow metric

Now let D denote a map in the plane all of whose borderlines are piecewise smooth curves. Assume that with each region of D one of the metrics $L_1$, $L_2$ is associated, see Figure 7i).

**Definition 4.2** Let $d_D(a,b)$ be the minimum Euclidean length of all curves connecting a and b that consist only of segments of the following types: straight line segments inside the $L_2$-regions, pieces of edges of D, and x-y staircases inside the $L_1$ regions.

Note that $d_D(a,b) = L_2(a,b)$ holds if a and b are on a line parallel to the x- or to the y-axis. Furthermore, $d_D$ is nice. But $d_D$-straight curves cannot always be prolonged, see Example 4.3 in [KlWo].

**Lemma 4.3**     i) *The $d_D$-circles have no holes.*

*ii) Each intersection of a line parallel to the x– or to the y–axis with a $d_D$–circle is connected.*

A sketch of the proof of i) is illustrated in Figure 7ii). $H$ denotes a closed curve in $C_r^{d_D}(V)$, $z$ lies inside $H$, and $\pi$ is a $d_D$–straight curve from $v$ to $w$. By Pythagoras' theorem, the straight line segment $\sigma$ between $u$ and $z$ is shorter than the segment $\rho$ of $\pi$ between $u$ and $w$. Thus, $d(v,z) \leq d(v,w) \leq r$. The proof of Lemma 4.3ii) is similar. An application of Theorem 3.2 and Theorem 3.3 yields

**Theorem 4.4** *The Voronoi diagram of $n$ points in $d_D$ can be computed within $O(n \log n)$ steps, with respect to the bisector complexity.*

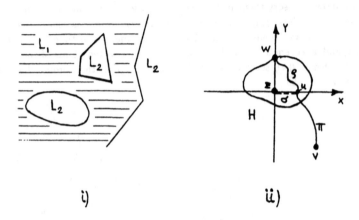

Figure 7: A composite metric

# 5   Concluding Remarks

We have presented an algorithm for contructing Voronoi diagrams in arbitrary "nice" metrics whose circles do not contain holes, provided that suitable separating curves exist. This divide–and–conquer algorithm runs in $O(n \log n)$ steps, relative to the bisector complexity. As an appliction, we have shown that Voronoi diagrams in the Moscow metric can be computed within optimal $O(n \log n)$ time. An interesting question is how to deal with metrics whose circles do contain holes. For the natural metric on the surface of a cone, for example, an $O(n \log n)$ sweepcircle algorithm was presented in [DeKl]. Another problem is how to compute shortest paths and bisectors of points for the class of metrics mentioned above, and for even more general classes.

# References

[Ar]     B. Aronov, "On the geodesic Voronoi diagram of point sites in a simple polygon", Proc. *3rd ACM Symposium on Computational Geometry*, Waterloo, 1987, pages 39–49.

[Br]     K. Q. Brown, "Voronoi diagrams from convex hulls", *Inf. Proc. Lett. 9*, pages 223–228, 1979.

[ChDr]   L. P. Chew and R. L. Drysdale, III, "Voronoi diagrams based on convex distance functions", Proc. *1st ACM Symposium on Computational Geometry*, Baltimore, 1985, pages 235–244.

[DeKl]    F. Dehne and R. Klein, "A sweepcircle algorithm for Voronoi diagrams", presented at the Workshop on Graph-Theoretic Concepts in Computer Science (WG 87), Staffelstein, 1987. To appear in LNCS.

[F]       S. Fortune, "A sweepline algorithm for Voronoi diagrams", *Algorithmica 2(2)*, 1987, pages 153–174.

[H]       F. K. Hwang, "An $O(n \log n)$ algorithm for rectilinear minimal spanning trees", *JACM 26*, 1979, pages 177–182.

[KlWo]    R. Klein and D. Wood, "Voronoi diagrams based on general metrics in the plane", in R. Cori and M. Wirsing (eds.), Proc. *5th Annual Symposium on Theoretical Aspects of Computer Science (STACS)*, Bordeaux, France, 1988, LNCS, pages 281–291.

[L]       D. T. Lee, "Two-dimensional Voronoi diagrams in the $L_p$ metric", *JACM 27*, 1980, pages 604–618.

[LePr]    D. T. Lee and F. P. Preparata, "Euclidean shortest paths in the presence of rectilinear barriers", *Networks 14(3)*, pages 393–410.

[LeWo]    D. T. Lee and C. K. Wong, "Voronoi diagrams in $L_1$ ($L_\infty$) metrics with 2-dimensional storage applications", *SIAM J. COMPUT. 9*, 1980, pages 200–211.

[MiPa]    J. S. B. Mitchell and Ch. H. Papadimitriou, "The weighted region problem", Proc. *3rd ACM Symposium on Computational Geometry*, Waterloo, 1987, pages 30–38.

[ShHo]    M. I. Shamos and D. Hoey, "Closest-point problems", Proc *16th IEEE Symposium on Foundations of Computer Science*, 1975, pages 151-162.

[WiWuWo]  P. Widmayer, Y. F. Wu, and C. K. Wong, "Distance problems in computational geometry for fixed orientations", Proc. *1st ACM Symposium on Computational Geometry*, Baltimore, 1985, pages 186–195.

# A sweep algorithm and its implementation:
# The all-nearest-neighbors problem revisited

Klaus Hinrichs[*], Jurg Nievergelt, Peter Schorn

Department of Computer Science, University of North Carolina, Chapel Hill, NC 27599-3175, USA

## Abstract
The 2-dimensional all-nearest-neighbors problem is solved directly in asymptotically optimal time O(n*log n) using a simple plane-sweep algorithm. We present the algorithm, its analysis, an optimization based on the concept of a clipped computation, and describe two robust realizations: a "foolproof" implementation which guarantees an exact result at the cost of using five-fold-precision rational arithmetic, and a robust floating point version.

## Keywords
Computational geometry, complexity, proximity problems, plane-sweep algorithms, robust implementation.

## Contents

---

[*] current address: Universität - Gesamthochschule - Siegen, Fachbereich 12,
Postfach 10 12 40, D - 5900 Siegen, West - Germany

# 1. Algorithms for the all-nearest-neighbors problem

We consider the 2-dimensional *all-nearest-neighbors problem*: Given a set S of n points in the plane, find a nearest neighbor of each with respect to the Euclidean metric.

It is well known that $\Omega(n*\log n)$ is a tight lower bound for this problem in the algebraic decision tree model of computation. Known algorithms with optimal worst case time complexity $O(n*\log n)$ are of two types:

1) Extract the solution (in linear time) from the answer to the more general problem of constructing the Voronoi diagram of the given points. The latter can be computed in time $O(n*\log n)$ both by a divide-and-conquer algorithm [SH 75], as well as by an intricate sweep based on transformed data [Fo 86].

2) Compute the desired result directly, without the additional information present in the Voronoi diagram. [Va 86] describes an algorithm which works for any number of dimensions in any $L_p$-metric. It maintains a growing set of shrinking boxes; when a box has shrunk to a single point, its associated information determines a nearest neighbor.

Algorithms of both types are complicated: their intricate logic and data structures are rarely specified in a sufficiently formal notation to allow the reader to assess the complexity of their implementation. We show how the 2-dimensional all-nearest-neighbors problem is solved effectively in asymptotically optimal time $O(n*\log n)$ using a plane-sweep algorithm [SH 76]. A vertical line (front, or cross section) sweeps the plane from left to right, stopping at every transition point (event) of a geometric configuration to update the cross section. All processing is done at this moving front, without any backtracking, with a look-ahead of only one point. Events to be processed are queued in an *x-queue*, the status of the cross section is maintained in a *y-table*. In the slice between two adjacent events, the relevant properties of the geometric configuration seen so far do not change; therefore the y-table needs to be updated only at transition points. Sweep algorithms have a simple structure typical of greedy algorithms:

```
initialize x-queue X;
initialize y-table Y;
while not emptyX do
    p := nextX;
    transition(p)
end
```

Sweep algorithms impose a left-to-right order on the data to be processed, thus distinguishing a *known past* from a *future yet to be discovered*. Since the nearest neighbor of a point may lie "in the future" as yet unknown we study a modified problem, *all-nearest-neighbors-to-the-left*, which is more easily solved by a sweep algorithm. A solution to the original *all-nearest-neighbors* problem is trivially obtained in linear time $O(n)$ from a solution to *all-nearest-neighbors-to-the-left* together with a solution to the analogous problem *all-nearest-neighbors-to-the-right*.

In the following figure an arrow points to the nearest neighbor of a point.

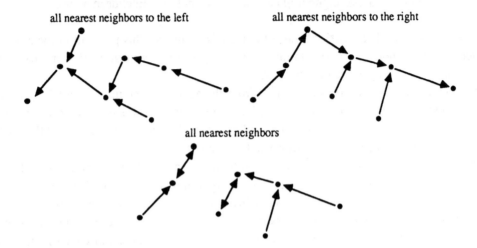

all nearest neighbors to the left

all nearest neighbors to the right

all nearest neighbors

When designing a new algorithm it is often convenient to assume a "non-degenerate data configuration", that is, to exclude at first sight all data with special properties that might cause trouble to the algorithm. In reading sections 2 to 4, we encourage the reader to assume that no two data points have equal x- or y-coordinates. The first assumption simplifies the ordering of events to be processed, the second one avoids vertical bisectors that might cause numerical problems. Both simplifying assumptions are removed in sections 6 and 7.

## 2. The sweep invariant

Let $S_{left}$ denote the set of those points of S that lie to the left of the sweep line: initially empty, $S_{left}$ grows one point at a time until at termination it equals S. During the sweep we maintain, as an invariant represented in the Y-table, the intersection $V \cap L$ of the Voronoi diagram V of $S_{left}$ with the sweep line L. For $q \in S_{left}$, let V(q) denote the Voronoi polygon of q with respect to $S_{left}$. The points q whose polygon V(q) intersects the sweep line are called *active points*, and the edges of V that intersect the sweep line are called *active bisectors*. V partitions the sweep line into disjoint intervals I(q), one for each active point $q \in S_{left}$. All points in I(q) have q as their nearest neighbor to the left. To each active point there corresponds an *upper* and a *lower* active bisector. The intersections of the active bisectors with the sweep line define a total order on the active points and on their bisectors.

The intersection $V \cap L$ of the Voronoi diagram $V$ of $S_{left}$ with the sweep line must be updated in either of the following cases:

1) $S_{left}$, and hence $V$, remain unchanged, but there is a topological change in $V \cap L$ as $L$ sweeps across a vertex of $V$. Type of transition: Intersection of two active bisectors. Action: An interval is deleted from $V \cap L$, its territory is distributed among its two neighbors.

2) $S_{left}$ changes, and thus $V$ changes, as $L$ sweeps across a new data point. Type of transition: A data point $p \in S - S_{left}$. Action: A new interval is created at the expense of existing intervals, some of which may disappear completely, and is inserted into $V \cap L$.

We discuss both types of transitions.

## 2.1 Intersection point

An intersection of two active bisectors is processed by removing an interval $I(p)$, then separating its two former adjacent intervals $I(r)$ and $I(s)$. As shown in the figure below, let $r$ be the lower neighbor of $p$, $s$ the upper neighbor of $p$. The two bisectors $bs(r, p)$ and $bs(p, s)$ are removed, the new bisector $bs(r, s)$ is inserted into $V \cap L$.

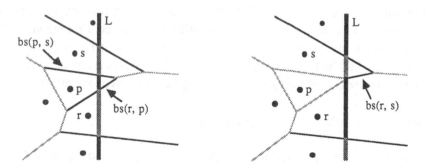

## 2.2 Data point

A new data point $p$ generates a new interval $I(p)$ created at the expense of intervals of currently active points. The interval $I(nn)$ of $p$'s nearest neighbor $nn$ will always lose some territory to $I(p)$. Some of its successor and predecessor intervals may also contribute to $I(p)$. We find those by starting with $nn$ and iterating upwards and downwards over active points. After computing $I(p)$, and possibly deactivating some points, $p$ becomes a new active point.

Updating the sweep invariant requires computing the new interval $I(p)$ by determining its lower and upper bisectors. We find the interval $I(nn)$ into which $p$ falls, and process it as the following figure shows: we remove from $I(nn)$ the subinterval of those points closer to $p$ than to $nn$, and initialize $I(p)$ to this subinterval. The intersection of bisector $bs(p, nn)$ of p and nn with L determines what needs to be done. If I(p) swallows up all of I(nn), nn is deactivated.

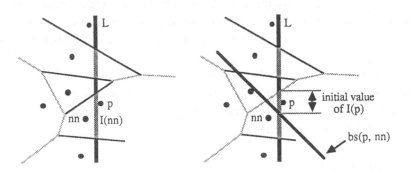

Having initialized I(p), we check whether I(p) needs to be extended upwards and/or downwards. We describe how the upper bisector of I(p) is computed; the lower is handled symmetrically. I(p) is extended upwards if nn has a successor, and bs(p, nn) intersects L outside the subinterval from p to the upper bisector of nn. The upward extension is the following loop:

Starting with q := nn, we immediately move upwards (q := successor(q)), and process each active point q in turn, until a termination condition T is satisfied. *Processing* q determines whether I(q) will lose some or all of its territory to I(p). The *termination condition* T is satisfied when T1: the bisector bs(p, q) of p and q intersects a test interval, namely the subinterval between p and the upper bisector of q, or T2: q has no successor.

*The processing step* depends on the termination test as follows:

P1: If q does not satisfy the termination test, q is deactivated, and I(q) is swallowed up by I(p).

P2: If q satisfies the termination test T1, I(q) loses some of its territory to I(p), namely everything below bs(p, q).

P3: If q does not satisfy the termination test T1, but satisfies T2, q is deactivated, and I(q), which extends to infinity, is swallowed up by I(p).

The next three pictures illustrate steps of this iteration. In the first, the termination test is not satisfied because bs(p, q) intersects L above the test interval that extends from p to the upper bisector of q: Step P1 deactivates q and extends I(p). In the second, termination test T1 is satisfied: I(q) loses some of its territory to I(p), a new bisector is inserted, and the final state is shown in the third figure.

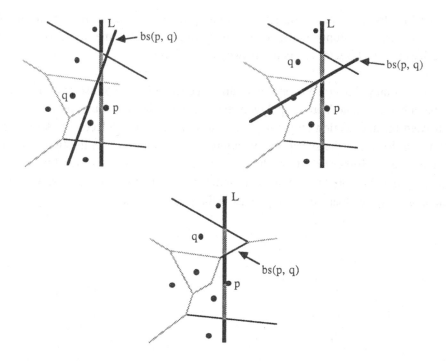

## 2.3 Embedding the invariant in the algorithm's data structures

The invariant developed in the previous section is embedded into a plane-sweep algorithm as the following figure suggests:

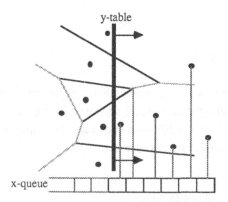

The x-queue stores the points of S and intersection points of bisectors sorted by x-coordinate. It is a priority queue which supports the operations insert and next, and is initialized with the points of S; during the sweep certain intersections of bisectors are inserted. Each intersection event points to its two bisectors.

The y-table stores the active bisectors sorted by their y-value on the sweep line. It is a dictionary which supports the operations insert, delete, member, successor and predecessor. Each bisector refers to the two points it bisects.

An intersection point in a Voronoi diagram is generated by two bisectors (its "parents") that belong to the same Voronoi polygon, and thus are adjacent. Thus a new active bisector is checked for intersection only against its successor and its predecessor. An intersection that lies to the left of the sweep line is ignored, one to the right (in the future) is inserted into the x-queue. Either or both of the parents may cease to be active (and thus be deleted from the y-table) before the intersection event they spawned is processed. Any intersection event in the x-queue that lacks two parents in the y-table is ignored.

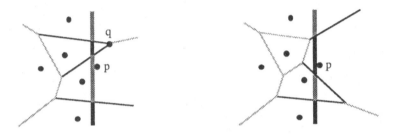

The figure above shows an example where an intersection q, although created, does not survive long enough to be processed as a transition when the sweep reaches q. Its parents are deactivated by the new point p. q could be deleted from the x-queue when p is processed; it is easier to leave it in the queue, and only process intersections whose parents are still active.


## 3. Program structure

We have implemented this all-nearest-neighbors algorithm as a MacPascal program. It lets the user enter points with the mouse and animates the algorithm's execution. The algorithm proper takes about 400 lines; additional code handles graphics and table management. The following procedures are called to process an event.

`findnnY(p, nn, lowerS, upperS)` returns the nearest neighbor nn to the left of a given point p, and its lower and upper bisectors.

`inInterval(y, b1, b2)` returns true if y lies in the interval bounded by $b_1$ and $b_2$, otherwise false.

If direction = $+\infty$ , `nextP(bs, direction)` returns the upper of the two points which determine bs, otherwise the lower one.

If direction $= +\infty$ , nextY(bs, direction) returns the successor of bs in the y-table, otherwise the predecessor.

memberY(bs) returns true if the bisector bs is contained in the y-table, otherwise false.

bisector(p₁, pᵤ) returns the bisector of two points $p_l$ and $p_u$.

deleteY(bs) removes a bisector from the y-table.

insertAndCheckY(bs) inserts a bisector bs into the y-table and calls another procedure which checks whether bs intersects its predecessor or successor on the right side of the sweep line. Any intersections found are inserted into the x-queue.

Two fictitious bisectors bound an interval extending toward $+\infty$ or $-\infty$.
infinity(bs) determines whether bs is such a sentinel.

The procedure transition incorporates all the work to be done in processing a transition point.

```
procedure transition(p);
begin
  if dataPoint(p) then
```

```
    findnnY(p, nn, lowerS, upperS);
    updateY(p, nn, upperS, +∞);
    updateY(p, nn, lowerS, -∞)
  elsif intersection(p) then
```

```
    if memberY(upperS) and memberY(lowerS) then
      bs := bisector(p₁, pᵤ);
      deleteY(upperS);
      deleteY(lowerS);
      insertAndCheckY(bs)
    end
  end
end transition;
```

If a new data point p lies in the interval I(q) bounded by nextbs, the invariant in direction
(+∞ or -∞) of nextbs is updated as explained in section 2:

```
procedure updateY(p, q, nextbs, direction);
begin
  bs := bisector(p, q);
  while not inInterval(bsy, py, nextbsy)
```

nextbs ⟍ │ ── nextbs$_y$ │● ── p $_y$ p

```
        and not infinity(nextbs) do
    q := nextP(nextbs, direction);
    n := nextY(nextbs, direction);
    deleteY(nextbs);
    nextbs := n;
    bs := bisector(p, q)
  end;
  if not infinity(nextbs) or
       inInterval(bsy, py, direction) then
```

bs ⟍ │ ── bs$_y$ │● ── p $_y$ p

```
    insertAndCheckY(bs)
  else
    Ytabledirection := p
  end
end updateY;
```

## 4. Analysis

Each time a new point is encountered during the sweep at most two new bisectors are
inserted into the y-table. The two bisectors may intersect each other, the upper bisector
may intersect its upper neighbor in the y-table, and the lower bisector may intersect its
lower neighbor. Therefore at most three intersection events are inserted into the x-queue.
Each time an active intersection point is encountered a new bisector is inserted into the y-
table. This bisector may intersect its upper and its lower neighbor in the y-table, and
therefore at most two new intersection events may be generated. Since at each active
intersection point an interval I(p) is removed, i.e. a point $p \in S$ is deactivated, there may be
at most n active intersection points. Therefore at most 5n intersection events are inserted
into the x-queue, and the x-queue contains no more than 6n events at any time.

If the x-queue is implemented as a heap the operations nextX and insertX can be performed in time $O(\log n)$. Initializing the x-queue with the points of S takes $O(n)$ time.

The y-table is implemented by a balanced binary tree. The operations deleteY and memberY can be performed in time $O(\log n)$. The procedure insertAndCheckY inserts a new bisector into the y-table and at most two intersection events into the x-queue. Therefore it takes $O(\log n)$ time. The nearest neighbor nn of a data point p is determined by searching in the y-table for the bisectors bounding the interval $I(nn)$; this can also be done in time $O(\log n)$.

The cost for updateY is $O((k+1)*\log n)$, where k is the number of points deactivated in a call of updateY. Since at most n points can be deactivated the total cost for all calls of updateY is $O(n*\log n)$. Hence the total cost for processing new points is $O(n*\log n)$. Since at most 5n intersection events are generated, the total cost for processing all intersection events is $O(n*\log n)$. Therefore the total cost of the plane-sweep algorithm for the all-nearest-neighbors problem is $O(n*\log n)$.

The constant factor in front of the $(n*\log n)$-term of the cost can be obtained from the total number of $O(\log n)$-operations performed on the heap and the balanced tree:
$$5n * \text{InsertInHeap} + 6n * \text{NextFromHeap}$$
$$+ \ 11n * \text{SearchInTree} + 3n * \text{InsertInTree} + 3n * \text{DeleteFromTree}$$


## 5. Clipped computation

The algorithm as presented in section 3 computes intersections of bisectors that may be parallel. In seeking a robust implementation, we first tackle the geometric problem of handling a point at infinity defined as the intersection of two parallel lines. A good solution to this problem will automatically take care of points defined as intersections of nearly parallel lines, too far away to be represented by coordinates in the number system we are using, i.e. points in the plane whose numerical evaluation causes overflow.

Plane sweep applied to the all-nearest-neighbors problem (and several other problems, such as the line-intersection-problem) avoids dealing with far-away points thanks to a simple and effective technique: any point constructed outside the smallest rectangle that encloses the data points is simply thrown away, as it will never be used. We present the geometric justification of this clipped computation, and its possible use as an optimization technique.

## 5.1 The clipping box

Enclose the set S of given points into a clipping box, any rectangle aligned with the axes that contains all data points. Let the left, right, bottom, and top sides of the box be given by $x_l$, $x_r$, $y_b$, and $y_t$, respectively. All input data is limited to this box, but intermediate results (intersection points) may lie outside: above, below, and to the right, but never to the left of the sweep line. Any point outside may be discarded, as it is never used.

Justification: The y-table is used to locate the interval into which the current point falls. As any data point to be processed is known to have a y-coordinate in the range $[y_b,y_t]$, parts of the table below $y_b$ or above $y_t$ are never used, and thus need not be maintained. We amend the invariant of section 2 as follows: The y-table contains the intersection of the Voronoi diagram of the points to the left *clipped to the interval $[y_b,y_t]$*. Let's look at the details:

The y-table stores formulas of straight lines. During the sweep, these are evaluated under two circumstances, giving rise to intermediate points of two different types:
a) Intersection of a bisector with the sweep line: This occurs when searching for the interval I(nn) into which the current data point p falls. Such a point is used only once and need not be stored.
b) Intersection of two bisectors. Such a point is normally stored in the x-queue, to trigger the replacement of these bisectors by a third one at some later time. The following figure shows that two parent bisectors bs(r, s) and bs(s, t) that intersect outside the box have a child bisector bs(r, t) that lies entirely outside the box, and thus will never participate in locating an interval for any future point to be processed.

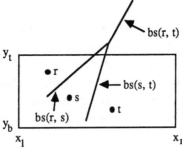

## 5.2 Robustness and optimization

Clipping leads to a practical optimization with respect to space and time. As intersections outside the clipping box are ignored, they need not be inserted into the x-queue. Thus we replace the line intersection procedure used in insertAndCheckY, which claimed to compute an intersection point for any pair of lines, by a more efficient one which only computes an intersection if the latter falls within the box:

```
function intersectInBox(s1, s2: line; var p: point): boolean;
```

And we modify the calling sequence to do nothing if the in-box test returns false:

```
if intersectInBox(s1, s2, p) then process(p);
```

It is instructive to reflect on the nature of the exceptionally well-behaved computation we are facing. The input data determines a box, and all computation can be confined within. This is not too surprising in view of the fact that the output data for this problem consists merely of indices in the range 1 .. n, indicating which point is a nearest neighbor to the left; even if we include the distance to the nearest neighbor in the specification of the required output, this quantity is confined to the length of the diagonal of the box. The approach via the Voronoi diagram, which organizes the entire plane, is once again revealed as an overkill: by introducing intermediate points that may be arbitrarily far away, it creates an unnecessary danger of overflow.

## 6. An exact implementation

We have implemented this algorithm on the Macintosh, using MacPascal, in such a way that it handles all degenerate configurations and achieves an exact result. The degeneracies to be dealt with are events of equal x-coordinate, including intersection events that coincide (equal x- and y-coordinates). If input coordinates are integers, five-fold precision rational arithmetic guarantees exact results.

### 6.1 Disambiguating degenerate configurations

Degeneracies are handled by extending the order on events in the x-queue and bisectors in the y-table to the case of equal x- and/or y-coordinates.

We define the order on events of the form $(x, y, t)$, $t \in \{$intersection, dataPoint$\}$ as follows:

$$(x_1, y_1, t_1) < (x_2, y_2, t_2) \iff (x_1 < x_2)$$
$$\lor \; (x_1 = x_2) \land (y_1 < y_2)$$
$$\lor \; (x_1 = x_2) \land (y_1 = y_2) \land (t_1 = \text{intersection}) \land (t_2 = \text{dataPoint})$$

Two distinct intersection events which coincide in the plane can be processed in any order - we need not break the tie in this case.

The vertical order on bisectors of the form $a\,x + b\,y = c$, $b \geq 0$, is defined by:

$((a_1 x + b_1 y = c_1) < (a_2 x + b_2 y = c_2))$ at $x_0$ $\Leftrightarrow$

$(b_1 > 0) \wedge (b_2 > 0) \wedge ((c_1 - a_1 x_0) / b_1 < (c_2 - a_2 x_0) / b_2)$

$\vee\ (b_1 > 0) \wedge (b_2 > 0) \wedge ((c_1 - a_1 x_0) / b_1 = (c_2 - a_2 x_0) / b_2) \wedge (-a_1 / b_1 < -a_2 / b_2)$

$\vee\ (b_2 = 0) \wedge (b_1 > 0)$

$\vee\ (b_2 = 0) \wedge (b_1 = 0) \wedge (a_1 > a_2)$

## 6.2 Bounding the computed quantities

Rational arithmetic yields exact results if both: 1) input data, and 2) intermediate and final results, are bounded in size and can be represented exactly. We meet assumption 1) by requiring that all data points lie on a finite integer grid; i.e. there exists an integer M such that any data point has integer coordinates $(x, y)$ with $|x| < M$ and $|y| < M$. By representing any number as a pair (numerator, denominator) we avoid division. Given a sufficiently high precision the operations "+", "-", "*" and "/" can be carried out error-free on bounded operands in rational representation. We meet requirement 2) by examining all expressions used by our algorithm and bounding all numerators and denominators that occur during the computation. Four kinds of calculations occur:

1) Computing the coefficients in the equation of the bisector of two data points. The equation of a bisector of two points $(x_1, y_1)$ and $(x_2, y_2)$ is $a x + b y = c$ where
   $a = 2 (x_1 - x_2)$, $b = 2 (y_1 - y_2)$ and
   $c = (x_1 - x_2) (x_1 + x_2) + (y_1 - y_2) (y_1 + y_2) = x_1^2 - x_2^2 + y_1^2 - y_2^2$.
   These expressions can be bounded by: $|a| < 4 M$, $|b| < 4 M$, and $|c| < 2 M^2$.

2) Calculating the intersection point of two bisectors. The intersection point $(x, y)$ of the bisectors generated by the pairs $(x_1, y_1)$, $(x_2, y_2)$ and $(x_2, y_2)$, $(x_3, y_3)$ are represented as $(X/D, Y/D)$ where
   $X = (x_1^2 + y_1^2) (y_2 - y_3) + (x_2^2 + y_2^2) (y_3 - y_1) + (x_3^2 + y_3^2) (y_1 - y_2)$
   $Y = (x_1^2 + y_1^2) (x_3 - x_2) + (x_2^2 + y_2^2) (x_1 - x_3) + (x_3^2 + y_3^2) (x_2 - x_1)$
   $D = 2 (x_1 y_2 - x_2 y_1 + x_2 y_3 - x_3 y_2 + x_3 y_1 - x_1 y_3)$
   From these formulas we conclude: $|X| < 12 M^3$, $|Y| < 12 M^3$ and $|D| < 12 M^2$.

3) Determining the order of two bisectors in the y-table. We have to compare the bisectors $a_1 x + b_1 y = c_1$ and $a_2 x + b_2 y = c_2$ at $x_0 = p / q$ (assume $q > 0$). The nontrivial part arises when $b_1 > 0$ and $b_2 > 0$.
   $(a_1 x + b_1 y = c_1) < (a_2 x + b_2 y = c_2)$ at $x = p / q$ $(b_1 > 0, b_2 > 0$ and $q > 0)$
   $\Leftrightarrow$
   $(c_1 - a_1 p / q) / b_1 < (c_2 - a_2 p / q) / b_2$ if $(c_1 - a_1 p / q) / b_1 \neq (c_2 - a_2 p / q) / b_2$
   and $- a_1 / b_1 < - a_2 / b_2$ if $(c_1 - a_1 p / q) / b_1 = (c_2 - a_2 p / q) / b_2$
   $\Leftrightarrow$
   $b_2 (q c_1 - p a_1) < b_1 (q c_2 - p a_2)$ if $b_2 (q c_1 - p a_1) \neq b_1 (q c_2 - p a_2)$
   and $a_1 b_2 > a_2 b_1$ if $b_2 (q c_1 - p a_1) = b_1 (q c_2 - p a_2)$

We can bound the quantities which have to be compared by using the fact that x is either a given point (in which case $|p| < M$ and $q = 1$) or x is an intersection point of two bisectors (in this case it follows from 2) that $|p| < 12 M^3$, $|q| < 12 M^2$). Together with the inequalities from 1) for the $a_i$, $b_i$, $c_i$ ($i = 1, 2$) we obtain:

$|b_2 (q c_1 - p a_1)| < 288 M^5$, $|b_1 (q c_2 - p a_2)| < 288 M^5$ and $|a_1 b_2| < 16 M^2$, $|a_2 b_1| < 16 M^2$.

4) Determining the order of two events in the x-queue. Here we have the three cases *data point against data point*, *data point against intersection* and *intersection against intersection*. Obviously only the last case may give rise to large numbers. In this case we have to compare two fractions whose absolute value of the numerator is bounded by $12 M^3$ and whose absolute value of the denominator is bounded by $12 M^2$. Since the comparison is done by cross multiplication we have to compare numbers bounded by $144 M^5$.

Conclusion: Let m be the number of bits used to represent the input data. Rational arithmetic using integers represented by $\lceil \log_2 288 \rceil + 5 m = 9 + 5 m$ bits yields results exactly.

A similar result was obtained in [Be 84]: exact computation in a different plane-sweep algorithm requires about five-fold precision. Rational arithmetic can be used to compute exact results for almost all geometric problems, as the great majority of these problems involve only rational expressions of the given data. By analyzing the formulas used it is usually possible to bound the precision required. This is because geometric computations, unlike ordinary numerical computations, rarely use iterative processes in calculating the results.

## 7. An approximate floating point implementation

The numbers declared as "real" in the program of section 3 participate in a floating point computation which is necessarily only approximate. Although it is difficult to enforce rigorous non-trivial bounds on the errors, the sweep algorithm leads to an exceptionally well-behaved floating point computation, which can be made practically foolproof by adding some precautions. The following points justify this assessment.

1) Roundoff errors do not propagate nor accumulate - the only lines ever computed are bisectors of the originally given data points. Constructed points (intersections of bisectors) serve as alarm clocks to trigger computations, but do not affect the result of these computations.

2) The equation $a x + b y = c$ for the bisectors given in section 6.2 shows that the coefficients a and b are bounded by the length and height of the clipping box,

respectively, whereas $c = x_1^2 - x_2^2 + y_1^2 - y_2^2$ is a sum of squares of coordinates of data points. Thus a, b, and c are represented nearly exactly in double precision, as is the intersection of a bisector with the sweep line and the borders of the bounding box. A double precision implementation of all bisector computations yields results that are close to least-significant-bit accurate.

3) Trapping overflow lets the computation proceed. In section 5 we have shown that intermediate points need to be computed only if they lie within the clipping box, and that any point that ends up outside can simply be thrown away. This fact can be used to defuse overflow due to intersection of (nearly) parallel lines, in any language or system that returns control to the application program after an overflow has been trapped. Execute the code of section 3, written under the fictitious assumption that the computation proceeds in the mathematical system of real numbers, rather than in a finite number system. Add an overflow-handling procedure, which is called whenever an overflow has been trapped; it identifies the pair of lines whose intersection caused the overflow to occur, avoids recomputing this intersection by incrementing the relevant state variables (counters or pointers), and returns control to the application program. Thus the computation proceeds after the offensive point has been effectively thrown out. Whereas in most computations overflow is taken as an indicator of failure, here we face the exceptional case where it is normal.

## 8. Conclusions

The algorithm presented is relatively simple both from the point of view of theory and implementation: it calls upon standard techniques in algorithm design (plane-sweep) and data structures (priority queues and dictionaries), and has been programmed to animate algorithm execution in just a few hundred lines of Pascal. It is plausible to assume that other optimal algorithms which first compute the Voronoi diagram of the given set S of points are less efficient, in practice, since the Voronoi diagram provides much more information about S than is needed to solve the problem.

Two issues need to be considered before a geometric program can be regarded as practical: 1) can it handle degenerate configurations, e.g. three lines that intersect in the same point? and 2) is it numerically robust, e.g. can errors be bounded? Geometric algorithms are rarely studied from this point of view. Our implementations address both problems. Degenerate cases are classified into two categories: those whose order is irrelevant, as in the case of coinciding intersections, and those that must be ordered consistently. Numerical issues are considered from two points of view: five-fold precision rational arithmetic yields exact results, assuming that the data points lie on an integer grid; single- or double-precision floating point yields a well-behaved approximate computation.

## Acknowledgement

This is an expanded and revised version of the authors' paper "A sweep algorithm for the all-nearest-neighbors problem", to appear in "Computational Geometry and its Applications" (Proceedings CG '88), H. Noltemeier (ed.), Springer Lecture Notes, 1988. It was supported in part by the National Science Foundation under grant DCR 8518796.

## References

[Be 84]   G. B. Beretta:
          An implementation of a plane-sweep algorithm on a personal computer, Ph. D. Thesis Nr. 7538, ETH Zurich, 1984.

[Fo 86]   S. Fortune:
          A Sweepline Algorithm for Voronoi Diagrams,
          Proc. 2nd Ann. Symp. on Computational Geometry, ACM, 313-322, 1986.

[SH 75]   M. Shamos, D. Hoey:
          Closest-Point Problems,
          16th Annual IEEE Symposium on Foundations of Computer Science,
          151 - 162 (1975).

[SH 76]   M. Shamos, D. Hoey:
          Geometric intersection problems,
          17th Annual IEEE Symposium on Foundations of Computer Science,
          208 - 215 (1976).

[Va 86]   P. Vaidya:
          An Optimal Algorithm for the All-Nearest-Neighbors Problem,
          Proc. 27th IEEE Symp. Foundations of Computer Science, 117-122, 1986.

## List of Participants
## 14th Int. Workshop on Graph-Theoretic Concepts

Alt, H.
Bache, R.
Bakker, E.M.
Bauderon, M.
Berg, M.T. de
Bodlaender, H.L.
Courcelle, B.
Emde Boas, P. van
Grech, W.
Habel, A.
Hartel, P.H.
Hinrichs, K.
Huckenbeck, U.
Icking, C.
Iwanowski, S.
Jayakumar, R.
Jørgensen, L.K.
Kaeslin, H.
Kersten, M.
Klein, R.
Knödel, W.
Kranakis, E.
Kreveld, M.J. van
La Poutré, J.A.
Lavault, C.
Leeuwen, J. van
Lingas, A.
Lipeck, U.W.
Marchetti-Spaccamela, A.
Meijer, H.
Mohanty, H.
Montonen, E.
Moser, T-J.
Müller, H.
Nagl, M.
Nedunuri, S.
Noltemeier, H.
Nurmi, O.
Ostfeld, Z.
Peper, F.
Plaxton, C.G.
Preilowski, W.
Sack, J-R.
Santoro, N.

Savage, J.E.
Schimmler, M.
Schmidt, U.
Schneider, H.J.
Schoone, A.A.
Seipel, D.
Smid, M.
Stewart, I.A.
Stiefeling, H.
Takaoka, T.
Tan, R.B.
Torenvliet, L.
Veldhorst, M.
Vidyasankar, K.
Whitty, R.W.
Wiegers, M.
Wloka, M.G.
Wöginger, G.

# AUTHOR INDEX